Advances in Dielectrics

Series Editor

Friedrich Kremer, Fakultät für Physik und Geowissenschaften, Peter-Debye-Institut für Physik der weichen Materie, Universität Leipzig, Leipzig, Germany

Aims and Scope

Broadband Dielectric Spectroscopy (BDS) has developed tremendously in the last decades. For dielectric measurements it is now state of the art to cover typically 8-10 decades in frequency and to carry out the experiments in a wide temperature and pressure range. In this way a wealth of fundamental studies in molecular physics became possible, e.g. the scaling of relaxation processes, the interplay between rotational and translational diffusion, charge transport in disordered systems, and molecular dynamics in the geometrical confinement of different dimensionality – to name but a few. BDS has also proven to be an indispensable tool in modern material science; it plays e.g. an essential role in the characterization of Liquid Crystals or Ionic Liquids and the design of low-loss dielectric materials.

It is the aim of "Advances in Dielectrics" to reflect this rapid progress with a series of monographs devoted to specialized topics.

Target Group

Solid state physicists, molecular physicists, material scientists, ferroelectric scientists, soft matter scientists, polymer scientists, electronic and electrical engineers.

More information about this series at https://link.springer.com/bookseries/8283

Andreas Schönhals · Paulina Szymoniak
Editors

Dynamics of Composite Materials

Springer

Editors
Andreas Schönhals
Bundesanstalt für Materialforschung
und -prüfung (BAM)
Berlin, Germany

Paulina Szymoniak
Bundesanstalt für Materialforschung
und -prüfung (BAM)
Berlin, Germany

ISSN 2190-930X ISSN 2190-9318 (electronic)
Advances in Dielectrics
ISBN 978-3-030-89722-2 ISBN 978-3-030-89723-9 (eBook)
https://doi.org/10.1007/978-3-030-89723-9

This Springer imprint is published by the registered company Springer Nature Switzerland AG
The registered company address is: Gewerbestrasse 11, 6330 Cham, Switzerland

Preface

Historically, to tune the properties of a polymer or more general soft matter systems by a second phase is not a new concept and dates back to the 40s of the last century. Beside some successes, the improvement of the properties remained somehow limited. The expectations of the enhancement of the properties of composites changed by the developments of Toyota Central research in the 1990s. It was shown that the incorporation of 5 vol% exfoliated layers of a clay system into a polymer leads to a strong improvement of the mechanical and thermal properties. This discovery stimulated a broad research interest of both fundamental and applied character. Today, polymer-based nanocomposites have reached a billion-dollar global market. The corresponding applications span from components for transportation, commodity plastics with enhanced barrier and/or flame retardancy characteristics, to polymers with electrical properties for shielding, electronics, sensors, and solar cells as well as to live science. Important fields are filled rubbers, reinforced thermoplastics, or thermosets for automotive, aircraft/space and marine industries, but also membranes for separation processes as well as barrier layers, just to mention a few.

For a variety of applications, the molecular mobility in nanocomposites is of great importance. This concerns the molecular mobility needed to form a percolating filler network in rubbers used in tires or in composites employed in electric shielding applications. In general, it is also essential for processing polymer-based nanocomposites. Furthermore, separation processes in composite materials for membranes require a certain molecular mobility. This also concerns nanodielectrics used in electrical applications or sensors where the mobility of charge carriers can be related to the fluctuations of molecular groups etc. Finally, the molecular mobility can be taken as probe for structure on a molecular scale.

Broadband dielectric spectroscopy is a powerful tool to investigate the molecular mobility in polymer systems. It is due to the extremely broad frequency and sensitivity range that can be covered by this technique. Information about localized and cooperative molecular fluctuations, polarization effects at interfaces, as well as charge transport processes can be deduced. Therefore, this book focuses on broadband dielectric spectroscopy of composite materials. Moreover, the dielectric studies are accompanied by mechanical spectroscopy, advanced calorimetry, NMR techniques, as well as transmission electron microscopy and X-ray scattering investigations.

Besides a brief introduction to (nano) composites, the book aims to address fundamental aspects of the molecular mobility in this innovative group of materials. Selected examples with scientific interest and some cases with higher industrial impact were chosen. Due to the breadth of the subject, unfortunately not all topics could be addressed in detail, such as processing for instance.

Berlin, Germany
July 2021

Andreas Schönhals
Paulina Szymoniak

Contents

(Nano)Composite Materials—An Introduction ... 1
Andreas Schönhals, Martin Böhning, and Paulina Szymoniak

Fundamentals

Fundamentals of Dielectric Spectroscopy in Polymer Nanocomposites ... 35
Ivan Popov and Alexei P. Sokolov

Dynamics of Polymer Bridges in Polymer Nanocomposites Through a Combination of Dielectric Spectroscopy and Rheology ... 63
Shiwang Cheng

Molecular Mobility in Nanocomposites Based on Renewable Semicrystalline Polyesters ... 87
Panagiotis A. Klonos, Dimitrios N. Bikiaris, and Apostolos Kyritsis

Dynamics in Polymer Nanocomposites—From Conventional to Self-suspended Hybrid Systems ... 123
Emmanuel U. Mapesa, Sara T. Hamilton, Dayton P. Street, Nelly M. Cantillo, Thomas A. Zawodzinski, S. Michael Kilbey, Ah-Hyung Alissa Park, and Joshua Sangoro

Special Systems

Polymer Composites with Molecular Fillers: Microscopic Views into Supramolecular Reinforcement ... 163
Kay Saalwächter

Dynamics of Hyperbranched Polymers Under Severe Confinement in Intercalated Nanocomposites ... 187
Kiriaki Chrissopoulou and Spiros H. Anastasiadis

Industrial Applications

Interfacial Effects on the Dielectric Properties of Elastomer Composites and Nanocomposites 225
Bo Li, Georgios Polizos, and Evangelos Manias

Microstructure and Segmental Dynamics of Industrially Relevant Polymer Nanocomposites 251
Julian Oberdisse and Anne-Caroline Genix

Filler Networks of Carbon Allotropes of Different Shapes and Dimensions in a Polymer Matrix 291
Ingo Alig, Konrad Hilarius, Dirk Lellinger, and Petra Pötschke

Epoxy-Based Nanocomposites—What Can Be Learned from Dielectric and Calorimetric Investigations? 335
Paulina Szymoniak and Andreas Schönhals

Index 369

(Nano)Composite Materials—An Introduction

Andreas Schönhals, Martin Böhning, and Paulina Szymoniak

Abstract The chapter gives a brief introduction to (nano)compositecomposite materials having the focus on polymer-based nanocomposites. The different dimensionalities of nanoparticles are introduced, along with their distribution in the matrix. Different application fields of polymer-based nanocomposites, like flame retardancy, filled rubbers, nanofilled thermosets and thermoplastics, separation membranes and nanodielectrics, are considered in greater detail.

Keywords Polymer-based nanocomposites · Nanoparticles · Distribution of nanoparticles · Filled rubbers · Filled thermosets and plastics · Separation membranes · Nanodielectrics

Abbreviations

α	Permselectivity
σ_{DC}	DC conductivity
$\varepsilon^* = \varepsilon' - i\varepsilon''$	Complex dielectric function
ε_0	Permittivity of vacuum
ε_s	Static permittivity
CNT	Carbon nanotubes
C, c	Concentration
D_{eff}	Effective diffusion coefficient
E_{BD}	Electrical breakdown field strength
f	Frequency
LDH	Layered Double Hydroxide
L_{D}	Thickness of the Debye layer
MWS	Maxwell/Wagner/Sillars
P	Permeability

A. Schönhals (✉) · M. Böhning · P. Szymoniak
Bundesanstalt für Materialforschung und –prüfung (BAM), Unter den Eichen 87, 12205 Berlin, Germany
e-mail: Andreas.Schoenhals@bam.de

A. Schönhals and P. Szymoniak (eds.), *Dynamics of Composite Materials*,
Advances in Dielectrics, https://doi.org/10.1007/978-3-030-89723-9_1

POSS	Polyhedral Oligomeric Silsesquioxane
S	Solubility
T_g	Glass transition temperature
TEM	Transmission electron microscopy
$\tan \delta$	Loss tangent, dissipation factor
U	Energy density

1 Introduction

The improvement of the properties of a material, e.g., a polymer, by a second phase is not a new concept in material science. Well-known examples of this approach are, for instance, metal alloys. Also, the improvement of the properties of polymers is considered already for a longer time. One concept widely used to tailor the properties of a polymeric material is to mix one polymer with another yielding a blend system [1]. This approach has numerous applications in industry. A second concept is to modify the properties of polymer systems by an inorganic phase. First examples for that approach date back to the 1940s. Glass fibres, carbon black, or carbon fibers can be considered as archetype of filler (see, for instance, [2]). Today, such materials cover a broad range of application in construction, automotive, and aerospace industries. Engineering tools have been developed to minimize delamination and other failure possibilities. In contrast to the widely established field of polymer-based composites with conventional fillers, i.e., from the millimeter to the micrometer scale, developments and application based on property improvements due to smaller filler structures, i.e., in the sub-micrometer to nanometer scale, remains limited.

This field of research was rediscovered due to developments of Toyota Central Research in the 1990s. It was shown that the exfoliation of the layers of clay particles of montmorillonite into a semicrystalline polyamide (Nylon-6) increases the modulus of the composite by a factor of approximately three and the heat deflection temperature was enhanced by 80 K [3]. It is worth to note that these achievements in terms of property improvements were accomplished with remarkable low filler contents such as 5 vol%. A review of these early works can be found elsewhere, also with amorphous polymers as matrices [4].

The discovery that polymers filled with nanoparticles, i.e. filler particles with at least one spatial dimension smaller than 100 nm can have strong property improvements compared to the unfilled system, stimulated a broad range of research activities with fundamental and applied character. Today, polymer-based nanocomposites have reached a billion-dollar global market with applications ranging from components for transportation, commodity plastics with enhanced barrier and/or flame retardancy characteristics to polymers with electrical properties for shielding, electronics, sensors, and solar cells as well as in live science. Typical applications of polymer nanocomposites are, for instance, filled rubbers, reinforced thermoplastics, or thermosets for automotive, aircraft/space and ship industries, as membranes for separation processes as well as barrier layers and in dielectrics for shielding, capacitors, and insulation. A recent review discusses both, theoretical aspects as well as possible applications of nanocomposites [5].

2 Nanoparticles, Surface Modification, and Dispersion

Various nanoparticles have been considered as fillers to enhance the properties of polymers and/or to enable special functions like electrical conductivity or antibacterial effects. They can have different shapes and dimensionalities. Different classification schemes for their dimensionalities exist in the literature. Here, the nanoparticles are characterized according to their geometrical shape. This concerns, for instance, spherical-shaped nanoparticles (3D nanoparticles) which can be organic or inorganic in nature. Some examples are metals, oxides or salts like silica, titanium, Fe_3O_4, ZnS, just to name a few. For an overview about metal oxide nanoparticles and their applications in nanotechnology see, for instance, [6]. Examples of organic spherical nanoparticles are fullerenes (C_{60}) (see, for instance, [7]) and single-chain polymer nanoparticles [8]. Polyhedral Oligomeric Silsesquioxane (POSS) are a kind of spherical organic–inorganic nanoparticles (see, for instance, [9–11]). From the chemical point of view POSS is a molecule with a central silica cage and organic substituents at its edges. It may be considered as the smallest silica nanoparticle with a well-defined organic shell. Due to the versatility of the chemistry of POSS, its interaction with a polymer matrix can be tuned. Besides the discussed classes of spherical nanoparticles quantum dots (for example, CdSe/ZnS) can be also considered as 3D nanofillers (see, for instance, [12]). (In other classifications quantum dots are called nanoparticles of zero dimension.)

Clay minerals are examples for 2D nanoparticles. The layers of the clay materials have lateral dimensions up to a few micrometers but with thicknesses in the lower nanometer range. Dispersed in a polymeric matrix these clay minerals exhibit morphologies between two edge cases, a completely exfoliated or an intercalated morphology (see Fig. 1). The former one comprises completely delaminated layers of the crystal that are present in the matrix as separated single sheets. In the latter case the polymer chains are intercalated into the intergalleries of the layered clay mineral keeping its overall crystal structure more or less intact but increasing the interlayer distances. Clays can be either cationic—like montmorillonite, hydro-micas, or kaolinite (for examples, see [1, 4, 13–17])—or anionic in nature. Examples for the latter case are Layered Double Hydroxides (LDH, for an overview see [18–20]). LDHs are natural minerals but they can be also synthesized with a broad range of chemical compositions. The most common naturally occurring LDH is hydrotalcite. LDHs are well known for their catalytic activities [21] and due to the large amount of tightly bound water as well as synergistic effects they are able to enhance the flame retardancy of polymer-based nanocomposites [22, 23]. In a sense, graphene oxide and/or graphene can be also regarded as 2D nanofiller materials [24]. Especially graphene has attracted an enormous scientific interest due to its exceptionally high electron

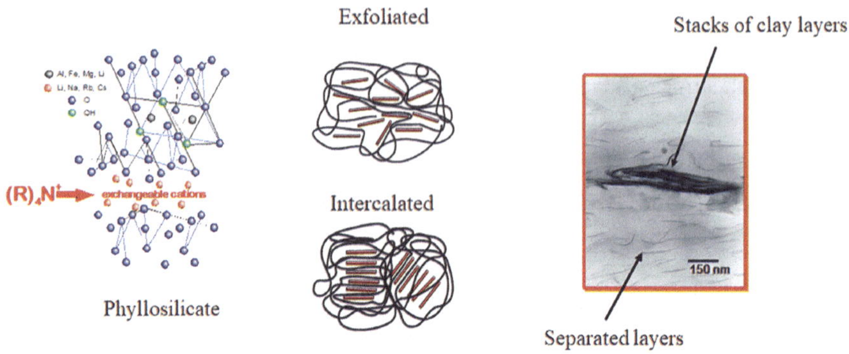

Fig. 1 Edge morphologies for a nanocomposite with layered clay material as nanofiller

transport ability, mechanical properties, and high surface area. Due to the atomically thin graphitic carbon layers, significantly improved properties can be expected already at quite low loadings when graphene is properly incorporated into a polymer matrix.

Tube-like materials can be considered as one-dimensional (1D) nanomaterials. They are characterized by a high aspect ratio. Examples are for instance Carbon Nanotubes (CNTs) [25–28]. CNTs can be single-walled or multiwalled. They are further characterized by a reasonably high electrical conductivity. Corresponding nanocomposites can be employed as electrical shielding materials, e.g., to reduce electro smog. Besides CNTs Halloysite Nanotubes (HNTs) gain an increasing attention [29] as they are cheaper, less toxic, and more environmentally friendly compared to CNTs. Halloysite is an aluminosilicate, and HNTs have a hollow tubular structure with an outer surface being chemically like SiO_2 [30]. Further types of one-dimensional nanofillers are nanorods and nanowires (see, for instance, [31, 32]). These nanofiller materials are mostly prepared from metals or metal oxides. Organic 1D nanofillers are, for instance, whiskers based on cellulose (see [33]) or nanocellulose fibers [34, 35]. Moreover, electrospun nanofibers made from various polymers can be also employed as nanofiller [36].

One aspect responsible for the very effective improvement of the properties of a nanocomposite compared to the unfilled polymer is the small size of the nanoparticles. Reducing the spatial size of the filler particles from that of a colloidal (for instance, colloidal carbon) to nanosized fillers leads to an increase of the surface area by about six orders of magnitude. This means that the amount of polymer segments located in the interfacial regions between the filler particles and the polymer matrix is high. This is sketched according to Ref. [37] in Fig. 2. Hence, for the performance of the nanocomposites firstly a homogenous dispersion of the nanoparticles on a molecular level is expected to be of essential importance. This aspect also points to a direct tailoring of the interface between the nanoparticles and the polymer matrix for a further improvement of the properties of nanocomposites. Secondly, the improvement of the mechanical properties of an elastomer used for tires or the electrical

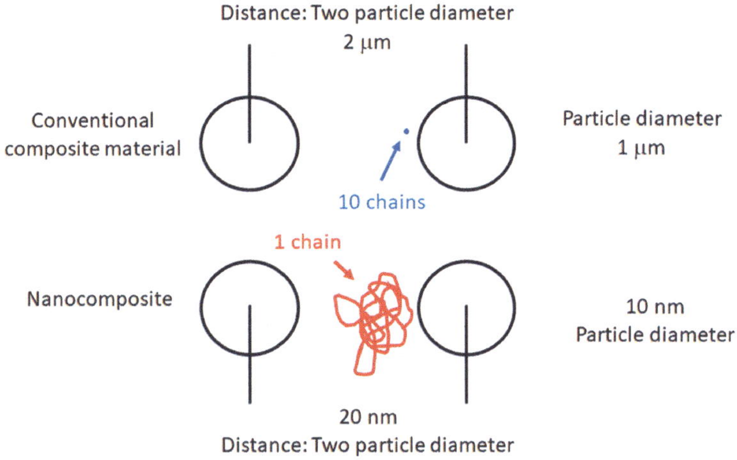

Fig. 2 Simplified scheme demonstrating the different size relationship of the polymer with regard to the filler particles for a conventional micronized filler particle and a nanoparticle. Modified from Ref. [37]

properties of polymers used for shielding are due to the percolation of the nanoparticles forming a network which can be considered as a hierarchical superstructure (see for example, [28, 38–48]). Especially for electrically conducting nanocomposites a contact between the surfaces of adjacent nanoparticles forming a continuous network is required (see Refs. [48–51]). This means that at least for electrical applications a homogenous distribution of the nanoparticles in the composite will not always lead to a maximum of its performance. This point is illustrated in Fig. 3a where the real part of the complex conductivity is plotted versus frequency for polycarbonate filled with 2 wt% multiwalled carbon nanotubes [52]. For the as received samples with a good dispersion of the nanotubes on a molecular scale, the DC conductivity σ_{DC}, represented by the frequency independent plateau value, is relatively low. The increase of σ_{DC} with the concentration of the nanotubes is smooth, rather than showing the expected step-like percolation behavior (see Fig. 3b). After a first heating run, where the samples were heated to temperatures well above the glass transition temperature (T_g) in the rubbery state (190 °C), the conductivity of a subsequent measurement at room temperature increases by more than two orders of magnitude. Also, the concentration dependence of the DC conductivity changes to a more percolation-like transition (Fig. 3b). The CNTs are not completely miscible with the polymer in a thermodynamical sense. Therefore, at temperatures above T_g the carbon nanotubes segregate into nanoclusters [48] and the conductivity is due to a percolating network of these nanoclusters. The existence of such clusters was also confirmed by gas sorption measurements (Fig. 3c), confirming a substantial rearrangement of the nanofiller. The concentration of sorbed carbon dioxide for the freshly prepared samples is similar to the unfilled polycarbonate and much higher

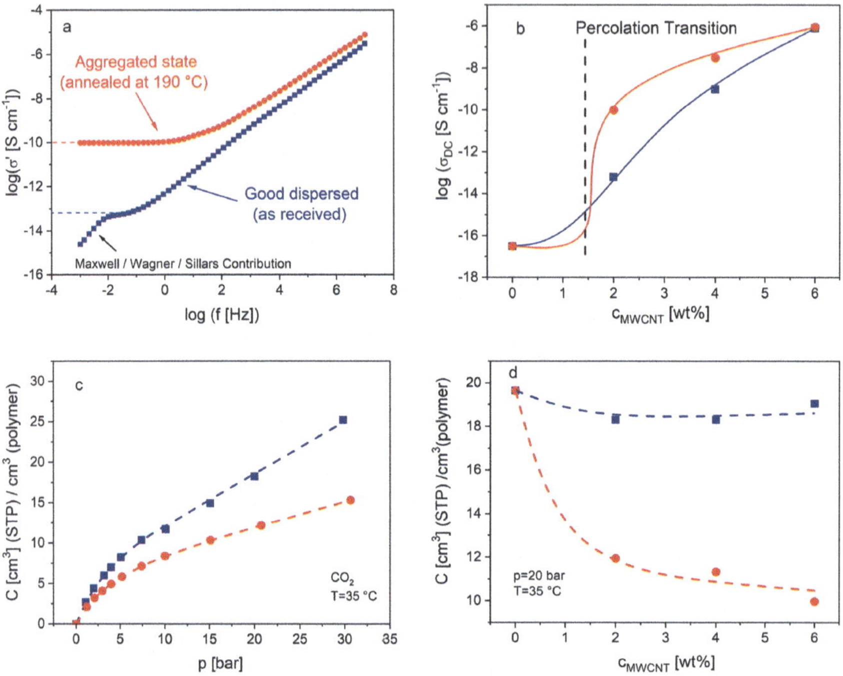

Fig. 3 Influence of the state of dispersion of the nanoparticles on properties of polycarbonate with multiwalled carbon nanotubes: Blue squares—as received in a good dispersed state, red circles—after annealing above the glass transition temperature at 190 °C: **a** Real part of the complex conductivity σ' versus frequency f at 25 °C for 2 wt% of multiwalled CNT. **b** DC conductivity σ_{DC} versus concentration of the multiwalled CNT. Lines are guides to the eyes; **c** Sorption isotherms of carbon dioxide at 35 °C for 2 wt% of multiwalled CNT. Lines are guides for the eyes. **d** Concentration of sorbed carbon dioxide at 20 bar gas pressure versus concentration of the multiwalled CNT at 35 °C. Lines are guides for the eyes. Unpublished results

than that of the annealed system because the gas cannot be solved in the clusters of the CNTs (see Fig. 3d).

Thirdly, to further tune the properties of polymer-based nanocomposites a control of the distribution of the nanoparticles from a homogenous dispersion on the nanoscale to self-assembled structures might be important. Such self-assembled structures can be strings, sheets, or different cellular structures, in some cases inspired by nature (for examples, see [53–57]). Nevertheless, such an advanced control of the distribution of the nanofiller remains challenging.

To improve the interaction with the polymer matrix, the surface modification of the nanoparticles is one technologically effective approach. Both methods, grafting-from and grafting-to, have proven useful to attach polymer chains to the surface of nanoparticles with different grafting density ranges [58–61]. Due to the steric constraints limiting the diffusion of polymer chains to the surface of the nanoparticles, the grafting-to approach is restricted to low grafting densities [62–64]. Higher chain grafting densities in the range from 0.01 to 0.8 chains/nm^2 can be obtained by grafting-from techniques.

These grafting-from approaches are based on a modification of the surface of a nanoparticle by a so-called coupling agent [62] which has a certain chemical functionality. For instance, silica surfaces can be modified by a silane as coupling agent carrying −OH, −COOH, $-HN_2$, or $-CH{=}CH_2$ groups [61–66]. These groups can then be further involved in different polymerization reactions, like atom-transfer radical polymerization [67], reversible addition-fragmentation chain transfer [68], or nitroxide-mediated polymerization [69]. These techniques allow to control many properties like chain density, molar mass distribution, as well as molecular architecture. These parameters are important to tune the dispersion state of nanoparticles and their interaction with the matrix, and therefore the final properties of the nanocomposites. Besides for silica based nanofillers (see, for example, [65]), similar approaches have been developed to modify other nanofillers like titanium nanoparticles, barium titanate or CNTs, just to mention a few [26, 70–72]. In addition to silane coupling agents, other approaches have been developed which are based on phosphate/phosphonate, amines or carboxylic compounds [63]. Furthermore, Azo- and Thiol-based strategies have been considered [73, 74] where the latter approach is mainly used for gold nanoparticles.

Clay minerals as 2D nanofiller contain counterions in the intergalleries between the clay layers. These small ions can be exchanged by bulky organic ions. Due to this exchange, the distance between the clay layers is increased which will enable an intercalation of polymer chains into the galleries or support the delamination of the layers leading to an exfoliated structure of the nanocomposite. Examples for bulky organic ions are, for instance, alkylammonium or alkylphosphonimum ions with long alkyl chains employed for cationic minerals like montmorillonite, hectorite or saponite [13] or sodium alkyl sulfonates for LDH [18]. An example for the widening the interlayer distances by ion exchange is given in Fig. 4 for the modification of an LDH material by sodium dodecylbenzensulfonate (for details see [75]).

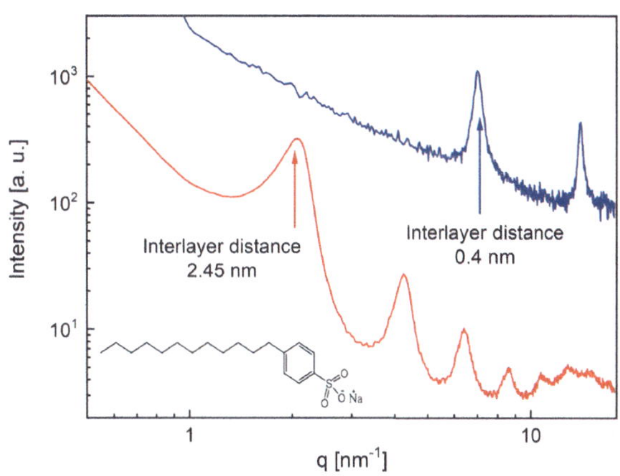

Fig. 4 SAXS pattern of unmodified (blue) and organically modified ZnAl-LDH (red). The inset gives the chemical formular of sodium dodecylbenzene sulfonate. Adapted with permissions from Ref. [75]. Copyright (2011) American Chemical Society

In some cases, the ions in the intergalleries are exchanged by ions carrying a chemical functionality. Due to possible functional groups of the exchanged ion, it can react with the polymeric matrix to improve the interaction of the nanofiller with the polymer by covalent bonds. One example is discussed for an epoxy-based nanocomposite filled with a taurine modified LDH [76]. Due to the amine groups of taurine the nanofiller is chemically linked to the epoxy.

3 Characterization of the Dispersion of the Nanofiller

The dispersion of nanoparticles in a polymer matrix or melt is determined by both entropic and enthalpic factors. In principle thermodynamic approaches can be used to predict the morphology of the nanocomposites. Nevertheless, in practical cases these approaches are limited by kinetic factors such as viscosity etc. In addition, external forces like mechanical (shear) or electrical fields can be employed to reach or to modify the distribution of a nanofiller in a polymeric matrix. This point concerns also the industrial processing of nanocomposites like extrusion (see, for example, [77, 78] and references cited).

A classical method employed to directly visualize the distribution of nanoparticles in a polymeric matrix is transmission electron microscopy (TEM). Two examples for TEM images of polymer-based nanocomposites are depicted in Fig. 5. As first example Fig. 5a and b give TEM images of a nanocomposite based on polylactide and NiAl-LDH at different magnifications [79]. The TEM pictures show larger aggregates but also smaller objects and exfoliated layers. The second example in Fig. 5c shows silver and/or silver-ion particles in a layer of a copolymer of acrylic acid and CO_2 [80], whereas Fig. 5d depicts the histogram of the Feret diameter of the nanoparticles.

The characterization of the distribution of nanoparticles by TEM is not accurate enough because it is space-resolved in nature and might be subjectively biased. For a detailed integral characterization scattering tools like X-ray or neutron scattering are more appropriate. In the case of low concentrations of spherical nanoparticles, the scattering pattern shows a strong increase with the scattering vector q. From the structure factor a mean number of particles in clusters can be extracted when the clusters are well dispersed in the nanocomposites. For higher concentrations of nanoparticles, models of the structure factor like the Percus–Yevick model for hard spheres has to be involved. Some examples for spherical nanoparticles can be found in Refs. [81–83]. An empirical multi-scale fractal approach was developed by Beaucage et al. to describe the interaction in the aggregates [84].

As one recent example for a more complex behavior, a detailed analysis of small- and wide-angle X-ray scattering (SAXS and WAXS) data was carried out for an epoxy thermoset nanofilled with boehmite [85]. From a linear combination of the data of the pure components in the WAXS region the concentration of the nanofiller in the composites was confirmed (see Fig. 6a). The concentration of the nanofiller in the composites can be also estimated from the SAXS data using the scattering length of both compounds and a Monte Carlo-based analysis (McSAS [86]) which

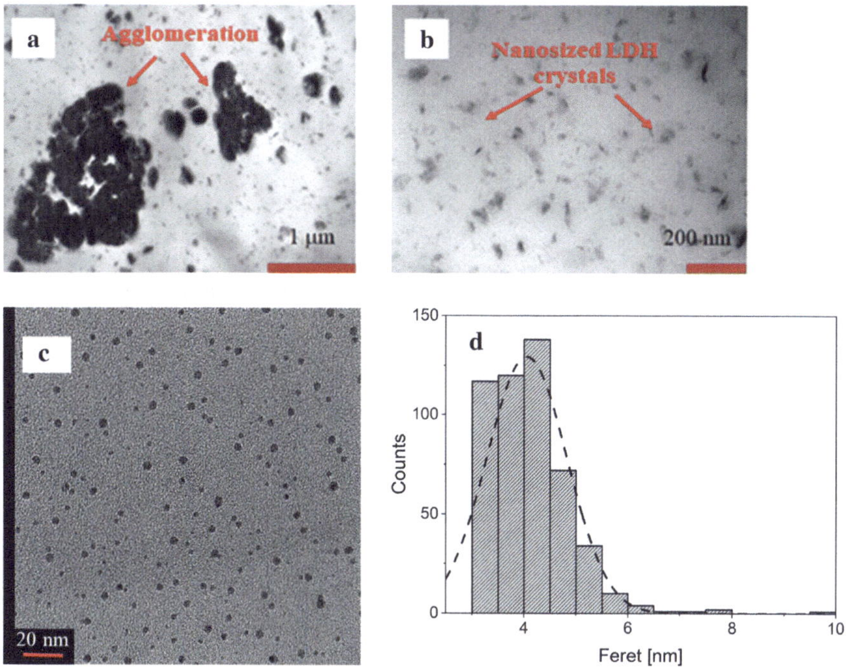

Fig. 5 Characterization of nanocomposites by transmission electron microscopy (TEM). **a** TEM image for nanocomposite based on PLA and NiAl-LDH with a concentration of 3 wt% of LDH at a low magnification. **b** TEM image for nanocomposite based on PLA and NiAl-LDH with a concentration of 3 wt% of LDH at a high magnification. Details of the nanocomposite system can be found in Ref. [79]. **c** Silver and/or silver-ion particles in a in a layer of a copolymer of acrylic acid and CO_2. **d** Feret distribution of the size of the silver or silver-ion nanoparticles. Images c and d were adapted with permission form Ref. [80]. Copyright (2020) American Chemical Society

provides comparable results to the linear combination of the WAXS data (see Fig. 6a). Employing a structure factor for the primary boehmite particles the mass fractions of the primary particles (<30 nm) and of the particles in nanodomains (30 nm < d < 1 µm) could be estimated (see Fig. 6b). Finally, the mass fraction of aggregates (>1 µm) could be extracted which is found to be low (see Fig. 6c). A related approach was also applied to a different epoxy thermoset with LDH as nanofiller [76] and for a thermoset filled with halloysite nanotubes [87].

Using a microfocus beamline, SAXS measurements can also be employed to investigate the homogeneity of nanocomposites on a macroscopic length scale. For that purpose, X-ray patterns are measured at different spatial positions of the sample and overlaid. When different Bragg peaks of the nanofillers for the nanocomposite overlap also in intensity, it can be concluded that the nanocomposites are homogeneous at a macroscopic length scale [75, 88]. An example is given in Fig. 7 for a nanocomposite system based on polypropylene and LDH as nanofiller.

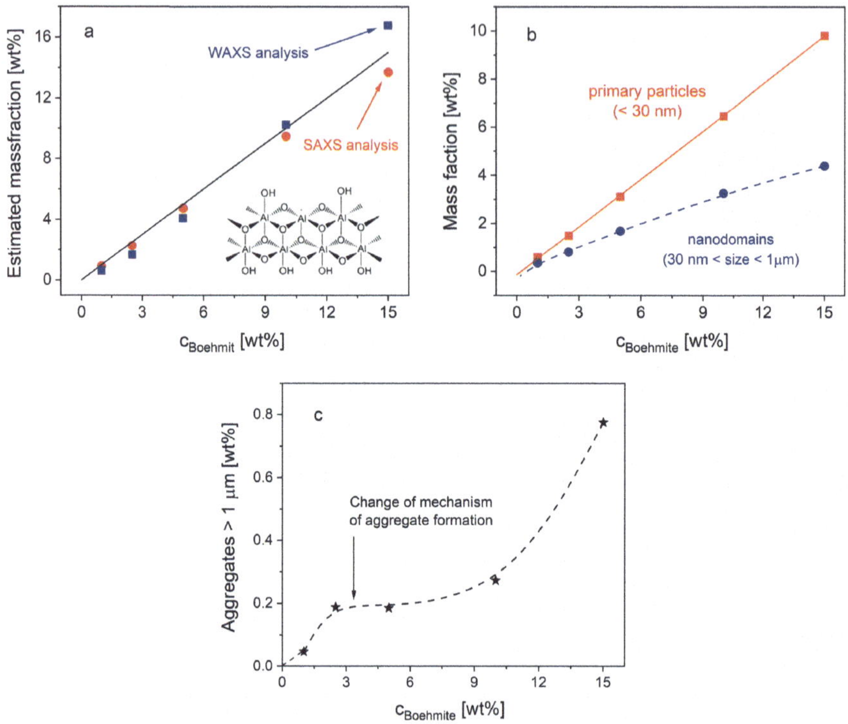

Fig. 6 Analysis of X-ray measurements for epoxy-based nanocomposite with boehmite as nanofiller. For details see Ref. [85]. **a** Concentration of the nanofiller in the nanocomposite versus the formulated concentration. Blue–from an analysis of WAXS data, red–from an analysis of SAXS. The line is the theoretical expectation. The inset gives the chemical structure of boehmite. **b** Mass fraction of primary particles (red, smaller than 30 nm) and nanodomains (blue, 30 nm ≤ size < 1 μm versus the concentration of the nanofiller. The Lines are guides for the eyes. **c** Mass fraction of aggregates (size larger than 1 μm) versus the concentration of the nanofiller. The line is a guide for the eyes. The figure is adapted from Ref. [85] with permission from The Royal Society of Chemistry

In some cases, the distribution of nanoparticles can be also evaluated by dielectric spectroscopy. This approach is based on the Maxwell/Wagner/Sillars (MWS) polarization. For a nanocomposite based on grafted polypropylene and clay, in addition to relaxation processes, a process related to a MWS polarization is observed [89] (see Fig. 8a). For the discussed nanocomposite system, the MWS polarization is due to the blocking of charge carriers at the clay layers. For well-dispersed nanoparticles, the distance between the clay layers is small and time constant for the corresponding MWS process is shorter, compared to the case when the nanofiller is poorly dispersed (see Fig. 8b). This behavior can be discussed in the framework of a simplified model considering an electrical double layer. In that model the time scale of the corresponding MWS process can be expressed as [89]

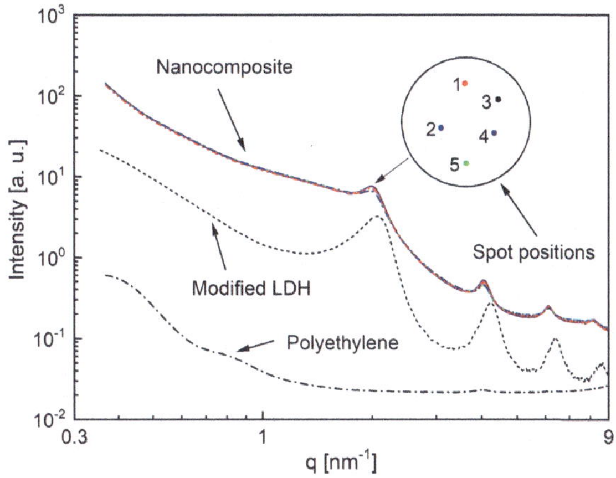

Fig. 7 Overlay of five SAXS patterns for a nanocomposite based on polyethylene and organically modified ZnAl-LDH (solid lines). Dashed line—organically modified ZnAl-LDH. Dashed-dotted line—polyethylene. The curves are shifted along the y-scale for clarity. Adopted with permissions from Ref. [75]. Copyright (2011) American Chemical Society

$$\tau_{MWS} = \frac{\varepsilon_s \varepsilon_0}{\sigma_{DC}} \frac{d}{2L_D}. \tag{1}$$

ε_0 is the permittivity of vacuum, ε_s is the static permittivity of the material, σ_{DC} is the DC conductivity, L_D is the thickness of the Debye layer, and d the average distance between two clay layers ($d >> L_D$). Assuming that the time constant τ_{MWS} is proportional the average distance between separated clay layers in the nanocomposite, the following relation holds:

$$\frac{\tau_1}{\tau_2} = \frac{f_2}{f_1} \approx \frac{d_1}{d_2} \tag{2}$$

The comparison of the rates of the MWS process for the nanocomposite with well and poorly dispersed clay reveals that the rate is 1.5 orders of magnitude higher for the former one (see Fig. 8b). This line of argumentation is supported also by the concentration dependence of the rate of the MWS process for a polypropylene/clay system (see Fig. 8c) with a good dispersion of the nanoparticles. If the concentration of the exfoliated clay layers is increased by a factor of four, the rate also increases by a factor of four.

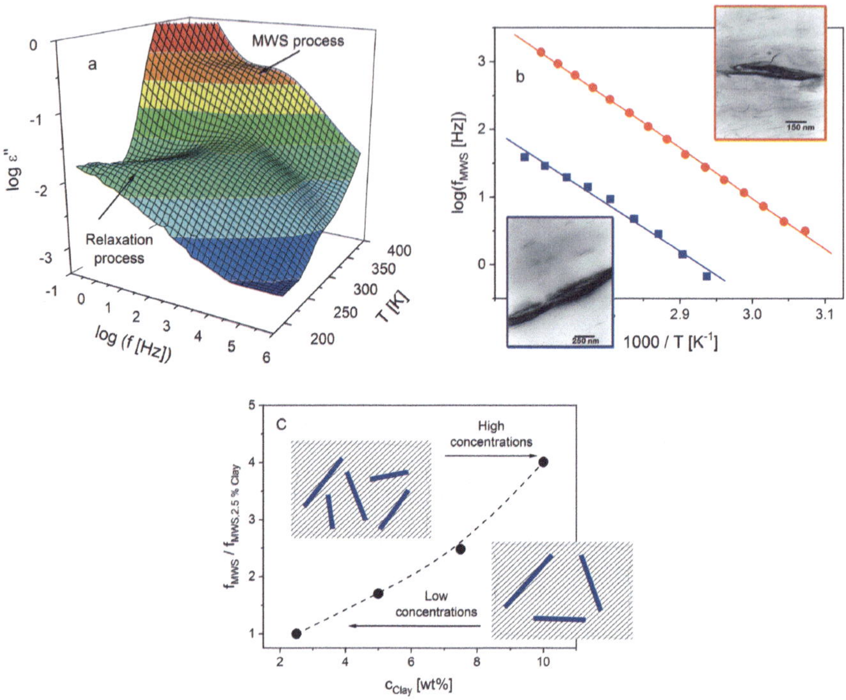

Fig. 8 **a** Dielectric loss versus frequency and temperature for a nanocomposite of anhydride grafted polypropylene nanofilled with well-dispersed 2.5 wt% clay. **b** Rate of the Maxwell/Wagner/Sillars process f_{MWS} versus inverse temperature: blue—poor dispersion, red—good dispersion. The insets show the TEM images of the nanocomposites: Red framed—well-dispersed clay. Blue framed—poorly dispersed clay. **c** Concentration dependence of the rate of the Maxwell/Wagner/Sillars process normalized to that of 2.5 wt% measured at 373 K. The line is a guide for the eyes. Unpublished results

4 Applications

As discussed above, nanocomposites can have a broad range of applications, for instance, as materials with enhanced flame retardancy, as filled rubbers used in tires, as construction materials, as filled plastics or thermosets, mixed matrix membranes for separation and barriers, dielectric materials for capacitors and insulation and active materials in sensors or solar cells, just to mention a few. Some of these applications will be briefly discussed in this section.

4.1 Nanocomposites with Enhanced Flame Retardancy

In spite of the great importance, flame retardancy of nanocomposites is not in the central scope of this chapter. Therefore, here only some examples for review literature are given [90–97]. The mechanism to reduce the flammability by the incorporation of nanoparticles is mainly to reduce the heat release rate due to a delay of the thermal or thermo-oxidative decomposition. Moreover, the active sites at the nanofiller can have a catalytic activity and may influence the decomposition process. Also catalyzed dehydrogenation and crosslinking will lead to a higher flame retardancy. Furthermore, the nanofiller will also stimulate the formation of a carbonaceous char which will suppress the decomposition and reduce the dripping of flaming polymer also by crosslinking with the nanoparticles.

4.2 Rubber Nanocomposites

Rubbers are crosslinked/network-forming polymeric materials having a glass transition temperature well below their application temperatures. Tires are probably the most relevant application for nano-filled rubbers (see, for instance, [38–40, 98–102] because the amount of literature is vast). The performance of tires is characterized by the magic triangle of rolling resistance, wear (abrasion) resistance and wet traction. The rolling resistance should be as low as possible to increase the fuel efficiency, while the wear resistance and the wet traction should be high. All these factors must be considered in the engineering of tires, keeping its performance high, while weight as well as the production costs should be low. The first filler particles for rubbers used in tire applications were colloidal silica and carbon black. It was found that an improved dispersion of the filler particles but also a percolating network of more or less aggregated filler particles is important to improve the properties of tires. Moreover, research showed that a decrease in the size of the aggregates of filler particles during the processing is a key issue to further improve the mechanical properties of filled rubbers [103, 104].

Furthermore, it was stressed that the formation of a bound layer is most relevant in the mechanical reinforcement (see, for examples, [105–109]). It should be noted that the formation of a bound layer can be due to adsorption and/or chemical bonds. From the latter point silane coupling agents play an essential role. Moreover, the improvement of the mechanical properties of filled rubbers is also due the nanoparticles by a kind of secondary effect. The nanoparticles will modify the polymeric network and the properties related to it, i.e., the Payne effect [101]. The Payne effect is found to be critical for the nano-filled rubbers (see, for instance, [38–40, 110, 111]). In conclusion, it must be noted that till now neither the effect of the bound layer nor that of the nanoparticles is completely understood. This also concerns the scaling up of procedures to control the dispersion developed at the lab scale to industrial processes.

4.3 Nanocomposites with Thermoplastics and Thermosets as Matrix

In contrast to rubbers, the application temperatures of thermoplastics and thermosets are below their glass transition temperatures. It should be noted that the first nanocomposite with a practical application as a belt cover developed by Toyota Central Research was a thermoplastic polyamide (Nylon-6) nanofilled with a clay system [112]. This application was further extended using such nanofillers in bumpers as well as in fuel tanks. For these applications mainly the enhancement of the mechanical properties and/or the thermal characteristics (deflection temperature, flame retardancy) are in the focus of the practical interest. This point concerns also the applications of epoxies in high performance applications (see, for instance, [5, 113–120]). Epoxy-based thermosets are characterized by glass transition temperatures much higher than room temperature. They have a high hardness, good adhesion, and high mechanical stability but they are relatively brittle materials. Due to their characteristics, epoxy have applications in automotive, aerospace and shipbuilding industries, but also as coatings, laminates, adhesives, and in electrical insulations [116, 121]. For instance, as adhesive materials epoxy materials can be applied between a substrate and electronic chip to minimize mechanical stress due to the differences in the thermal expansion coefficients. The brittleness of epoxy thermosets can be reduced by the incorporation of nanoparticles but by trade-off relationships the elongation to break or the yield stress might decrease. For instance, it was shown for an epoxy filled with copper oxide nanoparticles that mechanical properties have a maximum at 2.5 vol% [122]. At higher concentrations the nanoparticles start to agglomerate leading to a decrease in the material performance. A similar result was also found epoxy systems nanofilled with boehmite [85].

Most of the industrial composite systems used today are based on thermoplastics and thermosets having filler particles within the micrometer scale. Examples for that are continuous fiber reinforced composites (CFRC). The idea for further improvements in that field is not to replace the filler with sizes on the micrometer scale completely but to add nanoparticles to enhance the properties of CFRC. This approach is mainly realized using nanoparticles with a high aspect ratio to attain compatibilization with the fibers, i.e. clay materials (see, for example, [123, 124]) or CNTs (see, for example, [27, 28, 125]). After the discovery of graphene also this nanofiller was employed for that purpose [24]. By this concept an improvement of the properties of CFRC by 10–35 % has been reached at low filler concentrations up to 5 wt%. Nevertheless, the processing of these materials is found to be challenging due to the high viscosity, the agglomeration of the nanofiller and possible void formation. A surface modification of both the micronized filler particles, as well as the nanoparticles might be able to resolve these problems.

A further important area of the application of nano-filled thermoplastics is in packaging and food industries [126]. These applications take advantage of two effects, the extension of the diffusion pathways d_w (tortuosity effect) of gas molecules or other small molecules, and the interface between the nanoparticle and the polymer matrix.

The transport of small molecules through dense polymers is usually described in terms of the solution diffusion model [127]. This means that a gas molecule is solved in the free volume of the polymer and then diffuses through the membrane. The diffusion is characterized by the diffusion coefficient D which is mainly triggered by the molecular mobility of the polymer matrix. The solubility S of the gas molecules is determined by the chemical structure of the polymer and the amount and distribution of free volume. The permeability P is then given by

$$P = D \cdot S. \tag{3}$$

For barrier applications platelet-like nanofillers with a high aspect ratio like clay materials are preferred. For a nanocomposite with delaminated clay layers with a thickness w_{Clay} and a length l_{Clay} the tortuosity τ_{Clay} is given by

$$\tau_{Clay} = \frac{d_w}{d_{w,clay}} = \frac{l_{Clay}}{2w_{Clay}} f_s. \tag{4}$$

Here, f_s is a factor related to the concentration of the nanoparticles, the degree of delamination, and the orientation of the clay layers. The permeability P_{Clay} is then given by

$$P_{Clay} = \frac{(1 - f_s)}{\tau_{Clay}} P. \tag{5}$$

The effect of the tortuosity is shown in Fig. 9a where the permeability is depicted for a well-dispersed nanocomposite of a modified polypropylene with montmorillonite clay [128]. For details of the systems see [89]. For the three gasses, carbon dioxide, oxygen, and nitrogen, the permeability is reduced by ca. 50 % by the incorporation of only 5 wt% of the nanofiller.

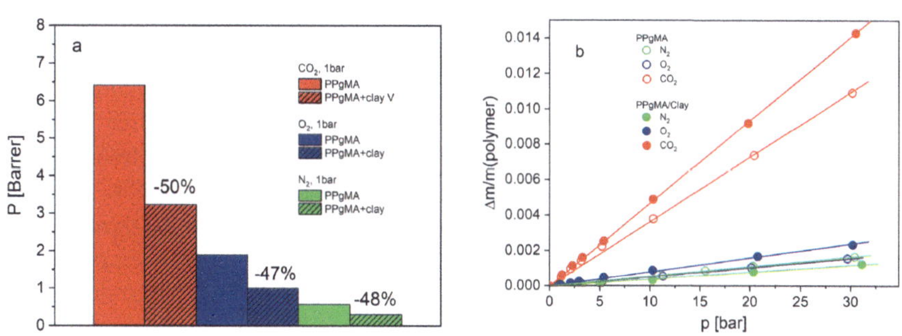

Fig. 9 **a** Permeability P at 308 K measured for different gasses as indicated for grafted polypropylene and the corresponding nanocomposite. **b** Relative mass change versus pressure at 308 K for the pure grafted polypropylene and the corresponding nanocomposite as indicated. Lines are linear regression to the corresponding data. Unpublished results

The presence of an interfacial layer between the nanofiller will mainly influence the solubility of the small molecules in the nanocomposite. This effect is shown in Fig. 9b. For the nanocomposite the amount of sorbed gas is increased by 27 % for carbon dioxide and by 54 % for oxygen in comparison to the unfilled polymer. For nitrogen the amount of sorbed gas is decreased by 30 %. The change in the sorption process for the nanocomposite compared to the unfilled polymers is a consequence of the interface between the nanoparticle and the polymer.

4.4 Nanocomposites for Membranes in Separation Processes

The separation and processing of mixtures of gasses or other small molecules are integral parts of many processes in chemical industry. Examples are the purification of natural gas, the upgrading of biogases, carbon capture, separation membranes in fuel cells (see, for instance, [129]) as well as water treatment (see, for examples, [130, 131]). Membrane processes are considered as essential future technology because they have a high potential in energy saving, compared to conventional materials in separation operations. Membrane-based processes are usually more cost and energy efficient, compared to other separation technologies.

In the following paragraph, the focus is on membranes for gas separation processes. One decisive performance characteristic of membranes for gas separation is the permeability P defined by Eq. (3). The second characteristic of gas separation membranes is the permselectivity α. The ideal permselectivity of a membrane is defined as the ratio of the gas permeabilities $P1$ and $P2$ of a gas pair (product purification). Generally, the permeability and the selectivity are subjected to a trade-off relationship—easily discernible when the permselectivity of a gas pair is plotted versus the permeability of the more permeable gas. The trade-off relationship implies that the selectivity decreases with increasing permeability and vice versa as described by the empirical Robeson upper bound [132]. The selectivity is basically controlled by the difference in the diffusion coefficients and solubility coefficients of a pair of gases. While the diffusion is mainly controlled by kinetic factors, such as the mobility of the polymer, the sorption is predominantly determined by thermodynamics. For glassy polymers both, diffusivity and solubility are affected by the free volume, i.e., its amount and its size distribution. Therefore, manipulating the free volume is a useful way to improve the performance of gas separation membranes. The free volume is either assigned to nanoscopic holes (voids) or to the molecular dynamics in the polymers [133, 134]. Synthetic efforts have been undertaken to create polymers with a high fractional free volume in the glassy state. Examples are polyimides (see, for examples, [135–137]) or polyacetylenes [138, 139]. Some more recent developments in the field concern the new class of polymers of intrinsic microporosity (PIMs) [140–143] and microporous polynorbornenes with bulky side groups [144, 145]. One common issue of all these (super) glassy polymers is that they undergo a strong physical aging process, related to the molecular mobility, which decreases their performance as active separation layers in gas separation membranes.

It was shown by several researchers that nanoparticles can change the free volume of the matrix and thus change the gas transport properties of the system in favorable way [146]. Freeman and coworkers reported an increase of both the permeability and the selectivity for the mixture n-butane/methane. This result obtained for the high free volume polymer poly(4-methyl-1-pentene) [147] filled with silica nanoparticles was quite unexpected because according to the simplified picture of macroscopic tortuosity discussed above (see Eq. 5) the permeability should decrease (see Fig. 9a) while the selectivity should remain unchanged. It can be discussed that the increase of both the permeability and the selectivity results from the unfavorable interactions between the polymer and the filler particles leading to an increase in free volume [148]. Nowadays these membranes are known as mixed matrix membranes [149].

In Ref. [150], the gas transport properties of CO_2 through polycarbonate membranes filled with Phenethyl-POSS have been investigated (see Fig. 10a). It

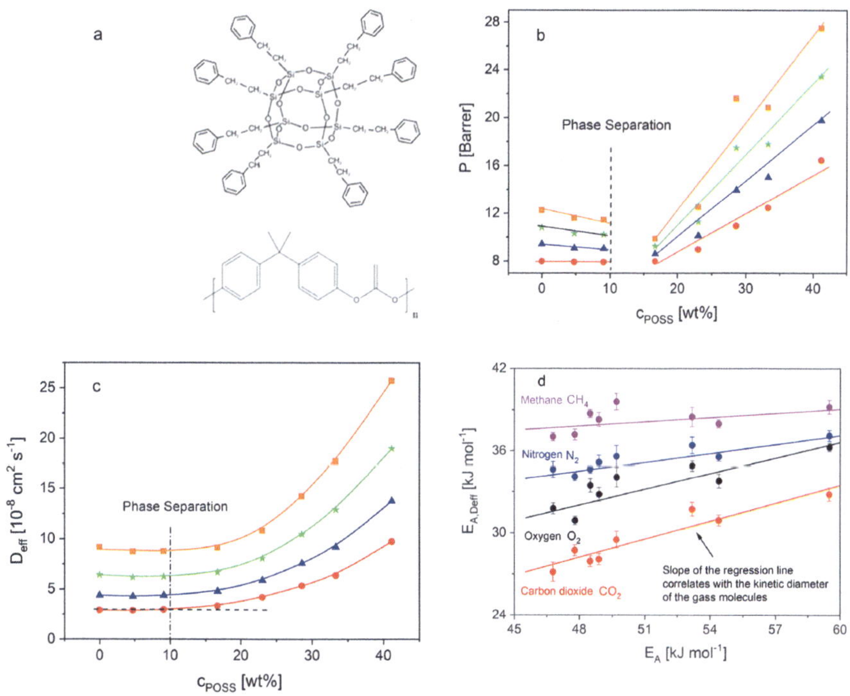

Fig. 10 **a** Chemical structures of Phenethyl-POSS and polycarbonate. **b** Permeability versus POSS concentration at 10 bar: Red circles—308 K; blue–triangle—318 K, green–asterisk—328 K, yellow squares—338 K. Lines are guide for the eyes. The figure was adopted from Ref. [150]. **c** Effective diffusion coefficient versus POSS concentration at 10 bar. Red circles—308 K; blue–triangle—318 K, green–asterisk—328 K, yellow squares—338 K. Lines are guides for the eyes. The figures **b** and **c** were adopted with permissions from Ref. [150]. Copyright (2010) American Chemical Society. **d** Activation energy for gas diffusion versus activation energy of the localized β-relaxation measured by dielectric spectroscopy for the indicated gases. Lines are linear regressions to the corresponding data. Data was taken from Ref. [152]

was shown by dielectric investigations that the system is miscible on a molecular scale up to a concentration of ca. 7 wt% Phenethyl-POSS [151]. For higher concentrations of POSS, the membranes undergo a nanophase separation in POSS-rich and polycarbonate-rich domains. Fig. 10b depicts the permeability of the POSS/polycarbonate versus the concentration of Phenethyl-POSS for different temperatures. Up to ca. 7 wt% the permeability decreases with increasing filler concentration as expected from the macroscopic tortuosity arguments or a decrease of the solubility. This behavior changes in the nanophase separated state where the permeability increases with further increase of the concentration of POSS. From time-lag measurements also an effective diffusion coefficient D_{eff} is estimated (see Fig. 10c). D_{eff} is approximately constant up to 7 wt% of POSS but increase with the POSS concentration in the nanophase separated state. As the solubility normalized to the amount of polymer is independent of the POSS concentration in the nanophase separated state [150] the increase of the permeability for POSS concentration higher than 7 wt% can be solely related to the increase of the diffusion coefficient which can be related to an increase of the free volume introduced by the nanophase separation. Furthermore, gas sorption measurements (sorption isotherms) reveal also the importance of the interfacial layer evidenced by dielectric spectroscopy [151] for the gas transport properties. Finally, from the temperature dependence of the diffusion coefficient D_{eff} an effective activation energy $E_{A,D_{eff}}$ can be deduced. $E_{A,D_{eff}}$ decreases up to 10 wt% of POSS and remains constant in the nanophase separated state. It could be shown that $E_{A,D_{eff}}$ correlates to the activation energy of the β-relaxation E_A estimated by dielectric spectroscopy not only for CO_2 but also for other gasses like oxygen, nitrogen, or methane (see Fig. 10d) [152]. This correlation evidences the importance of localized molecular fluctuations for the gas transport properties. This picture is supported further as the slope of the regression lines correlates with the kinetic diameter of the gas molecules. The discussed example underlines the interplay of structure and molecular mobility of a polymer for its performance as gas separation membranes.

Further progress in the development of mixed matrix membranes is made by adding nanoparticles having a certain porosity like zeolites, metal–organic frameworks (MOF), polymer-organic frameworks (POF), and/or Zeolitic-Imidazolate Frameworks (ZIF). For a recent review see, for instance, [153]. While solid nanoparticles are able to influence dynamics and free volume of a polymer matrix, nanoparticles exhibiting a well-defined pore structure itself may additionally contribute to selectivity due to the corresponding sieving properties. The chemical versatility of these fillers allows to tailor the selectivity for specific applications.

It was shown that nanoparticles generally might be able to reduce the physical aging in glassy polymers, i.e., the gradual decrease in free volume (see, for instance, [154]). This approach can also be adapted to stabilize their gas transport properties. A prominent example is the incorporation of nanoparticles to super-glassy polymers and PIMs [155].

4.5 Nanocomposites for Dielectric and Sensor Applications

Inorganic filler particles can be employed to tune the electric/dielectric properties of polymers to optimize them for various applications in electrical engineering. An overview can be found elsewhere [156]. The corresponding applications can comprise transistors, high-voltage insulation, or capacitors in power electronics. These applications include the development of materials with a low or high permittivity (low and high k-materials). Besides permittivity, the dielectric loss, the conductivity, or the DC breakdown strength can be tuned by using polymer-based composites [157]. For instance, silicon carbide, boron nitride, silica, aluminium oxide, and titania oxide have been employed as fillers for high-voltage applications [158]. These filler materials are particularly suited for applications which require higher temperatures and/or electrical stress, like the cabling/insulation for transformers, generators or motors. For high-voltage applications, the dielectric permittivity and the dielectric loss (for engineering purpose characterized by dissipation factor tan δ) should be as low as possible. It is further worth to note that inorganic filler particles can not only improve the electric properties but also the thermal conductivity, flame retardancy and mechanical characteristics of the nanocomposites employed for high-voltage applications. Moreover, an appropriate selection and arrangement of the nanoparticles might be also employed to tune the thermal expansion of insulation parts of a device improving the matching with metallic conductors, like the metal wire and the insulation coating of a cable. Especially for cables, but also for other applications (water) treeing is known to reduce the DC breakdown strength E_{BD}. Plate-like nanoparticles like mica platelets are used to improve the long-time performance and reliability of the cables by preventing the propagation of the trees and the associated electrical breakdown (see, for instance, [159]). Another point to be mentioned is that nanoparticles can be used to design anisotropic matrix/filler structures with anisotropic electrical/dielectric properties. Such composites are important for devices which require anisotropic resistivities, for instance, for ground insulation [160]. Due to their low dielectric permittivity values polyolefins are widely used as matrix for such composites (see, for instance, [161]). Besides polyolefins also epoxies are often employed for these purposes (see, for example, [162, 163]).

For high-voltage applications such as in capacitors which are mainly thin film based, other properties besides the permittivity become important, too. This concerns first an ultrafast energy uptake and delivery at times cales in the range of 10^{-3}–10^{-6} s, required in various applications [157]. One might think about electrically driven vehicles, power units of aircrafts and space vehicles, high frequency filtering or power-conditioning devices. Besides the as high as possible permittivity, the energy dissipation should be as low as possible to maximize the energy density U, related in the linear case to

$$U \sim \varepsilon_0 \varepsilon_s E_{BD}^2. \tag{6}$$

These applications are based on more or less conventional thermoplastics due to the necessity of a defect free large area production. Presently, in most cases unfilled thermoplastics are used but also materials with micron-sized fillers are employed. It was shown that the incorporation of nanoparticles can enhance the polarization dynamics and the charge decay rate as well as lower tan δ and space charge effects [158, 164]. In addition to these property improvements, the breakdown strength E_{BD} is increased leading to an increased service time of the corresponding components. It was discussed that these benefitable improvements are due to a refinement of the morphology of the composites and the interfacial area leading to improved filler/matrix interaction. Moreover, the size of the nanoparticles, and concentrations are key issues [157, 165]. Also, a reduction of production-related defects has been considered.

For all applications, the addition of nanofillers will change the distribution of trapping sites within the dielectric material. This is mainly due to strongly increased amount of the interfacial area between the nanoparticles and the polymeric matrix. The change in the distribution of the trapping sites will strongly influence the charge transport and its mechanism. For high-voltage applications, as it was shown, for example, for a polyimide filled with 10 vol% modified silica, that the discharge resistance increased four times [166]. This means that the time to failure is also strongly increased. Nevertheless, this increase in the time to failure is limited to low filler concentrations. For filler contents higher than 10 vol% E_{BD} will decrease [157]. This trade-off relationship is caused by a percolation of the nanoparticles which changes the trapping process, the local polarization, and charge relaxation. This is also observed for nanocomposites employed in applications for high voltages capacitors, where a slightly increased energy density is found for low concentrations of the nanoparticles accompanied with a decrease in the discharge efficiency.

To improve the properties of polymer nanocomposites for electrical applications and to overcome the discussed trade-off relationships it was tried mainly to optimize the interface between the nanoparticles and the matrix. The idea behind these works is to tailor on the one hand the distribution of the electrical field in the nanocomposite to avoid high local field strengths. On the other hand, an improvement of the interface between the nanoparticles and the matrix can hinder the charge transport to increase the discharge efficiency. As examples this concept is, for instance, employed in core–shell structures (see, for instance, [167–169]), gradient interfaces [170], or surface modifications [171, 172].

To optimize dielectric materials for a certain application, it might be useful to predict the permittivity of composites by models. The simplest model is a linear combination of the permittivity values of the filler particles and the matrix. What is typically observed in cases when a filler with a high permittivity is dispersed in polymeric matrix with a low permittivity is a measured ε_s lower than predicted by this simple mixing rule. This behavior is due to local shielding effects of the electrical field by the particles. Therefore, advanced theories have been developed to model the permittivity of random heterogeneous materials. Generally, they are based on effective medium approaches. The most well-known ones are the Maxwell–Garnett and the Bruggeman model [173–175]. For applications see also [157, 176–178]. It should be noted that these models are purely static which do not include charges and their distribution which might contribute to the polarization. Furthermore, interfaces including the relaxation dynamics are also not considered by these models.

The effective medium approaches have been extended by core/shell models [179–181] to model and to understand the role of the interfaces. Also, the frequency dependence of the complex permittivity has been considered in some cases. Nevertheless, the quantitative comparison of the model predictions with the experimental data remains poor. The weak predicting strength of the models requests an improvement of the theoretical approaches including simulations.

For the applications of nanocomposites in sensors or devices for energy conversion like solar cells, the nanoparticles must be electrically conductive [182]. The charge transport in these systems might be related to the molecular mobility of the matrix in these systems. Examples for conductive nanoparticles for these applications are graphene, graphene oxides, or reduced graphene oxides imbedded in a conductive matrix like poly(3-hexylthiophene) or polyaniline. For a recent review see, for instance, [183]. A second class of important fillers for these applications are fullerenes [184]. Moreover, besides conductivity the nanoparticles can act as donors or acceptors which are important for the function of these materials. Like for the applications discussed above also here the interface between the polymer matrix and the nanoparticles plays an important role to reach a homogeneous distribution of the nanoparticles in the matrix. This is mainly due to the fact that a recombination of charge carriers is taking place on a length scale of few nanometers. Therefore, the modification of the surface of the nanoparticles is also a way to improve the properties. One example is sulfonated graphene which can be copolymerized with poly(3,4-ethyldioxythiophene) [185].

Besides photovoltaics conductive nanocomposites can be employed in organic light emission or electrochromic devices.

5 Molecular Dynamics in Nanocomposites

For a variety of applications of nanocomposites their molecular mobility is of great importance. This issue was already addressed in the sections above like for nanocomposites employed in the field of separation membranes, in electrical shielding materials, where a formation of a percolated filler network is crucial, or for filled rubbers. One can also think of applications where an electrical conductivity is required, e.g., in nanodielectrics or sensors. In general, a certain molecular mobility is required for the processing of nanocomposites. Furthermore, molecular mobility can be used to probe the structure of nanocomposites on a molecular scale.

There is a broader agreement between the researchers that the molecular mobility is modified by the nanoparticles. Most polymers show two dielectrically active relaxation processes due to fluctuations of molecular dipoles [186]. At low temperatures (higher frequencies) a so-called β-relaxation is observed, which is due to localized fluctuations. For the most polymer-based nanocomposites the β-relaxation is only scarily affected by the nanofiller. This is due to the localized character of the relevant fluctuations. This might not apply to thermosets, where the nanoparticles influence the curing process and therefore the morphology of the matrix [187].

The second relevant process for polymers is the segmental dynamics (glassy dynamics, dynamic glass transition), denoted as α-relaxation. In the literature, there is a broad agreement that the segmental dynamics is related to the glass transition observed by differential scanning calorimetry (DSC). This was proven, for instance, by a comparison of the results obtained by broadband dielectric spectroscopy and temperature modulated DSC. The question how the segmental dynamics is influenced by the nanofiller depends on the interaction between the segments and the nanoparticles. In some cases, in the literature the segmental dynamics of the matrix is independent of nanofiller, whereas in others it is acerated or slowed down, compared to that of the matrix. Examples are discussed in several chapters of this book.

As discussed in detail in the sections above, for almost all polymer-based nanocomposite systems an interfacial, bound, or sometimes called adsorbed layer is found. The dynamics in this interfacial layer is widely discussed in the literature. For instance, for some nanocomposites the molecular dynamics in the adsorbed layer is more or less similar to the segmental dynamics but with slower relaxation times or rates. Example for that are composites of poly(vinyl acetate) filled with silica nanoparticles (see, for instance, [188, 189]) and poly(2-vinylpyridine) also filled with silica (see, for example, [190]). In other cases, no segmental dynamics, and no devitrification is observed for the interfacial layer in the temperature range where the polymer is stable. One example for such a composite is poly(methyl methacrylate) nanofilled with silica [191]. Commonly this adsorbed or bound layer is called rigid amorphous fraction (RAF). Here, the term "rigid" means that it is immobilized with respect to the segmental dynamics (see, for instance, [191–194]). Currently, it is intensively discussed whether the RAF could devitrify or not. Nevertheless, it should be noted that localized fluctuation with length scales smaller than that responsible for glassy dynamics are possible within RAF.

The chain dynamics of polymers can be only in seldom cases studied by broadband dielectric spectroscopy. To investigate these processes is the domain of mechanical spectroscopy. This is discussed in several chapters of this book (see Chapter 3 Dynamics of polymer bridges in polymer nanocomposites through a combination of dielectric spectroscopy and rheology).

6 Conclusions

Nanocomposites with a polymer or organic material as matrix have a wide variety of applications ranging from filled rubber materials for tires, nanofilled thermosets or thermoplastics as construction materials, barrier materials, as well as encapsulants, separation membranes, nanodielectrics. But also, in high tech applications such as sensors or solar cells they can be found. Only the application of nanofillers in rubbers may be regarded mature and with large scale production. Despite their importance most of the other fields need further development which require a full understanding of the structure properties relationship from the experimental, as well as from the theoretical side. For all application areas, the interfacial area between the nanoparticle and the matrix is crucial. This is due to the small size of the nanoparticles (one spatial dimension <100 nm). The resulting high surface to volume ratio leads to a high amount of interfacial area which determines the properties of a nanocomposite to a great extent.

For a variety of applications, homogeneous distribution of the nanoparticles on a molecular level is required. This applies for instance to applications in sensors, solar cells, or separation membranes. For other applications, e.g., in tires or conductive nanocomposites, a certain agglomeration of the nanoparticles is required to optimize the properties the nanocomposites.

It is concluded that still more research efforts are needed to fully understand the structure–property relationships of nanocomposites. This comprises fundamental and applicative research, including an optimization of processing condition, as well as theoretical approaches and simulations.

References

1. Utracki LA, Wilkie CA (eds) (2014) Polymer blends handbook. Springer, The Nethlands
2. Jones RM (1999) Mechanics of composite materials, 2nd edn. Taylor & Francis Inc., Philadelphia
3. Kojima Y, Usuki A, Kawasumi M, Okada A, Fukushima Y, Kurauchi T, Kamigaito O (1993) Mechanical-properties of Nylon 6-Clay hybrid. J Mater Res 8:1185–1189
4. Giannelis EP, Krishnamoorti R, Manias E (1999) Polymer-silicate nanocomposites: model systems for confined polymers and polymer brushes. Adv Polym Sci 138:107–147
5. Kumar SK, Benicewicz BC, Vaia R, Winey KI (2017) 50th annoversary prespective: are nanocomposites practical for applications? Macromolecules 50:714–731
6. Chavali MS, Nikolova MP (2019) Metal oxide nanoparticles and their applications in nanotechnology. SN Appl Sci 1:607
7. Innocenzi P, Brusatin G (2001) Fullerene-based organic–inorganic nanocomposites and their applications. Chem Mater 13:3126–3139
8. Pomposo JA (ed) (2017) Single-chain polymer nanoparticles: synthesis, characterization simulations, and applications. Wiley–VCH Verlag and Co KGaA, Weinheim
9. Lee A, Lichtenhan JD (1998) Viscoelastic responses of polyhedral oligosilsesquioxane reinforced epoxy systems. Macromolecules 31:4970–4974
10. Kuo SW, Chang FC (2011) POSS related polymer nanocomposites. Prog Polym Sci 36:1649–1696
11. Ayandele E, Sarkar B, Alexandridis P (2011) Polyhedral oligomeric silesesquioxane (POSS)-containing nanocomposites. Nanomaterials 2:455–475
12. Ahmed AM, Morsi RMM (2017) Polymer nanocomposites dielectric and electric properties with quantum dots nanofillers. Mod Phys Lett B 31:1,750,278
13. Ray SS, Okamoto K, Okamoto M (2003) Structure–property relationship in biodegradable poly(butylene succinate)/layered silicate nanocomposites. Macromolecules 36:2355–2367
14. LeBaron PC, Wang Z, Pinnavaia TJ (1999) Polymer-layered silicate nanocomposites: an overview. Appl Clay Sci 15:11–29
15. Ray SS, Okamoto M (2003) Polymer/layered silicate nanocomposites: a review from preparation to processing. Prog Polym Sci 28:1539–1641

16. Ke Y, Stroeve P (2005) Polymer-layered silicate and silica nanocomposites. Elsevier, Amsterdam, The Netherlands
17. Winey KI, Vaia RA (2007) Polymer nanocomposites. MRS Bull 32:314–322
18. Costa FR, Saphiannikova M, Wagenknecht U Heinrich G (2008) Layered double hydroxide based polymer nanocomposites. Adv Polym Sci 210:1–100
19. Kovanda F, Jindova E, Lang K, Kubat P, Sedlakova Z (2010) Preparation of layered double hydroxides intercalated with organic anions and their application in LDH/poly(butyl methacrylate) nanocomposites. App Clay Sci 48:260–270
20. Naseem S, Leuteritz A, Wagenknecht U (2019) Polymer nanocomposites based on layered double hydroxides (LDHs). In Kenig S (ed) Processing of polymer nanocomposites. Hanser
21. Tichit D, Coq B (2003) Catalysis by hydrotalcites and related materials. CATTECH 7:206–217
22. van der Ven L, van Gemert MLM, Batenburg LF, Keern JJ, Gielgens LH, Koster TPM, Fischer HR (2000) On the action of hydrotalcite-like clay materials as stabilizers in polyvinylchloride. Appl Clay Sci 17:25
23. Wang D-Y, Das A, Costa FR, Leuteritz A, Wagenknecht U, Heinrich G (2010) Synthesis of organo cobalt–aluminum layered double hydroxide via a novel single-step self-assembling method and its use as flame retardant nanofiller in PP. Langmuir 26:14162
24. Kim H, Abdala AA, Macosko CW (2010) Graphene/polymer nanocomposites. Macromolecules 43:6515–6530
25. Moniruzzaman M, Winey KI (2006) Polymer nanocomposites containing carbon nanotubes. Macromolecules 39:5194–5205
26. Mittal V (2014) Polymer nanotubes nanocomposites: synthesis, properties and applications. Wiley, Hobogen
27. Liu TX, Phang IY, Shen L, Chow SY, Zhang WD (2004) Morphology and mechanical properties of multiwalled carbon nanotubes reinforced nylon-6 composites. Macromolecules 37:7214–7222
28. Du FM, Scogna RC, Zhou W, Brand S, Fischer JE, Winey KI (2004) Nanotube networks in polymer nanocomposites: rheology and electrical conductivity. Macromolecules 37:9048–9055
29. Yuan P, Tan D, Annabi-Bergaya F (2015) Properties and applications of halloysite nanotubes: recent research, advances and further prospects. App Clay Sci 112–113:75–93
30. Liu M, Du M, Guo B, Jia D (2010) Newly emerging applications of halloysite nanotubes: a review. Polym Int 59:574–582
31. Wang X, Li Z, Yu Y (2014) One-dimensional titanium dioxide nanomaterials: nanowires, nanorods and nanobelts. Chem Rev 19:9346–9384
32. Khanal BP, Zubarev ER (2020) Gold nanowires from nanorods. Langmuir 36:15,030–15,038
33. Favier V, Chanzy H, Cavaille JY (1995) Polymer nanocomposites reinforced by cellulose whiskers. Macromolecules 28:6365–6367
34. Klemm D, Kramer F, Moritz S, Lindström T, Ankerfors M, Gray D, Dorris A (2011) Nanocelluloses: a new family of nature-based materials. Angew Chem Int Ed 50:5438–5466
35. Zhu H, Luo W, Ciesielski PN, Fang Z, Zhu JY, Henriksson G. Himmel ME, Hu L (2016) Wood-derived materials for green electronics, biological devices, and energy applications. Chem Rev 116:9305–9374
36. Wang G, Yu D, Kelkar AD, Zhang L (2017) Electrospun nanofiber: emerging reinforcing filler in polymer matrix composite materials. Prog Polym Sci 75:73–107
37. Jancar J, Douglas JF, Starr FW, Kumar SK, Cassagnau P., Lesser AJ, Sternstein SS, Buehler MJ. Current issues in the research on structure–property relationships in polymer nanocomposites. Polymer 51:3321–3343
38. Klüppel M (1997) Structure and properties of fractal filler networks in rubber. Kautsch Gummi Kunstst 50:282–291
39. Heinrich G, Klüppel M, Vilgis TA (2002) Reinforcement of elastomers. Curr Opin Solid State Mater Sci 6:195–203
40. Klüppel M (2003) The role of disorder in filler reinforcement of elastomers on various length scales. Adv Polym Sci 164:1–86

41. Sarvestani AS, Picu CR (2004) Network model for the viscoelastic behavior of polymer nanocomposites. Polymer 45:7779–7790
42. Kashiwagi T, Du FM, Douglas JF, Winey KI, Harris RH (2005) Shields JR (2005) Nanoparticle networks reduce the flammability of polymer nanocomposites. Nat Mater 4:928–933
43. Allegra G, Raos G, Vacatello M (2008) Theories and simulations of polymer-based nanocomposites: from chain statistics to reinforcement. Prog Polym Sci 33:683–731
44. Jouault N, Vallat P, Dalmas F, Said S, Jestin J, Boue F (2009) Well-dispersed fractal aggregates as filler in polymer-silica nanocomposites: long-range effects in rheology. Macromolecules 42:2031–2040
45. Chevigny C, Jouault N, Dalmas F, Boue F, Jestin J (2011) Tuning the mechanical properties in model nanocomposites: influence of the polymer-filler interfacial interactions. J Polym Sci, Part B: Polym Phys 49:781–791
46. Nusser K, Schneider GJ, Pyckhout-Hintzen W, Richter D (2011) Viscosity decrease and reinforcement in polymer-silsesquioxane composites. Macromolecules 44:7820–7830
47. Kaur J, Lee JH, Bucknall DG, Shofner ML (2012) Enabling nanoparticle networking in semicrystalline polymer matrices. ACS Appl Mater Interfaces 4:3111–3121
48. Alig I, Skipa T, Lellinger D, Pötschke P (2008) Destruction and formation of a carbon nanotube network, in polymer melts: rheology and conductivity spectroscopy. Polymer 49:3524–3532
49. Gubbels F, Blacher S, Vanlathem E, Jerome R, Deltour R, Brouers F, Teyssie P (1995) Design of electrical conductive composites - key role of the morphology on the electrical-properties of carbon-black filled polymer blends. Macromolecules 28:1559–1566
50. Ma D, Siegel RW, Hong JI, Schadler LS, Martensson E, Onneby C (2004) Influence of nanoparticle surfaces on the electrical breakdown strength of nanoparticle-filled low-density polyethylene. J Mater Res 19:857–863
51. Ma D, Hugener TA, Siegel RW, Christerson A, Martensson E, Onneby C, Schadler LS Influence of nanoparticle surface modification on the electrical behaviour of polyethylene nanocomposites. Nanotechnology 16:724–731
52. Schönhals A, Böhning M unpublished results.
53. Kang H, Detcheverry FA, Mangham AN, Stoykovich MP, Daoulas KC, Hamers RJ, Müller M, de Pablo JJ, Nealey PF (2008) Hierarchical assembly of nanoparticle superstructures from block copolymer-nanoparticle composites. Phys Rev Lett 100:148303
54. Darling SB (2007) Mechanism for hierarchical self-assembly of nanoparticles on scaffolds derived from block copolymers. Surf Sci 601:2555–2561
55. Wang JF, Cheng QF, Tang ZY (2012) Layered nanocomposites inspired by the structure and mechanical properties of nacre. Chem Soc Rev 41:1111–1129
56. Hudson ZM, Boott CE, Robinson ME, Rupar PA, Winnik MA, Manners I (2014) Tailored hierarchical micelle architectures using living crystallization-driven self-assembly in two dimensions. Nat Chem 6:893–898
57. Mu M, Walker AM, Torkelson JM, Winey KI (2008) Cellular structures of carbon nanotubes in a polymer matrix improve properties relative to composites with dispersed nanotubes. Polymer 49:1332–1337
58. Chen J, Liu M, Chen C, Gong H, Gao C (2011) Synthesis and characterization of silica nanoparticles with well-defined thermoresponsive PNIPAM via a combination of RAFT and click chemistry. ACS Appl Mater Inter 3:3215–3223
59. Balamurugan SS, Soto-Cantu E, Cueto R, Russo PS (2010) Preparation of organosoluble silica—polypeptide particles by "Click" chemistry. Macromolecules 43:62–70
60. von Werne T, Patten TE (2001) Atom transfer radical polymerization from nanoparticles: a tool for the preparation of well-defined hybrid nanostructures and for understanding the chemistry of controlled/"Living" radical polymerizations from surfaces. J Am Chem Soc 123:7497–7505
61. Bartholome C, Beyou E, Bourgeat-Lami E, Chaumont P, Lefebvre F, Zydowicz N (2005) Nitroxide-mediated polymerization of styrene initiated from the surface of silica nanoparticles. In Situ generation and grafting of alkoxyamine initiators. Macromolecules 38:1099–1106

62. Barbey R, Lavanant L, Paripovic D, Schüwer N, Sugnaux C, Tugulu S, Klok H (2009) Polymer brushes via surface-initiated controlled radical polymerization: synthesis, characterization, properties, and applications. Chem Rev 109:5437–5527
63. Wu L, Glebe U, Böker A (2015) Surface-initiated controlled radical polymerizations from silica nanoparticles, gold nanocrystals, and bionanoparticles. Polym Chem 6:5143–5184
64. Oyerokun FT, Vaia RA (2012) Distribution in the grafting density of end-functionalized polymer chains adsorbed onto nanoparticle surfaces. Macromolecules 45:7649–7659
65. Li C, Han J, Ryu CY, Benicewicz BC (2006) A versatile method to prepare RAFT agent anchored substrates and the preparation of PMMA grafted nanoparticles. Macromolecules 39:3175–3183
66. Hong C-Y, Li X, Pan C-Y (2007) Grafting polymer nanoshell onto the exterior surface of mesoporous silica nanoparticles via surface reversible addition-fragmentation chain transfer polymerization. Eur Polym J 43:4114–4122
67. Matyjaszewski K (2012) Atom transfer radical polymerization (ATRP): current status and future perspectives. Macromolecules 45:4015–4039
68. Moad G, Chong YK, Postma A, Rizzardo E, Thang SH (2005) Advances in RAFT polymerization: the synthesis of polymers with defined end-groups. Polymer 46:8458–8468
69. Nicolas J, Guillaneuf Y, Lefay C, Bertin D, Gigmes D, Charleux B (2013) Nitroxide-mediated polymerization. Prog Polym Sci 38:63–235
70. Fan X, Lin L, Messersmith PB (2006) Surface-initiated polymerization from TiO_2 nanoparticle surfaces through a biomimetic initiator: a new route toward polymer−matrix nanocomposites. Compos Sci Technol 66:1198–1204
71. Qiao Y, Yin X, Wang L, Islam MS, Benicewicz BC, Ploehn HJ, Tang C (2015) Bimodal polymer brush core−shell barium titanate nanoparticles: a strategy for high-permittivity polymer nanocomposites. Macromolecules 48:8998–9006
72. Baskaran D, Mays JW, Bratcher MS (2004) Polymer-grafted multiwalled carbon nanotubes through surface-initiated polymerization. Angew Chem, Int Ed 43:2138–2142
73. Prucker O, Ruehe J (1998) Synthesis of poly(styrene) monolayers attached to high surface area silica gels through self-assembled monolayers of Azo initiators. Macromolecules 31:592–601
74. Li D, Cui Y, Wang K, He Q, Yan X, Li J (2007) Thermosensitive nanostructures comprising gold nanoparticles grafted with block copolymers. Adv Funct Mater 17:3134–3140
75. Purohit PJ, Huacuja-Sánchez JE, Wang D-Y, Emmerling F, Thünemann A, Heinrich G, Schönhals A (2011) Structure-properties relationsships of nanocomposites based on polypropylene and layered double hydroxides. Macromolecules 44:4342–4354
76. Szymoniak P, Qu X, Abasi M, Pauw BR, Henning S, Zhi L, Wang D-Y, Schick C, Saalwächter K, Schönhals A (2021) Spatial inhomogeneity, interfaces and complex vitrification kinetics in a network forming nanocomposite. Soft Matter 17:2775–2790
77. Sanes J, Sánchez C, Pamies R, Avilés M-D, Bermúdez M-D (2020) Extrusion of polymer nanocomposites with graphene and graphene derivative nanofillers: an overview of recent developments. Materials 13:549
78. Shearer JDM, Barry CMF, Mead J (2014) Continuous extrusion for preparation of polymer nanocomposites. NIST-Nanotech 3:174–177
79. Leng J, Kang N, Wang D-Y, Falkenhagen J, Thünemann A, Schönhals A (2017) Structure–property relationships of nanocomposites based on polylactide and layered double hydroxides—comparison of MgAl and NiAl LDH as nanofillers. Macomolec Chem Phys 218:1700232
80. Fahmy A, Agudo Jácome L, Schönhals A (2020) Effect of Silver nanoparticles on the dielectric properties and the homogeneity of plasma poly(acrylic acid) thin films. J Phys Chem C 124:22817–22826
81. Chevigny C, Jestin J, Gigmes D, Schweins R, Di-Cola E, Dalmas F, Bertin D, Boue F (2020) "Wet-to-Dry" conformational transition of polymer layers grafted to nanoparticles in nanocomposite. Macromolecules 43:4833–4837
82. Robbes AS, Cousin F, Meneau F, Chevigny C, Gigmes D, Fresnais J, Schweins R, Jestin J (2012) Controlled grafted brushes of polystyrene on magnetic gamma-Fe_2O_3 nanoparticles via nitroxidemediated polymerization. Soft Matter 8:3407–3418

83. Chevigny C, Dalmas F, Di Cola E, Gigmes D, Bertin D, Boue F, Jestin J (2011) Polymer-grafted-nanoparticles nanocomposites: dispersion, grafted chain conformation, and rheological behavior. Macromolecules 44:122–133
84. Beaucage G (2012) Combined Small-Angle Scattering for Characterization of Hierarchically Structured Polymer Systems over Nano-to-Micron Meter. In Matyjaszewski K, Möller M (eds) Polymer science: a comprehensive reference, vol 2. Elsevier BV, Amsterdam
85. Szymoniak P, Pauw BR, Qu X, Schönhals A (2020) Competition of nanoparticle-induced mobilization and immobilization effects on segmental dynamics of an epoxy-based nanocomposite. Soft Matter 16:5406–5421
86. Bressler I, Pauw BR, Thünemann AF (2015) McSAS: software for retrieval of model parameter distributions from scatter patterns. J Appl Crys 48:962–969
87. Omar H, Smales GJ, Henning S, Zhi L, Wang D-Y, Schönhals A, Szymoniak P (2021) Calorimetric and dielectric investigations of epoxy-based nanocomposites with halloysite nanotubes as nanofillers. Polymers 13:1634
88. Purohit PJ, Wang D-Y, Emmerling F, Thünemann A, Heinrich G, Schönhals A (2012) Arrangement of layered double hydroxide in a polyethylene matrix studied by a combination of complementary methods. Polymer 53:2245–2254
89. . Böhning M, Goering H, Fritz A, Brzezinka KW, Turky G, Schönhals A, Schartel B (2005) Dielectric study of molecular mobility in poly(propylene-graft-maleic anhydride)/clay nanocomposites. Macromolecules 38:2764–2774
90. Morgan AB, Wilkie CA (eds) (2007) Flame retardant polymer nanocomposites. Wiley, Hoboken
91. Laoutid F, Bonnaud L, Alexandre M, Lopez-Cuesta JM, Dubois P (2009) New prospects in flame retardant polymer materials: from fundamentals to nanocomposites. Mater Sci Eng R-reports 63:100–125
92. Kiliaris P, Papaspyrides CD (2010) Polymer/layered silicate (clay) nanocomposites: an overview of flame retardancy. Progr Polym Sci 35:902–958
93. Morgan AB (2006) Flame retarded polymer layered silicate nanocomposites: a review of commercial and open literature systems. Polym Adv Tech 17:206–217
94. Wang X, Kalali EN, Wan JT, Wang D-Y (2017) Carbon-family materials for flame retardant polymeric materials. Prog Polym Sci 67:77–125
95. Zhang WJ, Camino G, Yang RJ (2017) Polymer/polyhedral oligomeric silsesquioxane (POSS) nanocomposites: an overview of fire retardance. Prog Polym Sci 67: 77–125
96. Pinnavaia TJ, Beall GW (2011) Polymer–clay nanocomposites. Wiley, Hoboken
97. Wilkie CA, Schartel B (eds) (2006) Polymers for advances technologies, vol 16, no 4
98. Tsagaropoulos G, Eisenberg A (1995) Dynamic-mechanical study of the factors affecting the 2 glass-transition behavior of filled polymers—similarities and differences with random ionomers. Macromolecules 28:6067–6077
99. Li Y, Wang L, Natarajan B, Tao P, Benicewicz BC, Ullal C, Schadler LS (2015) Bimodal "matrix-free" polymer nanocomposites. RSC Adv 5:14,788-14,795
100. Li Y, Tao P, Viswanath A, Benicewicz BC, Schadler LS (2013)Bimodal surface ligand engineering: the key to tunable nanocomposites. Langmuir 29:1211−1220
101. Payne AR (1965) Effect of dispersion on dynamic properties of filler-loaded rubbers. J Appl Polym Sci 9:2273–2284
102. Kumar SK, Krishnamoorti R (2010) Polymer nanocomposites: structure, thermodynamics and properties. Annu Rev Chem Biomol Eng 1:37–58
103. Bouty A, Petitjean L, Chatard J, Matmour R, Degrandcourt C, Schweins R, Meneau F, Kwasniewski P, Boue F, Couty M, Jestin J (2016) Interplay between polymer chain conformation and nanoparticle assembly in model industrial silica/rubber nanocomposites. Faraday Discuss 186:325–343
104. Baeza GP, Genix A-C, Paupy-Peyronnet N, Degrandcourt C, Couty M, Oberdisse J (2016) Revealing nanocompositefiller structures by swelling and small-angle X-ray scattering. Faraday Discuss 186:295–309

105. Berriot J, Montes H, Lequeux F, Long D, Sotta P (2002) Evidence for the shift of the glass transition near the particles in silica filled elastomers. Macromolecules 35:9756–9762
106. Papon A, Guy TCL, Saalwächter K, Oberdisse J, Merabia S, Long D, Sotta P, Frielinghaus HH, Radulescu A, Deme B, Noirez L, Montes H, Lequeux F (2013) Studying model samples to understand mechanical properties of filled elastomers. Kautsch Gummi Kunstst 66:52–58
107. Jouault N, Moll JF, Meng D, Windsor K, Ramcharan S, Kearney C, Kumar SK (2013) Bound polymer layer in nanocomposites. ACS Macro Lett 2:371–734
108. Berriot J, Lequeux F, Monnerie L, Montes H, Long D, Sotta P (2002) Filler-elastomer interaction in model filled rubbers, a H1 NMR study. J Non-cryst Solids 307:719–724
109. Heinrich G, Klüppel H (2004) The role of polymer-filler interphase in reinforcement of elastomers. Kautsch Gummi Kunstst 57:452–454
110. Gusev AA (2006) Micromechanical mechanism of reinforcement and losses in filled rubbers. Macromolecules 39:5960–5962
111. Chazeau L, Brown JD, Yanyo LC, Sternstein SS (2000) Modulus recovery kinetics end other insights into the Payne effect for filled elastomers. Polym Compos 21:202–222
112. https://mag.toyota.co.uk/toyota-brought-nanocomposite-materials-world/
113. Piscitelli F, Lavorgna M, Buonocore GG, Verdolotti L, Galy J, Mascia L (2012) Macromol Mater Eng 298:896–909
114. Irzhak TF, Irzhak VI (2017) Epoxy nanocomposites. Polym Sci Ser A 59:791–825
115. Kinjo N, Ogata M, Nishi K, Kaneda A (1989) Epoxy molding compounds as encapsulation materials for microelectronic devices. Adv Polym Sci 88:1–48
116. Kalali EN, Wang X, Wang D-Y (2015) Functionalized layered double hydroxide-based epoxy nanocomposites with improved flame retardancy and mechanical properties. J Mater Chem A 3:681
117. Yu AP, Ramesh P, Itkis ME, Bekyarova E, Haddon RC (2007) Graphite nanoplatelet-epoxy composite thermal interface materials. J Phys Chem C 111:7565–7569
118. Gojny FH, Wichmann MHG, Fiedler B, Schulte K (2005) Influence of different carbon nanotubes on the mechanical properties of epoxy matrix composites—a comparative study. Compos Sci Technol 65:2300–2313
119. Sinapius M, Ziegmann G (eds) (2021) Acting principle of nano-scaled matrix additives for composite structures. Springer, Berlin
120. Wetzel B, Haupert F, Zhang MQ (2003) Epoxy nanocomposites with high mechanical and tribological performance. Compos Sci Technol 63:2055–2067
121. Tseng CH, Hsueh BH, Chen CY (2007) Effect of reactive layered double hydroxides on the thermal and mechanical properties of LDHs/epoxy nanocomposites. Compos Sci Technol 67:2350
122. Sunny AT, Adhikari R, Mathew S, Thomas S (2016) Copper oxide nanoparticles in an epoxy network: microstructure, chain confinement and mechanical behaviour. Phys Chem Chem Phys 18:1955–21967
123. Brown JM, Curliss D, Vaia RA (2000) Thermoset-layered silicate nanocomposites. quaternary ammonium montmorillonite with primary diamine cured epoxies. Chem Mater 12:3376–3384
124. Koerner H, Misra D, Tan A, Drummy L, Mirau P, Vaia R (2006) Montmorillonite-thermoset nanocomposites via cryo-compounding. Polymer 47:3426–3435
125. Lin Y, Zhou B, Fernando KAS, Liu P, Allard LF, Sun YP (2003) Polymeric carbon nanocomposites from carbon nanotubes functionalized with matrix polymer. Macromolecules 36:7199–7204
126. Amrame A, Rajendran S, Sharoba AS (eds) (2020) Nanotechnology in the beverage industry. Elsevier, Amsterdam
127. Wijmans JG, Baker RW (1995) The solution diffusion model: a review. J Membr Sci 107:1–21
128. Böhning M, Schönhals. A unpublished resulst
129. Zaidi J, Matsuura T (eds) (2009) Polymer membranes for fuel cells. Springer, New York
130. Pandey N, Shuhkla SK, Singh NB (2017) Water purification by nanocomposites: an overview. Nanocomposites 3:47–66

131. Qadir D, Mukhtar H, Keong LK (2017) Mixed matrix membranes for water purification applications. Separ Prurificat Rev 46:62–80
132. Robson LM (2008) The upper bound revisited. J Membr Sci 320:390–400
133. Gusev AA, Müller-Plathe F, Vangunsteren WF, Suter UW (1994) Dynamics of small molecules in bulk polymers. Adv Polym Sci 116:207–247
134. Gusev AA, Suter UW (1993) Dynamics of small molecules in dense polymers subject to thermal motion. J Chem Phys 99:2228–2234
135. Bos A, Pünt IGM, Wessling M, Strathmann H (1998) Sep Purif Technol 14:27–39
136. Simmons JW, Ekiner OS (1993) Polyimide and polyamide-imide gas separation membranes. US patent US5232472 A.
137. Konnertz N, Böhning M, Schönhals A (2016) Dielectric investigations of nanocomposites based on matrimid and polyhedral oligomeric phenethyl-silsesquioxanes (POSS). Polymer 90:89–101
138. Nagai K, Masuda T, Nakagawa T, Freeman BD, Pinnau I (2001) Poly[1-(trimethylsilyl)-1-propyne] and related polymers: synthesis, properties, and functions. Prog Polym Sci 26:721–798
139. Sakaguchi T, Shiotsuki M, Masuda T (2004) Synthesis and properties of Si-containig poly(diarylacetylene)s and their desilylated polymer membranes. Macromolecules 37:4104–4108
140. Budd PM, Ghanem BS, Makhseed S, McKeown NB, Msayib KJ, Tattershall CE (2004) Polymers of intrinsic microporosity (PIMs): robust, solution-processable, organic nanoporous materials. Chem Commun (Cambridge, UK) 2004:230–231
141. Budd PM, Msayib KJ, Tattershall CE, Ghanem BS, Reynolds KJ, McKeown NB, Fritsch D (2005) Gas separation membranes from polymers of intrinsic microporosity. J Membr Sci 251:263–269
142. Swaidan R, Ghanem B, Pinnau I (2012) Fine-Tuned intrinsically ultramicroporous polymers redefine the permeability/selectivity upper bounds of membrane-based air and hydrogen separations. ACS Macro Lett 4:947–951
143. Yampolskii Y (2012) Polymeric gas separation membranes. Macromolecules 45:3298–3311
144. Gringolts M, Bermeshev M, Yamploskii Y, Starannikova LE, Stantarovich V, Finkelshtein ES (2010) New high permeable addition poly(tricyclononenes) with $Si(CH_3)_3$ side groups. Synthesis, gas permeation parameters, and free volume. Macromolecules 43:7165–7172
145. Chapala PP, Bermeshev M, Starannikova LE, Belov NA, Ryzhikh VE, Stantarovich V, Lakhtin VG, Gavrilova NN, Yamploskii Y, Finkelshtein ES (2015) A novel, highly gas-permeable polymer representing a new class of silicon-containing polynorbornenes as efficient membrane matcrials. Macromolcculcs 48:8055–8061
146. Merkel TC, Freeman BD, Spontak RJ, He Z, Pinnau I, Meakin P, Hill AJ (2002) Ultrapermeable, reverse-selective nanocomposite membranes. Science 296:519–522
147. Lopez LC, Wilkes GL, Stricklen PM, White SA (1992) Synthesis, Structure and Properties of Poly(4-Methyl-1-pentene). J Macrom Sci, Part C 32:301–406
148. Takahashi S, Paul DR (2006) Gas permeation in poly(ether imide) nanocomposite membranes based on surface-treated silica. Part 1: without chemical coupling to matrix. Polymer 47:7519–7534
149. Chung TS, Jiang LY, Li Y, Kulprathipanja S (2007) Mixed matrix membranes (MMMs) comprising organic polymers with dispersed inorganic fillers for gas separation. Prog Polym Sci 32:483–507
150. Hao N, Böhning M, Schönhals A (2010) CO_2 gas transport properties of nanocomposites based on polyhedral oligomeric phenethyl-silsesquioxanes and poly(bisphenol A carbonate). Macromolecules 43:9417–9425
151. Hao N, Böhning M, Schönhals A (2007) Nanocomposites of polyhedral oligomeric phenethyl-silsesquioxanes and poly(bisphenol A carbonate) as investigated by dielectric spectroscopy. Macromolecules 40:2955–2964
152. Böhning M, Hao N, Schönhals A (2013) Correlation of activation energies of gas diffusivity and local matrix mobility in polycarbonate/POSS nanocomposites. J Polym Sci, Polym Phys 51:1593–1597

153. Hamid MRA, Jeong H-K (2018) Recent advances on mixed-matrix membranes for gas separation: opportunities and engineering challenges. Korean J Chem Eng 35:1577–1600
154. Priestley RD, Ellison CJ, Broadbelt LJ, Torkelson JM (2005) Structural relaxation of polymer glasses at surfaces, interfaces and in between. Science 309:456–459
155. Lau H, Nguyen PT, Hill MR, Thornton AW, Konstas K, Doherty CM, Mulder RJ, Bourgeois L, Liu ACY, Sprouster DJ, Sullivan JP, Bastow TJ, Hill AJ, Gin DL, Noble RD (2014) Ending aging in super glassy polymer membranes. Angew Chem, Int Ed 53:5322–5326
156. Nelson JK (ed) (2010) Dielectric polymer nanocomposites. Springer, US
157. Thakur VK, Gupta RK (2016) Recent Progress on ferroelectric polymer-based nanocomposites for high energy density capacitors: synthesis, dielectric properties, and future aspects. Chem Rev 116:4260–4317
158. Cao Y, Irwin PC, Younsi K (2004) The future of nanodielectrics in the electrical power industry. IEEE Trans Dielectr Electr Insul 11:797–807
159. Vogelsang R, Farr T, Frohlich K (2006) The effect of barriers on electrical tree propagation in composite insulation materials. IEEE Trans Dielectr Electr Insul 13:373–382
160. Stone GC, Culbert I, Boulter EA, Dhirani H (2014) Electrical insulation for rotating machines: design, evaluation, aging, testing, and repair. 2nd edn. Blackwell Science Publ, Oxford
161. Syatirah MN, Muhamad NA, Kharul Anwar AH, Zakariya MZ, Anuar MNK, Zaidi AAH (2020) A review: polymer-based insulation material for HVDC cable application. IOP Conf Ser: Mater Sci Eng 932:012064
162. Andraschek N, Wanner AJ, Ebner C, Riess G (2016) Mica/epoxy-composites in the electrical industry: applications, composites for insulation, and investigations on failure mechanisms for prospective optimizations. Polymers 8:201
163. Lv Z, Rowland S, Chen S, Zheng H, Iddrissu I (2017) Evolution of partial discharges during early tree propagation in epoxy resin. IEEE Trans Dielectr Electr Insul 24:2995–3003
164. Roy M, Nelson JK, MacCrone RK, Schadler LS, Reed CW, Keefe R, Zenger W (2005) Polymer nanocomposite dielectrics—the role of the interface. IEEE Trans Dielectr Electr Insul 12:629–643
165. Plesa I, Notingher PV, Schlogl S, Sumereder C, Muhr M (2016) Properties of polymer composites used in high-voltage applications. Polymers 8:173
166. Irwin PC, Cao Y, Bansal A, Schadler LS (2003) Thermal and mechanical properties of polyimide nanocomposites. IEEE, New York, pp 120–123
167. Fredin LA, Li Z, Ratner MA, Lanagan MT, Marks TJ (20129 Enhanced energy storage and suppressed dielectric loss in oxide core-shell-polyolefin nanocomposites by moderating internal surface area and increasing shell thickness. Adv Mater 24: 5946–5953
168. Huang XY (2015) Jiang PK (2015) Core-shell structured high-k polymer nanocomposites for energy storage and dielectric applications. Adv Mater 27:546–554
169. Virtanen S, Krentz TM, Nelson JK, Schadler LS, Bell M, Benicewicz B, Hillborg H, Zhao S (2014) Dielectric breakdown strength of epoxy bimodal-polymer-brush-grafted core functionalized silica nanocomposites. IEEE Trans Dielectr Electr Insul 21:563–570
170. Dang ZM, Yuan JK, Yao SH, Liao RJ (2013) Flexible nanodielectric materials with high permittivity for power energy storage. Adv Mater 25:6334−6365
171. Wang GY, Huang XY, Jiang PK (2015) Tailoring dielectric properties and energy density of ferroelectric polymer nanocomposites by high-k nanowires. ACS Appl Mater Interf 7:18017–18027
172. Tang HX, Sodano HA (2013) Ultra high energy density nanocomposite capacitors with fast discharge using Ba0.2Sr0.8-TiO_3 nanowires. Nano Lett 13:1373–1379
173. Markel VA (2016) Introduction to the Maxwell Garnett approximation: tutorial. J Opt Soc America A 33:1244–1256
174. Bruggeman DAG (1935) Berechnung verschiedener physikalischer Konstanten von heterogenen Substanzen. I. Dielektrizitätskonstanten und Leitfähigkeiten der Mischkörper aus isotropen Substanzen. Ann Phys 416:636–664
175. Bruggeman DAG (1935) Berechnung verschiedener physikalischer Konstanten von heterogenen Substanzen. I. Dielektrizitätskonstanten und Leitfähigkeiten der Mischkörper aus isotropen Substanzen (Schluß). Ann Phys 416:665–679

176. Calame JP (2006) Finite difference simulations of permittivity and electric field statistics in ceramic-polymer composites for capacitor applications. J Appl Phys 99:084101
177. Kim P, Doss NM, Tillotson JP, Hotchkiss PJ, Pan MJ, Marder SR, Li JY, Calame JP, Perry JW (2009) High energy density nanocomposites based on surface-modified $BaTiO_3$ and a ferroelectric polymer. ACS Nano 3:2581–2592
178. Sihvola AH (1999) Electromagnetic mixing formulas and applications. IET Digital library
179. Tanaka T, Kozako M, Fuse N, Ohki Y (2005) Proposal of a multicore model for polymer nanocomposite dielectrics. IEEE Trans Dielectr Electr Insul 12:669–681
180. Katayama J, Ohki Y, Fuse N, Kozako M, Tanaka T (2013) Effects of nanofiller materials on the dielectric properties of epoxy nanocomposites. IEEE Trans Dielectr Electr Insul 20:157–165
181. Lewis TJ (1994) Nanometric dielectrics. IEEE Trans Dielectr Electr Insul 1:812–825
182. Zhan C, Yu G, Lu Y, Wan L, Wujcik E, Wie S (2017) Conductive polymer nanocomposites: a critical review of modern advanced devices. J Mater Chem C 7:1569–1585
183. Díez-Pascual AM, Sánchez JAL, Capilla RP, Díez PG (2018) Recent developments in graphene/polymer nanocomposites for applications in polymer solar cells. Polymers 10:217
184. Etxebarria I, Ajuria J, Pacios R (2015) Polymer: Fullerene solar cells: materials, processing issues, and cell layouts to reach power conversion efficiency over 10%, a review. J Photo Energ 5:057214
185. Xu Y, Wang Y, Liang J, Huang Y, Ma Y, Wan X, Chen Y (2009) A hybrid material of graphene and poly(3,4-ethyldioxythiophene) with high conductivity, flexibility, and transparency Nano Res 2:343–348
186. Schönhals A (2003) Molecular mobility in polymer model systems. In: Kremer F, Schönhals A (eds) Broadband dielectric spectroscopy. Springer, Berlin
187. Omar H, Smales GJ, Henning S, Li Z, Wang D-Y, Schönhals A, Szymoniak P (2021) Calorimetric and dielectric investigations of epoxy-based nanocomposites with halloysite nanotubes as nanofillers. Polymers 13:1634
188. Füllbrandt M, Purohit PJ, Schönhals A (2013) Combined FTIR and dielectric investigation of poly(vinyl acetate) adsorbed on silica particles. Macromolecules 46:4626–4632
189. Carroll B, Cheng S, Sokolov A (2017) Analyzing the interfacial layer properties in polymer nanocomposites by broadband dielectric spectroscopy. Macromolecules 50:6149–6163
190. Holt AP, Griffin PJ, Bocharova V, Agapov AL, Imei AE, Dadmun MD, Sangoro JR, Sokoloy AP (2014) Dynamics at the polymer/nanoparticle interface in poly(2-vinylpyridine)/silica nanocomposites. Macromolecules 47:1837–1843
191. Sargsyan A, Tonoyan A, Davtyan S, Schick C (2007) The amount of immobilized polymer in PMMA SiO_2 nanocomposites determined from calorimetric data. Euro Polym J 43:3113–3127
192. Wurm A, Ismail M, Kretzschmar B, Pospiech D, Schick C (2010) Retarded crystallization in polyamide/layered silicates nanocomposite caused by an immobilized interphase. Macromolecules 43:1480–1487
193. Klonos PA, Papadopoulos L, Papageorgiou GZ, Kyritsis A, Pissis P, Bikiaris DN (2020) Interfacial interaction, crystallization, and molecular dynamics of renewable poly(propylene furanoate) in situ filled with initial and surface functionalized carbon nanotubes and graphene oxide. J Phys Chem C 124:10220–10234
194. Leng J, Szymoniak P, Kang N-J, Wang D-Y, Wurm A, Schick C, Schönhals A (2019) Influence of interfaces on the crystallization behavior and the rigid amorphous phase pf poly(L-lactide)-based nanocomposites with different layered doubled hydroxides as nanofiller. Polymer 184:121919

Fundamentals

Fundamentals of Dielectric Spectroscopy in Polymer Nanocomposites

Ivan Popov and Alexei P. Sokolov

Abstract Properties of Polymer Nanocomposites (PNCs) are strongly modulated by the properties of the interfacial polymer layer surrounding the nanofillers. This interfacial region in PNCs experiences significant structural and dynamic changes in comparison to the bulk polymer. In this chapter, we first discuss experimental data to demonstrate how the presence of the interfacial layer can be detected in the dielectric spectra of different PNCs. We emphasize the impact of nanoparticles on segmental and chain dynamics in PNCs. Then, we present various theoretical approaches to describe the dielectric spectra of the PNCs, explain the behavior of the dielectric parameters, and describe how to extract the characteristics of the interfacial layer. We emphasize that the dielectric signal in heterogeneous systems is not additive, and the interfacial layer model provides the most accurate description of the dielectric data in PNCs. We also emphasize the often overlooked experimental observation—a strong decrease in the dielectric strength of segmental relaxation in the interfacial layer.

Keywords Polymer nanocomposites · Interfacial layer · Broadband Dielectric Spectroscopy

Abbreviations

APL	Adsorbed Polymer Layer
BDS	Broadband Dielectric Spectroscopy
DSC	Differential Scanning Calorimetry
GNPs	Grafted Nanoparticles
GSNC	Glycerol/Silica Nanocomposite

I. Popov (✉) · A. P. Sokolov
Department of Chemistry, University of Tennessee, Knoxville, TN 37996, USA
e-mail: ipopov@utk.edu

A. P. Sokolov
Chemical Science Division, Oak Ridge National Laboratory, Oak Ridge, TN 37831, USA

A. Schönhals and P. Szymoniak (eds.), *Dynamics of Composite Materials*, Advances in Dielectrics, https://doi.org/10.1007/978-3-030-89723-9_2

HN	Havriliak–Negami
IL	Interfacial Layer
ILM	Interfacial Layer Model
MD	Molecular Dynamic
MWS	Maxwell–Wagner–Sillars
NMR	Nuclear Magnetic Resonance
PNCs	Polymer Nanocomposites
PI	Polyisoprene
PPG	Poly(propylene glycol)
PVAc	Poly(vinyl acetate)
P2VP	Poly(2vinyl pyridine)
SAXS	Small Angle X-ray Scattering
SANS	Small Angle Neutron Scattering
TMDSC	Temperature-Modulated Differential Scanning Calorimetry
TPM	Two Phase Model

1 Introduction

Polymer Nanocomposites (PNCs) are actively used in many technologies, thanks to their light weight, particular mechanical properties and flexibility, and ability to tune macroscopic properties in an extremely broad range [1–19]. PNCs properties depend on fillers used, their dispersion, the loading and the interactions between polymer and the fillers. Despite their wide use, the fundamental understanding of many underlying phenomena controlling macroscopic properties of PNCs remains rather limited. In recent years, it becomes clear that nanoparticles affect the structure and dynamics of the polymer matrix around the nanoparticles strongly, forming the so-called Interfacial Layer (IL) [17, 20–24]. It has been demonstrated that interfacial layer plays a critical role in mechanical and viscoelastic properties of PNCs [6, 14, 25–27].

We want to emphasize that the interfacial layer is often confused in literature with polymer bound layer. The latter is the layer formed by polymer chains adsorbed to the nanoparticle surface, and its thickness is defined by the polymer radius of gyration, R_g (unless polymer is confined by a other interface on a shorter distance), i.e., it varies with molecular weight (Fig. 1) [27–36]. In contrast, the interfacial layer is defined as the polymer region with properties (e.g., density, segmental dynamics, mechanical modulus, etc.) significantly different from that of bulk polymer. Its thickness is usually estimated in the range ~2–6 nm rather independent of molecular weight (Fig. 1), and depends on polymer rigidity, nanoparticles size and in addition it increases upon cooling [17, 20–24, 37]. Although the thickness of the interfacial layer is relatively small, it usually occupies significant volume fraction of the polymer matrix due to large number of nanoparticles. As a result, the interfacial layer plays a critical role in many macroscopic properties of composites.

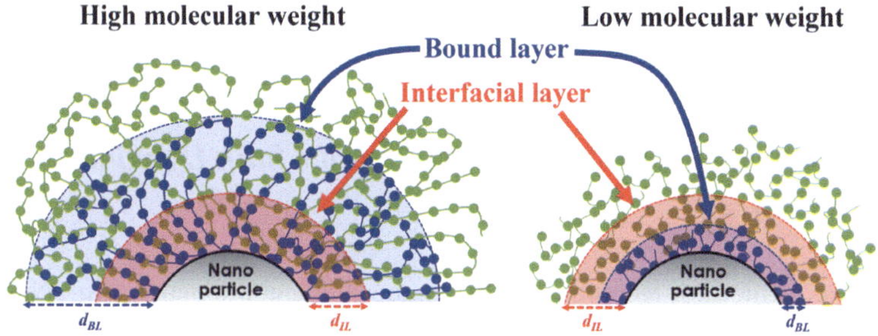

Fig. 1 Schematic presentation of bound polymer layer and of the interfacial polymer layer [36]. [Reprinted with permission from [36]: Popov, I.; et al. Macromolecules 2020, 53, 4126–4135. Copyright 2020 American Chemical Society]

The interfacial layer can be detected using Small Angle X-ray Scattering (SAXS) and Small Angle Neutron Scattering (SANS) techniques due to change in density of the polymer surrounding the nanoparticles. Surprisingly, several studies revealed a decrease in density in the interfacial layer for high molecular weight polymers with $2R_g$ larger than the distance between nanoparticles surfaces [26, 38, 39]. These results were ascribed to a frustration in the chain packing caused by chain confinements [38, 39]. Differential Scanning Calorimetry (DSC) measurements, and detailed Nuclear Magnetic Resonance (NMR) studies all revealed presence of the interfacial layer with slower segmental dynamics (higher T_g) in PNCs [24, 39–42]. All these experimental studies estimated the thickness of the interfacial layer to be just a few nm. Many coarse-grained simulations of composite materials also revealed the interfacial layer with a structure and segmental dynamics different from the bulk polymer, and its thickness was just a few segments size [43–45]. These results are very similar to the interfacial layer known for supported polymer films [46–63].

Among many experimental techniques used to study PNCs, Broadband Dielectric Spectroscopy (BDS) plays very important role because it covers an enormous frequency range, enabling studies of the polymer segmental and, in some cases, chain dynamics in broad time and temperature ranges. BDS studies were instrumental in developing a fundamental understanding of the parameters controlling interfacial layer in PNCs, as the role of chain rigidity and temperature [37, 39, 42]. This chapter considers fundamentals of the BDS data analysis in PNCs, presenting an overview of many fundamental studies. Usually studies focus on the change of the time scale of relaxation processes in the interfacial layer. We emphasize here another often overlooked but important parameter—amplitude of segmental fluctuations. The presence of the interface, the interaction with the nanoparticle surface, and the chain stretching strongly restrict the amplitude of the segmental reorientation in the interfacial layer on the time scale of segmental relaxation. This amplitude decreases regardless of whether the dynamics is getting slower or faster than in the bulk material. In this chapter, we consider major BDS experimental results and basic models proposed for their interpretations. Then, we briefly summarize knowledge accumulated on the interfacial layer in PNCs.

2 Dielectric Properties of the Interfacial Layer. Experimental Results

2.1 Dielectric Spectra for Segmental and Chain Dynamics in PNCs

BDS is one of the best techniques to study the segmental dynamics in PNCs [20, 53, 55, 64–67]. Broad overlapped relaxation peaks might cover several decades in frequency but vast operating frequency range of BDS allows to explore polymer dynamic without involving time–temperature superposition approach.

Figure 2 presents typical spectra of PNCs [39, 68]. The valuable information is usually extracted from imaginary part of the dielectric permittivity $\varepsilon''(\omega)$ and the derivative of the real part of the dielectric permittivity, $\varepsilon'_{der}(\omega) = -\pi/2 \times \partial\varepsilon'(\omega)/\partial \ln \omega$ [69]. Dielectric spectra of most of the polymers do not show chain modes, and the segmental relaxation dominates the spectra. In such polymers, e.g., poly(vinyl acetate) (PVAc) or poly(2vinyl pyridine) (P2VP), adding nanofillers reduces the amplitude of the segmental peak (α-process), strongly broadens its low-frequency side, and induces an additional process, the so-called Maxwell–Wagner–Sillars (MWS) polarization process (Fig. 2) [70]. The latter is related to charge accumulation and reorientation at interfaces between materials with different dielectric properties. It is usually covered by conductivity tail in $\varepsilon''(\omega)$, and appears clearly only in $\varepsilon'_{der}(\omega)$(see, e.g., Fig. 5d). The amplitude and position of the MWS process depends on the amount of interfaces and the conductivity (e.g., impurities) in the polymer matrix, the shape of nanofillers, and might provide some information on the nanofiller-polymer interface [71]. But to the best of our knowledge, it doesn't interfere much with the analysis of the IL. The decrease of the amplitude of segmental relaxation and its low-frequency broadening are directly related to the segmental dynamics in the IL. The low frequency tail changes of the α-process are more clearly visible in $\varepsilon'_{der}(\omega)$ (Fig. 2b). The dielectric loss spectra can be converted into a relaxation time distribution spectra, $\Delta\varepsilon \times g(\ln \tau)$, using he formal equation

$$\varepsilon^*(\omega) = \varepsilon_\infty + \frac{\sigma_{DC}}{i\omega\varepsilon_0} + \int \Delta\varepsilon \times \frac{g(\ln \tau)}{(1 + i\omega\tau)} \mathrm{d} \ln \tau, \quad (1)$$

with applied normalization condition $\int g(\ln \tau)\, d \ln \tau = 1$. To evaluate $\Delta\varepsilon \times g(\ln \tau)$ a regularization method can be utilized [72]. In that case the interfacial segmental process appears as a clear peak denoted as the α'-process (Fig. 2c).

The increase of nanofillers concentration (φ) leads to an intensity increase of the α'-process and an intensity decrease of the α-process (Fig. 3a). This is related to a decrease of bulk-like polymer fraction and an increase of the interfacial polymer layer fraction. Surprisingly, similar apparent effect is observed with a decrease in polymer molecular weight (Fig. 3b) [39]. However, the origin of this effect is different. Detailed analysis revealed that density of the interfacial layer and its mechanical

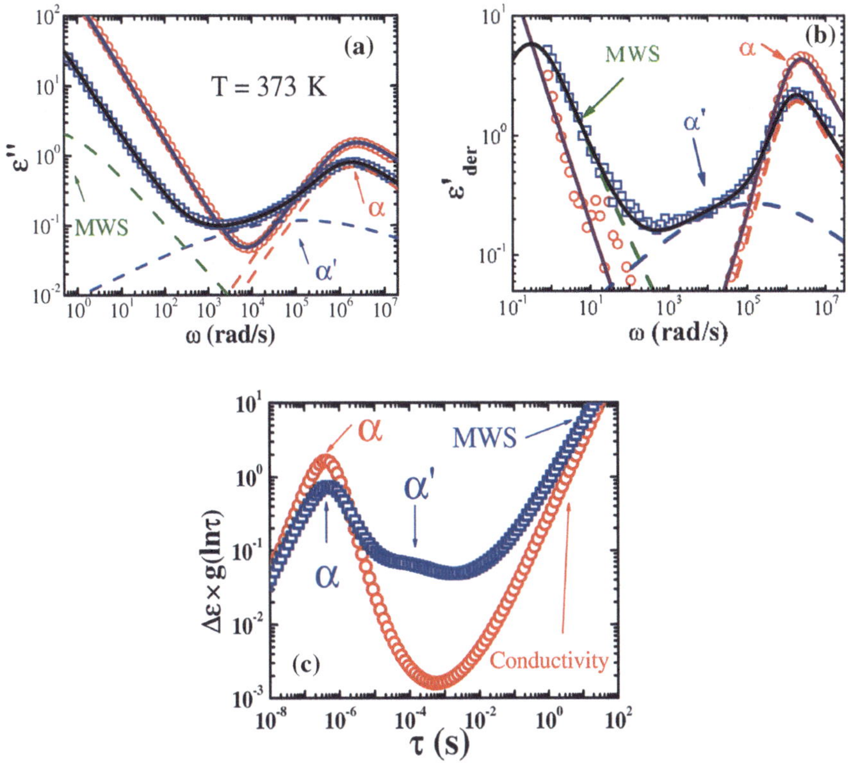

Fig. 2 **a** Imaginary part of the dielectric permittivity, **b** derivatives of the real part of dielectric permittivity $\varepsilon'_{der}(\omega)$, and **c** spectra of time distribution $\Delta\varepsilon \times g(\ln \tau)$ for PVAc/SiO_2 PNC with 20.5 vol% (blue squares) and neat PVAc (red circles). Black lines are the fit of the Havriliak-Negami function plus dc-conductivity for the neat PVAc, and the ILM model for PVAc/SiO2-20.5 vol%. Dashed lines in (**a**) and (**b**) show the contributions from bulk-like (α, red line) and interfacial (α', blue line) polymers. MWS marks tail of the Maxwell–Wagner–Sillars polarization [39]. [Figures (a) and (c) were reprinted with permission from [39]: Cheng, S et al. Physical Review Letters 2016, 116: 038302. Copyright 2016 by the American Physical Society. Figure (b) were reprinted from [68]: Cheng, S. et al. The Journal of Chemical Physics 2017, 146, 203,201, with the permission of AIP Publishing]

properties vary strongly with the molecular weight at the same volume fraction of nanofillers [26, 38, 39]. The density of the IL increases relative to the bulk polymer density in low molecular weight PNCs due to attractive polymer–nanoparticles interactions [39]. However, for longer chains, when 2Rg becomes larger than the distance between nanoparticles surfaces, confinement of polymer chains between nanoparticles leads to a frustration in the chain packing and a reduction in polymer density in the interfacial layer [26, 39]. As a result, change in the segmental dynamics of the interfacial layer decreases with the increase in molecular weight (Fig. 3b).

In several polymers, e.g., poly(propylene glycol) (PPG) and polyisoprene (PI), also the chain relaxation (normal mode) appears in the dielectric spectra. In these

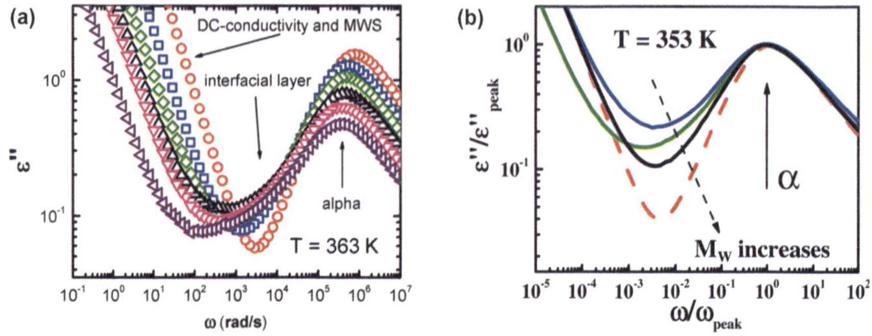

Fig. 3 **a** Imaginary part of the dielectric permittivity for the PVAc/SiO2 PNCs with different nanoparticle volume concentrations: 0 vol% (red circles), 6 vol% (blue squares), 12 vol% (olive diamonds), 20 vol% (black upper triangles), 23 vol% (pink down triangles), 32 vol% (purple left triangles). Data are taken from [20] [Reprinted with permission from [20]: Carroll, B. et al. Macromolecules 2017, 50, 6149–6163. Copyright 2017 American Chemical Society] **b** Normalized dielectric loss spectra of PNCs with 3 selected molecular weights, M_W, in comparison to the neat polymer PVAc spectrum. The dashed red line is neat PVAc with $M_W = 85$ kg/mol; the solid lines are PNCs with 20 vol% of silica, and PVAc with $M_W = 12$ kg/mol (blue line); PVAc with $M_W = 85$ kg/mol (green line); PVAc with $M_W = 253$ kg/mol (black line) [39]. [Figure were reprinted with permission from [39]: Cheng, S. et al. Physical Review Letters 2016, 116, 038302. Copyright 2016 by the American Physical Society]

polymers the dipole moment has a non-zero projection along the chain, which accumulates, enabling measurements of the chain end-to-end relaxation [64]. Studies of these polymers revealed that adding nanoparticles affects dielectric spectra of segmental and chain relaxations (Fig. 4) [73]. Although the analysis of the low frequency side of the segmental mode in this case is more difficult due to the overlap with the normal mode, a decrease in the amplitude of segmental relaxation is clearly observed. Furthermore, after removing the free none-adsorbed polymer chains, the PNC spectra reveals an acceleration and a strong broadening of both the normal and the segmental modes (Fig. 4). This effect increases with cooling [73].

BDS studies of glycerol/silica nanocomposite (GSNC) [74] present an interesting model example of nanocomposite free from the influence of chain confinement and entanglement effects, as well as the nonequilibrium chain adsorption phenomena. Reduction of the amplitude and low-frequency broadening of the α-process in the dielectric spectra is clearly observed also in this case (Fig. 5). Thus, adding nanoparticles to a molecular liquid with attractive molecule–nanoparticle interactions also leads to a formation of substantial IL with a slower α-dynamics. However, the detailed analysis of the BDS data revealed [39, 42, 68, 74–77] an important difference between PNC and molecular nanocomposites, as will be discussed below.

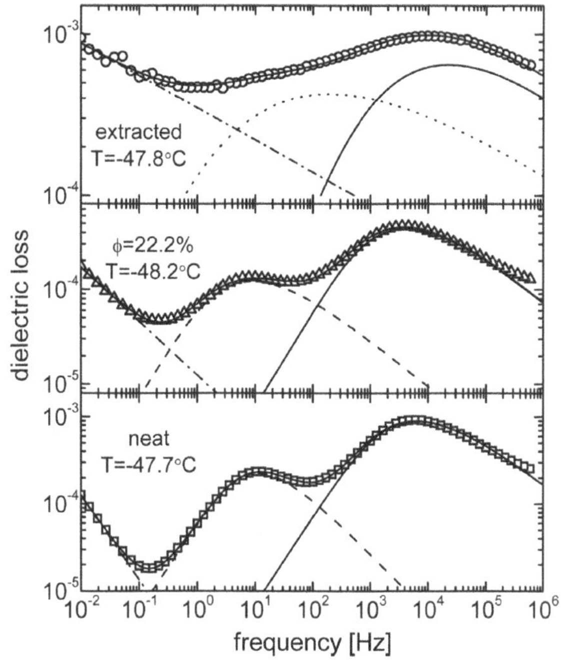

Fig. 4 Dielectric spectrum of neat PPG (bottom); PPG-22 % silica PNC (middle) and PNC after extraction of unattached polymer (upper). The lines represent fits of the Kohlrausch function to each of the two dispersions, along with a power law at low frequencies due to dc conductivity. Data from [73]. [Reprinted with permission from [73]: Casalini, R. et al. Macromolecules 2016, 49, 3919–3924. Copyright 2016 American Chemical Society]

2.2 *Segmental Relaxation Time in the Interfacial Layer in PNCs*

As mentioned above, the segmental process in the dielectric spectra of most PNCs appears broadened at low frequency side of the peak (Figs. 2 and 3). This clearly indicates slowing down of the segmental relaxation in the interfacial layer, as expected for attractive polymer–nanoparticles interactions [20, 23, 24, 39, 68, 78]. The latter is required for a good nanoparticles dispersion. Quantitative estimates of the slowing down depend on model approach chosen for analysis of the BDS data [20], and we will discuss several models below.

The quantitative analysis of BDS data for many regular PNCs and PNCs formed by polymer Grafted Nanoparticles (GNPs) revealed that the segmental dynamics in the IL is always slower than in the neat polymer, and thus $\tau_{IL}/\tau_{neat} > 1$ [23, 36] (Fig. 6a). This ratio is often ~ 10–30 [36]. However, τ_{IL} is usually estimated from the maximum of dielectric loss, and reflects some averaged relaxation time in the interfacial layer. In reality, the interfacial layer has strong gradient in segmental relaxation time [43–45]. The latter leads to the broad dielectric relaxation spectrum that extends 4–5 orders below the maximum in dielectric loss (Fig. 2).

A different scenario, an existence of the interfacial layer without glass-like dynamics was proposed in [79] based on the analysis of pol(methyl methacrylate) (PMMA)/silica DSC data. The data revealed a much smaller step in specific heat at Tg

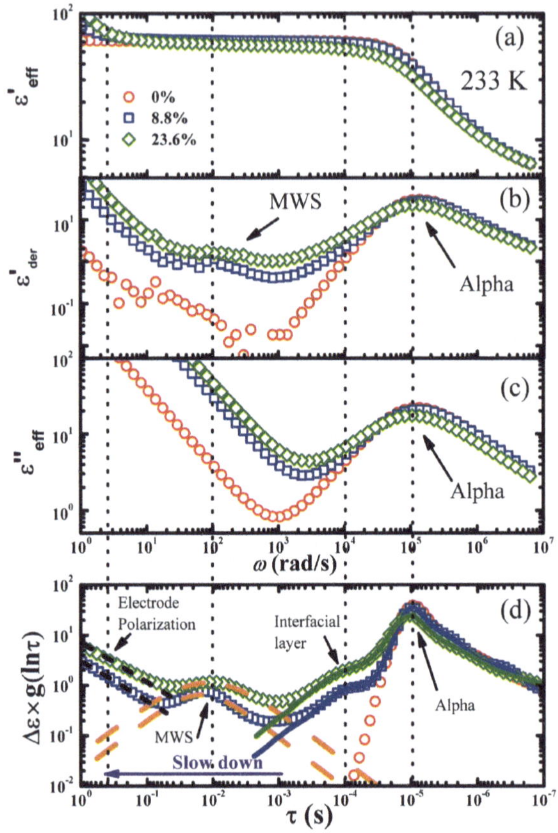

Fig. 5 Glycerol/silica nanocomposite dielectric spectra in different presentations for two silica concentrations and pure glycerol: **a** Real part of the dielectric permittivity and **b** its derivative, **c** Imaginary part of the dielectric permittivity and **d** Relaxation time distribution using Eq.(1). Data from [74]. [Reprinted from [74]: Cheng, S. et al. The Journal of Chemical Physics 2015, 143, 194,704, with the permission of AIP Publishing]

than expected from the decrease of the polymer weight fraction, and no second step at higher T. Similar effect in DSC has been also reported for polyamide, poly(dimethyl siloxane) (PDMS) and polylactic acid based PNCs [80–82]. A direct comparison of the results of DSC and BDS revealed a contradiction between these two techniques: the interfacial fraction appears as immobile in DSC, while it appears retarded in BDS [80]. Based on these results, the authors suggested that relaxation in the interfacial layer might differ from glassy dynamics [79, 80]. In contrast, studies of PPG, P2VP, and PVAc-based PNCs revealed a long tail in specific heat jump with the fraction of the polymer in agreement with that estimated from BDS [42]. This controversy in thermodynamic behavior of different PNCs remains unresolved.

An interesting result was observed when the free none-adsorbed chains are removed from PNCs. These kinds of samples have no bulk-like polymer left, and the

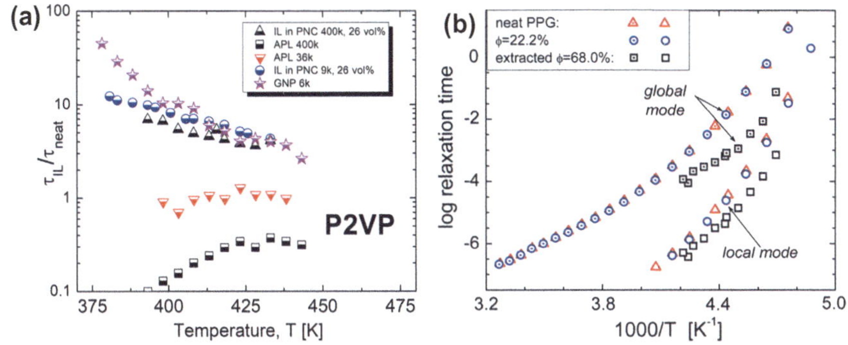

Fig. 6 **a** Temperature dependence of the ratio of the segmental relaxation times in the interfacial/adsorbed layer (IL), τ_{IL}, to that in the neat polymer, τ_{neat} [36]. [Reprinted with permission from [36]: Popov, I. et al. Macromolecules 2020, 53, 4126–4135. Copyright 2020 American Chemical Society]. **b** Segmental or local (dotted symbols) and normal or global mode (open symbols) relaxation times for the neat PPG (triangles), a nanocomposite (circles), and the nanocomposite after extraction of the unbound polymer (squares) [73]. [Reprinted with permission from [73]: Casalini, R. et al. Macromolecules 2016, 49, 3919–3924. Copyright 2020 American Chemical Society]

entire polymer fraction is essentially the interfacial layer. The segmental relaxation in these samples appears to be faster than in the neat polymer (Fig. 6) [36, 73]. It is clearly observed in PPG/silica nanocomposites after extraction of the unbound polymer (Fig. 6b) [73]. Moreover, faster dynamics in PPG occur for both normal and segmental modes. Similar analysis of P2VP/silica PNCs with removed free chains revealed extremely broad segmental relaxation spectra with the maximum shifted to higher frequency, relative to the neat polymer (Fig. 7 a). This surprising result clearly indicates a faster characteristic relaxation time in the interfacial layer with removed

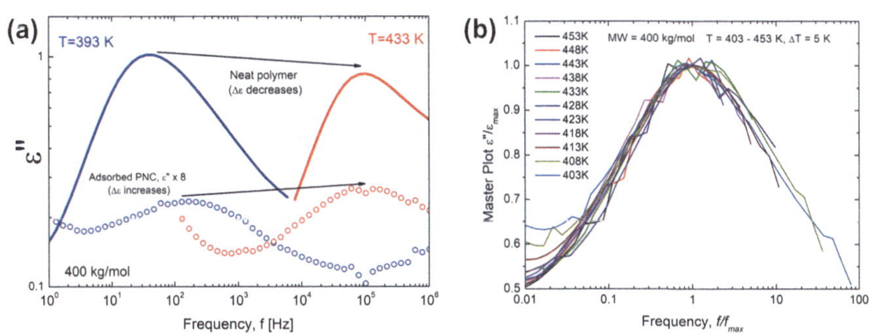

Fig. 7 **a** Dielectric loss spectra of neat P2VP (lines) and PNC with only Adsorbed Polymer Layer (APL) (symbols) at T = 393 K (blue) and T = 433 K (red). For convenience, the spectra of APL are multiplied by a factor of 8 [36]. **b** Master plot of the dielectric losses for segmental relaxation in APL reveals no significant change in the peak shape with temperature [36]. [Reprinted with permission from [36]: Popov, I. et al. Macromolecules 2020, 53, 4126–4135. Copyright 2020 American Chemical Society]

free chains, and was ascribed to a frustration in the chain packing in this layer [36]. It was also noted that the broad shape of the loss peak reflects the broad distribution of relaxation times, and shows no significant changes with temperature (Fig. 7b).

This acceleration of segmental dynamics in polymer adsorbed to an attractive surface is unusual, but has been also reported in other systems such as PMMA/silica PNCs [83], PMMA/Ag PNCs [84] and also in thin polymer films [48, 59, 60, 63]. In PNCs with only adsorbed chains this effect is more pronounced for higher molecular weight systems (Fig. 6a). These counter-intuitive results were ascribed to an additional free volume in these systems created by frustration in the packing of the adsorbed chains [36, 83–85]. The extremely broad relaxation spectrum (Fig. 7) suggests an extremely strong gradient of the relaxation times in the adsorbed layer.

2.3 Dielectric Strength of Structural Relaxation in the Interfacial Layer in PNCs

Another interesting experimental observation is the change in the dielectric strength of the segmental relaxation and its temperature dependence. The dielectric strength, $\Delta\varepsilon$, of the segmental relaxation in a neat polymer decreases with temperature increase, and can be approximately described as follows: $\Delta\varepsilon \propto 1/T$ (Fig. 8) [64]. However, in the PNC the dielectric strength of the segmental relaxation not only decreases with adding nanoparticles, but its temperature dependence may differ strongly from that of the neat polymer (Fig. 8) [20]. For example, in PVAc with 23 vol% of silica nanoparticles, $\Delta\varepsilon$ is almost independent of temperature (Fig. 8). This effect has been ascribed to a decrease in the volume fraction of the bulk-like polymer due to an increase in the interfacial layer thickness upon cooling [20].

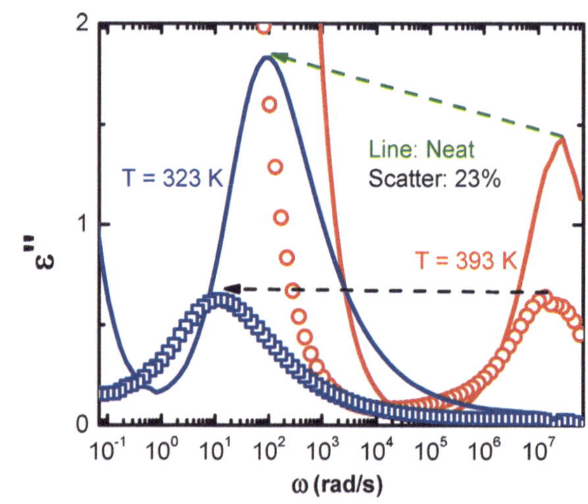

Fig. 8 Imaginary parts of the permittivity spectra of the neat PVAc (line) and PVAc/silica-23 vol% PNC (symbols) at two selected temperatures; the amplitude changing is shown by arrows [20]. [Reprinted with permission from [20]: Carroll, B. et al. Macromolecules 2017, 50, 6149–6163. Copyright 2017 American Chemical Society]

This effect is even more pronounced in PNCs with removed free none-adsorbed chains (Fig. 7a) [36]. The analysis of the dielectric spectra of the segmental relaxation in these samples reveals two important results (Fig. 7a) [36]: (i) significant (~30 times) drop in the dielectric strength of the segmental relaxation peak $\Delta\varepsilon$, despite that the weight fraction of the polymer is still ~40 % (i.e., normally one would expect only ~2.5 times drop in $\Delta\varepsilon$); and (ii) surprising decrease of the dielectric strength upon cooling. Thus, the dielectric strength of the segmental relaxation in the interfacial polymer layer decreases much stronger than expected from the reduction of the polymer content, and its temperature dependence also differs. This is an important experimental result that is often overlooked in many experimental studies, because researchers are usually focusing on changes of the characteristic relaxation time without paying attention to a change in the dielectric strength. Similar effect, strong reduction of the segmental dielectric strength was also reported for thin polymer films [47–50, 52–63], and we will discuss below models proposed for the explanation of this effect.

3 Theoretical Models for Description of the Dielectric Data in PNCs

3.1 Additive Approach

Several model approaches have been used to analyze the measured dielectric spectra in PNCs. The simplest approach assumes additive contributions of the bulk-like polymer and of the interfacial region, and fits the spectra by two Havriliak– Negami (HN) functions [24, 66, 80, 86]

$$\varepsilon^*(\omega) = \varepsilon_\infty + \frac{\sigma_0}{i\omega\varepsilon_0} + \frac{\Delta\varepsilon_{IL}}{\left[1 + (i\omega\tau_{\mathrm{HN}}^{IL})^{\alpha_{IL}}\right]^{\beta_{IL}}} + \frac{\Delta C_{\mathrm{bulk}}}{\left[1 + (i\omega\tau_{\mathrm{HN}}^{\mathrm{bulk}})^{\alpha_{\mathrm{bulk}}}\right]^{\beta_{\mathrm{bulk}}}}, \quad (2)$$

where $\Delta\varepsilon_{\mathrm{bulk}}$ and $\Delta\varepsilon_{IL}$ are the dielectric strength associated with the bulk polymer and the interfacial layer, respectively. $\tau_{\mathrm{HN}}^{\mathrm{bulk}}$ bulk and τ_{HN}^{IL} are the respective HN relaxation times, and the α_{bulk}, β_{bulk} and α_{IL}, β_{IL} are the corresponding HN broadening parameters. To avoid fitting uncertainties and to limit the amount of the free fit parameters, the bulk polymer process spectral shape (broadening parameters α_{bulk} and β_{bulk}) is usually assumed to be identical to the spectral shape of the neat polymer, and only its dielectric strength and relaxation time are allowed to vary. Then, assuming that the dielectric strength per volume is the same in the bulk and in the interfacial layer, the two HN functions approach provides an estimate of the volume fraction of interfacial layer as $\varphi_{IL} = (1 - \varphi_{\mathrm{NP}})\Delta\varepsilon_{IL}/(\Delta\varepsilon_{IL} + \Delta\varepsilon_{\mathrm{bulk}})$. Knowing the volume fraction and radius of nanoparticles, R_{NP}, the thickness of the interfacial layer can be calculated to

$$l_{IL} = R_{NP}\left[\left(\frac{\varphi_{IL} + \varphi_{NP}}{\varphi_{NP}}\right)^{1/3} - 1\right]. \tag{3}$$

This simple approach involves several assumptions that are not valid in general. First, the dielectric responses in heterogeneous systems are not additive, and interference terms should be explicitly taken into account. Second, the dielectric strength in the interfacial region can differ significantly from that in the bulk (see the Sect. 2.3).

The Two Phase Model (TPM) and the Interfacial Layer Model (ILM) take interference terms explicitly into account in analysis of the dielectric spectra of PNCs.

3.2 Two Phase Model

The two phase model (Fig. 9a) considers only The polymer matrix and nanoparticle, and for spherical nanoparticles it predicts [20, 39, 64, 74]

$$\begin{aligned} \varepsilon^*_{PNC}(\omega) &= (1-\varphi_{NP})\varepsilon^*_{bulk}(\omega) + \varphi_{NP}\varepsilon^*_{NP}(\omega) - (1-\varphi_{NP})\varphi_{NP}\varepsilon^*_{cross}(\omega), \\ \varepsilon^*_{cross}(\omega) &= \frac{\left[\varepsilon^*_{bulk}(\omega) - \varepsilon^*_{NP}(\omega)\right]^2}{(1-\varphi_{NP})\varepsilon^*_{NP}(\omega) + (2+\varphi_{NP})\varepsilon^*_{bulk}(\omega)}. \end{aligned} \tag{4}$$

Here $\varepsilon^*_{bulk}(\omega)$ and $\varepsilon^*_{NP}(\omega)$ are the complex dielectric permittivities of the polymer matrix and the nanoparticles, respectively; φ_{NP} is volume fraction of the nanoparticles, and $\varepsilon^*_{\text{cross}}(\omega)$ presents the interference term. This term always reduces the dielectric signal (Eq. 4). The TPM contains no free parameters and demonstrates the inaccuracy of the simple additive approach in describing the dielectric response of heterogeneous materials, because the interference term is not negligible and should be explicitly taking into account for detailed quantitative analysis.

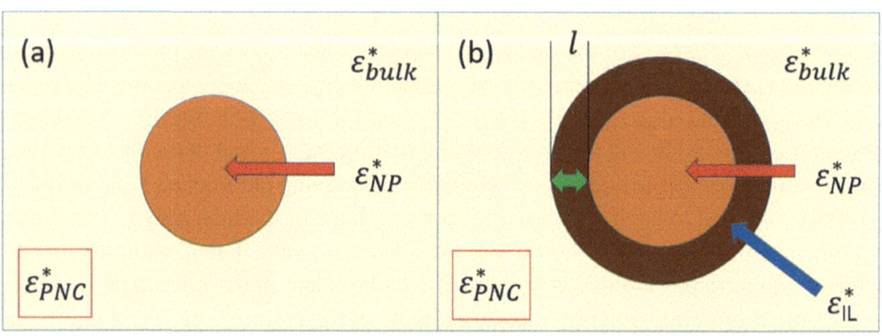

Fig. 9 Sketch for the geometry of **a** the Two Phase Model (TPM) and **b** Interfacial Layer Model (ILM) [20] [Reprinted with permission from [20]: Carroll, B. et al. Macromolecules 2017, 50, 6149–6163. Copyright 2017 American Chemical Society]

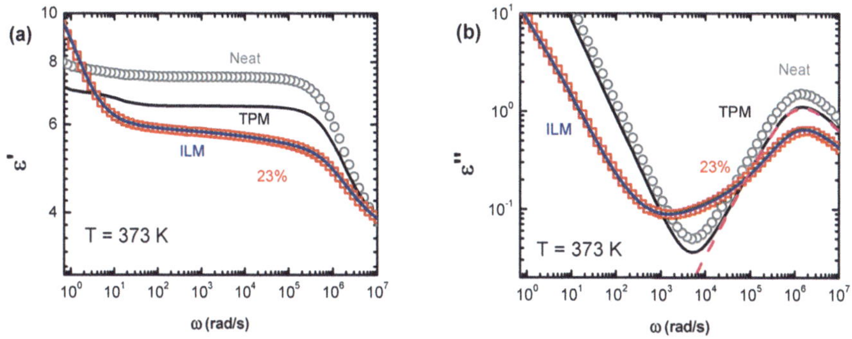

Fig. 10 Application of TPM to the dielectric spectra of PVAc/silica PNC (symbols) (black line) and fitting by the ILM (blue line). **a** real part, **b** imaginary part Unlike the ILM, the TPM cannot describe the broadening of segmental relaxation at low frequencies [20]. [Reprinted with permission from [20]: Carroll, B. et al. Macromolecules 2017, 50, 6149–6163 Copyright 2017 American Chemical Society]

It has been shown that the TPM cannot describe the spectra of PNCs (Fig. 10), and an interfacial layer with dielectric properties different from that of the bulk polymer should be taken explicitly into account [20]. However, the TPM analysis provides an expected estimation of the total dielectric strength of the nanocomposite spectra, because it does not have any free fit parameters. The analysis of PVAc/silica nanocomposites clearly revealed that $\Delta\varepsilon_{PNC}/(1-\varphi_{NP})$ decreases significantly stronger than predicted by the TPM (Fig. 11a) [20]. However, the same analysis of the dielectric spectra of glycerol/silica nanocomposites demonstrated good agreement with the TPM predictions (Fig. 11b) [20, 74]. It suggests that the dielectric strength of the structural relaxation of the interfacial layer of GSNC is essentially the same as in the bulk glycerol, while the dielectric strength in the polymer interfacial layer is strongly

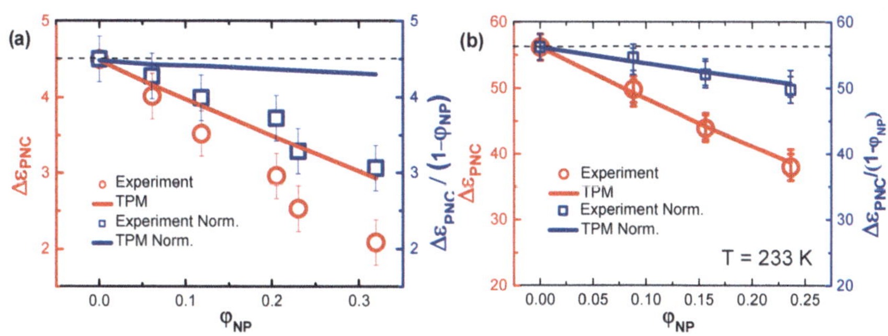

Fig. 11 Dielectric strength as a function of NP loading for the PVAc/SiO_2 (**a**) and glycerol/SiO_2 (**b**) nanocomposites. The experimental data (red circles) and the data normalized to the polymer volume fraction $(1-\varphi_{NP})$ (blue squares) and the two-phase model predictions (lines) [20, 74] are presented. Data were replotted from [20]. [Reprinted with permission from [20]: Carroll, B. et al. Macromolecules 2017, 50, 6149–6163. Copyright 2017 American Chemical Society]

reduced. This reduction was obvious without model data analysis for the adsorbed polymer layer (Fig. 7a). Thus, the ratio of the dielectric signals from the interfacial polymer and from the bulk-like polymer, $\Delta\varepsilon_{IL}/\Delta\varepsilon_{bulk}$, cannot be used for an accurate estimates of the interfacial volume fraction.

3.3 *Interfacial Layer Model*

The Interfacial Layer Model (ILM) assumes the existence of an additional layer around nanoparticle with different dielectric properties than in remaining bulk-like polymer (Fig. 9b). This model introduces the volume fraction of the IL, φ_{IL}, and predicts [20, 87]:

$$\varepsilon^*_{PNC}(\omega) = \frac{\varphi_{NP}\varepsilon^*_{NP}(\omega) + \varphi_{IL}\varepsilon^*_{IL}(\omega)R^* + \varphi_{bulk}\varepsilon^*_{bulk}(\omega)S^*}{\varphi_{NP} + \varphi_{IL}R^* + \varphi_{bulk}S^*},$$

$$S^* = \frac{\left(\varepsilon^*_{IL}(\omega) + 2\varepsilon^*_{bulk}(\omega)\right)\left(\varepsilon^*_{NP}(\omega) + 2\varepsilon^*_{IL}(\omega)\right) + 2d\left(\varepsilon^*_{IL}(\omega) - \varepsilon^*_{bulk}(\omega)\right)\left(\varepsilon^*_{NP}(\omega) - \varepsilon^*_{IL}(\omega)\right)}{9\varepsilon^*_{IL}(\omega)\varepsilon^*_{bulk}(\omega)},$$

$$R^* = \frac{2\varepsilon^*_{IL}(\omega) + \varepsilon^*_{NP}(\omega)}{3\varepsilon^*_{IL}(\omega)}, \quad d = \frac{\varphi_{NP}}{\varphi_{NP} + \varphi_{IL}}, \quad \varphi_{NP} + \varphi_{IL} + \varphi_{bulk} = 1. \tag{5}$$

Here, $\varepsilon^*_{IL}(\omega)$ is the complex permittivity of the IL, which can be approximated by HN function. Thus, this model has additional fit parameters to describe function $\varepsilon^*_{IL}(\omega)$ and parameter φ_{bulk}, while the shape parameters for the bulk-like polymer spectra are assumed to be the same as in the neat polymer. The demonstrates that ILM provides good description of the PNCs dielectric spectra (Fig. 10). Moreover, the amplitude of the bulk-like polymer contribution to the dielectric spectra provides estimates of the bulk-like polymer fraction, and correspondingly the interfacial layer fraction $\varphi_{IL} = 1\text{-}\varphi_{NP}-\varphi_{bulk}$ [20]. Using this approach, the IL thicknesses was estimated to be l_{IL} ~2–6 nm, depending on type of the polymer (Fig. 12a) and the temperature (Fig. 12b) [42]. These values are comparable to results obtained by other methods, such as SAXS and Temperature-Modulated Differential Scanning Calorimetry (TMDSC) [42].

Furthermore, ILM provides a direct estimates of the segmental relaxation time in the interfacial layer$_{IL}$. As we discussed in the Sect. 2.2, τ_{IL} is slower than τ_{bulk} in most cases, although it appears faster in the PNCs with removed free chains (Fig. 6). In addition, the ILM approach also enables to estimate the dielectric strength of the IL process. The temperature dependence of the dielectric strength of the segmental relaxation in the IL for the regular PNCs, GNPs, and PNCs with removed free none-bound polymers, i.e., Adsorbed Polymer Layer (APL), in comparison with that of the neat polymer are presented in Fig. 13. Assuming APL consists of the interfacial polymer only, a single HN function was used to estimate the dielectric

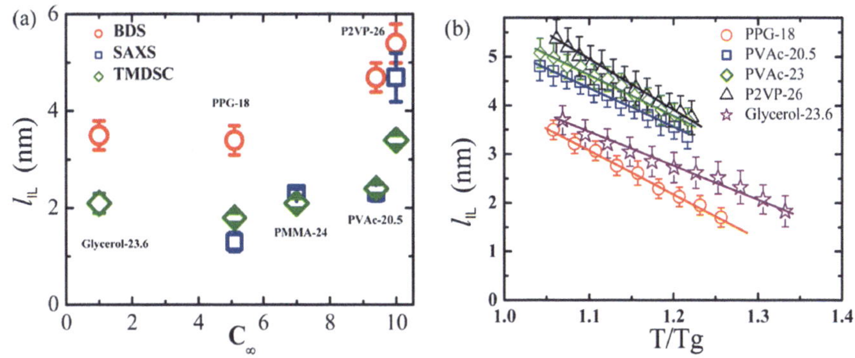

Fig. 12 **a** Thickness of the IL estimated using three different methods versus characteristic ratio C_∞ (rigidity of the polymer chain) for several polymers and glycerol [42]. **b** Temperature dependence of the IL thickness estimated from dielectric spectra [42] [Reprinted with permission from [42]: Cheng, S. et al. Macromolecules 2017, 50, 2397–2406. Copyright 2017 American Chemical Society]

strength. The detailed analysis revealed strong suppression of the dielectric strength of the segmental relaxation in the PNC's interfacial layer, in comparison to expected reduction of dielectric strength due to decreases of polymer content (Fig. 13) [36]. This is consistent with the earlier results (Figs. 7a, 10, and 11). This effect is more pronounced in regular PNCs with short chains and in APLs. Moreover, the temperature dependence of the $\Delta\varepsilon$ in IL clearly differs from that of the neat polymers (Fig. 13). In some PNCs and APLs the temperature behavior of $\Delta\varepsilon(T)$ becomes the opposite to that of bulk polymer (Figs. 7a, 13). A strong suppression of the dielectric strength of segmental relaxation has been observed also in thin polymer films

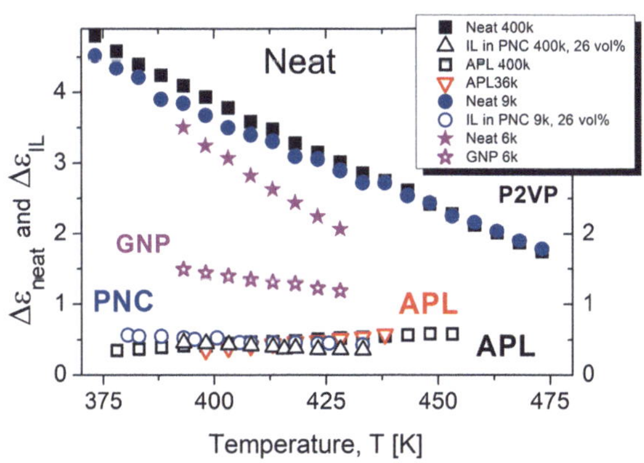

Fig. 13 Temperature dependence of the dielectric strength of the neat polymer, $\Delta\varepsilon_{neat}$ (closed symbols), and of the interfacial/adsorbed layer in regular PNCs, Grafted Nanoparticles (GNP) and Adsorbed Polymer Layer (APL), $\Delta\varepsilon_{IL}$ (open symbols). Data are taken from [36]. [Reprinted with permission from [36]: Popov, I. et al. Macromolecules 2020, 53, 4126–4135. Copyright 2020 American Chemical Society.]

[54–58]. However, glycerol/silica nanocomposites do not show a decrease in the dielectric strength of structural relaxation in the interfacial layer (see, e.g., Fig. 11b) [20, 74]. Thus this reduction appears to be a polymer specific property. A possible mechanism of this unexpected observation will be discussed in the next section.

3.4 Reduction of the Dielectric Strength in the Interfacial Layer

In the previous section, it was demonstrated that regardless of whether the segmental dynamics in thin films is getting faster or slower than in a bulk polymer (e.g., T_g decreases or increases), dielectric strength of segmental relaxation always decreases [47–50, 52–63]. The same appears in the interfacial layer in PNCs: while the averaged segmental relaxation in IL can be slower (PNC and GNP) or faster (APL) than that in the bulk polymer (Fig. 7a), the dielectric strength of segmental relaxation in IL is always lower than in the bulk polymer (Fig. 13).

The idea of "dead layer" was proposed to describe this effect in thin polymer films [58, 62, 88]. In a simple approximation, the dielectric strength of a polymer can be described as [89]:

$$\Delta\varepsilon_{neat} = \frac{N\,\overline{\mu}}{VE} = \frac{gN\mu^2}{3k_B TV} \tag{6}$$

where N is a number of dipoles in a volume V, μ is a segmental dipole moment and g is the Kirkwood–Fröhlich factor that accounts for dipole–dipole correlations. The dead layer approach assumes that a significant number of segmental dipoles is frozen, and this leads to a strong suppression of the dielectric response in the interfacial layer. The idea of "dead" (or glassy) layer has been widely used to explain various experimental results in PNCs and thin polymer films [46, 58, 79, 88, 90, 91]. However, many experimental and simulations data contradict the idea of a "dead" layer and emphasize that the dynamics in the interfacial layer is not frozen, although it can be much slower than in the bulk [20, 24, 39, 42]. The significant suppression of the dielectric signal is most clearly observed in the APL PNCs (Fig. 7a). However, no significant change in the spectral shape of the APL's segmental peak with temperature was detected (Fig. 7b). This contradicts the idea of a "dead" (frozen) layer, because a significant additional broadening would be expected on the low-frequency side of the segmental peak with "unfreezing" dipoles from the dead layer. However, no such low frequency broadening has been observed. Moreover, the averaged segmental dynamics in APL appears to be faster than in the bulk (Fig. 6a). That is completely inconsistent with the idea a of "dead" layer. The same inconsistency is observed in thin polymer films where segmental dynamics can be faster (T_g is lower) than in the bulk polymer, while $\Delta\varepsilon$ strongly decreases [47–50, 52–63].

An alternative explanation of the decreased dielectric signal in IL is based on the assumption of a restricted amplitude of the segmental fluctuations on the average time scale of segmental dynamics [36]. To perform the quantitative analysis, we may consider the segmental dipole moment of a polymer chain in IL or adsorbed layer to be restricted in the conic potential well, with well height equal to U (Fig. 14). The parameter 2δ define the conic angle at the top of the conic potential well, where the segment can fluctuate without restriction. The potential characterizes restrictions on segmental reorientation caused by the presence of the interface, chain crowding and stretching in the interfacial layer.

In this case the Eq. (6) can be written as follows

$$\Delta\varepsilon_{\mathrm{Ads}} = \left(1 - \phi_{NP}^{m}\right)\frac{N\,\overline{\mu}_{Ads}}{VE} = \left(1 - \phi_{NP}^{m}\right)\frac{gN}{VE}\langle\mu\overline{\cos}\,\theta\rangle_{NP} \tag{7}$$

where $\langle\ldots\rangle_{NP}$ defines an averaging of the potential wells around nanoparticle; g is the Kirkwood-Fröhlich factor without potential well and $\overline{\cos\theta}$ is an average of the angle between the dipole moment and the electrical field inside the potential well

$$\overline{\cos}\,\theta = \int \cos(\mathbf{E}\mu)\, e^{\frac{(\mathbf{E}\mu)-U}{k_B T}}\, d\Omega \Big/ \int e^{\frac{(\mathbf{E}\mu)-U}{k_B T}}\, d\Omega. \tag{8}$$

In this approach the temperature dependence is introduced for the effective Kirkwood–Fröhlich factor $g_{eff} = g(T)\,\langle\overline{\cos}\,\theta\rangle_{NP}$, instead of number of mobile dipoles, N, as it was assumed in the "dead" layer approach. The calculation of Eq. 7 and Eq. 8 [36] provides a simple relationship in the case of $\mu E/k_B T << 1$

$$\Delta\varepsilon_{\mathrm{IL}} = \frac{\Delta\varepsilon_{\mathrm{Ads}}}{\left(1 - \phi_{NP}^{m}\right)} = \Delta\varepsilon_{Neat}\left(1 - f^2\right), \quad f = \frac{\sin^2\delta}{2(\coth(U/2k_B T) - \cos\delta)} \tag{9}$$

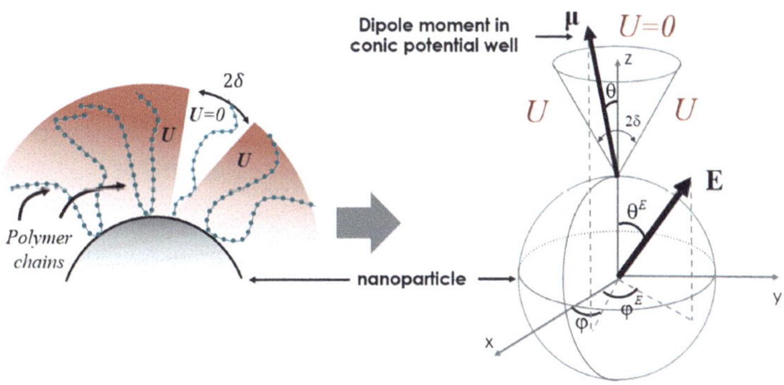

Fig. 14 Model for a polymer chain in the conic potential well in the interfacial layer around nanoparticle. Sketch is replotted from [36] [Reprinted with permission from [36]: Popov, I. et al. Macromolecules 2020, 53, 4126–4135. Copyright 2020 American Chemical Society]

The model predictions for $\Delta\varepsilon_{IL}$ (Eq. 9) are presented in the Fig. 15 at different values of the potential. The model predicts a change of the slope in $\Delta\varepsilon_{IL}(T)$ (Fig. 15), and at high potential energy barrier predicts increasing of the dielectric strength with temperature increase, in agreement with observed experimental results (Fig. 13). Furthermore, this model predicts a significant suppression of the dielectric signal in IL that depends on the amplitude of the free segmental fluctuations angle δ (Fig. 14), see the Eq. 9.

A high potential of the well restricts the fluctuation of the segmental dipole moments leading to a suppression of the dielectric response and, as a result, to a low value of the dielectric strength. Unlike the neat polymer, where temperature is a disordering factor, in the systems where dipole reorientation is restricted, the temperature helps to overcome the potential barrier leading to higher dielectric response at higher temperature. It explains the different temperature behavior of $\Delta\varepsilon(T)$ in the neat polymer and the IL. In turn, the average rate of relaxation dynamics depends on packing density. In the APL, where IL consists of strongly adsorbed polymer chains, the potential barrier is high, but density might be low, leading to a low $\Delta\varepsilon$ and a faster averaged dynamics in comparison to the neat polymer. However, in regular PNCs there are strongly and weakly adsorbed chains and even free chains, resulting in higher density of the IL. This density increase leads to slower averaged segmental dynamics in the IL, while presence of the interface and chain stretching still strongly suppress the amplitude of segmental fluctuations and $\Delta\varepsilon$. Thus, there is a decoupling between $\Delta\varepsilon$ and the averaged relaxation time in PNCs. We emphasize that in both cases, whether the averaged segmental relaxation time in the IL is slower or faster than in the bulk polymer, $\Delta\varepsilon$ in the IL always decreases. This result is inconsistent with the "dead" layer approach, because faster dynamics is incompatible with the idea of "glassy" layer with frozen segmental dynamics.

A similar qualitative explanation was proposed in [80]. The authors speculated that trains of polymer segments formed on the surface of nanoparticles which do not contribute to the dielectric spectrum because of their strong adsorption, while tails and loops have a reduced dielectric response, and this leads to the decrease of $\Delta\varepsilon$ [80].

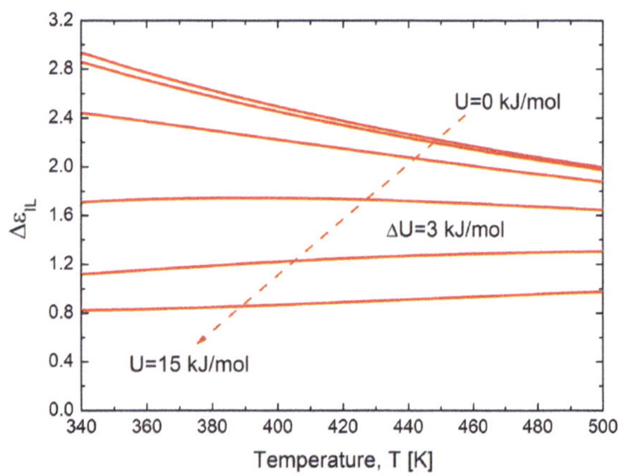

Fig. 15 Model predictions provided by the Eq. (9) for different value of the potential from 0 kJ/mol to 15 kJ/mol with the step 3 kJ/mol. The case U = 0 kJ/mol corresponds to bulk-like behavior

3.5 Dielectric Strength of the Interfacial Layer in PNCs Where the Polymer Has Normal Mode.

The model of confined polymer chains in potential well might help to explain another interesting effect observed for the normal mode in PNCs. A detailed analysis of the PPG/silica PNC (see Fig. 4) revealed that the dielectric strength of the segmental relaxation decreases with increasing silica concentration. However, this decrease significantly exceeds the one expected due to the reduction in the polymer content (Fig. 16) [73]. This is the same effect discussed in the section above.

However, the dielectric strength of the normal mode appears to be essentially independent of nanoparticle loading (Fig. 15), although the polymer fraction decreases [73].

To explain this effect we speculate that polymer chains in the IL are stretched. For the segmental mode, this effect does not play a significant role, but when polymer chain is stretched, its end-to-end distance increases resulting in an increased end-to-end dipole moment. Thus, there are two competitive effects for the normal mode: (1) reduction of the dielectric strength because of restricted fluctuation the of dipole moment, and (2) an increase of the mean dipole moment due to the chain stretching effect. Apparently in the case of PPG/silica PNCs presented in [73], the second effect is stronger and even compensates for the decrease expected due to the decrease of the polymer volume fraction.

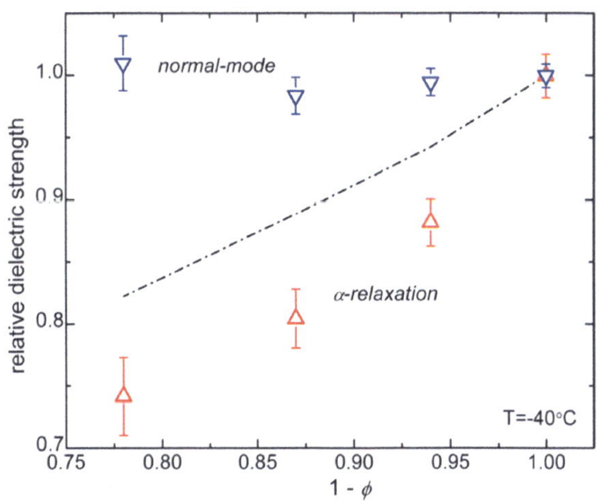

Fig. 16 Dielectric strength of the segmental mode (triangles) and normal mode (inverted triangles), normalized by the respective values for the neat PPG, as function of the polymer volume fraction. The dashed-dotted line represents the change expected due to the reduced polymer fraction in the PNCs. Data from [73]. [Reprinted with permission from [73]: Casalini, R. et al. Macromolecules 2016, 49, 3919–3924. Copyright 2016 American Chemical Society]

4 Parameters Controlling Thickness and Properties of the Interfacial Layer

BDS was critical in experimental studies of IL in PNCs, and in identifying structural and chemical parameters controlling thickness and properties of the IL [37, 39, 42, 68, 80, 84]. It has been demonstrated that the interfacial layer thickness, l_{IL}, decreases with decreasing of the nanoparticle size, and this effect was ascribed to the surface curvature (Fig. 17) [37]. Comparison of BDS data for PNCs with different polymers revealed that chain flexibility plays significant role, and the thickness of the l_{IL} increases with the increase of the polymer characteristic ratio C_∞ (rigidity of the polymer chain) (Fig. 12a) [42]. This result was confirmed also by calorimetric and SAXS measurements [42]. Detailed BDS studies also revealed that the thickness of the l_{IL} increases upon cooling (Fig. 12b) [42, 76]. Change of molecular weight has a rather weak effect on thickness of the interfacial layer, but affects strongly its properties when chain is strongly confined between nanoparticles, i.e., when 2Rg is larger than interparticle surface distance [26, 38, 39]. Apparently, this confinement leads to a frustration in chain packing, reducing the density in the interfacial layer. This reduction in density leads to weaker slowing down of the averaged segmental relaxation and smaller enhancement of the mechanical modulus in the interfacial layer [38, 39].

All these results are in agreement with many Molecular Dynamic (MD) simulations of PNCs [43–45]. However, simulations helped to unravel and disentangle more microscopic details of the interfacial layer. In particular, coarse-grained MD simulations allowed to vary independently nanoparticle–polymer interactions and polymer chain rigidity. These studies revealed that l_{IL} depends on the chain rigidity, but is rather insensitive to the nanoparticle–polymer interactions [43]. The latter strongly affect changes in segmental dynamics at the interface [43, 45]. Simulations also demonstrated that surface roughness of the nanoparticles plays an important role in the segmental dynamics [44].

We want to emphasize that in all these cases the thickness of the interfacial layer was estimated in the range of a few nm, i.e., ~1.5–6 nm [17, 20–24, 68], or a few

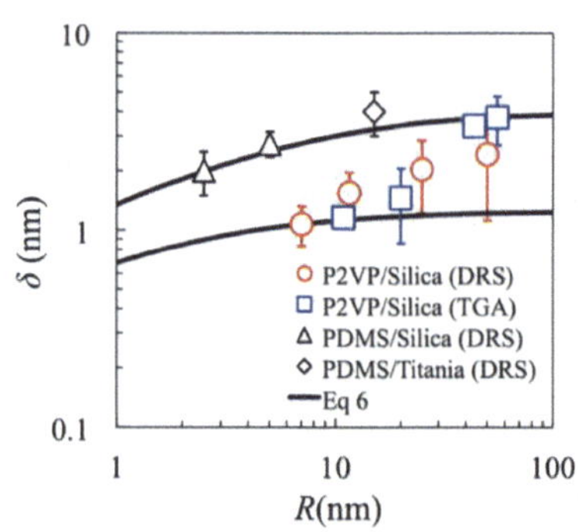

Fig. 17 Dependence of the interfacial layer thickness in PNCs on the radius of nanoparticles. Data from [37] [Reprinted with permission from [37]: Gong, S. et al. ACS Macro Letters 2014, 3, 773–777. Copyright 2014 American Chemical Society]

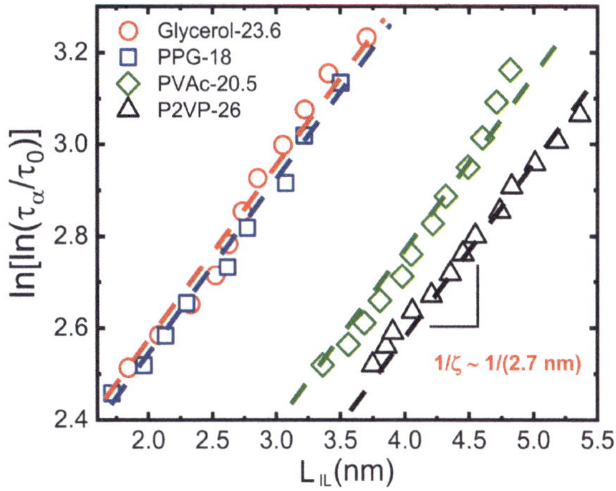

Fig. 18 Relationship between segmental relaxation time in a bulk system and the interfacial layer thickness, τ_0 is a relaxation time at infinitely high temperature Dashed line present model prediction from [45]. Data from [76]. [Reprinted from [76]: Cheng, S. et al. The Journal of Chemical Physics 2020, 152, 094,904, with the permission of AIP Publishing]

segment size in the case of coarse-grained simulations [43–45]. This length scale is comparable to the so-called dynamic heterogeneity length scale characteristic for liquids and polymers [68]. Although the specific microscopic details controlling the thickness of the interfacial layer remain unclear, some interesting suggestions relating them to the temperature dependence of segmental relaxation in bulk materials were discussed in (Fig. 18) [45, 76]. However, this discussion is out of scope of the current chapter.

5 Conclusions

The extremely broad frequency range covered by the dielectric spectroscopy makes this technique very efficient in studies of the dynamics in polymer nanocomposites. In particular, it enables detailed studies of segmental and chain dynamics in the interfacial polymer layer formed around nanofillers. In this chapter, we overviewed various approaches to analyze the dynamics in the interfacial layer, and stressed several important details of this analysis:

1. The dielectric signal in heterogeneous systems is not additive, and interference terms should be explicitly taken into account for accurate quantitative analysis.
2. Maxwell–Wagner–Sillar polarization will appear as an additional dielectric process and should not be confused with molecular relaxations.
3. The strength of the dielectric signal in the interfacial layer differs from that in the bulk. Thus, one cannot use the ratio of the interfacial and bulk dielectric signal to estimate the interfacial and the bulk-like volume fractions.
4. One can estimate the bulk-like volume fraction in PNCs from the amplitude of the bulk-like segmental dynamics. However, the interference terms in the

dielectric response should be accounted for, and the best approximation would be the Interfacial Layer Model.

5. Although the ratio of characteristic segmental relaxation time in the interfacial layer to that in the bulk is ~ 10–30, much stronger slowing down is expected at the interface. τ_{IL} estimated from the BDS data presents some averaged segmental time in the interfacial region. Dielectric spectra of the interfacial layer is usually very broad reflecting strong gradient in segmental dynamics.

The BDS data, as well as many other experimental and simulations results clearly demonstrate that the interfacial layer is not a dead or glassy layer. Segments are fluctuating in the interfacial layer. However, the amplitude of these fluctuations is strongly suppressed, most probably due to restrictions on segments reorientation in the vicinity of the interface. Detailed microscopic understanding of this effect remains unclear, although apparently this is a polymer specific effect, because it was not observed in glycerol/silica nanocomposites. Properties of the interfacial layer define many macroscopic properties of PNCs, especially when they form a percolating network at high nanoparticle loadings. In this sense, BDS might provide detailed information on volume fraction, segmental, and chain dynamics in the interfacial layer.

Acknowledgements This work was supported by the U.S. Department of Energy, Office of Science, Basic Energy Sciences, Materials Sciences and Engineering Division.

References

1. Jancar J, Douglas J, Starr FW, Kumar S, Cassagnau P, Lesser A, Sternstein SS, Buehler M (2010) Current issues in research on structure–property relationships in polymer nanocomposites. Polymer 51:3321–3343. https://doi.org/10.1016/j.polymer.2010.04.074
2. Balazs AC, Emrick T, Russell TP (2006) Nanoparticle polymer composites: where two small worlds meet. Science 314:1107–1110. https://doi.org/10.1126/science.1130557
3. Kumar SK, Benicewicz BC, Vaia RA, Winey KI (2017) 50th anniversary perspective: are polymer nanocomposites practical for applications? Macromolecules 50:714–731. https://doi.org/10.1021/acs.macromol.6b02330
4. Baker RW, Low BT (2014) Gas separation membrane materials: a perspective. Macromolecules 47:6999–7013. https://doi.org/10.1021/ma501488s
5. Croce F, Appetecchi G, Persi L, Scrosati B (1998) Nanocomposite polymer electrolytes for lithium batteries. Nature 394:456. https://doi.org/10.1038/28818
6. Crosby AJ, Lee JY (2007) Polymer nanocomposites: the "nano" effect on mechanical properties. Polym Rev 47:217–229. https://doi.org/10.1080/15583720701271278
7. Freeman BD (1999) Basis of permeability/selectivity tradeoff relations in polymeric gas separation membranes. Macromolecules 32:375–380. https://doi.org/10.1021/ma9814548
8. Leszczyńska A, Njuguna J, Pielichowski K, Banerjee J (2007) Polymer/montmorillonite nanocomposites with improved thermal properties: Part I. Factors influencing thermal stability and mechanisms of thermal stability improvement. Thermochim Acta 453:75–96. https://doi.org/10.1016/j.tca.2006.11.002
9. Mutiso RM, Winey KI (2015) Electrical properties of polymer nanocomposites containing rod-like nanofillers. Prog Polym Sci 40:63–84. https://doi.org/10.1016/j.progpolymsci.2014.06.002

10. Podsiadlo P, Kaushik AK, Arruda EM, Waas AM, Shim BS, Xu J, Nandivada H, Pumplin BG, Lahann J, Ramamoorthy A (2007) Ultrastrong and stiff layered polymer nanocomposites. Science 318:80–83. https://doi.org/10.1126/science.1143176
11. Ramanathan T, Abdala A, Stankovich S, Dikin D, Herrera-Alonso M, Piner R, Adamson D, Schniepp H, Chen X, Ruoff R (2008) Functionalized graphene sheets for polymer nanocomposites. Nat Nanotechnol 3:327. https://doi.org/10.1038/nnano.2008.96
12. Song J, Wang Y, Wan CC (1999) Review of gel-type polymer electrolytes for lithium-ion batteries. J Power Sources 77:183–197. https://doi.org/10.1016/S0378-7753(98)00193-1
13. Thakur VK, Gupta RK (2016) Recent progress on ferroelectric polymer-based nanocomposites for high energy density capacitors: synthesis, dielectric properties, and future aspects. Chem Rev 116:4260–4317. https://doi.org/10.1021/acs.chemrev.5b00495
14. Tjong SC (2006) Structural and mechanical properties of polymer nanocomposites. Mater Sci Eng R Rep 53:73–197. https://doi.org/10.1016/j.mser.2006.06.001
15. Merino S, Martin C, Kostarelos K, Prato M, Vazquez E (2015) Nanocomposite hydrogels: 3D polymer–nanoparticle synergies for on-demand drug delivery. ACS Nano 9:4686–4697. https://doi.org/10.1021/acsnano.5b01433
16. Merkel T, Freeman B, Spontak R, He Z, Pinnau I, Meakin P, Hill A (2002) Ultrapermeable, reverse-selective nanocomposite membranes. Science 296:519–522. https://doi.org/10.1126/science.1069580
17. Schadler LS, Kumar SK, Benicewicz BC, Lewis SL, Harton SE (2007) Designed interfaces in polymer nanocomposites: a fundamental viewpoint. MRS Bull 32:335–340. https://doi.org/10.1557/mrs2007.232
18. Song Q, Nataraj S, Roussenova MV, Tan JC, Hughes DJ, Li W, Bourgoin P, Alam MA, Cheetham AK, Al-Muhtaseb SA (2012) Zeolitic imidazolate framework (ZIF-8) based polymer nanocomposite membranes for gas separation. Energy Environ Sci 5:8359–8369. https://doi.org/10.1039/C2EE21996D
19. Xu H, Cheng L, Wang C, Ma X, Li Y, Liu Z (2011) Polymer encapsulated upconversion nanoparticle/iron oxide nanocomposites for multimodal imaging and magnetic targeted drug delivery. Biomaterials 32:9364–9373. https://doi.org/10.1016/j.biomaterials.2011.08.053
20. Carroll B, Cheng SW, Sokolov AP (2017) Analyzing the interfacial layer properties in polymer nanocomposites by broadband dielectric spectroscopy. Macromolecules 50:6149–6163. https://doi.org/10.1021/acs.macromol.7b00825
21. Senses E, Akcora P (2013) An interface-driven stiffening mechanism in polymer nanocomposites. Macromolecules 46:1868–1874. https://doi.org/10.1021/ma302275f
22. Fragiadakis D, Pissis P, Bokobza L (2006) Modified chain dynamics in poly (dimethylsiloxane)/silica nanocomposites. J Non-Cryst Solids 352:4969–4972. https://doi.org/10.1016/j.jnoncrysol.2006.02.159
23. Holt AP, Bocharova V, Cheng S, Kisliuk AM, White BT, Saito T, Uhrig D, Mahalik JP, Kumar R, Imel AE, Etampawala T, Martin H, Sikes N, Sumpter BG, Dadmun MD, Sokolov AP (2016) Controlling interfacial dynamics: covalent bonding versus physical adsorption in polymer nanocomposites. ACS Nano 10:6843–6852. https://doi.org/10.1021/acsnano.6b02501
24. Holt AP, Griffin PJ, Bocharova V, Agapov AL, Imel AE, Dadmun MD, Sangoro JR, Sokolov AP (2014) Dynamics at the polymer/nanoparticle interface in poly (2-vinylpyridine)/silica nanocomposites. Macromolecules 47:1837–1843. https://doi.org/10.1021/ma5000317
25. Cheng S, Bocharova V, Belianinov A, Xiong S, Kisliuk A, Somnath S, Holt AP, Ovchinnikova OS, Jesse S, Martin H, Etampawala T, Dadmun M, Sokolov AP (2016) Unraveling the mechanism of nanoscale mechanical reinforcement in glassy polymer nanocomposites. Nano Lett 16:3630–3637. https://doi.org/10.1021/acs.nanolett.6b00766
26. Genix A-C, Bocharova V, Kisliuk A, Carroll B, Zhao S, Oberdisse J, Sokolov AP (2018) Enhancing the mechanical properties of glassy nanocomposites by tuning polymer molecular weight. ACS Appl Mater Interfaces 10:33601–33610. https://doi.org/10.1021/acsami.8b13109
27. Baeza GP, Dessi C, Costanzo S, Zhao D, Gong S, Alegria A, Colby RH, Rubinstein M, Vlassopoulos D, Kumar SK (2016) Network dynamics in nanofilled polymers. Nat Commun 7:11368. https://doi.org/10.1038/ncomms11368

28. Griffin PJ, Bocharova V, Middleton LR, Composto RJ, Clarke N, Schweizer KS, Winey KI (2016) Influence of the bound polymer layer on nanoparticle diffusion in polymer melts. ACS Macro Lett 5:1141–1145. https://doi.org/10.1021/acsmacrolett.6b00649
29. Harton SE, Kumar SK, Yang H, Koga T, Hicks K, Lee H, Mijovic J, Liu M, Vallery RS, Gidley DW (2010) Immobilized polymer layers on spherical nanoparticles. Macromolecules 43:3415–3421. https://doi.org/10.1021/ma902484d
30. Jiang N, Endoh MK, Koga T, Masui T, Kishimoto H, Nagao M, Satija SK, Taniguchi T (2015) Nanostructures and dynamics of macromolecules bound to attractive filler surfaces. ACS Macro Lett 4:838–842. https://doi.org/10.1021/acsmacrolett.5b00368
31. Jiang N, Wang J, Di X, Cheung J, Zeng W, Endoh MK, Koga T, Satija SK (2016) Nanoscale adsorbed structures as a robust approach for tailoring polymer film stability. Soft Matter 12:1801–1809. https://doi.org/10.1039/C5SM02435H
32. Jimenez AM, Zhao D, Misquitta K, Jestin J, Kumar SK (2019) Exchange lifetimes of the bound polymer layer on silica nanoparticles. ACS Macro Lett 8:166–171. https://doi.org/10.1021/acsmacrolett.8b00877
33. Jouault N, Crawford MK, Chi C, Smalley RJ, Wood B, Jestin J, Melnichenko YB, He L, Guise WE, Kumar SK (2016) Polymer chain behavior in polymer nanocomposites with attractive interactions. ACS Macro Lett 5:523–527. https://doi.org/10.1021/acsmacrolett.6b00164
34. Jouault N, Moll JF, Meng D, Windsor K, Ramcharan S, Kearney C, Kumar SK (2013) Bound polymer layer in nanocomposites. ACS Macro Lett 2:371–374. https://doi.org/10.1021/mz300646a
35. Koga T, Barkley D, Nagao M, Taniguchi T, Carrillo J-MY, Sumpter BG, Masui T, Kishimoto H, Koga M, Rudick JG, Endoh MK (2018) Interphase structures and dynamics near nanofiller surfaces in polymer solutions. Macromolecules 51:9462–9470. https://doi.org/10.1021/acs.macromol.8b01615
36. Popov I, Carroll B, Bocharova V, Genix A-C, Cheng S, Khamzin A, Kisliuk A, Sokolov AP (2020) Strong reduction in amplitude of the interfacial segmental dynamics in polymer nanocomposites. Macromolecules 53:4126–4135. https://doi.org/10.1021/acs.macromol.0c00496
37. Gong S, Chen Q, Moll JF, Kumar SK, Colby RH (2014) Segmental dynamics of polymer melts with spherical nanoparticles. ACS Macro Lett 3:773–777. https://doi.org/10.1021/mz500252f
38. Genix A-C, Bocharova V, Carroll B, Lehmann M, Saito T, Krueger S, He L, Dieudonné-George P, Sokolov AP, Oberdisse J (2019) Understanding the static interfacial polymer layer by exploring the dispersion states of nanocomposites. ACS Appl Mater Interfaces 11:17863–17872. https://doi.org/10.1021/acsami.9b04553
39. Cheng S, Holt AP, Wang H, Fan F, Bocharova V, Martin H, Etampawala T, White BT, Saito T, Kang N-G, Dadmun MD, Mays JW, Sokolov AP (2016) Unexpected molecular weight effect in polymer nanocomposites. Phys Rev Lett 116:038302. https://doi.org/10.1103/PhysRevLett.116.038302
40. Kim SY, Meyer HW, Saalwächter K, Zukoski CF (2012) Polymer dynamics in peg-silica nanocomposites: effects of polymer molecular weight, temperature and solvent dilution. Macromolecules 45:4225–4237. https://doi.org/10.1021/ma300439k
41. Papon A, Montes H, Lequeux F, Oberdisse J, Saalwächter K, Guy L (2012) Solid particles in an elastomer matrix: impact of colloid dispersion and polymer mobility modification on the mechanical properties. Soft Matter 8:4090–4096. https://doi.org/10.1039/C2SM06885K
42. Cheng S, Carroll B, Lu W, Fan F, Carrillo J-MY, Martin H, Holt AP, Kang N-G, Bocharova V, Mays JW (2017) Interfacial properties of polymer nanocomposites: role of chain rigidity and dynamic heterogeneity length scale. Macromolecules 50:2397–2406. https://doi.org/10.1021/acs.macromol.6b02816
43. Carrillo J-MY, Cheng S, Kumar R, Goswami M, Sokolov AP, Sumpter BG (2015) Untangling the effects of chain rigidity on the structure and dynamics of strongly adsorbed polymer melts. Macromolecules 48:4207–4219. https://doi.org/10.1021/acs.macromol.5b00624
44. Hanakata PZ, Douglas JF, Starr FW (2014) Interfacial mobility scale determines the scale of collective motion and relaxation rate in polymer films. Nat Commun 5:4163. https://doi.org/10.1038/ncomms5163

45. Schweizer KS, Simmons DS (2019) Progress towards a phenomenological picture and theoretical understanding of glassy dynamics and vitrification near interfaces and under nanoconfinement. J Chem Phys 151:240901. https://doi.org/10.1063/1.5129405
46. Fukao K, Miyamoto Y (2000) Glass transitions and dynamics in thin polymer films: dielectric relaxation of thin films of polystyrene. Phys Rev E 61:1743–1754. https://doi.org/10.1103/PhysRevE.61.1743
47. Fukao K, Miyamoto Y (2001) Slow dynamics near glass transitions in thin polymer films. Phys Rev E 64:011803. https://doi.org/10.1103/PhysRevE.64.011803
48. Fukao K, Uno S, Miyamoto Y, Hoshino A, Miyaji H (2001) Dynamics of alpha and beta processes in thin polymer films: poly(vinyl acetate) and poly(methyl methacrylate). Phys Rev E 64:051807. https://doi.org/10.1103/PhysRevE.64.051807
49. Labahn D, Mix R, Schönhals A (2009) Dielectric relaxation of ultrathin films of supported polysulfone. Phys Rev E 79:011801. https://doi.org/10.1103/PhysRevE.79.011801
50. Madkour S, Szymoniak P, Heidari M, von Klitzing R, Schönhals A (2017) Unveiling the dynamics of self-assembled layers of thin films of poly(vinyl methyl ether) (PVME) by nano-sized relaxation spectroscopy. ACS Appl Mater Interfaces 9:7535–7546. https://doi.org/10.1021/acsami.6b14404
51. Madkour S, Yin H, Füllbrandt M, Schönhals A (2015) Calorimetric evidence for a mobile surface layer in ultrathin polymeric films: poly(2-vinyl pyridine). Soft Matter 11:7942–7952. https://doi.org/10.1039/C5SM01558H
52. Martínez-Tong DE, Vanroy B, Wübbenhorst M, Nogales A, Napolitano S (2014) Crystallization of poly(l-lactide) confined in ultrathin films: competition between finite size effects and irreversible chain adsorption. Macromolecules 47:2354–2360. https://doi.org/10.1021/ma500230d
53. Napolitano S, Capponi S, Vanroy B (2013) Glassy dynamics of soft matter under 1D confinement: How irreversible adsorption affects molecular packing, mobility gradients and orientational polarization in thin films. Eur PhysJ E 36:61. https://doi.org/10.1140/epje/i2013-13061-8
54. Napolitano S, Prevosto D, Lucchesi M, Pingue P, D'Acunto M, Rolla P (2007) Influence of a reduced mobility layer on the structural relaxation dynamics of aluminum capped ultrathin films of poly(ethylene terephthalate). Langmuir 23:2103–2109. https://doi.org/10.1021/la062229j
55. Napolitano S, Wübbenhorst M (2011) The lifetime of the deviations from bulk behaviour in polymers confined at the nanoscale. Nat Commun 2:260. https://doi.org/10.1038/ncomms1259
56. Rotella C, Napolitano S, De Cremer L, Koeckelberghs G, Wübbenhorst M (2010) Distribution of segmental mobility in ultrathin polymer films. Macromolecules 43:8686–8691. https://doi.org/10.1021/ma101695y
57. Rotella C, Napolitano S, Vandendriessche S, Valev VK, Verbiest T, Larkowska M, Kucharski S, Wübbenhorst M (2011) Adsorption kinetics of ultrathin polymer films in the melt probed by dielectric spectroscopy and second-harmonic generation. Langmuir 27:13533–13538. https://doi.org/10.1021/la2027779
58. Rotella C, Wübbenhorst M, Napolitano S (2011) Probing interfacial mobility profiles via the impact of nanoscopic confinement on the strength of the dynamic glass transition. Soft Matter 7:5260–5266. https://doi.org/10.1039/C1SM05430A
59. Serghei A, Mikhailova Y, Eichhorn KJ, Voit B, Kremer F (2006) Discrepancies in the characterization of the glass transition in thin films of hyperbranched polyesters. J Polym Sci Part B Polym Phys 44:3006–3010. https://doi.org/10.1002/polb.20929
60. Serghei A, Mikhailova Y, Huth H, Schick C, Eichhorn KJ, Voit B, Kremer F (2005) Molecular dynamics of hyperbranched polyesters in the confinement of thin films. Eur Phys J E 17:199–202. https://doi.org/10.1140/epje/i2005-10009-7
61. Serghei A, Tress M, Kremer F (2006) Confinement effects on the relaxation time distribution of the dynamic glass transition in ultrathin polymer films. Macromolecules 39:9385–9387. https://doi.org/10.1021/ma061290s
62. Serghei A, Tress M, Kremer F (2009) The glass transition of thin polymer films in relation to the interfacial dynamics. J Chem Phys 131:154904. https://doi.org/10.1063/1.3248368

63. Yin H, Napolitano S, Schönhals A (2012) Molecular mobility and glass transition of thin films of poly(bisphenol A carbonate). Macromolecules 45:1652–1662. https://doi.org/10.1021/ma202127p
64. Kremer F, Schönhals A (2002) Broadband dielectric spectroscopy. Springer, Berlin
65. Boucher VM, Cangialosi D, Alegrìa A, Colmenero J, González-Irun J, Liz-Marzan LM (2011) Physical aging in PMMA/silica nanocomposites: enthalpy and dielectric relaxation. J Non-Cryst Solids 357:605–609. https://doi.org/10.1016/j.jnoncrysol.2010.05.091
66. Füllbrandt M, Purohit PJ, Schönhals A (2013) Combined FTIR and dielectric investigation of poly(vinyl acetate) adsorbed on silica particles. Macromolecules 46:4626–4632. https://doi.org/10.1021/ma400461p
67. Holt AP (2016) The effect of attractive polymer-nanoparticle interactions on the local segmental dynamics of polymer nanocomposites. PhD Dissertation, University of Tennessee
68. Cheng S, Carroll B, Bocharova V, Carrillo J-M, Sumpter BG, Sokolov AP (2017) Focus: structure and dynamics of the interfacial layer in polymer nanocomposites with attractive interactions. J Chem Phys 146:203201. https://doi.org/10.1063/1.4978504
69. Wübbenhorst M, van Turnhout J (2000) Conduction-free dielectric loss ∂ε/∂ lnf–a powerful tool for the analysis of strong. Dielectrics Newsletter, NOVOCONTROL, GmbH, Hundsangen, pp 1–8
70. Boyd RH (1985) Relaxation processes in crystalline polymers: molecular interpretation—a review. Polymer 26:1123–1133. https://doi.org/10.1016/0032-3861(85)90240-X
71. Baeza GP, Oberdisse J, Alegria A, Saalwächter K, Couty M, Genix A-C (2015) Depercolation of aggregates upon polymer grafting in simplified industrial nanocomposites studied with dielectric spectroscopy. Polymer 73:131–138. https://doi.org/10.1016/j.polymer.2015.07.045
72. Roths T, Marth M, Weese J, Honerkamp J (2001) A generalized regularization method for nonlinear ill-posed problems enhanced for nonlinear regularization terms. Comput Phys Commun 139:279–296. https://doi.org/10.1016/S0010-4655(01)00217-X. (☆☆ This program can be downloaded from the CPC Program Library under catalogue identifier: http://cpc.cs.qub.ac.uk/summaries/ADOQ)
73. Casalini R, Roland CM (2016) Local and global dynamics in polypropylene glycol/silica composites. Macromolecules 49:3919–3924. https://doi.org/10.1021/acs.macromol.6b00354
74. Cheng S, Mirigian S, Carrillo J-MY, Bocharova V, Sumpter BG, Schweizer KS, Sokolov AP (2015) Revealing spatially heterogeneous relaxation in a model nanocomposite. J Chem Phys 143:194704. https://doi.org/10.1063/1.4935595
75. Cheng S (2021) Broadband dielectric spectroscopy of polymer nanocomposites. In: Broadband dielectric spectroscopy: a modern analytical technique. American Chemical Society
76. Cheng S, Sokolov AP (2020) Correlation between the temperature evolution of the interfacial region and the growing dynamic cooperativity length scale. J Chem Phys 152:094904. https://doi.org/10.1063/1.5143360
77. Cheng S, Xie S-J, Carrillo J-MY, Carroll B, Martin H, Cao P-F, Dadmun MD, Sumpter BG, Novikov VN, Schweizer KS, Sokolov AP (2017) Big effect of small nanoparticles: a shift in paradigm for polymer nanocomposites. ACS Nano 11:752–759. https://doi.org/10.1021/acsnano.6b07172
78. Holt AP, Bocharova V, Cheng S, Kisliuk AM, Ehlers G, Mamontov E, Novikov VN, Sokolov AP (2017) Interplay between local dynamics and mechanical reinforcement in glassy polymer nanocomposites. Phys Rev Mater 1:062601. https://doi.org/10.1103/PhysRevMaterials.1.062601
79. Sargsyan A, Tonoyan A, Davtyan S, Schick C (2007) The amount of immobilized polymer in PMMA SiO2 nanocomposites determined from calorimetric data. Eur Polymer J 43:3113–3127. https://doi.org/10.1016/j.eurpolymj.2007.05.011
80. Klonos P, Kulyk K, Borysenko MV, Gun'ko VM, Kyritsis A, Pissis P (2016) Effects of molecular weight below the entanglement threshold on interfacial nanoparticles/polymer dynamics. Macromolecules 49:9457–9473. https://doi.org/10.1021/acs.macromol.6b01931
81. Klonos P, Pissis P (2017) Effects of interfacial interactions and of crystallization on rigid amorphous fraction and molecular dynamics in polylactide/silica nanocomposites: a methodological approach. Polymer 112:228–243. https://doi.org/10.1016/j.polymer.2017.02.003

82. Wurm A, Ismail M, Kretzschmar B, Pospiech D, Schick C (2010) Retarded crystallization in polyamide/layered silicates nanocomposites caused by an immobilized interphase. Macromolecules 43:1480–1487. https://doi.org/10.1021/ma902175r
83. Chehrazi E, Taheri Qazvini N (2013) Nanoconfined segmental dynamics in miscible polymer blend nanocomposites: the influence of the geometry of nanoparticles. Iran Polym J 22:613–622. https://doi.org/10.1007/s13726-013-0160-4
84. Pandis C, Logakis E, Kyritsis A, Pissis P, Vodnik VV, Džunuzović E, Nedeljković JM, Djoković V, Rodríguez Hernández JC, Gómez Ribelles JL (2011) Glass transition and polymer dynamics in silver/poly(methyl methacrylate) nanocomposites. Eur Polymer J 47:1514–1525. https://doi.org/10.1016/j.eurpolymj.2011.06.001
85. Batistakis C, Lyulin AV, Michels MAJ (2012) Slowing down versus acceleration in the dynamics of confined polymer films. Macromolecules 45:7282–7292. https://doi.org/10.1021/ma300753e
86. Klonos P, Panagopoulou A, Kyritsis A, Bokobza L, Pissis P (2011) Dielectric studies of segmental dynamics in poly(dimethylsiloxane)/titania nanocomposites. J Non-Cryst Solids 357:610–614. https://doi.org/10.1016/j.jnoncrysol.2010.06.058
87. Steeman PAM, Maurer FHJ (1990) An interlayer model for the complex dielectric constant of composites. Colloid Polym Sci 268:315–325. https://doi.org/10.1007/bf01411674
88. Napolitano S, Wübbenhorst M (2007) Dielectric signature of a dead layer in ultrathin films of a nonpolar polymer. J Phys Chem B 111:9197–9199. https://doi.org/10.1021/jp072868i
89. Fröhlich H (1986) Theory of dielectrics: dielectrics constant and dielectric loss, 2nd edn. Clarendon Press, Oxford
90. Klonos PA, Goncharuk OV, Pakhlov EM, Sternik D, Deryło-Marczewska A, Kyritsis A, Gun'ko VM, Pissis P (2019) Morphology, molecular dynamics, and interfacial phenomena in systems based on silica modified by grafting polydimethylsiloxane chains and physically adsorbed polydimethylsiloxane. Macromolecules 52:2863–2877. https://doi.org/10.1021/acs.macromol.9b00155
91. Fragiadakis D, Pissis P, Bokobza L (2005) Glass transition and molecular dynamics in poly(dimethylsiloxane)/silica nanocomposites. Polymer 46:6001–6008. https://doi.org/10.1016/j.polymer.2005.05.080

Dynamics of Polymer Bridges in Polymer Nanocomposites Through a Combination of Dielectric Spectroscopy and Rheology

Shiwang Cheng

Abstract An important characteristic of the inclusion of nanoparticles (NPs) into soft polymer matrices, forming polymer nanocomposites (PNCs), is their strong modification to the dynamics of the matrix polymers, including the glassy dynamics and slow dynamics. This chapter focuses on one of the most prominent slow dynamics in PNCs with well-dispersed NPs, i.e. dynamics of polymer bridges, which is the key for thermomechanical properties of PNCs. Current theoretical and experimental understandings of the structure and dynamics of the adsorbed polymers and the polymer bridging network are summarized. The rheological way of quantifying the lifetime of the adsorbed polymers, the conformational rearrangement, and the desorption dynamics, is introduced. A recent proposed theoretical rationale for explaining the various dynamic features of polymer bridging networks is discussed. In addition, outlooks including the existent challenges and future directions on the slow dynamics of polymer nanocomposites are also presented.

Keywords Adsorbed polymers · Polymer bridges · Shift factors · Polymer nanocomposites · Broadband dielectric spectroscopy · Rheology

Symbols and Abbreviations

a_T^R	The shift factor from rheological measurements
$a_T^{R_{p-os}}$	The shift factor from rheological measurements after pre-deformation
a_T^B	The shift factor from dielectric measurements
BDS	Broadband dielectric spectroscopy
d_f	The fractal dimension of nanoparticle clusters
d_{IPS}	The average interparticle surface-to-surface distance
G'_{bulk}	The storage modulus of pristine polymer

S. Cheng (✉)
Department of Chemical Engineering and Materials Science, Michigan State University, East Lansing, MI 48824, USA
e-mail: chengsh9@egr.msu.edu

A. Schönhals and P. Szymoniak (eds.), *Dynamics of Composite Materials*, Advances in Dielectrics, https://doi.org/10.1007/978-3-030-89723-9_3

G_{lp}	Elastic modulus of polymer nanocomposites at low-frequency
G_p	Plateau modulus of a polymer
G'_{PN}	Storage modulus of the nanoparticle cluster
$G'_{polymer}$	Storage modulus of the polymer matrix in polymer nanocomposites
H-NMR	Proton nuclear magnetic resonance
$J(t)$	Creep compliance
l_{ad}	Thickness of the adsorbed polymer
M_b	Average molecular weight of a polymer bridge
M_w	Weight average molecular weight of a polymer
N_b	Average number of Kuhn segments in a polymer bridge
n_{char}	Average number of nanoparticles in a cluster
NP(s)	Nanoparticle(s)
SAOS	Small amplitude oscillatory shear
PNC(s)	Polymer nanocomposite(s)
PVAc	Poly(vinyl acetate)
R_g	The radius of gyration of a polymer chain
R_{NP}	The radius of a nanoparticle
SAOS	Small amplitude oscillatory shear
SCFT	Self-consistent field theory
t_a	Annealing time
T_g	Glass transition temperature
TTSP	Time-temperature superposition principle
τ_α	Segmental relaxation time
τ_{char}	Relaxation time of the nanoparticle cluster
τ_d	Terminal relaxation time of polymers
τ_{int}	Segmental relaxation time of the interfacial polymer
ω_{char}	Angular frequency of the longest relaxation time of a nanoparticle cluster
ω_d	Angular frequency of the terminal relaxation of a polymer chain
φ_c	Critical volume fraction of nanoparticles at the gel point
ΔE_a	Activation energy for polymer detachment

1 Introduction

The incorporation of inorganic nanoparticles (NPs) into a soft polymer matrix can significantly improve the mechanical, thermal, optical, and barrier properties of the formed polymer nanocomposites (PNCs) [1–6] compared to the polymer. The high performance along with lightweight makes PNCs attractive for various applications in energy, environment, healthcare, and infrastructure. For instance, in aerospace and automotive, high-strength PNCs are used as structural materials to reduce the total mass of vehicles and aircraft and increase fuel efficiency [7, 8]. In the energy sector, PNCs have been used as interface connectors, such as separators, O-rings, gaskets, and seals [9–11]. The modern healthcare system also requires the development of advanced PNCs for flexible electronics and sensors interfacing with intelligence [12–14]. In all these applications, the dynamical and mechanical properties of PNCs are among the first to consider.

Intensive work has been done in the past on understanding the dynamics and the mechanical properties of PNCs [2–4, 15–24]. An interfacial polymer layer with a thickness of 1–5 nm has been identified between the matrix polymer and the surface of NPs [4, 25–42]. Depending on the polymer-NP interactions, interfacial polymer layer with disturbed chain packing can experience a slower or a faster structural relaxation time than the bulk matrix polymer [4, 43, 44]. Recent measurements from broadband dielectric spectroscopy (BDS), H-NMR spectroscopy, and neutron scattering have witnessed significant progress in understanding the segmental dynamics of the interfacial layer, which have been summarized in recent review articles [4, 20]. Beyond the segmental dynamics, strong modifications to the sub-chain and chain dynamics of the adsorbed polymer were observed, leading to complex slow dynamics and intriguing linear and nonlinear viscoelastic properties of PNCs [45–47]. In terms of the viscoelastic properties of PNCs, the slow dynamics of the adsorbed or polymer bridges are the most relevant [47]. However, the complex geometry along with the multi-scale polymer-NP and NP-NP interactions makes it very challenging to investigate the structure and dynamics of the adsorbed polymer in PNCs [2]. The challenges are multifold: (i) the adsorbed polymers usually require an exceedingly large activation barrier to detach. (ii) The strong nano-confinement in PNCs promotes polymer bridges or polymer bridging networks. (iii) The dynamics and mechanical properties of PNCs are regulated strongly by the dispersion state of NPs. Due to these complications and their synergistic effects, our understanding of the structure and dynamics of polymer bridges in PNCs remains very limited. This chapter focuses on the highly challenging yet practically important subject of the dynamics of polymer bridges in PNCs. Salient features of the structure and dynamics of the polymer bridging network and their connections to the viscoelastic properties of PNCs will be reviewed and summarized. To avoid the complication from the physically connected nanoparticles (the NP network), we limited our discussions in PNCs with well-dispersed NPs where extensive NP networks and large NP clusters do not exist.

The chapter is organized as follows: Sect. 2 describes the theory and experiments on the structure of adsorbed polymers and polymer bridges. Section 3 introduces a recently proposed model on the structure of polymer bridging networks in PNCs. In Sect. 4, the dynamics of the polymer bridges will be discussed, including the segmental dynamics, the network dynamics, and the conformational rearrangement of the adsorbed polymers. A recently proposed rationale for the network dynamics will be discussed in Sect. 4. Section 5 will present the summary and outlook. Dielectric measurements were performed on Novocontrol Concept 80 or Concept 40 with an Alpha-A impedance analyzer and a Quatro Cryosystem temperature controller. The PNCs were all prepared through solution casting followed by vacuum drying. For dielectric measurements presented in this chapter, thin polymer film disks with 14 mm in diameter and 0.14 mm in thickness were sandwiched between two 20 mm in diameter gold electrodes. A Teflon spacer with a thickness of 0.14 mm was employed between two gold electrodes to avoid direct contact. The details of sample preparation and dielectric measurements were described in previous publications [27, 30]. The rheological measurements in Sect. 4 were carried out on an Anton Paar MCR302 on a pair of parallel plates with 4 mm or 8 mm in diameter. The creep compliance was obtained at a constant stress $\sigma = 1000$ Pa. The details of the rheological measurements were presented in Ref. [45].

2 The Formation of Polymer Bridges

2.1 Theoretical Predictions

A polymer bridge forms when a polymer chain attaches simultaneously to two neighboring nanoparticles. The competition of enthalpy and entropy at the interface gives birth to complex chain conformations of the adsorbed polymer, including trains, loops, and tails (Fig. 1) [48–53]. Trains are the segments consecutively adsorbed onto the surface. Loops are strands with two ends anchoring on the same surface. Tails are strands with only one end anchoring on the surface and the other free. Despite the wide acknowledgment of their existence, the current understanding of the adsorbed chain conformations is mostly based on theory and computer simulations as well as their comparison with experimental accessible macroscopic properties [54–60]. Two different theoretical approaches have been proposed on the structure of adsorbed polymers: (i) the self-consistent field theory (SCFT) approach pioneered by Scheutjens and Fleer; and (ii) the scaling theory (ST) approach by de Gennes. Details of these theories can be found in several review articles [50, 54–57, 61, 62].

According to de Gennes [50], the adsorbed polymer follows a self-similar grid structure (Fig. 2a). At any distance, z, from a planar plate, the local mesh size, ξ, of the adsorbed chain is proportional to z:

$$\xi(\phi(z)) \sim z \tag{1}$$

where $\phi(z)$ is the polymer concentration at a distance z from the plate. Since $\xi(\phi) \sim \phi^{-3/4}$ in a good solvent,

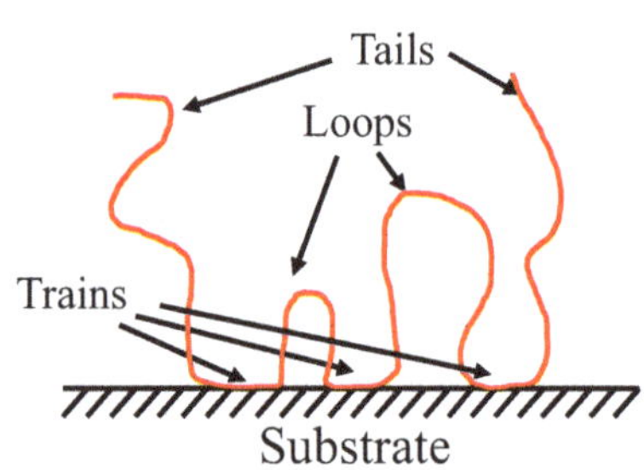

Fig. 1 The conformations of an adsorbed polymer chain

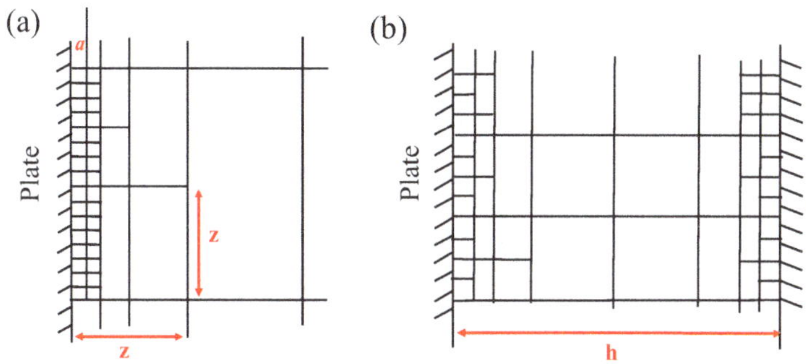

Fig. 2 **a** A sketch of the self-similar grid for adsorbed polymers. a is the size of a monomer and z is the distance away from the plate. **b** A sketch of the self-similar grid for polymer bridges at distance $h < z_{max}$. Reprinted from Ref. [50]. Copyright (1987), with permission from Elsevier

$$\phi(z) \sim z^{-4/3} \tag{2}$$

The lower limit of the adsorbed layer thickness is, $z_{min} = a$, and the upper limit of the diffusive layer is $z_{max} = aN^{0.6}$ in a good solvent with a being the monomer size and N the degree of polymerization. When two parallel plates with adsorbed polymers are approaching each other at a distance h, $a \ll h \ll aN^{0.6} = z_{max}$, polymer bridges start to form (Fig. 2b). The structure and the number density of the polymer bridges can be worked out through the self-similar grid approach. Specifically, the mesh size of the adsorbed polymer at the middle point is proportional to $\xi = h/2$. The force per unit area is then $F \sim \frac{k_B T}{\xi^3} \sim \frac{k_B T}{h^3}$ and the interaction energy, U, per unit area scales with $U \sim \frac{k_B T}{h^2}$ [50]. From these results, one can tell the number density of bridges of two parallel plates at distance h ($a \ll h \ll z_{max}$) is $n_b \sim \frac{1}{h^2}$.

2.2 *The Structure of Polymer Bridges*

In PNCs, polymer bridges form through a polymer chain simultaneously attaching onto two nanoparticles and can be divided into an interfacial portion and a non-interfacial portion (Fig. 3). In the interfacial portion, the structure and dynamics of the segments are strongly affected by NPs. A 3-dimensional polymer bridging network forms when polymer bridges percolate through NPs. This section focuses on the structure of the interfacial portion of polymer bridges.

Experimental efforts of quantifying the structure of adsorbed polymers have been performed in thin-film geometries, whose characteristic thickness were found to be comparable to the radius of gyration of the polymer, R_g [63]. Recent dynamic light scattering measurements demonstrated that the hydrodynamic size, R_h, of the physically adsorbed polymer at the surface of nanoparticles in dilute solution follows $R_h \sim N^{1/2}$ over an exceptionally wide range of molecular weight from $M \ll M_e$

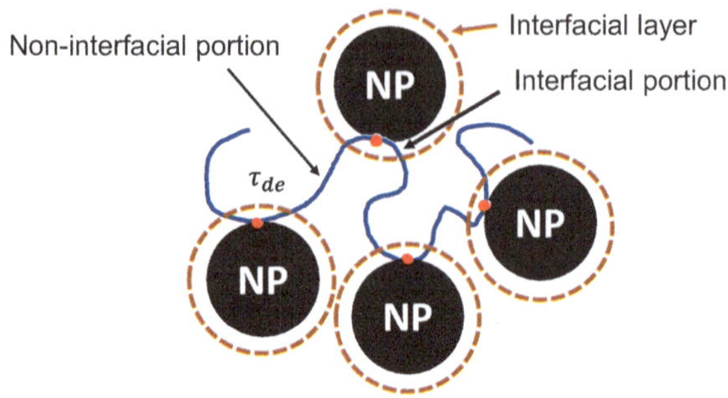

Fig. 3 A sketch of a polymer chain with multiple polymer bridge strands in polymer nanocomposites (PNCs), where an interfacial portion on the surface of nanoparticles (NPs) and a non-interfacial portion between adjacent NPs can be envisioned. τ_{de} is the detachment time of one adsorption site

to $M = 35M_e$ where M_e is the entanglement molecular weight of the polymer [64]. Although these experimental observations are not in perfect agreement with the scaling theory, a characteristic size of the adsorbed polymer comparable to the radius of gyration of the polymer, R_g, is generally accepted.

For PNCs, the structure of adsorbed polymer remains unknown. A two-layer structure of the adsorbed polymer was proposed by Koga and co-workers with a densely packed inner layer and a loosely packed outer layer [65, 66]. Recent measurements on the structures and dynamics of the interfacial polymer layer pointed to a slightly different perspective in bulk PNCs [30]. Specifically, a strong molecular weight dependence had been observed for the dynamics of the interfacial layer. As shown in Fig. 4, the dielectric broadening in segmental dynamics varies for PNCs with different molecular weights. The polymer matrix is poly (vinyl acetate) (PVAc) and the NP is SiO_2 with a radius of$R_{NP} = 12.5$ nm. The volume fraction of the NP was fixed at $\varphi_{NP} = 20\,\%$ in these measurements. As discussed in Sect. 2.1, the thickness of the adsorbed polymer, l_{ad}, should be proportional to R_g, $l_{ad} \sim R_g$. One would anticipate a larger dielectric broadening in PNCs with a higher molecular weight. However, the strongest dielectric broadening was from the PNC with the shortest matrix chain length, signifying an unexpected molecular weight effect in PNCs. Similar molecular weight effects of PNCs have been found for other measurements, such as the glass transition temperatures [30].

A frustrated chain packing mechanism (Fig. 5) was proposed to account for the unexpected molecular weight effects [30]. When $2R_g < d_{IPS}$ with $d_{IPS} = \left(\left(\frac{16}{\pi\varphi_{NP}}\right)^{1/3} - 2\right)$ where R_{NP} being the average interparticle surface-to-surface distance, adsorbed polymers of neighboring NPs did not interfere with each other. As a result, a dense polymer packing at the interface is formed and the dynamics of the interfacial polymer slowed down due to the strong polymer-NP attractive interactions. When $2R_g > d_{IPS}$, the adsorbed polymers of neighboring NPs overlapp with each other. A strong repulsive interaction existed between the adsorbed polymers, leading to frustrated chain packing. Thus, the segmental dynamics of the adsorbed polymer

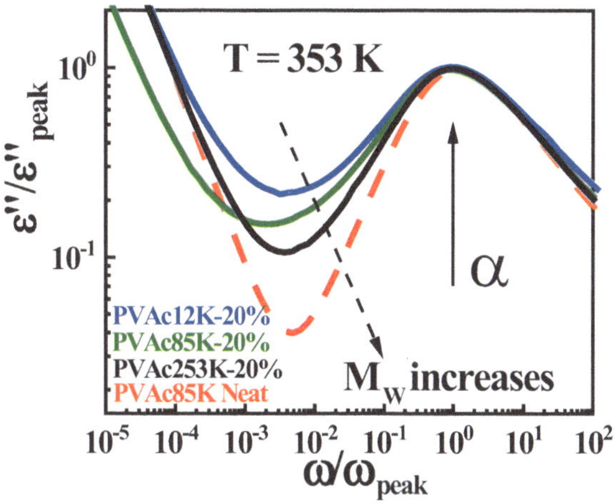

Fig. 4 Dielectric loss spectra of polymer nanocomposites with different molecular weights. The larger the molecular weight of the polymer matrix, the less the dielectric broadening for segmental relaxation (α-relaxation) peak. Reprinted with permission from Ref. [30] Copyright (2016) by the American Physical Society

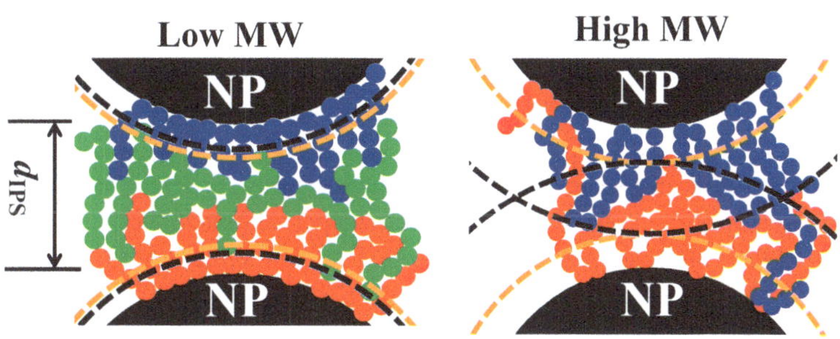

Fig. 5 A sketch for the interfacial polymer (the dashed dark-yellow line) and the bound polymer (the dashed black lines) for PNCs with different molecular weights. The red, blue, and green beads represent polymer segments with the same chemistry. The colors are used to better illustrate their adsorption state with the red chains mainly adsorbed onto the bottom NP and the blue chains mainly adsorbed onto the top NP. The green chains are non-adsorbed. A dense packing in the interfacial layer (left panel) was proposed for PNCs with $d_{IPS} > 2R_g$ and a frustrated chain packing (right panel) for PNCs with $d_{IPS} < 2R_g$. Reprinted with permission from Ref. [30]. Copyright (2016) by the American Physical Society

was affected simultaneously by the polymer-NP interaction that slows down the dynamics, and the frustrated interfacial chain packing that speeds up the dynamics. The two effects worked against each other and led to the unexpected molecular weight effect. The frustrated chain packing was further supported by small-angle scattering experiments and mass density measurements [30, 67–69]. Recent studies on the secondary dynamics of PNCs also supported the frustrated chain packing mechanism at the interface [26, 28]. X-ray photoelectron spectroscopy and sum-frequency generation found much less H-bonding at the polymer/NP interface with higher molecular weights, favoring the frustrated chain packing mechanism [70]. All these recent efforts highlight the emerging of frustrated chain packing at the polymer-nanoparticle interface that is regulated strongly by the detailed conformations of the adsorbed polymer.

As a separate note, the frustrated chain packing mechanism is valid only if large-scale conformational rearrangements of adsorbed polymers are forbidden. Indeed, long-time annealing experiments, $t_a \sim 10^{15}\tau_\alpha$, showed no sign of changes in the chain packing at the interface [30]. These observations were in agreement with the viewpoint of irreversible polymer adsorption for long-chain polymers and the long relaxation time for the conformational rearrangement of adsorbed polymers [71–73]. The extremely long lifetime of adsorbed polymers also pointed to the important role of the sample preparation on the structure and properties of PNCs.

3 The Structure of Polymer Bridging Network in Polymer Nanocomposites

The most prominent evidence for the polymer bridging network is the gel-like response, whose critical roles on the macroscopic properties of PNCs have been elaborated in the past [47, 74–76]. However, an explicit description of the structure of the polymer bridging network in PNCs is still missing.

For PNCs with well-dispersed NPs, Colby and co-workers [77] recently proposed that the polymer bridging network follows critical percolation. Specifically, polymer bridges and the connected NPs form clusters. The number density of a cluster of n NPs, $P(n)$, and the corresponding cluster size, $R(n)$, follow the static and dynamic scaling of critical percolation with

$$P(n) \sim n^{-\tau} \text{ for } 1 \le n \le n_{char} \tag{3}$$

and

$$R(n) \sim n^{1/d_f} \tag{4}$$

where the exponent $\tau = 2.2$ is for critical percolation, $n_{char} \sim |\varphi - \varphi_c|^{-1/0.45}$ is the characteristic size of the NP clusters with φ_c the volume fraction of NPs at the gel point, and the fractal dimension of the NP cluster is $d_f = 2.5$. The gel point, φ_c, can be defined through rheological measurements at $G' \sim G'' \sim \omega^{2/3}$. With these assumptions, the authors were able to calculate the linear viscoelastic spectra of polymer bridging networks. Figure 6a presents the model predictions and their comparison with experiments, where the $\tau_d = 1/\omega_d$ is the terminal relaxation time of

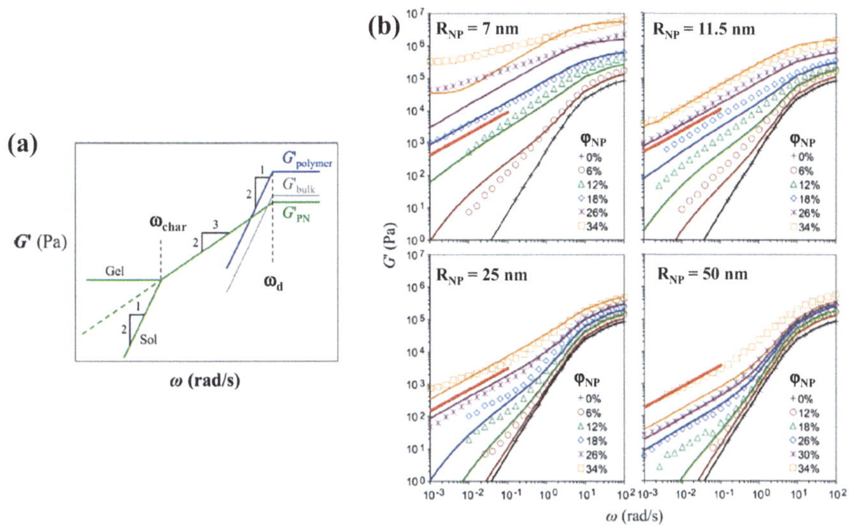

Fig. 6 **a** The prediction of the Colby-Kumar model on the viscoelastic properties of polymer bridges and polymer bridging network. $\tau_d = 1/\omega_d$ and $\tau_{char} = 1/\omega_{char} = \tau_d n_{char}^{-(2+d_f)/d_f}$ are the terminal relaxation time of the polymer and the relaxation time of an NP cluster of n_{char} particles. **b** The comparison between model predictions (the lines) and experiments (the symbols) for PNCs. The solid red line at each sub-panel in (**b**) dictates the viscoelastic response of PNCs at the gel point. Reprinted with permission from Ref. [77]. Copyright (2015) American Chemical Society

the polymer, G'_{bulk} the storage modulus of the pristine polymer, $G'_{polymer}$ the storage modulus of the polymer matrix, G'_{PN} the storage modulus of the NP clusters with polymer bridges. Figure 6b presents the comparison between experiments and model predictions for PNCs with different NP sizes and NP loadings, where the symbols are experiments and lines are model predictions. The solid red line at each sub-panel in Fig. 6b dictates the viscoelastic response of the polymer bridging network at the gel point.

4 Dynamics of the Polymer Bridging Network

4.1 Segmental Dynamics

The dynamics of a polymer bridge should be composed of two parts (Fig. 3): one is the interfacial portion, the other is the non-interfacial portion. The interfacial portion adopts dynamic features of the interfacial polymer that has been actively researched [4, 25, 27, 28, 32, 39, 78–83].

Figure 7a showed a comparison of the dielectric loss spectra, $\varepsilon''(\omega)$, of the pristine PVAc, the PVAc/SiO_2 nanocomposites with $\varphi_{NP} = 32\, vol\%$, and the polymer

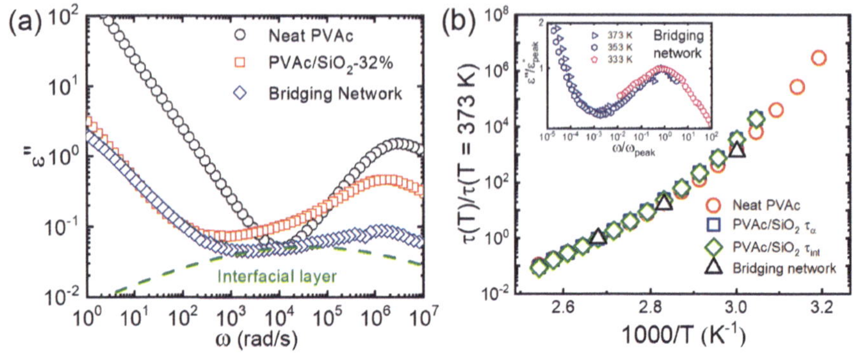

Fig. 7 **a** Dielectric loss spectra of the neat PVAc, the PVAc/SiO_2 nanocomposites with $\varphi_{NP} = 32\,\%$, and the polymer bridging network. The dashed green line is the dielectric loss function of the interfacial polymer of the PVAc/SiO_2 nanocomposite. **b** The dielectric shift factors of the neat PVAc, the matrix of the PVAc/SiO_2 nanocomposites, τ_α, the interfacial layer of the PVAc/SiO_2 nanocomposites, τ_{int}, and the relaxation time of the bridging network. The inset of **b** presents the normalized dielectric loss spectra of the polymer bridging network, where negligible shape changes with temperature were observed. Figure 7a is reprinted with permission from Ref. [28]. Copyright (2017) American Chemical Society

bridging network from PVAc/SiO_2 ($\varphi_{NP} = 32\,vol\%$). The polymer bridging network was prepared by separating the gel part of the PNC through solvent extraction [28]. Compared with the pristine polymer, a clear dielectric broadening in the low-frequency side of the segmental peak was identified in the PVAc/SiO_2 nanocomposites and the polymer bridging network. The low-frequency broadening of the segmental relaxation peak is more pronounced in the polymer bridging network than the PNC, suggesting a stronger contribution from the interfacial portion. A clear dielectric structural relaxation peak was also observed in the dielectric measurement, representing the non-interfacial portion of the polymer bridging network. Detailed analyses of the dielectric spectra have been published elsewhere [27, 28, 31].

Interestingly, the temperature dependence of the polymer bridging network is almost identical to that of the pristine polymer (Fig. 7b). Since the shape of the dielectric spectra of the polymer bridging network remains almost the same with temperature (inset of Fig. 7b), the conclusion of an identical temperature dependence between the polymer bridging network and the pristine polymer is well-supported. These results are also consistent with previous analyses of the dielectric spectra of bulk PNCs where the dynamics of the interfacial polymer and the bulk matrix are strongly correlated and follow almost the same temperature dependence [27, 28].

Therefore, the segmental dynamics of polymer bridges are composed of two different parts: one from the interfacial polymer, and the other from the non-interfacial portion. The segmental dynamics of polymer bridges follow an identical temperature dependence than neat polymer.

4.2 Network Dynamics

4.2.1 General Characteristics

Rheology has been actively employed to investigate the slow dynamics and viscoelastic properties of polymer bridging networks. Figure 8 presents master curves of the dynamic mechanical spectra of PNCs with well-dispersed NPs, where a strong mechanical reinforcement was observed. The master curves were constructed through the time–temperature superposition principle (TTSP). In Fig. 8 and *all* figures below, the same set of model PNCs will be frequently cited in the discussion. However, we want to emphasize that the revealed physics is general. The model PNCs are PVAc/SiO_2 nanocomposites with a polymer molecular weight (M_w) of 40 kg/mol (R_g = 5.7 nm), φ_{NP} = 7.5 % − 35 %, and R_{NP} = 7 ± 2 nm. The dispersion state of NPs was characterized by small-angle scattering and transmission electron microscopy [45, 84]. The mechanical reinforcement of PNCs at the rubbery plateau region remains a topic of active debate [85]. Recent studies demonstrate a strain redistribution surrounding the NPs, i.e. the classical hydrodynamic effect of particles, instead of the prevailing molecular overstaining as the molecular origin of the nano-reinforcement in PNCs [85]. For highly-filled PNCs with well-dispersed NPs, polymer bridging network dominates the rheological response at the long-time limit, resulting in the emergence of a low-frequency modulus plateau, G_{lp}, as shown in Fig. 8.

Linear rheological measurements provide the dynamic shift factors, a_T^R, representing the dynamic changes of PNCs. Figure 9 presents the comparison of dynamic

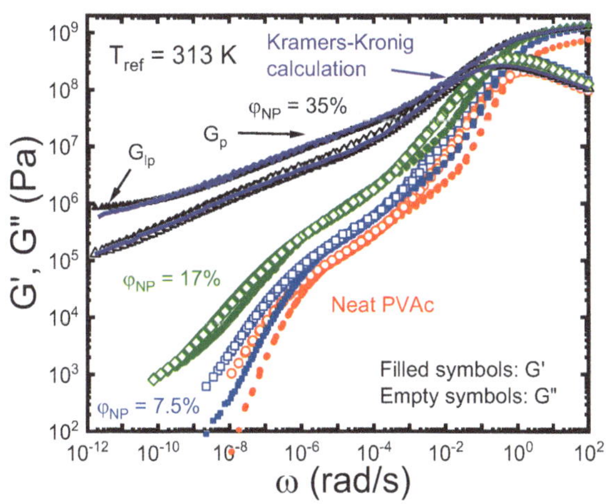

Fig. 8 Linear viscoelastic master curves of PNCs with different NP loadings at $T_{ref} = 313K$. The purple lines present the Kramers-Kronig calculation of the storage modulus, $G'(\omega)$, from the loss spectra, $G''(\omega)$. Reprinted with permission from Ref. [45]. Copyright (2020) American Chemical Society

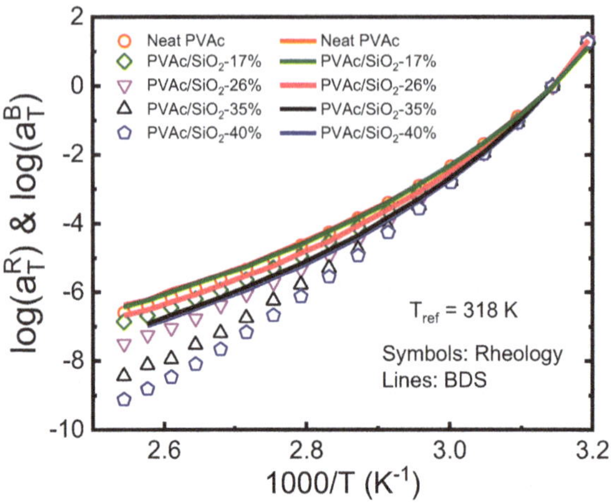

Fig. 9 The comparison between of shift factors of the dielectric measurements, $\log(a_T^B)$, and the rheology measurements, $\log(a_T^R)$, of the PVAc/SiO_2 nanocomposites. The lines are from dielectric measurements and the symbols are from rheology. Reprinted with permission from Ref. [45]. Copyright (2020) American Chemical Society

shift factors from rheology, a_T^R, and from dielectric measurements, $a_T^B = \frac{\tau_\alpha(T)}{\tau_\alpha(T_r)}$ with τ_α being the segmental relaxation time of PNCs and T_r the reference temperature. Several important features are worth noting: (i) At $T \sim T_g$, the a_T^B and the a_T^R show very similar temperature dependence for the neat polymer and the PNCs of different loadings, indicating a similar fragility index of the neat polymer and PNCs. (ii) At $T \gg T_g$, strong deviations between the a_T^R and the a_T^B were observed for PNCs at $\varphi_{NP} \geq 17\,\%$ ($d_{IPS} < 2R_g$). Almost identical a_T^R and a_T^B were observed (data not shown in Fig. 9) for the neat PVAc and PNCs at $\varphi_{NP} < 17\,\%$. Similar observations have been found in poly(propylene glycol) (PPG)/SiO_2 nanocomposites [45] and P2VP/SiO_2 nanocomposites [86]. Given the almost identical values of a_T^R and a_T^B for neat PVAc, the large separation between them in PNCs at $\varphi_{NP} \geq 17\,\%$ is most likely not due to the decoupling between the segmental dynamics and the chain dynamics of polymers [87].

4.2.2 The Origin of the Separation Between the a_T^R and the a_T^B in PNCs

Pre-deformation, which could destruct the polymer bridging network, was applied to investigate the origin of the separation between the a_T^R and the a_T^B in PNCs. As shown in Fig. 10, the creep compliance, $J(t)$, of PNCs exhibited a strong softening upon pre-deformation, confirming the network destruction by pre-deformation [45]. The inset of Fig. 10 presents the protocol of pre-deformation where oscillatory shear (OS) with medium or large strain amplitude, γ_1, was utilized. Detailed descriptions of the experiments were presented in a recent publication [45]. Interestingly, linear

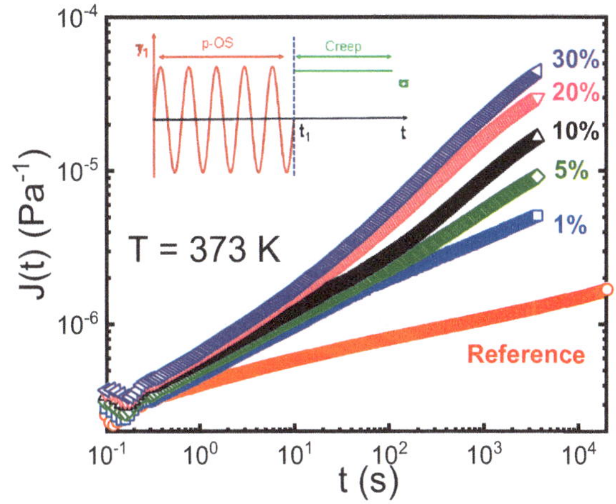

Fig. 10 Creep compliance, $J(t)$, of a PVAc/SiO_2 nanocomposite ($R_{NP} = 7 \pm 2nm, and\, \varphi_{NP} = 35\,\%$) after pre-deformation (p-OS) with different strain amplitudes, $\gamma_1 = 1\,\%, 5\,\%, 10\,\%, 20\,\%, and\, 30\,\%$. The reference is $J(t)$ of the PNC before pre-deformation. Clear softening was observed for PNCs after p-OS. The inset showed the experimental protocol, where a creep stress $\sigma = 1000Pa$ was employed. Reprinted with permission from Ref. [45]. Copyright (2020) American Chemical Society

viscoelastic measurement after pre-deformation yielded a completely different set of dynamic mechanical spectra (Fig. 11a). A master curve can be obtained from the linear viscoelastic spectra (Fig. 11b). Remarkably, the shift factors from the linear

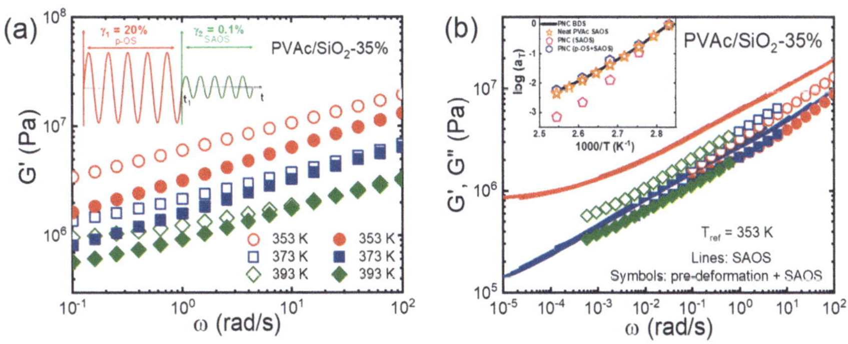

Fig. 11 **a** The storage modulus of the PNC before (empty symbols) and after (filled symbols) the pre-deformation with $\gamma_1 = 20\,\%$. The strain amplitude $\gamma_2 = 0.1\,\%$ was fixed in the small-amplitude oscillatory shear (SAOS). The PNC is PVAc/SiO_2 ($R_{NP} = 7 \pm 2\,\text{nm}, and \varphi_{NP} = 35\,\%$). The inset of (**a**) showed the experimental protocol. **b** Linear viscoelastic master curves of the PVAc/SiO_2 ($R_{NP} = 7 \pm 2\,\text{nm}, and \varphi_{NP} = 35\,\%$) before (lines) and after (symbols) pre-deformation. The inset shows the comparison of the a_T^B of the PNC (the black line) and the a_T^R of the neat PVAc (the orange stars), the PNC before deformation (the pink pentagons), and PNC after deformation (the purple hexagons). Reprinted with permission from Ref. [45]. Copyright (2020) American Chemical Society

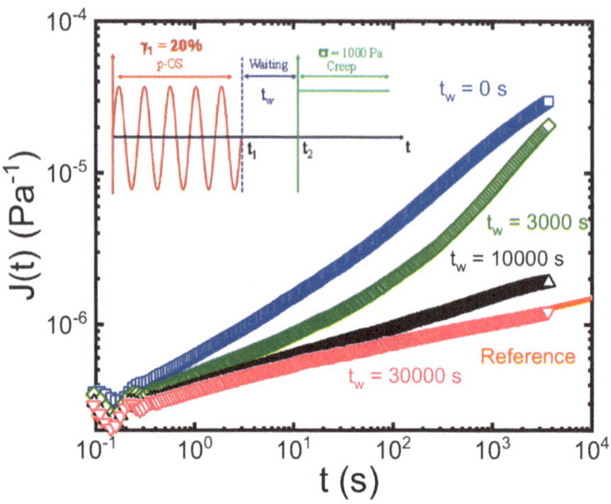

Fig. 12 Creep compliance, $J(t)$, of the PVAc/SiO$_2$ PNC ($R_{NP} = 7 \pm 2nm, and\varphi_{NP} = 35\,\%$) after pre-deformation (p-OS) at $\gamma_1 = 20\,\%$ at different annealing times, t_w. The stress applied for creep was $\sigma = 1000Pa$. The inset presents a sketch of the experimental protocol. Reprinted with permission from Ref. [45]. Copyright (2020) American Chemical Society

viscoelastic spectra after pre-deformation, $a_T^{R_{p-OS}}$, were identical with the a_T^B. Hence, the pre-deformation experiments revealed conclusively the emergence of the polymer bridging network as the origin of the separation between the a_T^R and the a_T^B.

Moreover, the pre-deformation test also enables monitoring of the kinetics of the reconstruction of polymer bridging networks. As shown in Fig. 12, $J(t)$ of the pre-deformed PNC was measured after different waiting times. A waiting time of $t_a =$ 30,000 s (~$10^{10}\tau_\alpha$) was required to achieve complete healing of the pre-deformed PNC [45]. The reconstruction of the polymer bridging network should involve conformational rearrangement of the adsorbed polymer that experiences a high activation energy barrier. Recent experiments on thin polymer film demonstrated an annealing time of 10^{10}–10^{12} τ_α is required to reach an equilibrium state for polymers at the interface [72]. The coincidence of the waiting time for healing of the polymer bridging network and the annealing time for conformational rearrangement in thin polymer films encourages further investigation.

4.2.3 The Influence of Network Dynamics on the Time–Temperature Superposition in PNCs

The strong separation between the a_T^R and the a_T^B imposes the question of the validation of the TTSP in PNCs that has been largely overlooked previously [45]. Experimentally, the linear viscoelastic spectra of PNCs often exhibit rheological simplicity and the constructed master curves followed the Kramer-Kronig relationship (solid purple lines in Fig. 8) [45]. The limited frequency window of conventional small amplitude oscillatory shear (SAOS) measurement also imposes a technical challenge to test the validation of the TTSP. Alternatively, creep was employed that

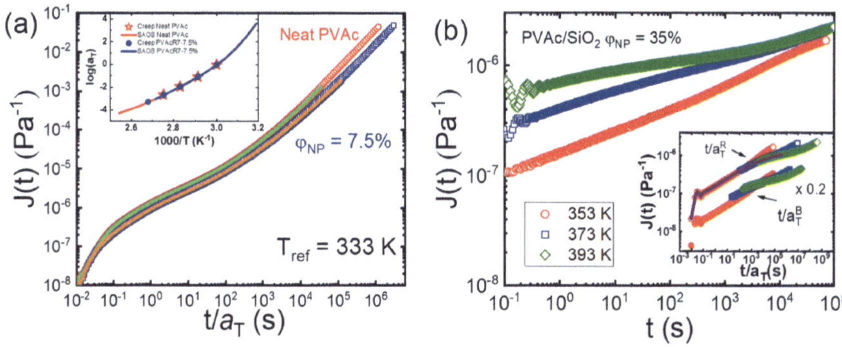

Fig. 13 **a** Creep compliance, $J(t)$, of neat PVAc and PVAc/SiO$_2$ PNC ($R_{NP} = 7 \pm 2nm, and\, \varphi_{NP} = 7.5\,\%$). Master curves can be constructed for $J(t)$ where the shift factors were identical to the SAOS measurements (inset of (**a**)). **b** Long-time creep measurements (~10^5 s) of a PVAc/SiO$_2$ PNC ($R_{NP} = 7 \pm 2\,\text{nm}, and \varphi_{NP} = 35\,\%$). Master curves could not be constructed as shown in the inset. The purple line in the inset of **b** is a master curve for truncated creep compliance at $t < 100\,s$. Reprinted with permission from Ref. [45]. Copyright (2020) American Chemical Society

allowed direct access to a wide dynamic range. Figure 13a, b show the long-time creep of PNCs from 10^{-2} to 10^5 s, where the solid lines in Fig. 13a represent a single creep test each. The inset of Fig. 13a summarized the shift factors of neat PVAc and PVAc/SiO$_2$ ($\varphi_{NP} = 7.5\,vol\%, d_{IPS} > 2R_g$) from $J(t)$ as well as their comparison with the SAOS measurements. The excellent agreement between the creep and SAOS suggests the validation of TTSP for PNCs in the absence of polymer bridging networks. On the other hand, strong deviations in $J(t)$ were observed for PNCs with polymer bridging networks (Fig. 13b). No master curves could be constructed for PVAc/SiO$_2$ ($\varphi_{NP} = 35\,vol\%, d_{IPS} \sim 0.5R_g$) as shown in the inset of Fig. 13b. An interesting observation was that if one truncated the time scale of the creep to $t < 100\,s$, which corresponds to an angular frequency range of $\omega > 10^{-2}$ rad/s in SAOS measurements, a master curve (the solid purple line in the inset of Fig. 13b) could be constructed. The shift factors were identical with a_T^R, exhibiting an "apparent" validation of the TTSP. Therefore, the TTSP is *valid* for PNCs in the absence of a polymer bridging network and *breaks down* when a strong polymer bridging network presents.

4.3 A Proposed Rationale

Polymer desorption at each anchoring site is an activated process with an energy barrier, ΔE_a. Thus, the lifetime of an adsorbed polymer site can be written as:

$$\tau_{de}(T) = \tau_\alpha^{int}(\text{T})\exp(\frac{\Delta E_a}{RT}) \quad (5)$$

with τ_α^{int} the segmental relaxation time of the interfacial polymer layer. The dynamics of the polymer bridging network should then follow the characteristics of the sticky-Rouse model (Fig. 3) [88]. In the long-time limit with $t < \tau_{de}$, the dynamics of a polymer chain with N_s anchoring sites follow

$$\tau(T) = \tau_{de}(T)N_s^2 = \tau_\alpha^{int}(\text{T})\exp\left(\frac{\Delta E_a}{RT}\right)N_s^2. \tag{6}$$

Note that $a_T^R = \frac{\tau(T)}{\tau(T_r)}$ and $a_T^B = \frac{\tau_\alpha(T)}{\tau_\alpha(T_r)} = \frac{\tau_\alpha^{int}(T)}{\tau_\alpha^{int}(T_r)}$, where the second equate sign of a_T^B is from the observation that the bulk polymer and the interfacial polymer have identical shift factors (Fig. 7b) [25, 27, 28, 32]. Rewriting Eq. 2, one has

$$a_T^R = \frac{\tau(T)}{\tau(T_r)} = \frac{\tau_\alpha^{ad}(T)\exp(\frac{\Delta E_a}{RT})N_s^2}{\tau_\alpha^{ad}(T_r)\exp(\frac{\Delta E_a}{RT_r})N_s^2} = a_T^B\exp\left(\frac{\Delta E_a}{RT} - \frac{\Delta E_a}{RT_r}\right). \tag{7}$$

Equation 7 explains the observed much stronger temperature dependence of the a_T^R than the a_T^B for PNCs with polymer bridging network.

More importantly, the above analyses offer a unique way to identify the activation energy,ΔE_a, for polymer desorption [45]. According to Eq. 7,

$$\log\left(\frac{a_T^R}{a_T^B}\right) = \left(\frac{\Delta E_a}{RT} - \frac{\Delta E_a}{RT_r}\right)loge \tag{8}$$

with $e = 2.71828\ldots$ being the Euler's number. Thus, a linear plot between $\log\left(\frac{a_T^R}{a_T^B}\right)$ and $1000/T$ should be anticipated, whose slope, $loge \times \Delta E_a/(1000R)$, gives the activation energy. Figure 14 shows the $\log\left(\frac{a_T^R}{a_T^B}\right) vs 1000/T$ and the inset of Fig. 14 presents the$\Delta E_a\, vs\, d_{IPS}/R_g$. The stronger the nano-confinement, the higher the activation barrier for the polymer desorption.

For the PVAc/SiO_2 PNC ($\varphi_{NP} = 35\,\%$), an activation energy $\Delta E_a = 56$ kJ/mol was observed. According to $\tau_{de}(T) \sim \tau_\alpha^{int}(\text{T})\exp(\frac{\Delta E_a}{RT})$ and $\frac{\tau_\alpha^{int}(\text{T})}{\tau_\alpha(T)} \sim 35$ for PVAc/SiO_2 nanocomposites [27, 28], a desorption time of $\tau_{de} \sim 10^9 - 10^{10}\tau_\alpha$ is predicted at $T = 373K$ that is very close to the annealing time ($10^{10}\tau_\alpha$) for the polymer bridge reconstruction (Fig. 12). A pseudo-permanent polymer bridging network can then be assumed. Assuming the conformation of the polymer bridge follows Gaussian distribution, the end-to-end distance of a polymer bridge strand with number of Kuhn segments, N_b, should be comparable to the d_{IPS}, $R_{ee} \sim N_b^{1/2} \sim d_{IPS}$. The average molecular weight of polymer bridge strand is $M_b \sim N_b \sim d_{IPS}^2$. According to de Gennes [50], the number density of the polymer bridge under confinement d_{IPS} is $n_b \sim d_{IPS}^{-2}$ (See Sect. 2.1). The modulus of the polymer bridge network thus follows $G \sim G_{lp} \sim \frac{n_b k_B T}{M_b} \sim d_{IPS}^{-4}$. Thus, the sticky-Rouse approach predicts the following scaling, $G_{lp} \sim d_{IPS}^{-4}$, for highly-filled PNCs. Figure 15 summarizes the available data in the literature, where excellent agreement

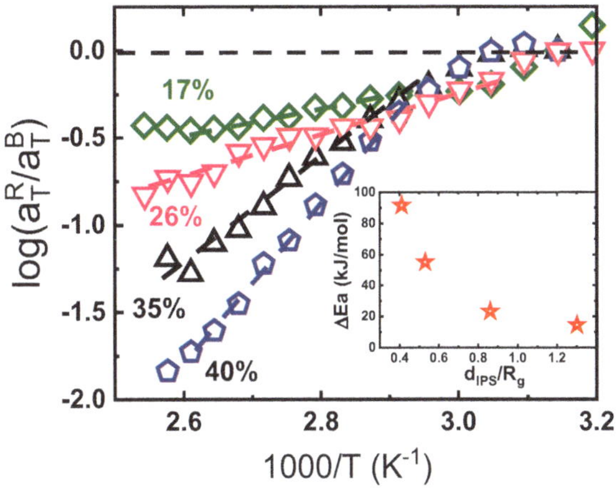

Fig. 14 Plot of $\log\left(\frac{a_T^R}{a_T^B}\right)$ versus $1000/T$ for PNCs (PVAc/SiO$_2$ ($M_W = 40$ kg/mol, $R_{NP} = 7 \pm 2$ nm) with $\varphi_{NP} = 17\,\%, 26\,\%, 35\,\%, and\ 40\,\%$). The inset shows the activation energy for detachment, where the $d_{IPS} = \left(\left(\frac{16}{\pi\varphi_{NP}}\right)^{1/3} - 2\right)R_{NP}$ is the characteristic interparticle surface-to-surface distance. The mainframe of Fig. 14 is adapted with permission from Ref. [45]. Copyright (2020) American Chemical Society

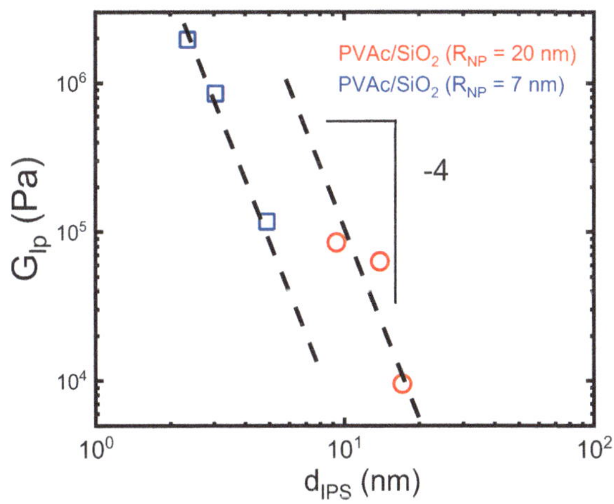

Fig. 15 The comparison between the model prediction and the experiments where $G_{lp} \sim d_{IPS}^{-4}$ was observed for PNCs with polymer bridging networks. The data of the $R_{NP} = 7$ nm is adapted from Ref. [45]

is found between experiments and the model prediction. Future experiments should be designed to examine the validity of the scaling.

5 Summary and Outlook

In summary, the chapter presented the current understanding of the structure and dynamics of polymer bridges in PNCs with well-dispersed NPs. Specifically, the average thickness of the adsorbed polymers and the critical condition for the polymer bridge formation were reviewed. Theoretical work suggested the percolation of the polymer bridges followed the critical percolation, leading to semi-quantitative predictions of linear viscoelastic spectra of PNCs [77]. Dynamic measurements of the polymer bridging network showed dominant contributions from the interfacial layer as well as distinct dynamics from the non-interfacial portion of the polymer bridge [28]. An important outcome was that the segmental dynamics of polymer bridges exhibited identical temperature dependence with the pristine polymer. Moreover, the emergence of the polymer bridging network was the molecular origin of the strong deviation in the a_T^R and the a_T^B. A sticky-Rouse dynamic model was proposed to understand the dynamics of polymer bridges, which enabled estimating the activation energy for polymer detachment and predicting a scaling between G_{lp} and d_{IPS}.

Despite the significant progress, our understanding of the structure and dynamics of polymer bridges is still at its very early stage. On the structure part, the Colby-Kumar model made strong assumptions that require future experimental validation. Experimental determination of the fractal dimension of NP clusters is limited. Future studies through small-angle scattering, advanced microscopy, and computer simulations should be able to provide helpful insights. On the dynamics side, a detailed analysis of the dielectric spectra of polymer bridges should be performed, which might offer crucial information on the dynamics of the interfacial portion and the non-interfacial portion of polymer bridges. Recent efforts demonstrated the powerful combination of rheology and dielectric measurements on the slow dynamics of polymer bridges, including desorption dynamics and the conformational rearrangement of the adsorbed polymers [45]. Given the technological importance of polymer adsorption and desorption and the lack of understanding of this topic [89–91], more effort should be put into the slow dynamics of adsorbed polymers.

Acknowledgements This work was supported by the College of Engineering at Michigan State University.

References

1. Ray SS, Okamoto M (2003) Polymer/layered silicate nanocomposites: a review from preparation to processing. Prog Polym Sci 28:1539–1641. https://doi.org/10.1016/j.progpolymsci.2003.08.002
2. Jancar J, Douglas JF, Starr FW, Kumar SK, Cassagnau P, Lesser AJ, Sternstein SS, Buehler MJ (2010) Current issues in research on structure-property relationships in polymer nanocomposites. Polymer 51:3321–3343. https://doi.org/10.1016/j.polymer.2010.04.074
3. Kumar SK, Benicewicz BC, Vaia RA, Winey KI (2017) 50th Anniversary perspective: are polymer nanocomposites practical for applications? Macromolecules 50:714–731. https://doi.org/10.1021/acs.macromol.6b02330
4. Cheng S, Carroll B, Bocharova V, Carrillo J-M, Sumpter BG, Sokolov AP (2017) Focus: Structure and dynamics of the interfacial layer in polymer nanocomposites with attractive interactions. J Chem Phys 146:203201. https://doi.org/10.1063/1.4978504
5. Yang J, Yang W, Chen W, Tao X (2020) An elegant coupling: freeze-casting and versatile polymer composites. Prog Polym Sci 109:101289. https://doi.org/10.1016/j.progpolymsci.2020.101289
6. Njuguna J, Pielichowski K, Desai S (2008) Nanofiller-reinforced polymer nanocomposites. Polym Adv Technol 19:947–959. https://doi.org/10.1002/pat.1074
7. Patil A, Patel A, Purohit R (2017) An overview of polymeric materials for automotive applications. materials today: proceedings 4 (2, Part A):3807–3815. http://www.sciencedirect.com/science/article/pii/S2214785317304881
8. Njuguna J, Pielichowski K, Fan J (2012) 15—Polymer nanocomposites for aerospace applications. In: Gao F (ed) Advances in Polymer Nanocomposites. Woodhead Publishing, pp 472–539. https://doi.org/10.1533/9780857096241.3.472
9. Balasooriya W, Clute C, Schrittesser B, Pinter G (2021) A review on applicability, limitations, and improvements of polymeric materials in high-pressure hydrogen gas atmospheres. Polym Rev 62:175-209. https://doi.org/10.1080/15583724.2021.1897997
10. Lagadec MF, Zahn R, Wood V (2019) Characterization and performance evaluation of lithium-ion battery separators. Nat Energy 4:16–25. https://doi.org/10.1038/s41560-018-0295-9
11. Lee H, Yanilmaz M, Toprakci O, Fu K, Zhang X (2014) A review of recent developments in membrane separators for rechargeable lithium-ion batteries. Energy Environ Sci 7:3857–3886. https://doi.org/10.1039/C4EE01432D
12. Peng H, Sun X, Weng W, Fang X (2017) Polymer materials for energy and electronic applications. Elsevier Inc
13. Khan F, Tanaka M, Ahmad SR (2015) Fabrication of polymeric biomaterials: a strategy for tissue engineering and medical devices. J Mater Chem B 3:8224–8249. https://doi.org/10.1039/C5TB01370D
14. Liu Y, Pharr M, Salvatore GA (2017) Lab-on-skin: a review of flexible and stretchable electronics for wearable health monitoring. ACS Nano 11(10):9614–9635. https://doi.org/10.1021/acsnano.7b04898
15. Song Y, Zheng Q (2016) A guide for hydrodynamic reinforcement effect in nanoparticle-filled polymers. Crit Rev Solid State Mater Sci 41:318–346. https://doi.org/10.1080/10408436.2015.1135415
16. Song Y, Zheng Q (2016) Concepts and conflicts in nanoparticles reinforcement to polymers beyond hydrodynamics. Prog Mater Sci 84:1–58. http://www.sciencedirect.com/science/article/pii/S0079642516300597
17. Rueda MM, Auscher M-C, Fulchiron R, Périé T, Martin G, Sonntag P, Cassagnau P (2017) Rheology and applications of highly filled polymers: a review of current understanding. Prog Polym Sci 66:22–53. http://www.sciencedirect.com/science/article/pii/S0079670016301174
18. Kumar SK, Ganesan V, Riggleman RA (2017) Perspective: outstanding theoretical questions in polymer-nanoparticle hybrids. J Chem Phys 147:020901. https://doi.org/10.1063/1.4990501

19. Vogiatzis GG, Theodorou DN (2018) Multiscale molecular simulations of polymer-matrix nanocomposites. Arch Comput Meth Eng 25:591–645. https://doi.org/10.1007/s11831-016-9207-y
20. Bailey EJ, Winey KI (2020) Dynamics of polymer segments, polymer chains, and nanoparticles in polymer nanocomposite melts: a review. Prog Polym Sci 105:101242. http://www.sciencedirect.com/science/article/pii/S0079670020300356
21. Allegra G, Raos G, Vacatello M (2008) Theories and simulations of polymer-based nanocomposites: from chain statistics to reinforcement. Prog Polym Sci 33:683–731. https://doi.org/10.1016/j.progpolymsci.2008.02.003
22. Zeng QH, Yu AB, Lu GQ (2008) Multiscale modeling and simulation of polymer nanocomposites. Prog Polym Sci 33:191–269. http://www.sciencedirect.com/science/article/pii/S0079670007001049
23. Lyulin SV, Larin SV, Nazarychev VM, Fal'kovich SG, Kenny JM, (2016) Multiscale computer simulation of polymer nanocomposites based on thermoplastics. Polym Sci Ser C 58:2–15. https://doi.org/10.1134/S1811238216010082
24. Hall LM, Jayaraman A, Schweizer KS (2010) Molecular theories of polymer nanocomposites. Curr Opin Solid State Mater Sci 14:38–48. https://doi.org/10.1016/j.cossms.2009.08.004
25. Cheng S, Sokolov AP (2020) Correlation between the temperature evolution of the interfacial region and the growing dynamic cooperativity length scale. J Chem Phys 152:094904. https://doi.org/10.1063/1.5143360
26. Holt AP, Bocharova V, Cheng S, Kisliuk AM, Ehlers G, Mamontov E, Novikov VN, Sokolov AP (2017) Interplay between local dynamics and mechanical reinforcement in glassy polymer nanocomposites. Phys Rev Mater 1:062601. https://doi.org/10.1103/PhysRevMaterials.1.062601
27. Cheng S, Carroll B, Lu W, Fan F, Carrillo J-MY, Martin H, Holt AP, Kang N-G, Bocharova V, Mays JW, Sumpter BG, Dadmun M, Sokolov AP (2017) Interfacial properties of polymer nanocomposites: role of chain rigidity and dynamic heterogeneity length scale. Macromolecules 50:2397–2406. https://doi.org/10.1021/acs.macromol.6b02816
28. Carroll B, Cheng S, Sokolov AP (2017) Analyzing the interfacial layer properties in polymer nanocomposites by broadband dielectric spectroscopy. Macromolecules 50:6149–6163. https://doi.org/10.1021/acs.macromol.7b00825
29. Holt AP, Bocharova V, Cheng S, Kisliuk AM, White BT, Saito T, Uhrig D, Mahalik JP, Kumar R, Imel AE, Etampawala T, Martin H, Sikes N, Sumpter BG, Dadmun MD, Sokolov AP (2016) Controlling interfacial dynamics: covalent bonding versus physical adsorption in polymer nanocomposites. ACS Nano 10:6843–6852. https://doi.org/10.1021/acsnano.6b02501
30. Cheng S, Holt AP, Wang H, Fan F, Bocharova V, Martin H, Etampawala T, White BT, Saito T, Kang N-G, Dadmun MD, Mays JW, Sokolov AP (2016) Unexpected molecular weight effect in polymer nanocomposites. Phys Rev Lett 116:038302. https://doi.org/10.1103/PhysRevLett.116.038302
31. Cheng S, Mirigian S, Carrillo J-MY, Bocharova V, Sumpter BG, Schweizer KS, Sokolov AP (2015) Revealing spatially heterogeneous relaxation in a model nanocomposite. J Chem Phys 143:194704. http://scitation.aip.org/content/aip/journal/jcp/143/19/10.1063/1.4935595
32. Cheng S (2021) Broadband dielectric spectroscopy of polymer nanocomposites. In: Broadband dielectric spectroscopy: a modern analytical technique, vol 1375. ACS Symposium Series, vol 1375. American Chemical Society, pp 157–183. https://doi.org/10.1021/bk-2021-1375.ch007
33. Klonos P, Kyritsis A, Pissis P (2016) Interfacial and confined dynamics of PDMS adsorbed at the interfaces and in the pores of silica–gel: Effects of surface modification and thermal annealing. Polymer 84:38–51. http://www.sciencedirect.com/science/article/pii/S0032386115304584
34. Fragiadakis D, Pissis P, Bokobza L (2006) Modified chain dynamics in poly(dimethylsiloxane)/silica nanocomposites. J Non-Cryst Solids 352:4969–4972. http://www.sciencedirect.com/science/article/pii/S0022309306009197
35. Füllbrandt M, Purohit PJ, Schönhals A (2013) Combined FTIR and dielectric investigation of Poly(vinyl acetate) adsorbed on silica particles. Macromolecules 46:4626–4632. https://doi.org/10.1021/ma400461p

36. Gong S, Chen Q, Moll JF, Kumar SK, Colby RH (2014) Segmental dynamics of polymer melts with spherical nanoparticles. ACS Macro Lett 3:773–777. https://doi.org/10.1021/mz500252f
37. Holt AP, Sangoro JR, Wang Y, Agapov AL, Sokolov AP (2013) Chain and segmental dynamics of Poly(2-vinylpyridine) nanocomposites. Macromolecules 46:4168–4173. https://doi.org/10.1021/ma400418b
38. Papon A, Montes H, Hanafi M, Lequeux F, Guy L, Saalwächter K (2012) Glass-transition temperature gradient in nanocomposites: evidence from nuclear magnetic resonance and differential scanning calorimetry. Phys Rev Lett 108:065702. https://doi.org/10.1103/PhysRevLett.108.065702
39. Papon A, Montes H, Lequeux F, Oberdisse J, Saalwächter K, Guy L (2012) Solid particles in an elastomer matrix: impact of colloid dispersion and polymer mobility modification on the mechanical properties. Soft Matter 8:4090–4096. https://doi.org/10.1039/C2SM06885K
40. Berriot J, Montes H, Lequeux F, Long D, Sotta P (2002) Evidence for the shift of the glass transition near the particles in silica-filled elastomers. Macromolecules 35:9756–9762. https://doi.org/10.1021/ma0212700
41. Berriot J, Montes H, Lequeux F, Long D, Sotta P (2003) Gradient of glass transition temperature in filled elastomers. EPL (Europhysics Letters) 64:50. http://stacks.iop.org/0295-5075/64/i=1/a=050
42. Martín J, Krutyeva M, Monkenbusch M, Arbe A, Allgaier J, Radulescu A, Falus P, Maiz J, Mijangos C, Colmenero J, Richter D (2010) Direct observation of confined single chain dynamics by neutron scattering. Phys Rev Lett 104:197801. https://doi.org/10.1103/PhysRevLett.104.197801
43. Mackay ME, Dao TT, Tuteja A, Ho DL, Van Horn B, Kim HC, Hawker CJ (2003) Nanoscale effects leading to non-Einstein-like decrease in viscosity. Nat Mater 2:762–766. https://doi.org/10.1038/nmat999
44. Cheng S, Xie S-J, Carrillo J-MY, Carroll B, Martin H, Cao P-F, Dadmun MD, Sumpter BG, Novikov VN, Schweizer KS, Sokolov AP (2017) Big effect of small nanoparticles: a shift in paradigm for polymer nanocomposites. ACS Nano 11:752–759. https://doi.org/10.1021/acsnano.6b07172
45. Yang J, Melton M, Sun R, Yang W, Cheng S (2020) Decoupling the polymer dynamics and the nanoparticle network dynamics of polymer nanocomposites through dielectric spectroscopy and rheology. Macromolecules 53:302–311. https://doi.org/10.1021/acs.macromol.9b01584
46. Sun R, Melton M, Zuo X, Cheng S (2020) Nonmonotonic strain rate dependence on the strain hardening of polymer nanocomposites. ACS Macro Lett 9:1224–1229. https://doi.org/10.1021/acsmacrolett.0c00525
47. Vilgis TA, Heinrich G, Klüppel M (2009) Reinforcement of polymer nano-composites: theory, experiments and applications. Cambridge University Press, Cambridge. https://www.cambridge.org/core/books/reinforcement-of-polymer-nanocomposites/370003E0509451FF20CF7A4078CDD7F1
48. Scheutjens JMHM, Fleer GJ (1979) Statistical theory of the adsorption of interacting chain molecules. 1. Partition function, segment density distribution, and adsorption isotherms. J Phys Chem 83:1619–1635. https://doi.org/10.1021/j100475a012
49. Scheutjens JMHM, Fleer GJ (1980) Statistical theory of the adsorption of interacting chain molecules. 2. Train, loop, and tail size distribution. J Phys Chem 84:178–190. https://doi.org/10.1021/j100439a011
50. De Gennes PG (1987) Polymers at an interface; a simplified view. Adv Colloid Interface Sci 27(3–4):189–209. https://doi.org/10.1016/0001-8686(87)85003-0
51. Semenov AN, Joanny JF (1995) Structure of adsorbed polymer layers: loops and tails. EPL (Europhysics Letters) 29:279. http://stacks.iop.org/0295-5075/29/i=4/a=002
52. Milner ST (1990) Strong-stretching and Scheutjens-Fleer descriptions of grafted polymer brushes. J Chem Soc Faraday Trans 86:1349–1353. https://doi.org/10.1039/FT9908601349
53. Guiselin O (1992) Irreversible adsorption of a concentrated polymer solution. EPL (Europhysics Letters) 17:225. http://stacks.iop.org/0295-5075/17/i=3/a=007

54. Scheutjens JMHM, Fleer GJ (1982) Effect of polymer adsorption and depletion on the interaction between two parallel surfaces. Adv Colloid Interface Sci 16:361–380. https://doi.org/10.1016/0001-8686(82)85025-2
55. Stuart MAC, Fleer GJ (1996) Adsorbed polymer layers in nonequilibrium situations. Annu Rev Mater Sci 26:463–500. https://doi.org/10.1146/annurev.ms.26.080196.002335
56. Fleer GJ, Stuart MAC, Scheutjens JMHM, Cosgrove T, Vincent B (1998) Interactions in the presence of polymers. In: Polymers at interfaces. Springer Netherlands, Dordrecht, pp 431–457. https://doi.org/10.1007/978-94-011-2130-9_11
57. Fleer GJ (2010) Polymers at interfaces and in colloidal dispersions. Adv Colloid Interface Sci 159:99–116. http://www.sciencedirect.com/science/article/pii/S0001868610000795
58. K Binder, A Milchev a, Baschnagel J (1996) Simulation studies on the dynamics of polymers at interfaces. Annu Rev Mater Sci 26:107–134. https://doi.org/10.1146/annurev.ms.26.080196.000543
59. Pazmino Betancourt BA, Douglas JF, Starr FW (2013) Fragility and cooperative motion in a glass-forming polymer-nanoparticle composite. Soft Matter 9:241–254. https://doi.org/10.1039/C2SM26800K
60. Carrillo J-MY, Cheng S, Kumar R, Goswami M, Sokolov AP, Sumpter BG (2015) Untangling the effects of chain rigidity on the structure and dynamics of strongly adsorbed polymer melts. Macromolecules 48(12):4207–4219. https://doi.org/10.1021/acs.macromol.5b00624
61. Fleer GJ, Scheutjens JMHM, Stuart MAC (1988) Theoretical progress in polymer adsorption, steric stabilization and flocculation. Colloids Surf 31:1–29. https://doi.org/10.1016/0166-6622(88)80178-1
62. Cohen Stuart MA, Cosgrove T, Vincent B (1985) Experimental aspects of polymer adsorption at solid/solution interfaces. Adv Colloid Interface Sci 24:143–239. https://doi.org/10.1016/0001-8686(85)80030-0
63. Hu H-W, Granick S, Schweizer KS (1994) Static and dynamical structure of confined polymer films. J Non-Cryst Solids 172–174:721–728. https://www.sciencedirect.com/science/article/pii/002230939490569X
64. Jouault N, Moll JF, Meng D, Windsor K, Ramcharan S, Kearney C, Kumar SK (2013) Bound polymer layer in nanocomposites. ACS Macro Lett 2:371–374. https://doi.org/10.1021/mz300646a
65. Gin P, Jiang N, Liang C, Taniguchi T, Akgun B, Satija SK, Endoh MK, Koga T (2012) Revealed architectures of adsorbed polymer chains at solid-polymer melt interfaces. Phys Rev Lett 109:265501. https://doi.org/10.1103/PhysRevLett.109.265501
66. Jiang N, Endoh MK, Koga T, Masui T, Kishimoto H, Nagao M, Satija SK, Taniguchi T (2015) Nanostructures and dynamics of macromolecules bound to attractive filler surfaces. ACS Macro Lett 4:838–842. https://doi.org/10.1021/acsmacrolett.5b00368
67. Jouault N, Crawford MK, Chi C, Smalley RJ, Wood B, Jestin J, Melnichenko YB, He L, Guise WE, Kumar SK (2016) Polymer chain behavior in polymer nanocomposites with attractive interactions. ACS Macro Lett 5:523–527. https://doi.org/10.1021/acsmacrolett.6b00164
68. Genix A-C, Bocharova V, Carroll B, Lehmann M, Saito T, Krueger S, He L, Dieudonné-George P, Sokolov AP, Oberdisse J (2019) Understanding the static interfacial polymer layer by exploring the dispersion states of nanocomposites. ACS Appl Mater Interfaces 11:17863–17872. https://doi.org/10.1021/acsami.9b04553
69. Genix A-C, Bocharova V, Kisliuk A, Carroll B, Zhao S, Oberdisse J, Sokolov AP (2018) Enhancing the mechanical properties of glassy nanocomposites by tuning polymer molecular weight. ACS Appl Mater Interfaces 10:33601–33610. https://doi.org/10.1021/acsami.8b13109
70. Voylov DN, Holt AP, Doughty B, Bocharova V, Meyer HM, Cheng S, Martin H, Dadmun M, Kisliuk A, Sokolov AP (2017) Unraveling the molecular weight dependence of interfacial interactions in Poly(2-vinylpyridine)/Silica nanocomposites. ACS Macro Letters 6:68–72. https://doi.org/10.1021/acsmacrolett.6b00915
71. Napolitano S, Capponi S, Vanroy B (2013) Glassy dynamics of soft matter under 1D confinement: how irreversible adsorption affects molecular packing, mobility gradients and orientational polarization in thin films. Eur Phys J E 36:61. https://doi.org/10.1140/epje/i2013-13061-8

72. Napolitano S, Wubbenhorst M (2011) The lifetime of the deviations from bulk behaviour in polymers confined at the nanoscale. Nat Commun 2:260. https://doi.org/10.1038/ncomms1259
73. Napolitano S, Cangialosi D (2013) Interfacial free volume and vitrification: reduction in Tg in proximity of an adsorbing interface explained by the free volume holes diffusion model. Macromolecules 46:8051–8053. https://doi.org/10.1021/ma401368p
74. Zhu Z, Thompson T, Wang S-Q, von Meerwall ED, Halasa A (2005) Investigating linear and nonlinear viscoelastic behavior using model silica-particle-filled polybutadiene. Macromolecules 38(21):8816–8824. https://doi.org/10.1021/ma050922s
75. Sternstein SS, Zhu A-J (2002) Reinforcement mechanism of nanofilled polymer melts as elucidated by nonlinear viscoelastic behavior. Macromolecules 35:7262–7273. https://doi.org/10.1021/ma020482u
76. Zhang Q, Archer LA (2002) Poly(ethylene oxide)/Silica Nanocomposites: structure and rheology. Langmuir 18:10435–10442. https://doi.org/10.1021/la026338j
77. Chen Q, Gong S, Moll J, Zhao D, Kumar SK, Colby RH (2015) Mechanical reinforcement of polymer nanocomposites from percolation of a nanoparticle network. ACS Macro Lett 4 (4):398–402. https://doi.org/10.1021/acsmacrolett.5b00002
78. Giovambattista N, Buldyrev SV, Starr FW, Stanley HE (2003) Connection between Adam-gibbs theory and spatially heterogeneous dynamics. Phys Rev Lett 90:085506. https://doi.org/10.1103/PhysRevLett.90.085506
79. Simmons DS (2016) An emerging unified view of dynamic interphases in polymers. Macromol Chem Phys 217:137–148. https://doi.org/10.1002/macp.201500284
80. Lang RJ, Merling WL, Simmons DS (2014) Combined dependence of nanoconfined tg on interfacial energy and softness of confinement. ACS Macro Lett 3:758–762. https://doi.org/10.1021/mz500361v
81. Phan AD, Schweizer KS (2020) Theory of spatial gradients of relaxation, vitrification temperature and fragility of glass-forming polymer liquids near solid substrates. ACS Macro Letters 9:448–453. https://doi.org/10.1021/acsmacrolett.0c00006
82. Schweizer KS, Simmons DS (2019) Progress towards a phenomenological picture and theoretical understanding of glassy dynamics and vitrification near interfaces and under nanoconfinement. J Chem Phys 151:240901. https://doi.org/10.1063/1.5129405
83. Phan AD, Schweizer KS (2019) Influence of longer range transfer of vapor interface modified caging constraints on the spatially heterogeneous dynamics of glass-forming liquids. Macromolecules 52:5192–5206. https://doi.org/10.1021/acs.macromol.9b00754
84. Cheng S, Bocharova V, Belianinov A, Xiong S, Kisliuk A, Somnath S, Holt AP, Ovchinnikova OS, Jesse S, Martin H, Etampawala T, Dadmun M, Sokolov AP (2016) Unraveling the mechanism of nanoscale mechanical reinforcement in glassy polymer nanocomposites. Nano Lett 16:3630–3637. https://doi.org/10.1021/acs.nanolett.6b00766
85. Sun R, Melton M, Safaie N, Ferrier RC, Cheng S, Liu Y, Zuo X, Wang Y (2021) Molecular view on mechanical reinforcement in polymer nanocomposites. Phys Rev Lett 126:117801. https://doi.org/10.1103/PhysRevLett.126.117801
86. Baeza GP, Dessi C, Costanzo S, Zhao D, Gong S, Alegria A, Colby RH, Rubinstein M, Vlassopoulos D, Kumar SK (2016) Network dynamics in nanofilled polymers. Nat Commun 7:11368. https://doi.org/10.1038/ncomms11368
87. Ding YF, Sokolov AP (2006) Breakdown of time-temperature superposition principle and universality of chain dynamics in polymers. Macromolecules 39:3322–3326. https://doi.org/10.1021/ma052607b
88. Chen Q, Tudryn GJ, Colby RH (2013) Ionomer dynamics and the sticky Rouse model. J Rheol 57:1441–1462. https://doi.org/10.1122/1.4818868
89. Yu C, Guan J, Chen K, Bae SC, Granick S (2013) Single-molecule observation of long jumps in polymer adsorption. ACS Nano 7:9735–9742. https://doi.org/10.1021/nn4049039
90. Douglas JF, Johnson HE, Granick S (1993) A simple kinetic-model of polymer adsorption and desorption. Science 262:2010–2012. http://www.sciencemag.org/content/262/5142/2010
91. Johnson HE, Douglas JF, Granick S (1993) Topological influences on polymer adsorption and desorption dynamics. Phys Rev Lett 70:3267–3270. https://doi.org/10.1103/PhysRevLett.70.3267

Molecular Mobility in Nanocomposites Based on Renewable Semicrystalline Polyesters

Panagiotis A. Klonos, Dimitrios N. Bikiaris, and Apostolos Kyritsis

Abstract Broadband dielectric spectroscopy is employed as the basic tool to study molecular mobility (local, segmental) in various polyesters in the bulk and in the form of nanocomposites. In particular, the investigation involves three categories of renewable/ecofriendly polyesters, the traditional polylactides, poly(ε-caprolactone) and, the more advanced, poly(alkylene furanoate)s. The polymers are reinforced by various inorganic nanoparticles (NP), differing in geometry (nanotubes, nanoplatelets, spherical nanoparticles) and surface chemistry, including modifications. The polymer nanocomposites (PNCs) are investigated in two states, amorphous, to assess the direct filler effects on molecular mobility, and semicrystalline, for the indirect effects on molecular mobility due to modifications in the crystalline domains. Depending on the degree of interfacial interactions, the filler aspect ratio and the surface area, the segmental mobility (α relaxation) is either accelerated or decelerated, while at the same time the number of mobile polymer segments mainly drops. However, the most dominant effects on mobility are the indirect ones, arising from the formation of crystals which generally slow down dynamics and suppress the mobile chain segment fraction, or from lowering of molar mass. In the case of weak polymer-NP interactions, the inclusions offer new nucleation sites enhancing the crystallization process. Models involving localized changes in the free volume seem to apply well within the overall effects. In the cases of stronger interactions, the crystallization kinetics is severely hindered, whereas additional filler-related relaxations are recorded. Regarding local relaxation processes (β, γ), the fillers have no direct impact on their dynamics. However, the nano-inclusions affect indirectly the local mobility, since for most polyesters, β relaxation seems to sense the early stages of the overall matrix changes (glass transition, cold crystallization).

P. A. Klonos · A. Kyritsis (✉)
Department of Physics, National Technical University of Athens, Zografou Campus, 15780 Athens, Greece
e-mail: akyrits@central.ntua.gr

D. N. Bikiaris
Department of Chemistry, Aristotle University of Thessaloniki, 54124 Thessaloniki, Greece

A. Schönhals and P. Szymoniak (eds.), *Dynamics of Composite Materials*,
Advances in Dielectrics, https://doi.org/10.1007/978-3-030-89723-9_4

Keywords Broadband dielectric spectroscopy · Molecular mobility · Polymer nanocomposites · Semicrystalline matrix · Renewable polymers · Polylactide · Poly(ε-caprolactone) · Poly(alkylene furanoate)

Abbreviations

AR	Aspect Ratio
BDS	Broadband dielectric spectroscopy
CF	Crystalline fraction
CNT	Carbon nanotube
DMFD	2,5-dimethylfuran dicarboxylate
DSC	Differential scanning calorimetry
FTIR	Fourier transform infrared spectroscopy
GO	Graphene oxide
HN	Havriliak-Negami
HNT	Halloysite nanotube
MAF	Mobile amorphous fraction
MMT	Montmorillonite
MW	Molecular weight
MWCNT	Multi-walled carbon nanotube
NM	Normal Mode
NP	Nanoparticle
PAF	Poly(alkylene furanoate)
PBF	Poly(butylene furanoate)
PCL	Poly(ε-caprolactone)
PHF	Poly(hexylene furanoate)
PLM	Polarized light microscopy
PPF	Poly(propylene furanoate)
PDLA	Poly(D-lactic acid)
PDLLA	Poly(L,D-lactic acid)
PLA	Poly(lactic acid)
PLLA	Poly(L-lactic acid)
PNC	Polymer nanocomposite
RAF	Rigid amorphous fraction
$RAF_{crystal}$	RAF due to (around) the crystals
RAF_{NP}	RAF due to (around) the NPs
SAXS	Small angle X-ray scattering
VFT	Vogel-Fulcher-Tammann
WAXS	Wide angle X-ray scattering
α	(ref. to mobility) Main segmental relaxation
α or α'	(ref. to crystallinity) PLA crystal polymorph type
α^* or a_f	(ref. to mobility) filler induced/related relaxation
a_{HN}	Width shape parameter of HN equation

a_n	Alternative for the Normal Mode relaxation
a_s	Alternative for α segmental relaxation
β	Secondary (local) relaxation
β_{HN}	Asymmetry shape parameter of HN equation
γ	Secondary (local) relaxation
c_p	Specific heat capacity
$\Delta\varepsilon$	Dielectric strength
Δc_p	Change in specific heat capacity
ΔH	Enthalpy change
ε or ε^*	Dielectric permittivity
ε'	Real part of ε
ε''	Imaginary part of ε
E_{act}	Activation energy
f	Frequency
M_n	Molar mass
R_g	Radius of gyration
σ	Conductivity or conductivity relaxation
T_c	Crystallization temperature
t_c	Crystallization time
T_{cc}	Cold crystallization temperature
T_g	Glass transition temperature
$T_{g,cal}$	Calorimetric T_g
$T_{g,diel}$	Dielectric T_g
T_m	Melting temperature
X_c	Degree of crystallinity. Alternative for CF

1 Introduction

Polymer nanocomposites (PNCs) have been intensively investigated during the last decades because they offer the possibility of developing materials with enhanced, or even new properties due to the presence of the nanoparticles (NPs) or the modification of the polymer matrix [1–3]. Usually, the good dispersion of the nano-inclusions is a prerequisite for the achievement of these goals, and several strategies have been developed for that purpose. Among them, the establishment of favorable interactions between polymer segments and the embedded NPs seems to be the best choice. The existence of such interactions results in the appearance of a polymer fraction around the NPs with different structural and dynamic properties than the bulk [4]. It is recognized that this particular polymer fraction, usually called 'bound layer', is responsible for the unique properties of the PNCs and, therefore, it is in the focus of many research efforts for many years [4] (and references therein). This research aims to improve our understanding of various characteristics of this interfacial layer, such as its thickness and its temperature dependence [5], the spatial variation of the mobility across the layer [6, 7], its dependence on both, the size of the NP and of the polymeric chain (e.g., the radius of gyration, R_g) and on the dimensionality of the NP [8, 9], on the rigidity of the matrix polymer [10] or on the surface modification of the NP, e.g., by grafting polymer chains [11, 12].

Despite the different conceptual approaches regarding this layer, it is well appreciated that the polymeric chains at the interface adopt train, loop or tail conformations [13, 14] and that the relative portion of each conformation dictates the specific properties of this interfacial layer [15–17]. Overall, it is accepted now that the properties of this interfacial layer are controlled not only by the strength of the polymer-NP interaction but also on the 'degree' of these interactions [4, 8, 18], i.e., the number of polymer-NP contact points [15, 16], the strength of each individual bond [19–21] and the chain topology at the interfaces [12, 22]. On the other hand, there have been also prepared PNCs wherein the polymer-NP interactions are weak (or even of repulsive character). In this case, entropic depletion interactions may dominate leading to dissociation between polymer chains and NPs [23], while the size of the NPs seems to play a critical role [24].

With respect to the contribution of this interfacial layer to molecular fluctuations corresponding to the calorimetric glass transition of the polymer, the concept of a Rigid Amorphous Fraction (RAF) of polymers associated with the NPs, RAF_{NP} (it can be also found as RAF_{filler} in the literature), and not contributing to the corresponding heat capacity endothermic step at the glass transition, has been introduced [25, 26]. It is well established now that this RAF_{NP}, which is considered as immobile in calorimetric experiments, may exhibit reduced molecular mobility (compared to the bulk) as has been revealed by experimental techniques that probe molecular mobility at the adequate length and time scale [24, 27–29]. Broadband dielectric spectroscopy (BDS) [30] has been proved to be one of the most powerful tools with regard to such effects.

The detection of new molecular relaxation processes in PNCs with different time scales than the segmental process of bulk polymer matrix has been considered as an additional indication for the existence of this particular interfacial polymer layer [7, 27, 29]. Furthermore, for high NP loadings or for very small NPs, where the surface-to-surface distance is very small and polymer segments may be spatially confined between the NPs, confinement effects on the dynamics of polymer chains have also been reported [31]. NP effects on the matrix mobility are also reflected in changes of molecular mobility at both, longer length scales, e.g., the global chain mobility and the diffusion of the NPs in the melt [1, 18], and shorter length scales than the segmental fluctuations, e.g., in perturbations of the non-diffusive local mobility [32, 33].

In the PNCs based on a semicrystalline polymer matrix, the impact of the embedded NPs on the properties of the matrix becomes more complex due to the fact that in most cases the NPs affect also the crystallinity of the matrix ('direct' and 'indirect' effects being involved). Therefore, the final properties of the PNCs are controlled by both the NPs and the crystalline domains, with the latter being proved severe. More specifically, NPs may affect the heterogeneous nucleation process in the PNCs, by offering new nucleation sites leading, usually, to acceleration of the crystallization process and to an increase of the crystalline fraction. In the case where they are not acting as nucleation agents and, therefore, the crystals are formed away from the NPs, they may also affect the crystallization process, now by hindering the process of crystal growth due to polymer-nanoparticle interaction or by imposing topological steric hindrance. It is worth to notice here that in this case it has been suggested that the dispersion and ordering of the NPs may be controlled by the crystallization process [34, 35]. On the other hand, it has been shown that modified segmental dynamics in PNCs may also have an impact on the crystallization process [36, 37]. It is, thus, anticipated that the properties of the aforementioned interfacial layer in the PNCs affect significantly the role of the NPs in the crystallization process [38] and especially in the early stage of crystallization [39]. We may assume in this context that the crystallization behavior of the PNCs may be another source of information with respect to the distribution of the NPs and their specific interactions with the polymer matrix.

For semicrystalline polymers, the concept of another rigid amorphous polymer fraction has been introduced, i.e., the RAF due to the crystals that are connected with the crystalline domains (RAF_{cryst}) [26, 40]. This fraction consists of polymer chains that are connected with the crystallites and are located in the amorphous region among the crystallites, most probably, they are chains semi-immersed at the outer layers of the crystals. However, due to the strong coupling to the crystal, these polymer segments are not contributing to the supercooled liquid-like molecular mobility above the glass transition temperature, T_g. Due to the location of RAF_{cryst} at the interface between the crystals and the amorphous regions, this fraction depends strongly on the characteristics of the crystal surface area. Worth noting at this point is that the two RAFs mentioned so far (RAF_{NP} and RAF_{cryst}) are formed by different mechanisms and they are anticipated to exhibit distinct properties. Therefore, in semicrystalline PNCs the molecular mobility is controlled by the relative fractions of the amorphous part (mobile amorphous polymer fraction, MAF), the two RAF(s) and the crystalline domains.

With respect to segmental mobility in the amorphous polymer part, its manifestation in BDS is the so-called segmental α relaxation process [30, 41], which is the dielectric analogue of the calorimetric glass transition. The individual dynamics of NP-interfacial chains, whenever recorded, is usually denoted as α', α^* or α_{int} [16, 27, 29]. Usually in semicrystalline homopolymers, the presence of crystals tends to decelerate the chain mobility [41] acting as an obstacle for chain diffusion, whereas in some cases an additional process may be recorded, either together with α relaxation [20, 42] or not [43]. This is the so-called, α_c process. In extreme cases of tight semicrystalline morphologies, there can be recorded either elimination of mobile amorphous segment mobility [44], or even acceleration of mobility, for example, due to spatial confinement [36, 45]. When coming to PNCs based on semicrystalline polymers, the molecular mobility is affected by the interfacial filler-polymer interactions (retardation of dynamics, direct filler effect) and, more severely, by the imposed changes in crystalline fraction, CF and semicrystalline morphology (number, size, distribution of the crystals) [26, 46–48]. All the cases described above are visualized in Scheme 1, which was created considering recordings by various works from the literature.

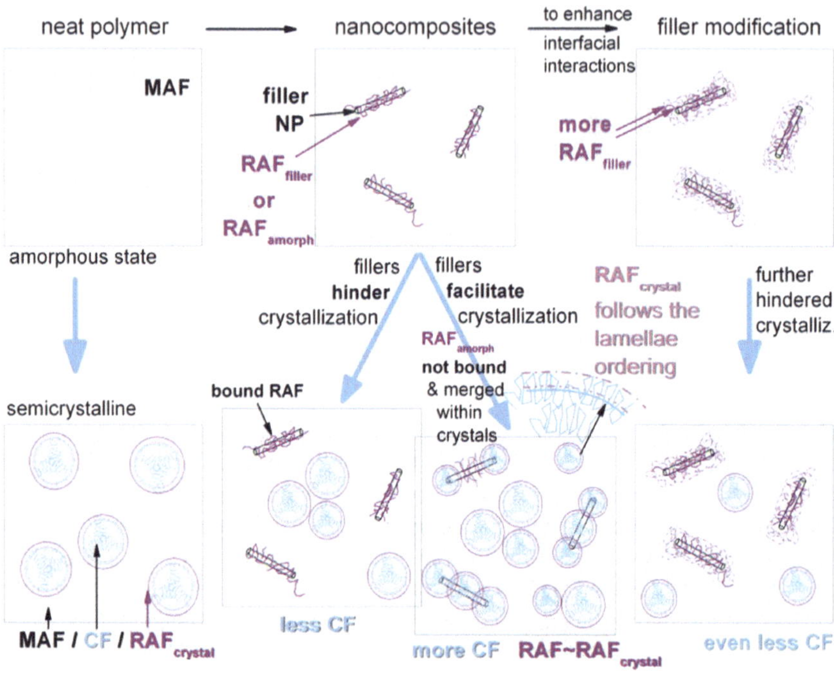

Scheme 1 Schematic view of the various polymer fractions (phases, MAF, CF, RAF) and NP filler distributions in neat semicrystalline matrices and nanocomposites at the (top) amorphous and (bottom) semicrystalline state and in the cases of fillers either serving or not as crystallization agents. RAF_{filler} and RAF_{amorph} refer to RAF_{NP}. The scheme has been created considering changes in the semicrystalline morphology upon filler addition [49, 50], probable filler rearrangements upon crystallization [35], differences in the interfacial polymer adsorption (interaction strength) on nucleation [51] (and references therein) and the severe differences in the structure between interfacial polymer around the fillers and the crystals [52–56]. (Reproduced with permission from Ref. [57]. Copyright 2021 Elsevier.)

Nowadays, there has been a serious concern for environmentally friendly materials, also in the modern frame of circular and green economy. These concerns have driven the scientific and industrial communities toward the research, development and application of renewable and non-toxic polymers, in particular, for large-scale and everyday uses. PNCs based on such polymers, belonging specifically in the class of semicrystalline polyesters, are discussed in this chapter with the focus being on the molecular mobility. The well-known polylactide or poly(lactic acid) (PLA) [58–60] and poly(ε-caprolactone) PCL [61] are firstly discussed. Last but not least, we study a quite new class of polyesters that are based on 2,5-furandicarboxylic acid [62–64] and called them poly(alkylene furanoate)s (PAF)s. The employment of such polymers in a wide range of applications arises not only from their renewable and non-toxic character but also from the wide potential to tune their performance (mechanical, permeability, etc.) by employing different thermal treatments and, more effectively, by the addition of properly chosen NPs. Within the latter, the employment of NPs as direct and indirect improving/reinforcing agents has been extensively explored [58, 64–66].

2 Experimental

2.1 Materials

With the exception of PAFs, the other two studied polymers, i.e. PLA and PCL, were commercial and of high purity. Details are given in previous publications [37, 47, 57, 67, 68]. The molecular structures of the polymers can be seen in Scheme 2.

The PLAs under investigation are employed in a variety of PNC series, wherein PLA differs mainly in the amount of D/L-isomers, namely 98/2, 4/96 and 0/100, and in the molecular weight M_n between ~30 and 700 kg/mol. Results on PCL-based PNCs are shown by one representative series, with the matrix PCL being of M_n~65 kg/mol. PAFs, being quite new in the field of polymer science, were synthesized by the two-stage polycondensation method [51, 69], within the Laboratory of Chemistry and Technology of Polymers and Dyes, Department of Chemistry, Aristotle University of Thessaloniki, Greece [62, 69–71]. Due to the method of synthesis, used also for the in situ PNC synthesis, PAFs are of relatively lower M_n, namely 10–20 kg/mol. The corresponding effects are demonstrated via many series of PNCs and basically three types of PAF, differing in the n-alkylene length (Scheme 2c) as n = 3, 4 or 6.

Regarding the used NPs, these consist of a variety of particles (Scheme 3), such as multi-walled carbon nanotubes (MWCNT or, simply, CNT), graphene (Gr) and graphene oxide (GO) shapes, montmorillonite (MMT) nanoclays, spherical fumed or mesoporous silicas (SiO_2), titania (TiO_2), silver (Ag), zinc oxide (ZnO) and Halloysite nanotubes (HNT).

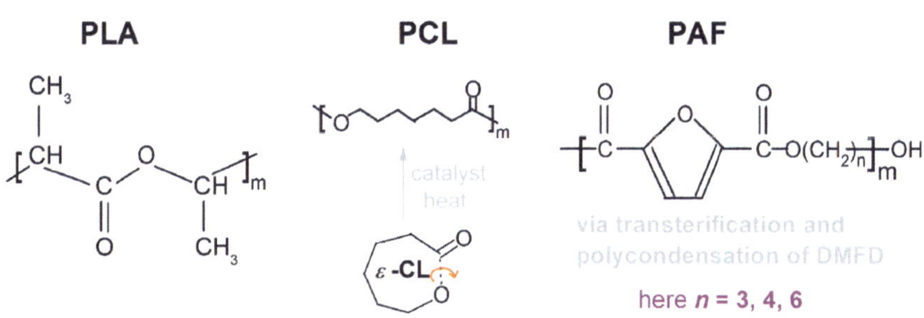

(a) (b) (c)

Scheme 2 Chemical structurs of the materials under investigation: **a** PLA, **b** PCL and **c** PAF

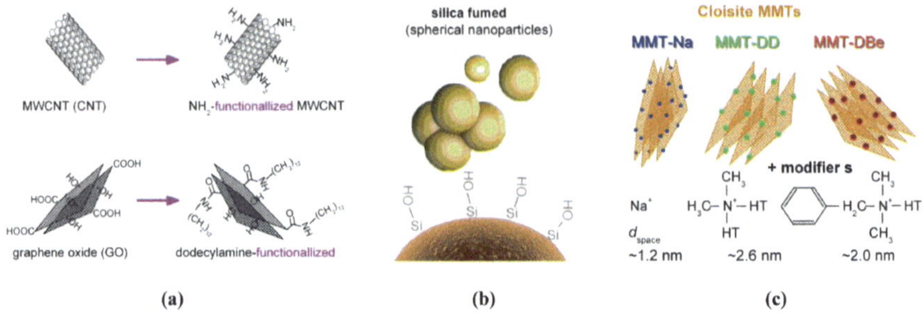

Scheme 3 Examples of nanofillers employed as reinforcing means for the PNCs under investigation: **a** CNTs and GO, initial and modified, **b** spherical silica NPs and **c** montmorillonite nanoclays, initial and modified (functionalized)

For selected cases of NPs and polymers, and in order to increase the polymer-NPs adhesion, for example, by manipulating the surface chemistry or the specific area of the NPs, the NPs were either surface-modified (Scheme 3a, c) [51, 70] or combined with each other, in the form of hybrid particles or mixtures [37, 47, 71].

More details are given in the following sections, along with the corresponding experimental results and discussions.

2.2 *Techniques*

2.2.1 Differential Scanning Calorimetry (DSC)

DSC is one of the most known and widely employed thermal analysis tools, for studying the thermal transitions of materials [72]. In the case of polymer-based systems, DSC enables the determination of the glass transition, crystallization, melting and mesomorphic transitions, in terms of characteristic temperatures (T_g, T_c, T_m and T_{MT}, respectively), specific heat capacity, c_p, changes (Δc_p), changes in enthalpy (ΔH) and entropy (ΔS).

In Fig. 1a shown below, we show typical DSC curves during cooling and heating of semicrystalline polymeric systems (such as those studied here).

From the estimation of the abovementioned physical properties, we may further evaluate the various fractions existing in the semicrystalline polymers and PNCs. Firstly, the crystalline fraction, CF, or else degree of crystallinity, X_c, is obtained by the following equation:

$$CF = X_c = \Delta H_{c,n} / \Delta H_{100\%} \qquad (1)$$

where $\Delta H_{c,n}$ is the measured enthalpy of crystallization, ΔH_c, normalized to the polymer fraction, w_{polym}, and $\Delta H_{100\%}$ is the enthalpy of fusion of the fully crystallized

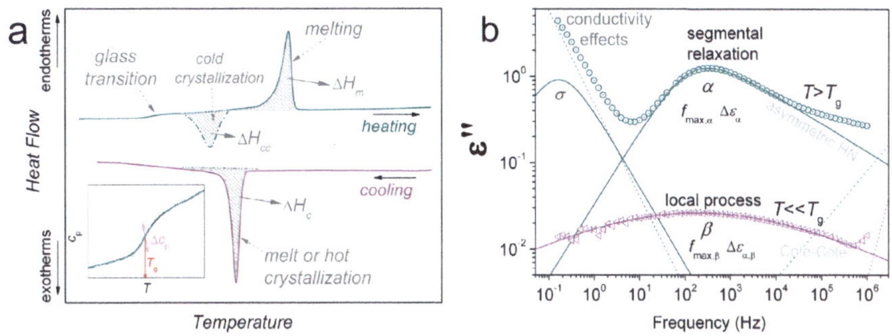

Fig. 1 **a** Typical DSC curves of a semicrystalline polymer during cooling and heating, with the main thermal transitions being indicated. The inset shows details of the glass transition region. **b** Typical isothermal BDS spectra of the imaginary part of dielectric permittivity (dielectric loss), ε'', versus frequency, at a low and a high temperature, indicating local and segmental relaxations, respectively. Along with the experimental data, examples of the analysis in terms of individual Havriliak-Negami (HN) and the Cole-Cole model function are shown

polymer which is mainly taken from the literature. Then from the measured change in the heat capacity, Δc_p, during glass transition, which is the measure of the mobile polymer fraction that contributes to the glass transition, we may estimate the mobile amorphous fraction, MAF, and, furthermore, upon assumptions, the rigid amorphous fraction(s), RAF(s) [25, 26] by the following equations [16, 51, 73]:

$$\Delta c_{p,n} = \Delta c_p / \left[w_{polym}(1 - CF) \right] \tag{2}$$

$$MAF = \Delta c_{p,n}(1 - CF)/\Delta c_{p,matrix}^{amorphous} \tag{3}$$

$$1 = CF + MAF + RAF = 1 + CF + MAF + \left(RAF_{NP} + RAF_{crystal} \right) \tag{4}$$

where $\Delta c_{p,matrix}^{amorphous}$ is the heat capacity change at the glass transition of the amorphous unfilled polymer matrix, while RAF_{NP} and $RAF_{crystal}$ are the RAFs produced by and located around the filler NPs and the crystals, respectively. Equation 4 describes the so-called '3 phase model' (RAF, MAF, CF) of polymer systems [25, 26].

It should be noted, from the methodological point of view, that for a more accurate determination of c_p, and Δc_p, more specialized DSC modes should be employed (e.g., temperature modulated or step-scan DSC) [26, 50, 74]. Also, to disentangle RAF_{NP} and $RAF_{crystal}$ from the total RAF (Eq. 4), certain and serious assumptions must be made [22, 26, 50] Moreover, we have proposed that for special cases of PNCs such disentanglement should not be attempted (Scheme 1) [46, 67, 73].

2.2.2 Broadband Dielectric Spectroscopy (BDS)

Isothermal measurements in the frequency domain are the most common mode of dielectric measurements and various terms, such as broadband dielectric spectroscopy (BDS) and impedance spectroscopy, are used for that. The latter is of particular interest here. The samples, in the form of thin disks or films, are usually placed between finely polished disk electrodes and form a sandwich-like capacitor. In some cases, the samples are melted and quenched inside the capacitor, with the use of thin silica spacers (~50 μm in thickness) to keep the distance between the electrodes fixed (~50 μm in thickness) and preclude the electrodes' contact. This capacitor is inserted between the parallel electrodes of a Novocontrol BDS-1200 sample cell and an alternate voltage is applied. The complex dielectric permittivity $\varepsilon^* = \varepsilon' - i\varepsilon''$ is recorded using a Novocontrol Alpha analyzer, isothermally as a function of frequency, f, in the broad range from 10^{-1} to 10^6 Hz, at temperatures from −150 to 130 °C in nitrogen atmosphere, on heating in steps of 2.5–10 K. The latter depends on the process under investigation.

Regarding molecular mobility, the most exploitable studied parameter is the imaginary part of dielectric permittivity, ε'', which represents the dielectric losses. Figure 1b shows a characteristic example for the recorded frequency dependence of ε'' and the corresponding method of analysis. To extract the wanted information for a relaxation process (i), i.e., the time scale ($\tau_{\mathrm{max,i}}$ or $f_{\mathrm{max,i}}$), strength ($\Delta\varepsilon_{\mathrm{i}}$) and relaxation time distribution, proper mathematical models are used to analyse the $\varepsilon''(f,T)$ spectra. The model that usually is adopted for the data referring to relaxation processes is the Havriliak-Negami (HN) equation [30, 75],

$$\varepsilon^*(f) = \varepsilon_\infty + \frac{\Delta\varepsilon}{\left[1 + (if/f_0)^{\alpha_{HN}}\right]^{\beta_{HN}}} \tag{5}$$

where ε_∞ describes the value of the real part of dielectric permittivity, ε', for $f >> f_0$, f_0 is a characteristic frequency related to the frequency of maximum dielectric loss (ε'') and α_{HN} and β_{HN} are the shape parameters of the relaxation function, determining the width and symmetry/asymmetry, respectively. A sum of HN terms, one for each of the relaxations present in the frequency window at the temperature of measurement (e.g., β and α, in Fig. 1b), is fitted to the experimental data and the HN parameters for each relaxation process are determined. The data by this analysis are generally shown in the form of the so-called 'Arrhenius plots' or 'dielectric maps' in terms of the reciprocal temperature (1000/T) dependence of the frequency of the ε'' peak maximum, f_{max}.

For local (secondary) relaxations, the time scale of the response is analyzed in terms of the Arrhenius equation [30, 76],

$$f(T) = f_{0,Arrh} \cdot e^{-\frac{E_{act}}{kT}} \tag{6}$$

whereas, the co-operative segmental (primary) α relaxation is analyzed in terms of the Vogel-Fulcher-Tammann (VFT) equation [77]:

$$f(T) = f_{0,VFT} \cdot e^{-\frac{B}{T-T_0}} \tag{7}$$

In Eq. (6), $f_{0,\,\mathrm{Arrh}}$ is a frequency constant (usually in the order of 10^{13} Hz, i.e., the inverse of quasi-lattice vibration time) and E_{act} is the activation energy of the relaxation. $f_{0,\mathrm{VFT}}$ in Eq. (7) is a frequency constant (in the range from 10^{12} to 10^{14} Hz), B is a material constant (=DT_0, where D is the so-called *fragility strength parameter*) [78], and T_0 is the Vogel temperature, all these parameters being material-dependent. After fitting Eq. (7) to the experimental data and fixing $f_{0,\mathrm{VFT}}$ to the phonon value 10^{13} Hz [16, 27, 30], we obtained values for T_0 and D. Then, the fragility index, m, is estimated according to Eq. (8):

$$m = 16 + 590/D \tag{8}$$

3 Results and Discussion

3.1 *Polylactide-Based Systems*

PLA is a biodegradable aliphatic polyester (Scheme 2), which can be derived from renewable resources, for example, corn starch. The PLA polymers synthesized from L-lactic acid or D-lactic acid are called poly(L-lactic acid) (PLLA) or poly(D-lactic acid) (PDLA), respectively. Commercial PLA for large-scale applications mainly contain predominant the L-monomer and small amounts of the D-monomer, i.e., poly(D-/L-lactic acid) (PDLLA) [79]. PLLA is crystallizable, whereas PDLA does not crystallize.

Prior to the discussion of the properties of PLA-based PNCs, we present the current knowledge with respect to PLA chain structure effects on the properties of neat PLA, namely on the **glass transition** of the amorphous part, the **crystallization** behavior and in the **molecular mobility** (in wide range of length and time scales). On the other hand, with the term chain structure, it means the **chain length** and in the **presence of small amounts of D-isomer** co-units in the PDLLA chain.

3.1.1 Amorphous PLA

Fully amorphous PLAs are the PDLA and PDLLA macromolecules with the content of D-isomer higher than about 10%. In addition, crystallizable PDLLA chains with low D-isomer content (<10%) and PLLA chains remain fully amorphous at temperatures below T_g by fast cooling from the melt (cooling rate higher than about 0.5 K/s) [80]. Therefore, the features of the glass transition and the related molecular mobility of fully amorphous PLA can be investigated by studying either non-crystallizable PDLA or crystallizable PLA chains that bypassed crystallization via fast cooling from the melt. The later systems exhibit usually cold crystallization during heating above T_g.

Table 1 gives the calorimetric T_gs, heat capacity increment at T_g, Δc_p, for amorphous PLA with various molar masses, M_n, and D-isomer contents. It can be observed that T_g values of PLLA with no or very low D-isomer content exhibit no M_n dependence (in the M_n range of Table 1). It is interesting to note that PDLA exhibits a lower T_g as compared to the amorphous PLLA and that PDLLA exhibits T_gs that vary in between 55 and 69 °C.

In Table 1, the $T_{g,\,diel}$ values, as estimated by dielectric measurements, are also included ($T_{g,diel}$ is defined as the temperature at which the time scale of the segmental α relaxation becomes 100–1000 s). We would like to mention here that there is a

Table 1 Calorimetric glass transition temperature, T_g, heat capacity change (strength), Δc_p and dielectric glass transition temperature, $T_{g,diel}$, *L-/D* content and molar mass M_n for various **amorphous** PLA samples

	L-lactide (%)	*D*-lactide (%)	M_n (g/mol)	$T_{g,cal}$ (°C)	Δc_p (J/gK)	$T_{g,diel}$ (°C)
PLLA [81]	100	0	34 k	57	N/A	58
PDLLA [82]	1.5	98.5	38 k	43	0.53	41
PDLLA [57]	96	4	42 k	59	0.52	48
PLLA [47]	100	0	42 k	58	0.55	53
PDLLA [83]	98	2	47 k	57	0.54	53
PDLLA [37]	96	4	50 k	58	0.53	48
PDLLA [84]	98	2	73 k	55	0.50	53
PLLA [85]	100	0	80 k	60	N/A	N/A
PLLA [50]	100	0	86 k	57	0.49	57
PLLA [86]	100	0	100 k	59	0.56	N/A
PDLLA [87]	96	4	116 k	49	0.48	N/A
PLLA [43]	100	0	269 k	69	N/A	N/A
PLLA [67, 73, 88]	100	0	700 k	60	0.56	58

discrepancy between the calorimetric and dielectric T_g values obtained for crystallizable PLLA (up to 3–4 K). However, this discrepancy does not exist for the non-crystallizable PDLA, while it is maximized for PDLLA with 4% D-isomer content (by ~10 K).

Regarding the molecular mobility in PLLA, we consider firstly the segmental mobility and then the mobility at shorter and longer time and length scales. For amorphous PLA, in general the α relaxation is an asymmetric relaxation peak ($\beta_{HN}<1$ in Eq. 5) of remarkable strength (Fig. 2a). Its dynamics follows the VFT behavior (Eq. 7), as expected for segmental processes, whereas its time scale (Fig. 2b) and fragility index (*m* from ~130 to ~190) are affected by the *L*/*D* ratio and (slightly) by M_n [89, 90]. The representative data in Fig. 2 indicate that the time scale of the α relaxation depends remarkably on the *D*-isomer content in the PLLA: α becomes faster with increasing *D*-isomer content in amorphous PLLA.

At temperatures below T_g, the local β relaxation of PLA is obsserved. β has been suggested to arise from localized fluctuations [89, 92] or local twisting motions [93] in the PLA chain. As a secondary (local) process, it obeys the Arrhenius law (Eq. 6) characterized by an average activation energy, E_{act}, varying in the narrow range from 39 to 48 kJ/mol (Fig. 2). No effects worthy of noting have been observed in its relaxation time width, with α_{HN} ~0.3–0.4 and $\beta_{HN} = 1$ (symmetric process, Eq. 5). Regarding its time scale, β does not seem to follow systematically M_n (Fig. 2a, b), however, within many works in the literature it seems that β 'senses' the early stages of glass transition, as recorded by a disturbance in the Arrhenius plots at the higher temperatures (e.g., Figure 2b) [43, 91, 94]. Bras et al. [94] have reported up to three processes contributing to the local dynamics range, depending on the amount and the method of the polymer crystallization.

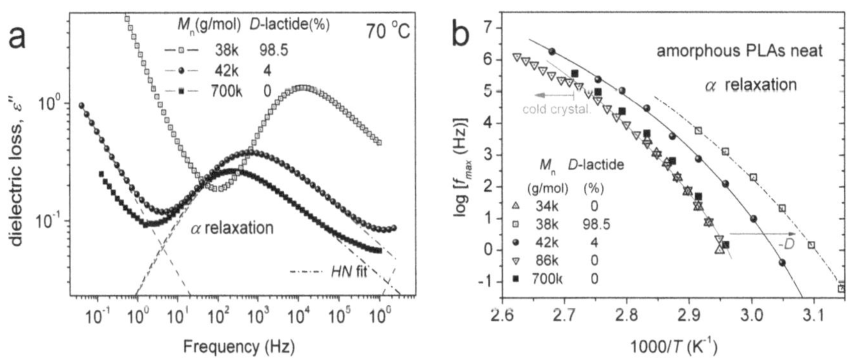

Fig. 2 BDS data for the segmental α relaxation for various amorphous PLA samples differing in M_n and the D-/L- contents, described on the plots, in the forms of **a** frequency dependence of ε'' at 70 °C along with analysis results in terms of Havriliak-Negami model and **b** activation diagram (Arrhenius plots). Source-literature description: up-pointing triangles [81], open squares [82] and solid spheres [57]. Down-pointing triangles [91] and solid squares [67]

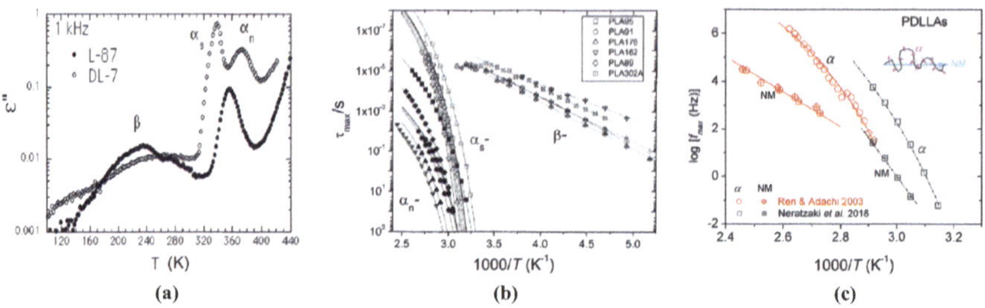

Fig. 3 **a** Isochronal ε'' curves at 1 kHz for (solid symbols, L-87) PLLA of $M_n = 87$ kg/mol and (open symbols) PDLLA of $M_n = 74$ kg/mol. (Reproduced with permission from from Ref. [89]. Copyright 2021 ACS.). **b** Arrhenius plot for local (β), segmental (α_S or α) and global chain dynamics (α_n or NM) dynamics in a series of amorphous PLAs with L:D ratio of 80:20 and M_n values from ~4 to 47 kg/mol (Reproduced with permission from Ref. [90]. Copyright 2021 Elsevier.). **c** Arrhenius plot for (open symbols) segmental α and (crossed symbols) Normal Mode relaxation in PDLLA samples of moderate [89] and high [82] *D*-lactide contents. (Adapted from Refs. [82, 89].)

Finally, in the literature [89, 90] the detection of a Normal Mode (NM)-like relaxation process in PLLA has been reported, i.e., a relaxation process that is related to the global chain mobility (Fig. 3). In a PDLA sample (fully amorphous, with M_n~38 kg/mol) [82], we have detected a relaxation process that resembles the NM relaxation (Fig. 3c). According to the published data, it seems that this relaxation process, of longer time and length scales than that of the segmental one, is more pronounced for low M_n.

3.1.2 Semicrystalline PLLA

PLLA is a crystallizable polymer which exhibits crystal polymorphism [95, 96]. Isothermal crystallization from the melt at temperatures higher than about 120 °C typically leads to orthorhombic α-type crystals with an equilibrium melting temperature of around 220 °C. Isothermal crystallization from the melt at temperatures lower than 120 °C or isothermal annealing at temperatures higher than T_g leads mainly to the formation of less ordered α'-crystals, in which molecule segments exhibit conformational disorder (Fig. 4a, b). It has been proposed that α'-crystals are larger than α-crystals. (Fig. 4c). Thus, the melting temperature of α'-crystals is lower than that of α-crystals, and the same trend follows also the bulk enthalpy of melting and the crystal density of α'-crystals. Slow heating of α'-crystals leads to a transformation into more stable α-crystals at around 150 °C, which then melt at 170–180 °C.

The kinetics of PLLA crystal nucleation under non-isothermal conditions has been reported in the literature [96–98]. It was found that the cooling rates required for suppressing crystallization and crystal nucleation were quite different. For PLLA with a molar mass of ~100 kg/mol, cooling the melt at rates higher than 0.5 K/s could avoid the occurrence of crystallization, while the rate for suppressing crystal nuclei formation was higher than 50 K/s [79]. Therefore, PLLA has very slow crystallization

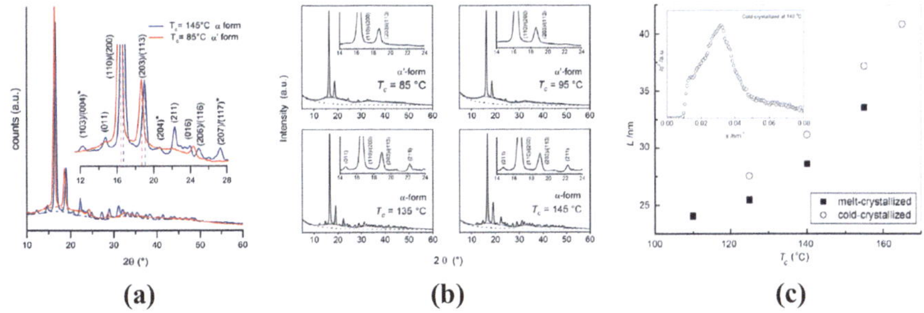

Fig. 4 **a**, **b** Monitoring of PLLA of $M_n = 100$ kg/mol crystallization by Wide Angle X-ray Scattering (WAXS) upon annealing at two distinct temperatures [100]. (Reproduced with permission from Refs. [86, 100]. Copyrights 2021 Elsevier.) **c** Monitoring of semicrystalline morphology for PLLA of $M_n = 180$ kg/mol by combining Small Angle X-ray Scattering (SAXS), WAXS and DSC, demonstrating the effects on the crystals' long period, L, by the crystallization temperature, T_c. (Reproduced with permission from Ref. [103]. Copyright 2021 Wiley.)

rate and a low degree of crystallinity in traditional processing such as injection molding. Furthermore, the dependence of the crystallization behavior of PLLA (ratio of α/α' forms [99], degree of crystallinity, size of spherulites) on the annealing temperature/time, the heating rate and the *D*-isomer content has been systematically investigated [87, 97, 98, 100, 101]. The manipulation of PLLA crystallization is usually achieved by adding nucleating agents and facilitating, thus, heterogeneous crystallization at low supercooling [102].

The impact of the crystallization process and crystal morphology on the properties of the MAF are also reported in the literature. Usually, confinement effects on the mobile chains induced by the crystals that overwhelm the polymer volume lead to a modification of the segmental relaxation process (α_c process) as in many semicrystalline polymers [41, 42, 104]. Additionally, different MAF mobility in different semicrystalline PLLA samples has been suggested to arise from different semicrystalline morphologies, and consequently different MAF region size [105] (and references 12, 26 and 27 therein). BDS is a powerful tool for the investigation of the impact of PLA crystallization on the mobility of the polymeric chains in the amorphous part. On the other hand, by probing the molecular mobility in the amorphous part, BDS may indirectly provide insight into the crystallization process. Indeed, BDS has been used for the monitoring of PLA cold crystallization. In Fig. 5, dielectric spectra recorded at various times during the cold crystallization process are shown [43, 47, 67]. For comparison, dielectric spectra recorded on PLA undergoing melt crystallization are also shown in Fig. 5c. It is demonstrated that the α process gradually transforms to a α_c process, either upon cold crystallization (almost all our data) or hot crystallization (fewer data), independently from *D*-isomer content (0, 2 or 4%). During the formation of crystals, only the α_c process is recorded which is symmetric with respect the distribution of relaxation times. Interestingly, reports that appeared in the literature regarding other semicrystalline polymers indicate that the α and α_c process may coexist (e.g., in polydimethylsiloxane [20, 42], polyether

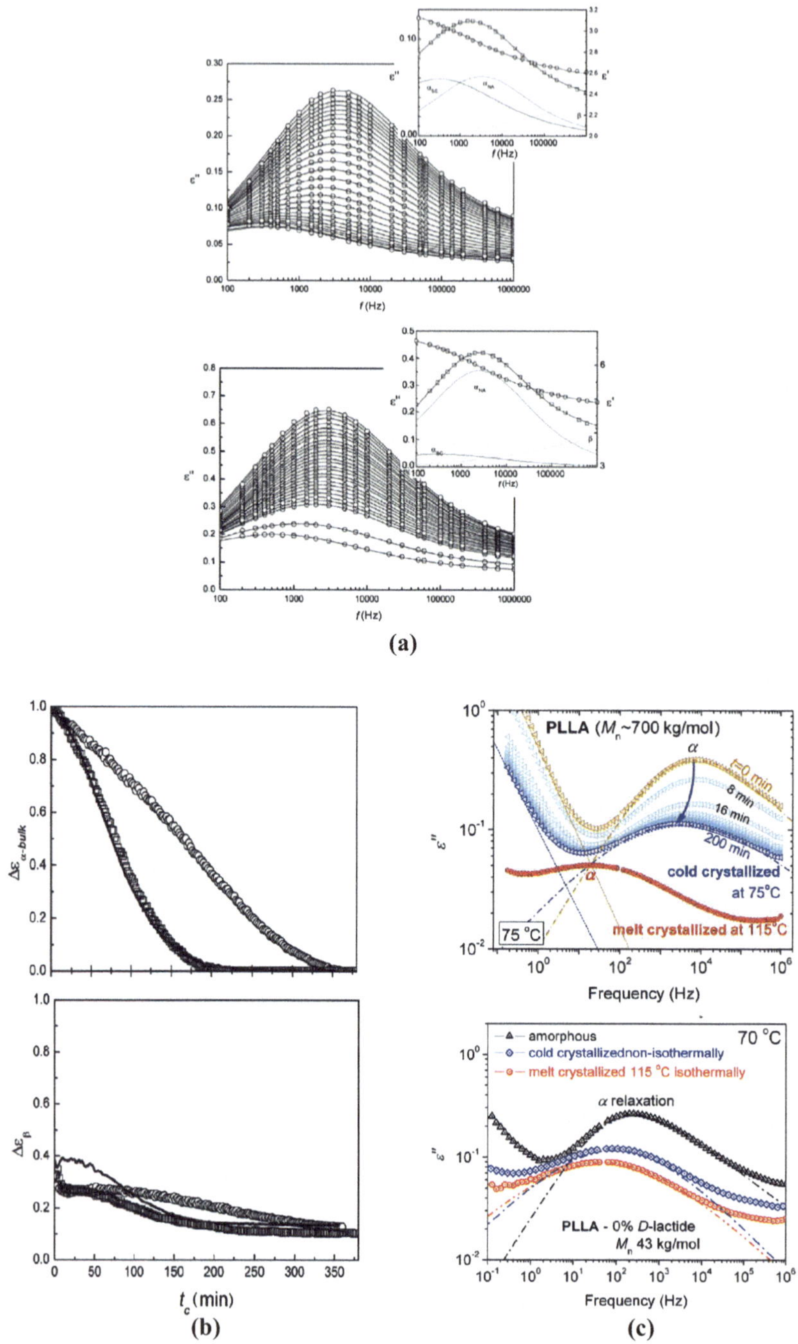
f (Hz)
(a)
$\Delta\varepsilon_{\alpha\text{-bulk}}$
$\Delta\varepsilon_{\beta}$
t_c (min)
(b)
PLLA (M_n~700 kg/mol)
t=0 min
8 min
16 min
200 min
cold crystallized
at 75°C
melt crystallized at 115°C
75 °C
Frequency (Hz)
amorphous
cold crystallizednon-isothermally
melt crystallized 115 °C isothermally
70 °C
α relaxation
PLLA - 0% D-lactide
M_n 43 kg/mol
(c)

◀**Fig. 5** Indirect monitoring of PLLA crystallization via versus the molecular mobility by Broadband Dielectric Spectroscopy in terms of ε'' against frequency, f. **a, b** In PLLA of $M_n = 269$ kg/mol during (a—top) cold and (a—bottom) melt crystallization at 80 °C. **b** Shows the corresponding results of **a** in terms of time evolution of dielectric strength for (b—top) the segmental α relaxation and (b—bottom) the local β relaxation, during cold (squares) and melt (circles) crystallization. Included in **b** are data for a PLLA of lower $M_n = 86$ kg/mol during cold crystallization (solid line). (Reproduced with permission from Ref. [43]. Copyright 2021 ACS.) (c—top) PLLA of M_n~700 kg/mol cold crystallized isothermally at 75 °C (Adapted from Ref. [67]) and (c—bottom) PLLA of M_n~43 kg/mol shown at 70 °C comparatively for an amorphous sample and upon cold and melt crystallization. (Adapted from Ref. [47].) In addition, **a, b** show two different routes for analyzing the $\varepsilon''(f)$ data in terms of fitting models

ether ketone [104], polyvinylidene fluoride [106]). For PLA, it seems that the case is different with only one exception, where the recorded process has been analyzed by two coexisting contributions (α and α_c) [43]. Finally, it is worth noting also that whenever we record it, α_c relaxation in the cold crystallized sample is, interestingly, faster (slightly or moderately) as compared to melt-crystallized sample, as can be seen in Fig. 5a, c [43, 47, 67].

Moreover, comparing with DSC, BDS (by probing the dynamics of the α relaxation) seems to be more sensitive to the evolution of crystallization at its earlier stages, e.g., via the rapid reduction of $\Delta\varepsilon$ [67, 73], responding faster than DSC in the crystallization of PLLA (Fig. 6). This feature of BDS has been observed also for other polymers [41, 107, 108].

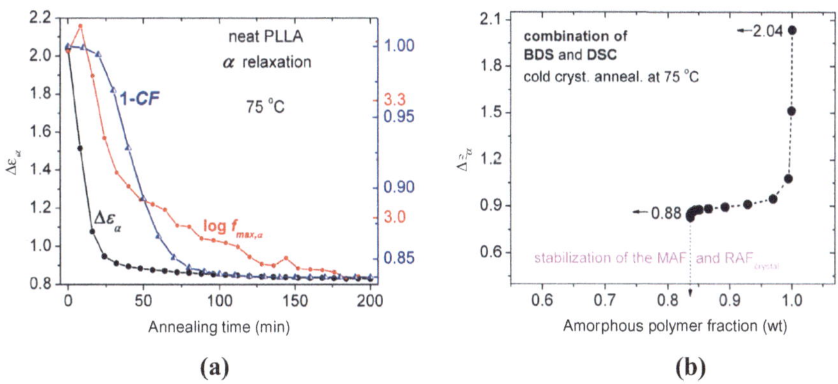

Fig. 6 **a** Comparative data for the amorphous polymer fraction, '1−CF' (=MAF + RAF), the dielectric strength of he α relaxation, $\Delta\varepsilon_\alpha$, and the frequency maximum of α relaxation, log$f_{max,\alpha}$, versus the annealing time, for initially amorphous PLLA during isothermal crystallization annealing at 75 °C. **b** $\Delta\varepsilon_\alpha$ versus the amorphous polymer fraction, '1−CF'. (Reproduced with permission from Ref. [67]. Copyright 2021 Elsevier.)

3.1.3 PNCs Based on PLA

In the case of the PLA, with a low crystallization rate, the corresponding PNCs exhibit a remarkably different crystallization behavior, mainly because the nano-inclusions promote the heterogeneous nucleation and the corresponding crystallization process. Thus, these effects of the NPs dominate the other modifications that may appear, e.g., changes in molecular mobility of polymer segments or the formation of RAF_{NP} and, consequently, the investigation of all effects independently becomes a very difficult task. Surely, remarkable changes in the crystallinity of a polymer affect in general the morphological and dynamical characteristics of the matrix. For this reason, we will discuss firstly the crystallization behavior of PLA-based PNCs and, then, its impact on the morphology and dynamics of the polymer.

Rate of crystallization, degree of crystallinity and morphology of crystals in PNCs

Regarding the non-isothermal crystallization from the melt of the PNCs, experimental DSC results are shown in Fig. 7a. Therein, the impact of NPs is severely additive on crystallization, in both the rate (elevation of crystallization temperatures) and in fraction (increase in CF). Exceptions to these behaviors are the cases where the NPs are aggregated/agglomerated (here in Ag and silica at high loadings). Since the non-isothermal crystallization involves an initially melted sample, the acceleration and enhancement of crystallinity are the first indications of the NPs, respectively, offering additional crystallization sites (nucleation, a ‘direct’ effect) and most probably favor easier and better (thicker) lamellae packings (an ‘indirect’ effect).

Results obtained during isothermal crystallization of PNCs with PLLA by employing DSC technique are shown in Fig. 7b, c. The acceleration of crystallization is confirmed by melt crystallization annealing via the suppression of crystallization

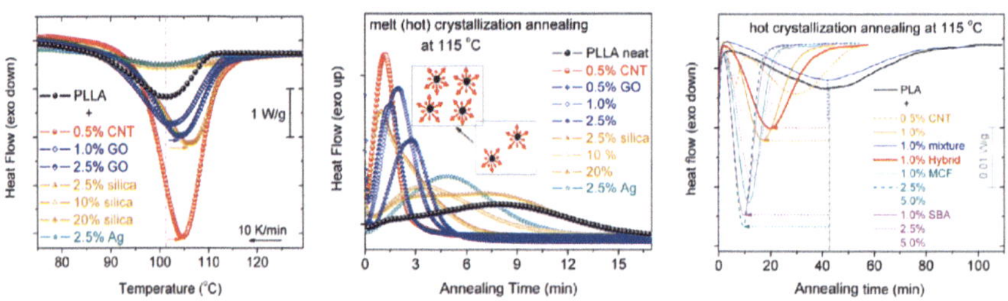

(a) (b) (c)

Fig. 7 DSC curves for **a** non-isothermal and **b** isothermal melt (hot) crystallization of PLLA of M_n~700 kg/mol in bulk and PNC form (descriptions on the plots). **c** Isothermal hot crystallization of PDLLA (4%-D) of M_n~42 kg/mol in bulk and PNC form [57] (NPs used: CNT, mixtures and hybrids of CNT and MMT, and 2 types of nanosilica, MCF and SBA). The latter PDLLA does not crystallize non-isothermally (i.e., during cooling) from the melt in any form, bulky or in PNCs. (a,b Adapted from Refs. [67, 73, 88] and **c** Reproduced with permission from Ref. [57]. Copyright 2021 Elsevier.)

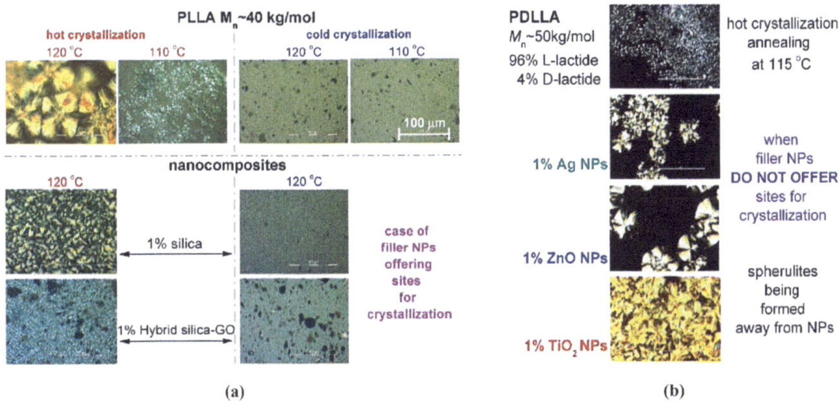

Fig. 8 Results by Polarized Light Microscopy on the effects of various filler nanoparticles on the crystallization of PLA, i.e., upon hot (melt) crystallization (direct filler effects) and cold crystallization. The polymers and fillers are described on the figures. (**a** Partially adapted from Ref. [47] and **b** partially reproduced with permission from Ref. [37] Copyright 2021 Elsevier, respectively.)

completion time. Interestingly, the maximum of CF achieved by melt crystallization is quite similar (~40–44 wt%) for the unfilled matrix and the PNCs, the same is not found for non-isothermal crystallization. Qualitatively similar data with respect to crystallization time, although from the opposite direction, are recorded in the case of isothermal cold crystallization at temperatures slightly above T_g (not shown) [67, 73]. When increasing the size of the filler particles (e.g., by aggregations), there seems to be the tendency for decelerating crystallization, moreover, in the amorphous state this has been accompanied by an increase in the T_g. These effects indicate different semicrystalline morphologies between the PLA matrices and the PNCs.

The isothermal and non-isothermal crystallization processes can be monitored, in terms of semicrystalline morphology, by employing Polarized Light Microcopy (PLM). Figure 8 shows representative images on the two distinct crystal morphologies, corresponding to the two roles of the inclusions in the crystallization process, namely in the cases of NPs (Fig. 8a) offering or (Fig. 8b) not-offering sites for crystallization of PLA. In Fig. 8a additionally, the impact of the crystallization initiating condition, namely hot crystallization from the melt and cold crystallization of a melt-quenched (~nucleated) sample is shown.

Based on the overall findings, we have concluded that in the cases when **large *D*-lactide content exists** and, additionally, when the **interfacial interactions are extensive** in terms of strength and of numbers of NPs-polymer contact points, the **NPs do not act as nucleation agents.** This is the case of silica and other metal oxide particles, especially when the filler surface is quite large [37, 73, 82] or very small particles (e.g., fumed silica) are aggregated [73].

3.1.4 Features of the Amorphous Part (Glass Transition, RAF and Molecular Mobility) in the PNCs

Changes in T_g, Δc_p and RAF

In the case where fillers are not acting as nucleation agents, which occurs for large *D*-lactide content and when the interfacial interactions are quite strong, T_g increases (Fig. 9a). This may happen due to the enhanced constraints in the segmental mobility [37, 73, 82]. In the cases of weaker (or absent) interactions in combination with non-aggregated fillers, there seems to be either an unchanged or accelerated segmental mobility, reflected by a constant or suppressed T_g by a few K (Fig. 9a) [47, 57, 67, 88]. In all the latter cases of finely dispersed fillers, which happens for low NP loadings (0.5–3 wt%), there seems to be a strong nucleating effect. The results of acceleration/retardation of segmental mobility, so far, have been rationalized in terms of an increase/decrease of the free volume hole diffusion as previously proposed for amorphous PNC [109]. Herein, there should to be a driving force for the free volume holes to redistribute closely or away from the NPs, and this should depend on the degree NP-interfacial chain interaction [37, 57] (co-represented by RAF, Fig. 9b). The discussed redistribution could be also the origins of the nucleating role of NPs for PLA (and other polymers) [46, 51, 110], this being again an indirect filler effect.

Regarding the strength of calorimetric glass transition, in all cases and without exception, in the amorphous samples, Δc_p (upon proper normalizations) [25, 26] is suppressed as compared to the unfilled amorphous PLLA (Fig. 9b). While this suppression is in general rationalized in terms of the formation of RAF_{NP}, this is true in the case for fillers not acting as additional nuclei, and the polymer-filler attraction is not severe. On the other hand, in the filler-nuclei case, the estimated amount of RAF seems not to have the meaning of 'bound RAF', but that of the initial polymer structuring/ordering at the early stages of crystallization. This has been reported for PNCs in [67, 88] for the first time and this scenario has been successfully checked for other PLA-based [47, 57] and other polymer-based systems [46, 51, 68, 110, 111].

Obviously, the shape and surface characteristics of NPs, either contributing or not in crystallization, should be related to the amount of RAF_{NP}. Figure 10a, b shows by complementary techniques that the specific surface area and the aspect ratio (AR) are the NP properties which represent well their facilitating role on RAF and crystallization/nucleation.

Changes in the features of α relaxation

The α process exhibits in general similar trends as the calorimetric glass transition (Table 1, Fig. 9), with, however, more pronounced changes in the time scale and strength. The α process may accelerate or decelerate (Fig. 11a) and this depends on the strength of NP-PLA interaction.

The changes in its shape as well as fragility (cooperativity) are moderate or weak, while the fragility index (*m*) tends to increase when the relaxation is accelerated. Models involving alternation in free volume and probably distributions of the free volume holes [109] seem to fit the amorphous bulk-like dynamics [37]. As in calorimetry, the $\Delta\varepsilon$ of α process drops upon the filler addition, and this suppression is evaluated and discussed in terms of RAF (i.e., RAF_{NP}, Fig. 12, for CF =

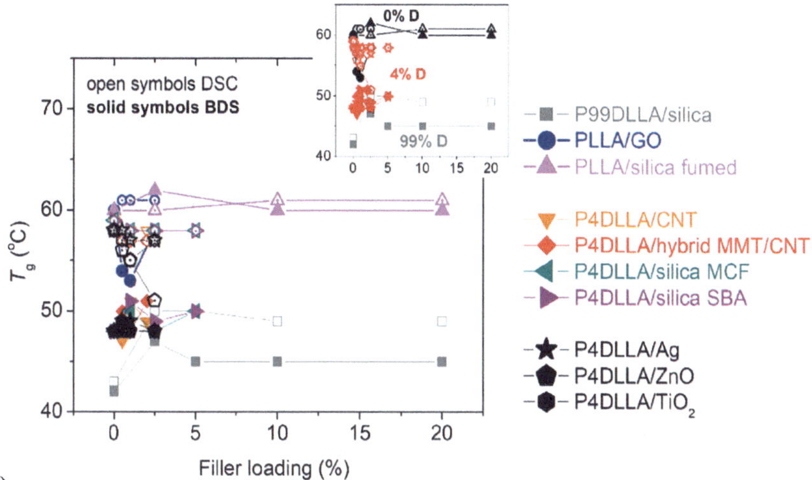

(a)

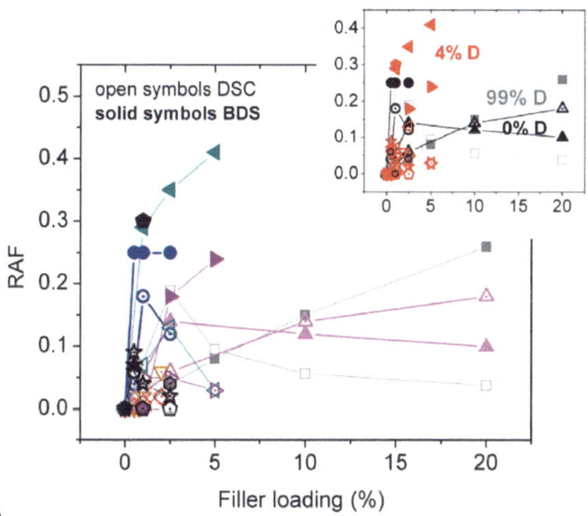

(b)

Fig. 9 Comparison between BDS and DSC in terms of recording **a** T_g and **b** RAF = RAF_{NP} in various amorphous PLA-based nanocomposites against the filler loading, samples being described on the plots. The data have been adapted from previous work from the literature [37, 57, 73, 82, 88]. The insets show the same data categorized with respect to the % *D*-content

0). Qualitatively, the results estimated for the RAF are similar to those obtained by calorimetry and other techniques, e.g., FTIR [29, 47] (Fig. 10). However, the data shown in Figs. 11b, 12 indicate that the most severe effects on the α relaxation arise from the strong force of crystallization.

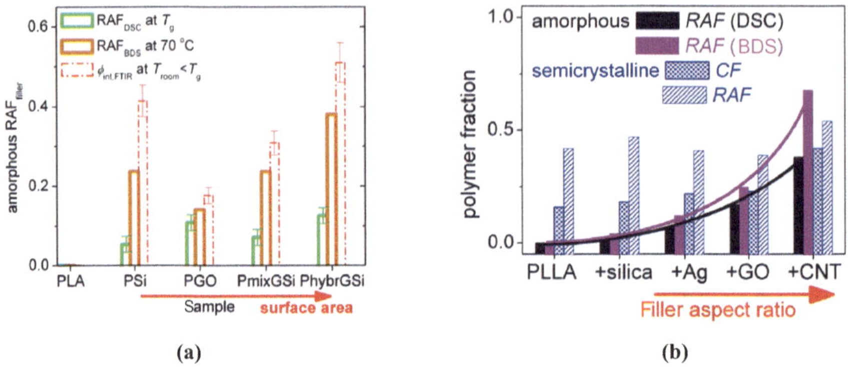

Fig. 10 **a** Interfacial fraction, RAF_{filler} (=RAF_{NP}), in amorphous PLLA filled with 1 wt% silica (PSi), graphene oxide (PGO), mixture and hybrid of the latter two (PmixGSi and PhybrGSi, respectively), as estimated by dielectric relaxation spectroscopy (BDS, DSC and Fourier transform infrared spectroscopy (FTIR). **b** Amorphous/semicrystalline RAF and CF of PLLA and its nanocomposites, filled with 1 wt% silica, Ag, graphene oxide (GO) and carbon nanotubes (CNT). **a**, **b** have been reproduced with permission from Refs. [47, 67], respectively. Copyright 2021 Elsevier.)

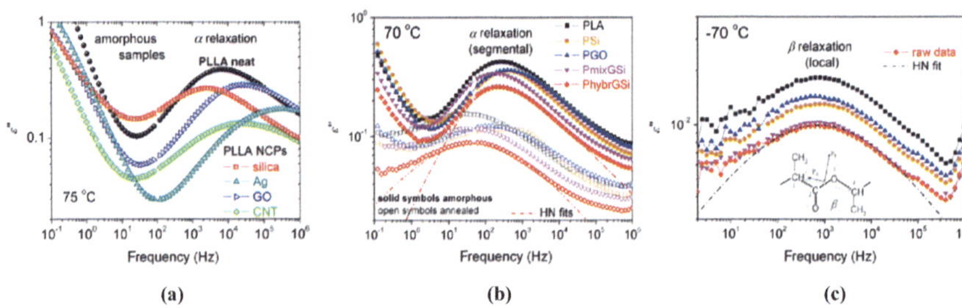

Fig. 11 **a**, **b** The α relaxation at fixed temperatures for two series of samples comparing between neat PLLA matrices and corresponding nanocomposites in the amorphous state and in **b** also in the semicrystalline state upon melt (hot) crystallization annealing. **c** Shows the data for β relaxation for the samples described in (**b**). (**a**–**c** have been reproduced with permission from Refs. (47, 67], respectively. Copyright 2021 Elsevier.)

Changes in the features of local fluctuations

The β process seems to change in time scale (Fig. 11c) only for small M_ns in the PNCs. It does not follow the changes in T_g. (Figure 13a, b). The effects of the fillers are indirect, as in some cases [43, 91] there seems to be a disturbance in its mobility at the early stages of the glass transition or of the cold crystallization, and both processes can change at filler presence (Fig. 13b). The filler-effects on the strength of β are again indirect, as β is slightly suppressed when the crystallinity is enhanced (Fig. 11c and 13b) [43, 47]. There are weak indications that β relaxation weakens in the PNCs due to interfacial interactions, however, clearing this point further is precluded as in many cases the fillers increase ε' to different extents (e.g., by filler-induced internal fields) [88].

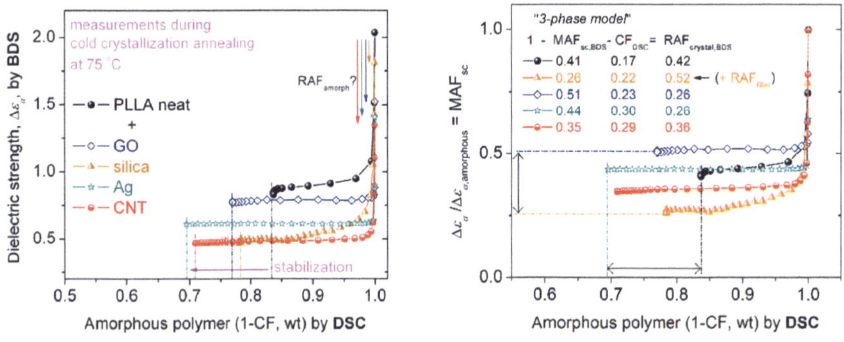

(a) (b)

Fig. 12 **a** Monitoring the amorphous polymer fractions, (AF = MAF + RAF = 1−CF) in PLLA and PLLA nanocomposites via the corresponding evolution (suppression) of the dielectric strength of the α relaxation, $\Delta\varepsilon_\alpha$. (Adapted from the free accessed Supplementary Material of Ref. [67].) In **b**, the data of **a** are shown as normalizations of $\Delta\varepsilon_\alpha$ to the same sample amorphous $\Delta\varepsilon_\alpha$. Included in **b** are the various calculated polymer fractions (MAF, RAF, CF)

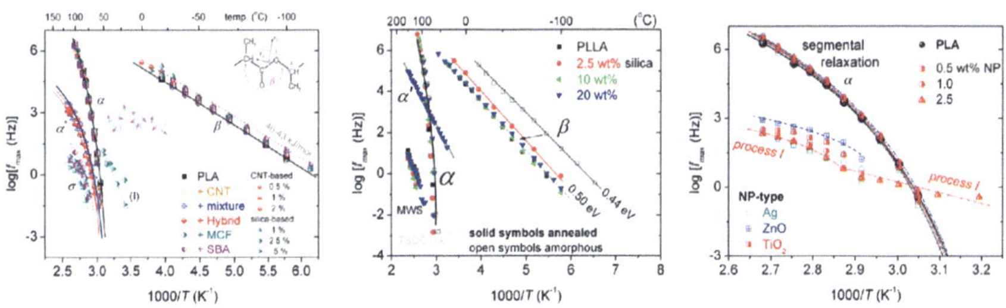

(a) (b) (c)

Fig. 13 Arrhenius plots of various PLA-based samples, described in the plots [37, 57, 73], showing in to the known α and β relaxations the additional processes **a** α^* [57], **b** α' [73] and **c** *process I* [37], which are most probably filler-related. (**a**, **b**, **c** have been reproduced with permission from Refs. [57, 73] and [37], respectively. Copyright 2021 Elsevier.)

Filler-related relaxation processes

Additional processes that are observed within nanocomposites are included in Fig. 13, namely denoted as α', $\alpha*$ process or process *I*, exhibiting a different temperature dependence, i.e., an Arrhenius- or VFT-like dependence. The intensities of the processes enhance with the filler addition and they are present in the case of the expected strong interfacial interactions, i.e., for silicas [37, 73], other metal oxides [37] and clays [47]. These processes are well fitted by the Cole-Cole equation and exhibit relatively wide ranges in their relaxation times. The relaxations do not exhibit noteworthy dependence on crystallization. Therefore, a possible scenario is their origination from interfacial polymer mobility, which is however a quite strong argument. Wu et al. [112] studied PNCs where PLLA chains were grafted on nanosilicas and recorded additional relaxations that were assigned to the interfacial PLLA mobility. More clear data on interfacial polymer relaxation in PNCs have been proved for other polymers, e.g., in rubbers [16, 27, 28, 113].

3.2 Poly(ε-Caprolactone)-Based Systems

We proceed with recordings on PCL-based systems [68]. The chemical structure of PCL is shown in Scheme 2. PCL is crystallizable, thus, qualitatively similar effects to those of PLA (Sect. 3.1) in its dynamics are expected. PCL was filled with low amounts (i.e., 0.5–2.5 wt%) of NPs with varied dimensionality in the nano-metric scale, i.e., silica and Ag NPs (spheres, 3D), CNTs (2D) and GO (nanosheets, 1D).

The filler NPs were found well dispersed in the PCL matrix, with the exception of Ag, where a significant aggregation was observed. The strong crystallization tendency of PCL could not be bypassed by conventional cooling, therefore, we could not separate the individual (direct) NPs' effects. Non-noteworthy effects of NPs on CF (0.47–0.56 wt) [68] were recorded (Fig. 14a). However, T_c was found to significantly increase for the PNCs. The order of increasing T_c (raw data in Fig. 14a and in terms of nucleation efficiency in Fig. 14b) [114] was GO>CNT>Ag>silica. This order is found again similar to the filler aspect ratio increasing and, thus, these results were explained by the model of nucleating action of the NPs (Scheme 1). Surprisingly, as in for PLA [67], scanning electron microscopy provided almost direct proofs that the NPs were embedded within the PCL crystals [68].

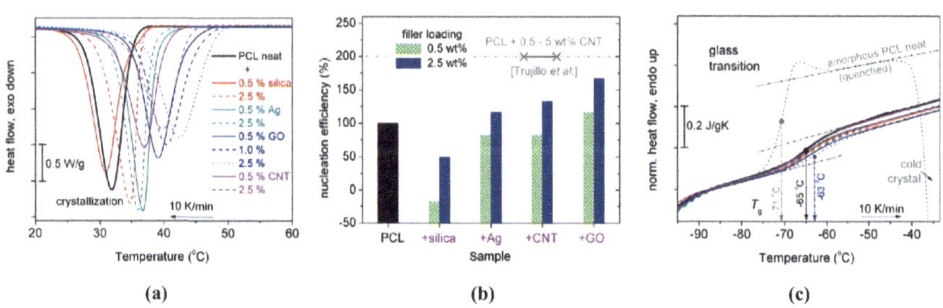

Fig. 14 **a** Crystallization of PCL and PCL-based PNCs during cooling. **b** Comparative column diagram for the nucleation efficiency in PCL-based PNCs. **c** DSC heating curves in the region of the glass transition for PCL-based samples in the semicrystalline state, included for comparison being a heating curve for initially amorphous neat PCL. (Reproduced with permission from Ref. [68]. Copyright 2021 Elsevier.)

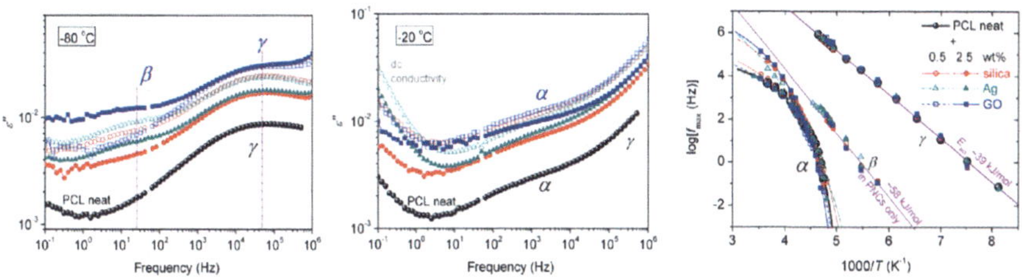

(a) (b) (c)

Fig. 15 **a**, **b** Comparative isothermal spectra of $\varepsilon''(f)$ for the PCL-based systems at the selected temperatures of **a** –80 °C for local dynamics and **b** –20 °C for segmental relaxation. **c** Dielectric map (Arrhenius plots) for neat PCL and PCL-based PNCs. The lines connecting the experimental points are fittings of the Arrhenius and the VFT equations. (Reproduced with permission from Ref. [68]. Copyright 2021 Elsevier.)

As expected, and similar to PLAs, T_g was found to correlate well with the crystalline fraction CF in neat PCL matrix and the PNCs (Fig. 14c). The overall results in combination with each other, and reports in the literature, indicate the effects of NPs on the semicrystalline morphology. BDS (Fig. 15) revealed a single α relaxation in the PNCs, similar to neat PCL, namely with time scale and fragility being unaffected by NP addition. The same was found true for the local dynamics (localized γ crankshaft motions and β relaxations) [115, 116].

3.3 *Poly(N-Alkylene Furanoate)-Based Systems*

The attention is now turned to the modern class of PAF-based PNCs. PAFs were synthesized by a known two-stage polycondensation method, whereas the PNCs were prepared in situ by the same polymerization method. The latter results in an excellent dispersion of the NPs. Moreover, in principle this method results in low M_n values for the neat PAFs [e.g., ~20 kg/mol for PPF ($n = 3$) and PBF ($n = 4$)], moreover and even lower in PNCs (from 19 down to 10 kg/mol) [48, 70, 71].

Neat PPF and PBF in the amorphous state exhibit a glass transition at 54 and 31 °C, respectively. These PAFs exhibit a difficulty to crystallize due to structural limitations, such as the quite rigid chain backbone [62, 69], originating from the existence of the ‘heavy’ furan ring and the relatively small chain lengths (low M_n values). The crystallinity is enhanced, in general, by increasing the n-alkylene sequence length, for example, when changing from $n = 2$ and 3 (PEF, PPF) to $n = 4$ and further to 6 (PBF and PHF, respectively) [62, 69, 71]. These properties depend, as expected, on the thermal-crystallization treatment. For almost all cases of PAFs where they do not crystallize during cooling from the melt, strong cold crystallization takes place during heating. Therefore, for the crystallization of PAFs, a strong supercooling is necessary [72]. This suggests that the main obstacle for the PAF crystallization arises from nucleation disability. A solution to that problem comes, expectedly, from the

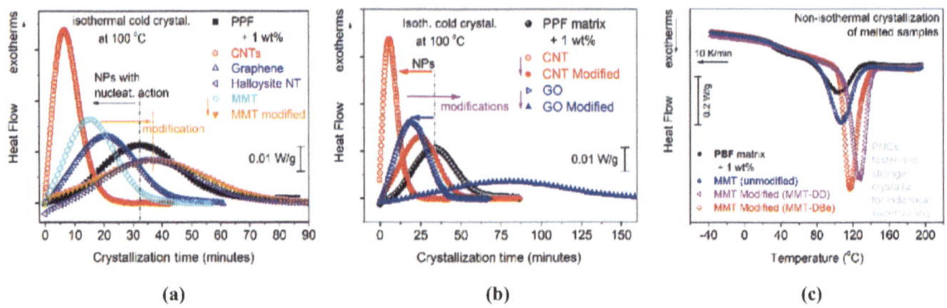

Fig. 16 **a**, **b** Isothermal DSC curves for initially amorphous PPF-based systems during cold crystallization annealing at 100 °C (i.e., >T_g). In general, the unmodified NPs, i.e., CNT, graphene, MMT (Cloisite Na^+) and GO accelerate crystallization, whereas NPs with surface modification, applied for enhancing polymer-NP adhesion, i.e., MMT modified (Cloiste-20A), CNT modified and GO modified show retardations in crystallization. **c** Non-isothermal crystallization of PBF and PBF + 1 wt% MMT of three types, initial and upon two modifications (MMT-DD and MMT-DBe), enhancing crystallization/nucleation. (**a**, **b**, **c** have been adapted from Refs. [48, 51, 117], respectively.)

introduction of NPs. The CF of PPF and PBF can be maximized upon isothermal annealing processes up to ~25–30 wt%, whereas by the NP introduction, CF may increase up to ~40 wt%.

PHF ($n = 6$) crystallizes more easily, furthermore, it cannot be kept amorphous by conventional cooling (10–100 K/min) from the melt [71]. PHF with a CF ~45 wt%, exhibits a T_g of 7 °C, while at the presence of NPs, CF increases, in general, up to ~50 wt%. It is interesting to report, from the methodological point of view, that the CF measured by DSC is systematically lower with a relatively small uncertainty, whereas the CF as estimated by WAXS is slightly larger, accompanied, nevertheless, by a larger uncertainty.

Besides the increase in CF by the NPs, more important is the facilitation of a faster crystallization, either this being developed isothermally (Fig. 16 a, b) or non-isothermally (Fig. 16c). It becomes undoubtable that, as in the previously discussed polyesters, in PAFs the NPs introduce a nucleating action. Again, the surface characteristics of the NPs seem to reflect their nucleating role (Fig. 17a) [48, 117]. To definitely check whether the interfacial interactions play a nucleation promoting role or not in PNCs, as discussed for PLA in the previous and in other polymers (e.g., rubbers [20, 46]), we have selected two of the most effective NPs, CNT and GO, and modified them (Scheme 3a) in order to enhance the NP-polymer adhesion. The latter enhancement was confirmed by DSC (RAF_{NP}) and checked by FTIR [51]. Strikingly, the modification and, subsequently, **the enhanced interfacial interaction resulted in a retardation of crystallization** (Fig. 16b).

Expectedly, depending on the thermal treatment and the amount and distribution of nucleating-acting NPs, semicrystalline morphology (number/size/distribution of crystals) differs in PNCs as compared to the unfilled matrices. This is confirmed also in PAFs by the direct technique PLM, an example being shown in Fig. 17b [48].

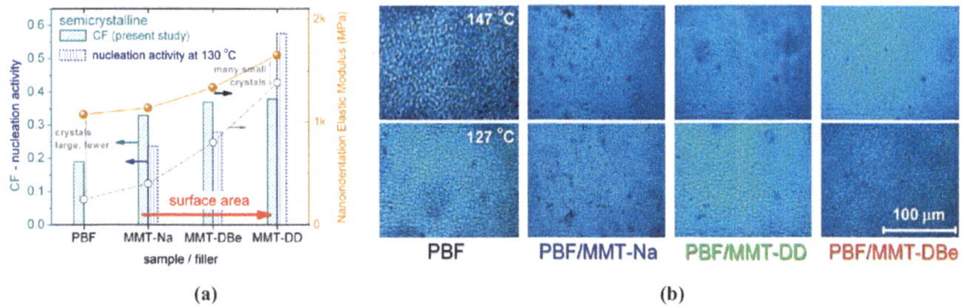

Fig. 17 **a**, left axis shows in the form of column diagrams the nucleating action/activity of MMT NPs of three modifications, leading to increased interlayer distances of MMT, thus, surface area, on PBF, in addition to imposing indirect mechanical reinforcement (right axis). **b** PLM micrographs of PBF-based PNCs upon melt crystallization at (top) 147 °C and (bottom) 127 °C. (Reproduced with permission from Ref. [48]. Copyright 2021 ACS.)

Coming to the molecular mobility, the PAFs-based systems have been investigated by BDS [118–121], thus, corresponding PNCs were studied by the same technique [48, 51, 71, 111, 117]. Selected and representative data are shown in the following, in the form of raw ε'' (f,T) results (Fig. 18) and upon analysis in terms of time scale [$\log f_{\max}$ versus (1000/T)] and $\Delta\varepsilon$ versus (1000/T) (Figs. 19 and 20).

As far as local dynamics is concerned, the β relaxation (α_{HN}~0.3–0.4, β_{HN} = 1) has been proposed to originate from crankshaft motions of the molecular group of PAF related to the chemical link between the ester carbon and the aromatic ring (inset to Fig. 18a) [118]. Moreover, it has been suggested that this process is active in the amorphous polymer regions [117, 122]. Similar to the previous polymers (PLA, PCL), this mobility seems to sense large-scale structure changes of the matrix, as revealed from the disturbance (Figs. 19c and 20) in its time scale and the sharp dielectric strength Increase (please compare $\Delta\varepsilon$ between Fig. 18a and b and in Fig. 20)

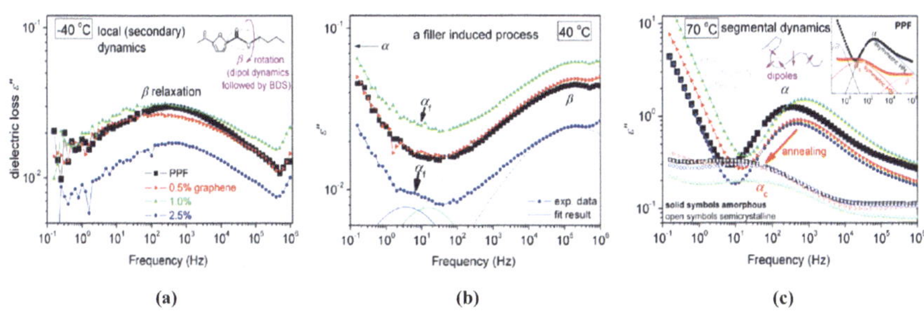

Fig. 18 Comparative $\varepsilon''(f)$ curves for PPF and PPF/graphene PNCs at selected temperatures for the (a, b) local and **c** segmental mobility. In (**c**), the data for amorphous samples (solid symbols) are compared with those for the semicrystalline ones (open symbols). Included are representative results for the performed analysis. (Reproduced with permission from Ref. [111]. Copyright 2021 Elsevier.)

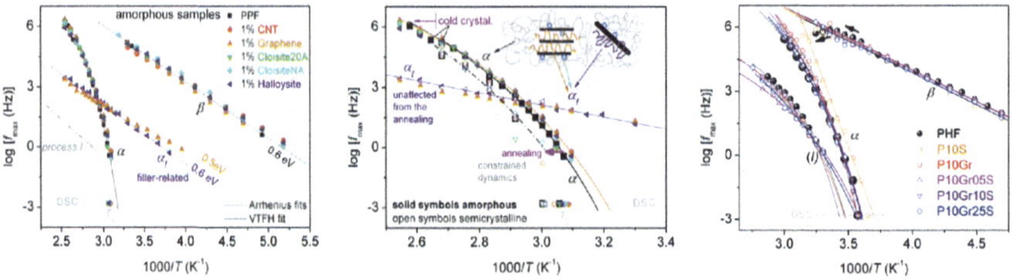

(a) (b) (c)

Fig. 19 Arrhenius plots for (a, b) PPF and PPF-based PNCs and **c** PHF and PHF-based PNCs filled with low amounts of silica (S), graphene (Gr) and mixtures of the latter. The added schemes to **b** demonstrate the origins of recorded dynamics. In (**a**), the E_{act} values have been added for the Arrhenius obeying processes, namely 0.5 and 0.6 eV, or else, 48 and 58 kJ/mol, respectively. (**a**, **b** and **c** have been reproduced with permission from Refs. [71, 117], Copyright 2021 Elsevier and MDPI, respectively.)

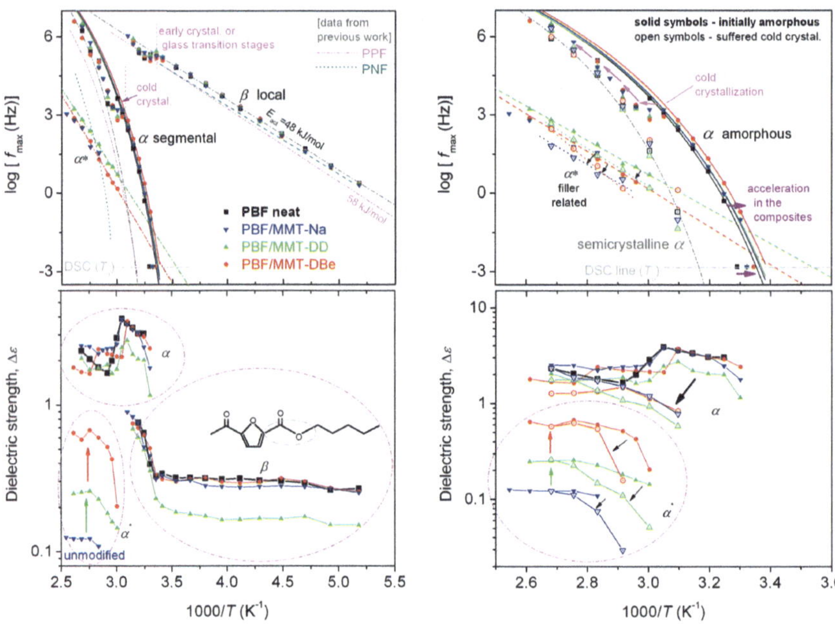

Fig. 20 (Top) Dielectric-calorimetric map (activation diagrams) and (bottom) $\Delta\varepsilon$, for PBF-based systems, in the temperature ranges of (left) both the local and segmental dynamics and (right) focusing on the segmental α and the filler-related α^* relaxations. Results from the recent literature for similar polymer structures, namely poly(propylene 2,5-furan-dicarboxylate) (PPF) [117] and poly(neopentyl 2,5-furanoate) (PNF) [118] have been added for comparison. (Reproduced with permission from Ref. [48]. Copyright 2021 ACS.)

as T approaches T_g or T_{cc} [48, 71]. Considering the proposed origins of the β relaxation process, being related to units of PAF backbone, it was rather expected that β process would exhibit such disturbance, as its time scale is quite close to that of the α process. This seems to be a common characteristic of the polyesters studied here. This dielectric behavior may originate from the fact that the dipolar units that contribute to β process are parts of the chain backbone of the polymers.

Coming to the segmental mobility, the dielectric α relaxation process of PAFs is clearly a strong and asymmetric process in the amorphous state (Fig. 18c), while upon crystallization the same relaxation changes to slower, weaker and symmetric one (Fig. 18c). The direct effects of the NPs on the time scale of α are systematically similar, as α tends to accelerate in all cases of the PNCs (Figs. 18, 19, 21). On the other hand, in PNCs occurs a remarkable suppression of the M_n values and this effect might be the main reason for the recorded acceleration of segmental mobility. Strong support to that is provided by the systematic suppression of the fragility index, m. Again, as in most semicrystalline polymers, the features of α relaxation process depend strongly on the developed crystallites, therefore, in the PNCs the NP effects on segmental dynamics are mainly indirect, since the NPs affect strongly the nucleation process and consequently the morphology of the crystallites. Strikingly, in the case of PAFs with both enhanced nucleation (filler-induced) and large CF, i.e., PHF filled with silicas (sample P10S in Fig. 19c), spatial confinement effects were recorded based the fast and non-cooperative dynamics of the α process [71]. The confinement therein was proposed to arise from the quite high crystallinity degree and the specific crystalline morphology.

Around the temperature/frequency region of α process, namely additional filler induced mobility has been recorded at slowr and faster time scales. We refer to α_f processes in PPF PNCs (Figs. 18c and 19a, b) [111, 117] and α^* process in PBF PNCs (Fig. 20) [48]. The $\Delta\varepsilon$ of these processes increases with the filler loadings, whereas their time scale and strength are changed by crystallization. For both processes, the time scale shows a non-cooperative character (Arrhenius behavior). These facts make o that the disccussed dynamics arises from the modified mobility of PAF chains located at the interfaces with NPs. Another interesting point refers to the PPF-based PNCs: α_f process exhibits both fast (lower temperatures) and slower (higher temperatures) modes as compared to bulk α process (Fig. 19a, b). A similar molecular behavior has been recorded before for other PNCs [31, 37, 110], however, also in neat polymers in bulk [123] and, in the form of supported thin films [124]. The additional processes have been proposed to originate either from confined polymer chains between nanoplatelets [31] or from the interfacial polymer [37, 110]. Therefore, in our opinion and to the best of our knowledge, the origins of the additional processes is a still open issue.

Last but not the least, in Fig. 21, in the PPF-based systems at temperatures above $T_{g,diel}$, the global chain dynamics was recorded (mainly resolved upon the critical analysis, inset to Fig. 21b), via the so-called Normal Mode (NM) relaxation [89, 90]. NM was fitted with relatively narrow (α_{HN} ~0.8–0.9) and symmetric ($\beta_{HN} = 1$) HN term, as expected. For our case of the low M_n values, NM was found to accelerate when M_n decreases in the PNCs (from 20 to 10 kg/mol, Fig. 21c).

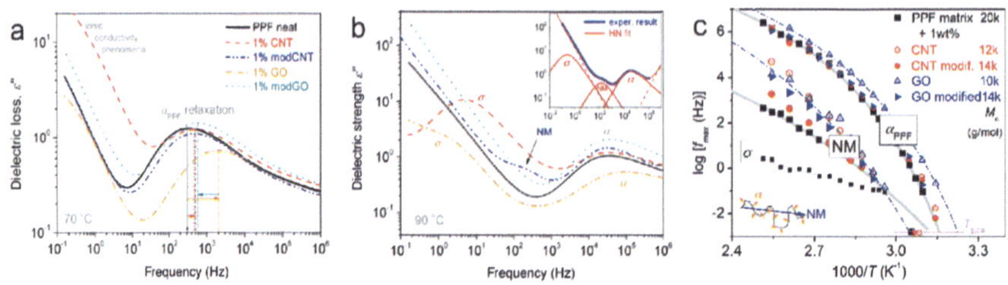

Fig. 21 Comparative $\varepsilon''(f)$ curves by BDS for unfilled PPF and PNCs containing 1 wt% CNTs and GO, initial and surface modified, at temperatures **a** slightly and **b** further above T_g, focusing on segmental (α process) and additional dynamics (NM and σ processes). The inset of **b** shows examples of the analysis by fitting models (HN). (Reproduced with permission from Ref. [51]. Copyright 2021 Elsevier.) **c** Arrhenius plots for the results on α, NM and σ processes for the data from **a**, **b** [Adapted from Ref. [51].)

4 Concluding Remarks

The results presented here and many reports that appeared in the literature suggest that for the PNCs based on semicrystalline polyesters, the crystallinity controls the molecular mobility, mainly in terms of crystalline fraction and crystal morphology. In this context, the NP effects on polymer mobility seem to be indirect, namely via altering crystallization. However, taking into account the overall picture, it is clear that the formation of a polymer fraction around the NPs with a different dynamics than the bulk, e.g., RAF or amorphous chains with modified mobility, affect significantly the crystallization behavior of the chains revealing, thus, the crucial role of the NPs in the dynamics and the morphology of the semicrystalline PNCs.

Below, a categorization of the NP effects on the molecular mobility depending on the extent of interfacial interactions is proposed.

For **weak or absent interfacial interactions**, the crystals grow onto the NPs (additional nucleation sites) and the meaning of RAF_{NP} in the semicrystalline PNCs is not any more that of the 'bound/rigid polymer' [46, 51, 67]. It correlates with the initial stage of crystallization and finally is merged within the crystals. The crystallization becomes faster during hot crystallization annealing and non-isothermal crystallization. During cold crystallization and besides the implemented large supercooling, PNCs exhibit slower crystallization rate, as T_g of the amorphous chains has been increased and, thus, the polymer chain diffusion (at temperatures close to T_g for cold crystallization) is severely reduced. When **interfacial interactions are moderately stronger**, i.e., increased number of polymer-NP contact points and enhanced strength of each individual bond, then the formation of RAF_{NP} (small amount) leads to less crystals (less filler-nucleation sites) than in the previous case via retarded or even absent non-isothermal crystallization. However, the presence of NPs may increase the amorphous free volume, and acceleration of hot isothermal crystallization is observed. The enhanced mobility of the amorphous chains can be explained by models such as that of diffusion of free volume holes (defects) [109, 125] (and references therein). Furthermore, findings on the formation of thicker lamellae by XRD and experiments on PNCs with fillers of different types and/or varying concentrations [37] support this scenario. Finally, in the case of **strong interfacial interactions (either ab initio or upon NP modification)**, the amount of RAF is quite high and the rigidity of interfacial polymer quite high. Then, the isothermal hot crystallization and non-isothermal crystallizations are severely hindered or completely precluded (loss of NPs nucleating action) [51, 57, 73]. In terms of free volume rearrangements, this may suggest that the diffusion of defects toward the interfaces is limited, contrary to the previous situation [16, 20, 26] (and references therein). This is the main case of PNCs studied in the literature via the concepts of RAF and interfacial dynamics.

Coming to **segmental mobility**, in the amorphous state, chain dynamics of the polymer matrix at the presence of NPs may be decelerated (strong interactions) or accelerated. The latter occurs when interfacial interactions are weak or/and M_n is in general low and drops in PNCs (case of PAFs). Within all studied semicrystalline polyesters, single α relaxations are observed. Upon crystallization and due to the simultaneous reduction of MAF, the strength of the α relaxation drops, while the relaxation transforms from a fast/asymmetric/strong (amorphous α process) to a slower/symmetric/weaker process (can be called crystalline α or α_c process). The cooperativity of α process exhibits a moderate suppression. As mentioned above, in the semicrystalline state, the effects of NPs seem to be mostly indirect, via the resulting semicrystalline morphology, namely for high CF and sparsely distributed large or smaller crystals the α process is moderately slower, respectively, as compared to amorphous α relaxation. For extreme cases of quite high crystalline fractions and particular crystal morphology the semicrystalline α process may become faster with a non-cooperative character (case of PHF) [71], suggesting crystal-induced spatial nano-confinement [71].

Regarding **local dynamics**, the overall results provide enough indications that local β relaxation in the PNCs is segmental mobility-dependent in terms of time scale and $\Delta\varepsilon$. These relaxations are most probably activated at the amorphous polymer regions. It seems that these characteristics are common for polyesters, most probably because β relaxation process originates from dipole moments of large molecular groups at the polymer backbone.

Finally, we have recorded in PLA and PAFs, **additional relaxations** (named α', α^* process or process I process) that seem to be filler-related/induced. The time scale of these processes is in similar range than that of α, whereas they exhibit either very weak or no changes by crystallization. For the time being, the adopted scenario for the origins of the additional processes is that they reflect the mobility of polymer chains in the NP-polymer interfacial layer. However, we cannot exclude previously proposed scenarios suggesting additional relaxation arising from the bulk-like polymer, for example, involving special chain conformations (helices) [123].

References

1. Bailey EJ, Winey KI (2020) Prog Polym Sci 105:101242
2. Paul DR, Robeson LM (2008) Polymer 49:3187
3. Kumar SK, Krishnamoorti R (2010) Annu Rev Chem Biomol Eng 1:37
4. Netravali AN, Mittal KL (2016) Interface/interphase in polymer nanocomposites. John Wiley & Sons Inc. and Scrivener Publishing LLC, Hoboken, New Jersey Beverly, Massachusetts
5. Jimenez AM, Zhao D, Misquitta K, Jestin J, Kumar SK (2019). ACS Macro Lett 8:166
6. Senses E, Narayanan S, Faraone A (2019) ACS Macro Lett 8:558
7. Holt AP, Griffin PJ, Bochatova V, Agapov AL, Imel AE, Dadmun MD, Sangoro JR, Sokolov AP (2014) Macromolecules 47:1837
8. Emamy H, Kumar SK, Starr FW (2020) Macromolecules 53:7845
9. Harton SE, Kumar SK, Yang H, Koga T, Hicks K, Lee HK, Mijovic J, Liu M, Vallery RS, Gidley DW (2010) Macromolecules 43:3415
10. Carrillo JMY, Cheng S, Kumar R, Goswami R, Sokolov AP, Sumpter BG, Macromolecules 48:4207
11. Bilchak CR, Jhalaria M, Huang Y, Abbas Z, Midya J, Benedetti FM, Parisi D, Egger W, Dickmann M, Minelli M, Doghieri F, Nikoubashman A, Durning CJ, Vlassopoulos D, Jestin J, Smith ZP, Benicewicz BC, Rubinstein M, Leibler L, Kumar SK (2020) ACS Nano 14:17174
12. Klonos PA, Goncharuk OV, Pakhlov EM, Sternik D, Deryło-Marczewska A, Kyritsis A, Gun'ko VM, Pissis P (2019) Macromolecules 52:2863
13. Scheutjens JMHM, Fleer GJ (1979) J Phys Chem 83:1619
14. Gin P, Jiang N, Taniguchi T, Akgun B, Satija SK, Endoh MK, Koga T (2012) Phys Rev Lett 109:265501
15. Klonos P, Kyritsis A, Pissis P (2015) Polymer 77:10
16. Klonos P, Kulyk K, Borysenko MV, Gun'ko VM, Kyritsis A, Pissis P (2016) Macromolecules 49:9457
17. Cheng S, Holt AP, Wang H, Fan F, Bocharova V, Martin H, Etampawala T, White BT, Saito T, Kang NG, Dadmun MD, Mays JW, Sokolov AP (2016) Phys Rev Lett 116:038302
18. Bailey EJ, Griffin PJ, Composto RJ (2019) Macromolecules 52:2181
19. Bokobza L, Diop AL (2010) eXPRESS Polym Lett 4:355
20. Klonos P, Panagopoulou A, Bokobza L, Kyritsis A, Peoglos V, Pissis P (2010) Polymer 51:5490
21. Lin Y, Liu L, Xu G, Zhang D, Guan A, Wu G (2015) J Phys Chem C 119:12956
22. Klonos P, Dapei G, Sulym IY, Zidropoulos S, Sternik D, Derylo-Marczewska A, Borysenko MV, Gun'ko VM, Kyritsis A, Pissis P (2016) Eur Polym J 74:64
23. Zhou Y, Yavitt BM, Zhou Z, Bocharova V, Salatto D, Endoh MK, Ribbe AE, Sokolov AP, Koga T, Schweizer KS (2020) Macromolecules 53:6984
24. Lin CC, Parrish E, Composto RJ (2016) Macromolecules 49:5755
25. Sargsyan A, Tonoyan A, Davtyan S, Schick C (2007) Eur Polym . 43:3113
26. Wurm A, Ismail M, Kretzschmar B, Pospiech D, Schick C (2010) Macromolecules 43:1480
27. Fragiadakis D, Pissis P, Bokobza L (2005) Polymer 46:6001
28. Fragiadakis D, Bokobza L, Pissis P (2011) Polymer 52:3175
29. Füllbrandt M, Purohit PJ, Schönhals A (2013) Macromolecules 46:4626
30. Kremer F, Schönhals A (2003) Broadband dielectric spectroscopy. Springer, New York
31. Elmahdy MM, Chrissopoulou K, Afratis A, Floudas G, Anastasiadis SH (2006) Macromolecules 39:5170
32. Holt AP, Roland CM (2018) Soft Matter 14:8604
33. Bailey EJ, Tyagi M, Winey KI (2020) J Polym Sci 58:2906
34. Jimenez AM, Krauskopf AA, Pérez-Camargo RA, Zhao D, Pribyl J, Jestin J, Benicewicz BC, Müller AJ, Kumar SK (2019) Macromolecules 52:9186
35. Khan J, Harton SE, Akcora P, Benicewicz B, Kumar SK (2009) Macromolecules 42:5741

36. Shi G, Guan Y, Liu G, Müller AJ, Wang D (2019) Macromolecules 52:6904
37. Pusnik Cresnar K, Fras Zemljič L, Papadopoulos L, Terzopoulou Z, Zamboulis A, Klonos PA, Bikiaris DN, Kyritsis A, Pissis P (2021) Mat Tod Commun 27:102192
38. Wen X, Zhao W, Su Y, Wang D (2019) Polymer Crystallization 2:e10066
39. Ning N, Fu S, Zhang W, Chen F, Wang K, Deng H, Zhang Q, Fu Q (2012) Prog PolymSci 37:1425
40. Suzuki H, Grebowicz J, Wunderlich B (1985) British Polymer J 17:1
41. Ezquerra T, Nogales A (2020) Crystallization as studied by broadband dielectric spectroscopy. Springer Nature, Switzerland
42. Lund R, Alegría A, Goitandia L, Colmenero J, Gonzalez MA, Lindner P (2008) Macromolecules 41:1364
43. Brás AR, Viciosa MT, Wang Y, Dionisio M, Mano JF (2006) Macromolecules 39:6513
44. Dobbertin J, Hensel A, Schick C (1996) J Therm Anal Calorim 47:1027
45. Martínez-Tong DE, Vanroy B, Wübbenhorst M, Nogales A, Napolitano S (2014) Macromolecules 47:2354
46. Klonos P, Sulym IY, Sternik D, Konstantinou P, Goncharuk OV, Deryło-Marczewska A, Gun'ko VM, Kyritsis A, Pissis P (2018) Polymer 139:130
47. Terzopoulou Z, Klonos PA, Kyritsis A, Tziolas A, Avgeropoulos A, Papageorgiou GZ, Bikiaris DN (2019) Polymer 166:1
48. Klonos PA, Papadopoulos L, Terzopoulou Z, Papageorgiou GZ, Kyritsis A, Bikiaris DN (2020) J Phys Chem B 124:7306
49. Schick C, Androsch R (2018) Polymer Crystallization 1:e10036
50. Leng J, Kang N, Wang DY, Wurm A, Schick C, Schönhals A (2017) Polymer 103:257
51. Klonos PA, Papadopoulos L, Papageorgiou GZ, Kyritsis A, Pissis P, Bikiaris DN (2020) J Phys Chem C 124:10220
52. Savage RC, Mullin N, Hobbs JK (2015) Macromolecules 48:6160
53. Coleman JN, Cadek M, Ryan KP, Fonseca A, Nagy JB, Blau WJ, Ferreira MS (2006) Polymer 47:8556
54. Hedge M, Samulski ET, Rubinstein M, Dingemans TJ (2015) Compos Sci Technol 110:176
55. Guan Y, Liu G, Ding G, Yang T, Müller AJ, Wang D (2015) Macromolecules 48:2526
56. Klonos PA, Peoglos V, Bikiaris DN, Kyritsis A (2020) J Phys Chem C 123:5469
57. Papadopoulos L, Klonos PA, Tezopoulou Z, Psochia E, Sanusi OM, Aït Hocine N, Benelfellah A, Giliopoulos D, Triantafyllidis K, Kyritsis A, Bikiaris DN (2021) Polymer 217:123457
58. Saeidlou S, Huneault MA, Li H, Park CB (2012) Prog Polym Sci 37:1657
59. Chronaki K, Korres D, Papaspirides CD, Vouyiouka S (2020) Polym Degrad Stabil 179:109283
60. Kontou E, Niaounakis M, Georgiopoulos P (2011) J Appl Polym Sci 122:1519
61. Gumede TP, Luyt AS, Müller AJ (2018) eXPRESS Polym Lett 12:505
62. Papageorgiou GZ, Papageorgiou DG, Terzopoulou Z, Bikiaris DN (2016) Eur Polym J 83:202
63. Sousa AF, Vilela C, Fonseca AC, Matos M, Freire CSR, Gruter GJM, Coelho JFJ, Silvestre AJD (2015) Polym Chem 6:5961
64. Terzopoulou Z, Papadopoulos L, Zamboulis A, Papageorgiou DG, Papageorgiou GZ, Bikiaris DN (2020) Polymers 12:1209
65. Sanusi OM, Benelfellah A, Bikiaris DN, Ait Hocine N (2021) 32:444
66. Armentano I, Bitinis N, Fortunati E, Mattioli S, Rescignano N, Verdejo R, Lopez-Manchado MA, Kenny JM (2013) Prog Polym Sci 38:1720
67. Klonos P, Terzopoulou Z, Koutsoumpis S, Zidropoulos S, Kripotou S, Papageorgiou GZ, Bikiaris DN, Kyritsis A, Pissis P (2016) Eur Polym J 82:16
68. Koutsoumpis S, Poulakis A, Klonos P, Kripotou S, Tsanaktsis V, Bikiaris DN, Kyritsis A, Pissis P (2018) Thermochimica Acta 666:229
69. Papageorgiou GZ, Papageorgiou DG, Tsanaktsis V, Bikiaris DN (2015) Polymer 62:28
70. Papadopoulos L, Terzopoulou Z, Vlachopoulos A, Klonos PA, Kyritsis A, Tzetzis D, Papageorgiou GZ, Bikiaris D (2020) Appl Clay Sci 190:105588

71. Sanusi OM, Papadopoulos L, Klonos PA, Terzopoulou Z, Aït Hocine N, Benelfellah A, Papageorgiou GZ, Kyritsis A, Bikiaris DN (2020) Polymers 12:1239
72. Schick, C (2009) Anal Bioanal Chem 395:1589
73. Klonos P, Pissis P (2017) Polymer 112:228
74. Xu H, Cebe P (2004) Macromolecules 37:2797
75. Havriliak S, Negami SA (1967) Polymer 8:161
76. Arrhenius SAZ (1889) Physik Chem 4:96
77. Tammann G, Hesse W (1926) Z Anorg Allg Chem 156:245
78. Boehmer R, Ngai K, Angell CA, Plazek DJ (1993) J Chem Phys 99:4201
79. Wang Y, Liu C, Shen C (2020) Polym Crystal https://doi.org/10.1002/pcr2.10171
80. Androsch R, Di Lorenzo ML, Schick C (2016) Eur Polym J 75:474
81. Mijovic J, Sy JW (2002) Macromolecules 35:6370
82. Neratzaki M, Prokopiou L, Bikiaris DN, Patsiaoura D, Chrissafis K, Klonos P, Kyritis A, Pissis P (2018) Thermochimica Acta 669:16
83. Chryssafi I, Pavlidou E, Christodoulou E, Vourlias G, Klonos PA, Kyritsis A, Bikiaris DN (2021) Thermochimica Acta 698:178883
84. Karava V, Siamidi A, Vlachou M, Christodoulou E, Zamboulis A, Bikiaris DN, Kyritsis A, Klonos PA (2021) Soft Matter 17:2439
85. Kortaberria G, Marieta C, Jimeno A, Arruti P, Mondragon I (2006) J Microscopy 224:277
86. Righetti MC, Gazzano M, Delpouve N, Saiter A (2017) Polymer 125:241
87. Delpouve N, Saiter A, Dargent E (2011) Eur Polym J 47:2414
88. Klonos P, Kritpotou S, Kyritsis A, Papageorgiou GZ, Bikiaris DN, Gournis D, Pissis P (2015) Thermochimica Acta 617:44
89. Ren J, Urakawa O, Adachi K (2003) Macromolecules 36:5180
90. Mierzwa M, Floudas G, Dorgan J, Knauss D, Wegner J (2002) J Non-Cryst Solids 307–310:296
91. Leng J, Purohit PJ, Kang N, Wang DY, Falkenhagen J, Emmerling F, Thünemann AF, Schönhals A (2015) Eur Polym J 68:338
92. Leng J, Kang N, Wang DY, Falkenhagen J, Thünemann AF, Schönhals A (2017) Macrom Chem Phys 218:1700232
93. Laredo E, Newman D, Pezzoli R, Müller AJ, Bello A (2016) J Polym Sci B Polym Phys 54:680
94. Bras AR, Malik P, Dionisio M, Mano JF (2008) Macromolecules 41:6419
95. Androsch R, Zhang R, Schick C (2019) Polymer 176:227
96. Androsch R, Naeem Iqbal HM, Schick C (2015) Polymer 81:151
97. Androsch R, Di Lorenzo ML, Schick C (2017) Eur Polym J 96:361
98. Androsch R, Zhuravlev E, Schick C (2014) Polymer 55:4932
99. Yin Y, Liu G, Song Y, Zhang X, de Vos S, Wang R, Joziasse CAP, Wang D (2016) Eur Polym J 82:46
100. Righetti MC, Gazzano M, Di Lorenzo ML, Androsch R (2015) Eur Polym J 70:215
101. Wang S, Yuan S, Wang K, Chen W, Yamada K, Barkley D, Koga T, Hong YI, Miyoshi T (2019) Macromolecules 52:4739
102. Song P, Sang L, Zheng L, Wang C, Liud K, Wei Z (2017) RSC Adv 7:27150
103. Wang Y, Funari SS, Mano JF (2006) Macromol Chem Phys 207:1262
104. Yu L, Cebe P (2009) J Polym Sci B Polym Phys 47:2520
105. Righetti MC, Mele E (2019) Thermochimica Acta 672:157
106. Kripotou S, Sovatzoglou S, Pandis S, Kulicek J, Micusik M, Omastova M, Kyritsis A, Konsta A, Pissis P (2016) Phase Transitions 89:717
107. Soccio M, Nogales A, Lotti N, Munari A, Ezquerra TA (2007) Phys Rev Lett 98:037801
108. Wurm A, Soliman R, Goossens JGP, Bras W, Schick C (2005) J Non-Cryst Solids 351:2773
109. Boucher VM, Cangialosi D, Alegría A, Colmenero J, Pastoriza-Santos I, Liz-Marzan LM (2011) Soft Matter 7:3607
110. Koutsoumpis S, Klonos P, Raftopoulos KN, Papadakis CM, Bikiaris D, Pissis P (2018) Polymer 153:548

111. Papadopoulos L, Klonos PA, Tzetzis D, Papageorgiou, GZ, Kyritsis A, Bikiaris DN (2020) Polymer 189:122172
112. Wu F, Zhang S, Chen Z, Zhang B, Yang W, Liu Z, Yang M (2016) Polymer 90:264
113. Klonos PA, Nosach LV, Voronin EF, Pakhlov EM, Kyritsis A, Pissis P (2019) J Phys Chem C 123:28427
114. Trujillo M, Arnal ML, Müller AJ, Mujica M, de Navarro CU, Ruelle B, Dubois P (2012) Polymer 53:832
115. Grimau M, Laredo E, Pérez Y MC, Bello A (2001) J Chem Phys 114:6417
116. Sabater i Serra R, Escobar Ivirico JL, Meseguer Duenas JM, Andrio Balado A, Gomez Ribelles JL, Salmeron Sanchez M (2007) Eur Phys J E 22:293
117. Klonos PA, Papadopoulos L, Tzetzis D, Kyritsis A, Papageorgiou GZ, Bikiaris DN (2019) Mat Tod Commun 20:100582
118. Genovese L, Soccio M, Lotti N, Munari A, Szymczyk A, Paszkiewicz S, Linares A, Nogales A, Ezquerra TA (2018) Phys Chem Chem Phys 20:15696
119. Soccio M, Martinez-Tong DE, Alegria A, Munari A, Lotti N (2017) Polymer 128:24
120. Poulopoulou N, Pipertzis A, Kasmi N, Bikiaris DN, Papageorgiou DG, Floudas G, Papageorgiou GZ (2019) Polymer 174:187
121. Soccio M, Martínez-Tong DE, Guidotti G, Robles-Hernández B, Munari A, Lotti N, Alegria A (2020) Polymers 12:1355
122. Nogales A, Denchev Z, Ezquerra TA (2000) Macromolecules 33:9367
123. Lupaşcu V, Picken SJ, Wübbenhorst M (2006) Macromolecules 39:5152
124. Madkour S, Szymoniak P, Heidari M, Von Klitzing R, Schönhals A (2017) ACS Appl Mater Interfaces 9:7535
125. Debot A, White RP, Lipson JG, Napolitano S (2019) ACS Macro Lett 8:41

Dynamics in Polymer Nanocomposites—From Conventional to Self-suspended Hybrid Systems

Emmanuel U. Mapesa, Sara T. Hamilton, Dayton P. Street, Nelly M. Cantillo, Thomas A. Zawodzinski, S. Michael Kilbey, Ah-Hyung Alissa Park, and Joshua Sangoro

Abstract The functionalization of nanoparticles has been broached as a possible means to ensuring their dispersion in polymer nanocomposites. We seek to draw a dynamic distinction between composites with "bare" nanoparticles and those with polymer-grafted nanoparticles. For clarity, the former and latter systems are referred to simply as polymer nanocomposites (PNCs) and polymer-grafted nanocomposites (PGNCs), respectively. Further, we explore dynamics in self-suspended nanoparticle organic hybrid materials (NOHMs) which retain liquid-like character.

E. U. Mapesa · N. M. Cantillo · T. A. Zawodzinski · S. M. Kilbey · A.-H. A. Park · J. Sangoro (✉)
Department of Chemical and Biomolecular Engineering, University of Tennessee, Knoxville, TN 37996, USA
e-mail: jsangoro@utk.edu

E. U. Mapesa
e-mail: euomapesa@gmail.com

N. M. Cantillo
e-mail: ncantill@vols.utk.edu

T. A. Zawodzinski
e-mail: tzawodzi@utk.edu

S. M. Kilbey
e-mail: mkilbey@utk.edu

A.-H. A. Park
e-mail: ap2622@columbia.edu

S. T. Hamilton
Department of Earth and Environmental Engineering, Columbia University, New York City, NY 10027, USA
e-mail: sth2128@columbia.edu

D. P. Street
Department of Chemistry, Columbia University, New York City, NY 10027, USA
e-mail: dstreet4@vols.utk.edu

S. M. Kilbey
Department of Chemistry, University of Tennessee, Knoxville, TN 37996, USA

A. Schönhals and P. Szymoniak (eds.), *Dynamics of Composite Materials*, Advances in Dielectrics, https://doi.org/10.1007/978-3-030-89723-9_5

Keywords Polymers · Nanocomposites · Nanoparticles · Dynamics · NOHMs · Dielectric spectroscopy

Abbreviations

BDS	Broadband dielectric spectroscopy
BPP	Bloembergen–Purcell–Pound model
DLS	Dynamic light scattering
DSC	Differential scanning calorimetry
GPC	Gel permeation chromatography
HN	Havriliak–Negami
MEA	Monoethanol amine
NMR	Nuclear magnetic resonance spectroscopy
NOHMs	Nanoparticleorganic hybrid materials
NOHM-I-PEI	PEI ionically bonded to the core
NPs	Nanoparticles
PDI	Polydispersity index
PEI	Polyethylenimine
PGNCs	Polymer-grafted nanocomposites
PMMA	Poly(methyl methacrylate)
PNCs	Polymer nanocomposites
PVAc	Polyvinyl acetate
SAXS	Small-angle X-ray scattering
THF	Tetrahydrofuran
VFT	Vogel–Fulcher–Tammann
WAXS	Wide-angle X-ray scattering

1 Introduction

1.1 Conventional Nanocomposites

The addition of very small amounts of nanoparticles into a polymer matrix is known to drastically alter the macroscopic optical, electrical, or mechanical properties, hence providing a route to synthesizing hybrid materials with desired superior attributes. Properties such as refractive index, tensile strength, and hardness can be tuned by the superposition of hard and soft materials into one composite [1–14]. While most applications of nanocomposites hinge on the increased mechanical strength in the hybrid materials, there are other uses in electrochemical energy storage, optics, photovoltaics, and electronics. Necessarily, one focus of many studies is centered around understanding the molecular origin of the drastic changes that occur in a polymer matrix upon introduction of nanoparticles. A key finding of this effort is the determination of the fact that interactions at the polymer/nanoparticle interface play an important role in influencing the macroscopic physical properties of the hybrid material [2, 13–35]. It is now evident that the microscopic details of the interfacial region—and ultimately the macroscopic properties of the nanocomposite—can be altered by tuning certain aspects such as rigidity [24, 31] and chain length [23, 25, 36] of the matrix, the nature of polymer/nanoparticle interactions [31, 37], and the shape [35] and size [22, 25, 38] of the nanoparticles. Although it has been suggested by some studies that dynamics within the interfacial region are completely frozen [36, 39–41], quite a number of other investigations have demonstrated that the segmental dynamics only get relatively slower than the corresponding relaxation in the bulk host matrix [31, 33, 34, 42]. For instance, in a study of poly(vinyl acetate) (PVAc)/silica nanocomposites, Füllbrandt et al. observed a slower relaxation, in addition to the structural process corresponding to the glass transition in bulk PVAc [34]. This slow process is attributed to segments in the interfacial zone that are slowed down due to favorable interactions with the surfaces of the nanoparticles. Similar findings have been made for other systems [30, 43].

We will differentiate between polymer nanocomposites having "bare" nanoparticles from those bearing functionalized (i.e., polymer-grafted) nanoparticles. For clarity, we refer to these systems, respectively, as polymer nanocomposites (PNCs) and polymer-grafted nanocomposites (PGNCs). The former systems have received the most attention in investigations of their dynamics than the latter [4, 12]. In order to avoid aggregation, surface modification of the nanoparticles is necessary; grafted chains, for instance, mitigate against the strong van der Waals forces that drive aggregation. In symmetric PGNCs where the matrix and the graft have the same chemical identity, the tethered chains convey a steric barrier thereby hindering the aggregation of the nanoparticles [44–47]. Accompanying these alterations that show up in terms of the organization of the nanoparticles, arising from changes to the range and nature of polymer/nanoparticle interactions, are changes in dynamics that demand in-depth understanding.

Using silica nanoparticles in poly(methyl methacrylate) (PMMA)-based PNCs and PGNCs, we probe molecular and interfacial dynamics, and draw a distinction between the two systems. While syndiotactic PMMA is not the go-to polymer for dielectric studies because of the prominent β-relaxation that largely overshadows the α-process, the fact that PMMA-based nanocomposites find wide industrial application means that further dynamic studies of the system are inevitable. Detailed analysis of the shape of the structural relaxation reveals that while slower dynamic modes are present in both PNCs and PGNCs because of an interphase comprising segments in the vicinity of the nanoparticles, faster modes too arise—in dependence on molecular weight and loading rate of the nanoparticles. A consistent and unifying view accounting for interactions and nanoscale organization in the interfacial zones is presented to explain both the slower and faster modes present in the investigated systems.

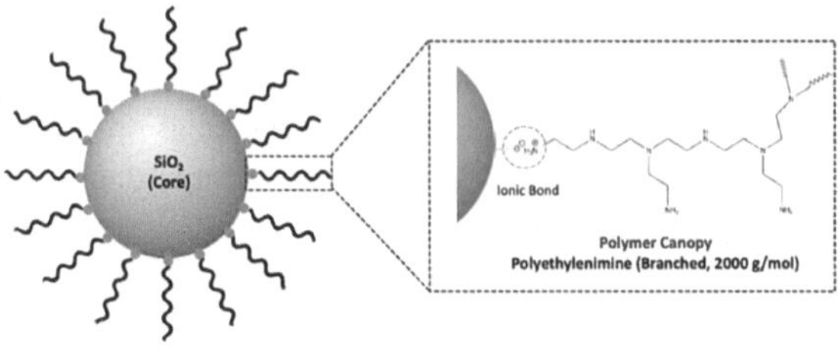

Fig. 1 A schematic representation of the structure of the hybrid material, NOHM-I-PEI. Core: silica nanoparticles; canopy: branched polyethylenimine with a molecular weight 2 of kg/mol

1.2 Self-suspended Nanocomposites

Self-suspended nanoparticle organic hybrid materials (NOHMs)—composites in which each particle carries its own suspending liquid of grafted organic species—are an emerging class of materials that hold great promise for a wide variety of applications (e.g., as conductive lubricants [48–52], electrolytes, magnetic fluids [53, 54], thermal control materials [55–61], novel reaction solvents [62–67], and nanocomposite materials [68, 69]) because of bearing useful properties such as enhanced thermal stability, negligible vapor pressure, and appreciable solvating ability [70–73]. NOHMs are made by either covalently or ionically grafting organic polymers or oligomers onto surface-modified inorganic nanoparticles. Because some NOHMs retain liquid-like properties at room temperature, they can be synthesized to combine the high mechanical modulus of the core nanoparticles with desired characteristics of liquids such as photoluminescence. One of the specific areas of application for which NOHMs are being investigated is carbon dioxide capture [74–80]. The development of efficient CO_2-capture techniques is now an urgent global issue given the fact that the atmospheric concentration of CO_2 is estimated to be growing at a rate of about 2 ppm per annum. To date, the most advanced methods for CO_2-capture are hinged on amine-based solvents like aqueous monoethanol amine (MEA) [81–84]. While MEA has a high CO_2-capture capacity and fast kinetics, its use demands an energy-intensive solvent regeneration process due to its high water content. In addition, MEA is highly volatile making its corrosive fume a concern when considering large-scale process design and operation [85, 86]. Consequently, novel approaches are being sought and liquid-like NOHMs stand a good chance of being the materials of choice for CO_2-capture in the near future.

The hybrid nature of NOHMs (see Fig. 1) has the implication that the structural parameter space that can be tweaked to obtain desirable materials is wide because it encompasses permutations of chemistry, composition, and structure details. One may, for example, tune the composition and molecular weight of the canopy, length, and chemistry of the corona, size, and shape of the core, among other properties. Therefore, the success of current engineering efforts to use NOHMs for specific applications depends on the availability of in-depth understanding of the role played by the hybrids' constituent components in determining overall properties. Among the major properties that need concise characterization are the molecular and interfacial dynamics since they dictate the temperature and size ranges in which desired innovations can be realized. Using nuclear magnetic resonance (NMR) spectroscopy, Jespersen et al. sought to probe the dynamics of the canopy in hybrids comprising silica nanoparticles (modified by an alkylsilane monolayer with sulfonic acid functionality) and a canopy of amine-terminated block (ethylene oxide/propylene oxide) copolymers [70]. Their work shows that the diffusion of the canopy is slowed down when compared to that of the neat polymer, but not to the degree predicted by theory for the diffusion of hard spherical nanoparticles. In their quest to understand the role played by constituent components, Bourlinos and co-workers studied ionically modified silica nanoparticles with large sulfonate or isostearate counterangions [87]. They showed that the glass transition in these systems is governed by the large counteranions while spatial correlation between the nanoparticles determines flow properties. In the present work, the dynamics in neat polyethylenimine (PEI) (used for the canopy) are compared to those of an NOHM material, comprising PEI oligomers ionically bonded to silica nanoparticles. The interplay between the structure and the local as well as collective (primary and secondary) dynamics in the two systems is probed, thereby availing fundamental understanding that is imperative for the advancement of this fast-emerging field of NOHMs.

2 Experimental Details

Most of the synthesis and experimental details in Sects. 2.1 and 2.2 are reproduced with permission from Ref. [95] Copyright (2020) American Chemical Society, and Ref. [123] Copyright (2021) American Chemical Society.

2.1 *Materials Synthesis*

2.1.1 Materials and Reagents

Methyl methacrylate (MMA, 99 %, Aldrich) was passed through basic alumina for purposes of removing the inhibitor. Azobis (isobutyronitrile) (AIBN, 98 %, Aldrich) was recrystallized from methanol and dried overnight in vacuum. The following materials were used as received: poly(methyl methacrylate) (PMMA) with reported average molecular weights of 15 and 120 kg/mol (Aldrich), (3-aminopropyl)dimethyethoxysilane (Gelest), 4-cyano-4- (phenylcarbonothioylthio) pentanoic acid (Aldrich), anhydrous tetrahydrofuran (THF, 99.9 %, Fisher), anhydrous benzene (99.8 %, Aldrich), and hexanes (95 %, Fisher). Unfunctionalized (also referred to in this work as "bare") silica nanoparticles (spherical, diameter 14 ± 4 nm) (received as a gift from Nissan Chemical Inc.) were isolated, collected, and dried prior to use as detailed in Ref. [88].

2.1.2 Synthesis of Amine-functionalized Silica Nanoparticles

Silica NPs were functionalized with (3-aminopropyl)dimethylethoxy silane as described in Ref. [89]. To accomplish this, 16 mL of Si NPs dispersed in MEK were added to a 100 mL round bottom flask containing 50 mL of THF and 0.5 mL (3-aminopropyl)dimethylethoxy silane. The mixture was then heated to reflux at 75 °C for 24 h. The reaction was then quenched by precipitating the amine-functionalized NPs into 400 mL of hexanes. The suspended amine-functionalized NPs were transferred to a centrifuge vial and centrifuged for 10 min at 4000 rpm. After decanting the supernatant, the amine-modified NPs amassed at the bottom of the tube were re-dispersed by adding THF (~20 mL). To ensure unattached silanes were removed, this cycle of precipitation–centrifugation–(re-)dispersion was completed at least three times [89, 90]. After purification, the amino-modified nanoparticles were collected and dispersed in THF until further use.

2.1.3 End-Functionalized PMMA Chains *via* RAFT Polymerization

Details of the protocols used can be found in the work of Li et al. [91]. PMMA graft chains containing a terminal mercaptothiazoline end group were polymerized by adding 9.0 g (89.9 mmol) of MMA into a 50 mL round bottom flask containing 66.1 mg (174.0 mmol) of MCPDB, 5.4 mg (32.9 mmol) of AIBN, and 20 mL of benzene. After the reagents were added, the mixture was sparged with argon for 10 min to remove oxygen. Next, the flask containing the reaction mixture was placed in an oil bath preheated to 65 °C for 18 h. Then, the polymerization was quenched by immersing the flask in liquid nitrogen. After thawing, the crude product mixture was precipitated into cooled hexanes. The precipitated polymer was collected, dissolved in THF, and then re-precipitated into hexanes again to remove residual monomer or reagents. Finally, the polymer was collected via filtration and dried *in vacuo* overnight at 40 °C. After drying, the PMMA was characterized via gel permeation chromatography (GPC). The molecular weight of PMMA chains was determined to

Table 1 Characteristics of PMMA matrices and PMMA-grafted nanoparticles (PMMA–g–NPs) prior to the synthesis of polymer nanocomposites

Sample details				
Sample name	M_n (g/mol)	PDI	T_g (°C)	σ (chains/nm^2)
PMMA matrix 15 kg/mol	13,880	1.77	88	N/A
PMMA matrix 120 kg/mol	86,260	1.55	103	N/A
PMMA graft 30 kg/mol	26,760	1.09	N/A	0.03

be 26,760 g/mol with a polydispersity index (PDI) of 1.09. Table 1 gives a summary of the characteristics of the samples prepared for this study.

2.1.4 Synthesis of PMMA-grafted Nanoparticles

As explained in Ref. [91], the mercaptothiazoline end group is an excellent leaving group that facilitates linking with primary amines, rather than aminolysis of thioesters. In addition, MCPDB is well-suited for a controlled polymerization of methacrylates. Therefore, MCPDB was used to graft PMMA chains to amine-functionalized nanoparticle surfaces, resulting in PMMA-grafted nanoparticles. The experimental protocol involved placing 0.3 g of amine-functionalized Si NPs in a 50 mL round bottom flask containing 0.52 g (0.019 mmol) of mercaptothiazoline-terminated PMMA (30 kg/mol) and 20 mL of THF. The flask containing this reaction mixture was placed in an oil bath preheated to 75 °C and refluxed for 48 h. Later, the reaction mixture was precipitated in 400 mL of cooled hexanes and the suspended polymer/polymer-grafted nanoparticle mixture was collected by centrifugation at 4000 rpm for 10 min. PMMA-grafted nanoparticles (PMMA–g–NPs) are separated from unattached (free) polymer chains by utilizing a mixed solvent precipitation procedure with THF and hexanes [92].

2.1.5 Generating Nanocomposites

The preparation of nanocomposites followed a procedure described in Ref. [93]: a blend of the inorganic additive, either bare Si NPs or PMMA–g–NPs, and a PMMA matrix was prepared by co-dissolving the PMMA–g–NPs and PMMA in THF and mixing at room temperature for 24 h. The protocol for generating PMMA nanocomposites containing PMMA–g–NPs at 5 wt% (by mass, based on the mass of silica only) is described as an example. TGA analysis showed that the PMMA–g–NPs had a weight loss of 25.6 %, so to create a nanocomposite that is 5 wt% based on the mass of silica, 6.73 mg of PMMA–g–NPs were used. Those PMMA–g–NPs were dissolved in 15 mL of THF, while a separate solution consisting of 93.27 mg of the PMMA matrix (either 15 or 120 kg/mol) in 15 mL of THF was also prepared. These solutions were sonicated for 30 min and then stirred for 12 h. Then, the two solutions were combined, sonicated for an additional 30 min, and stirred overnight. The PMMA nanocomposite solution was poured into PTFE dishes and the solvent was allowed to evaporate overnight in a fume hood. The nanocomposites were then annealed *in vacuo* at 120 °C for 24 h. The loading level of PMMA–g–NPs (based solely on wt% of silica) was probed by TGA before further characterizations.

2.1.6 Synthesis of NOHM-I-PEI

A colloidal silica suspension (7 nm diameter, Ludox SM–30 Sigma–Aldrich) was diluted to 3 wt% and a cation exchange column (Dowex Marathon C, Sigma–Aldrich) was employed to replace the sodium ions on the nanoparticle surface with protons. This surface-protonated silica suspension was mixed with a dilute solution of polyethylenimine, branched (Branched, M.W. 2000, Polyscience Inc.) to achieve the desired polymer canopy-to-silica core mass ratio of 80:20. The PEI has a molecular weight of 2,000 g/mol and a density of 1.05 g/cm^3, as reported by the manufacturer. It contains primary, secondary, and tertiary amines in an approximate ratio of 40/36/24, as reported. Dynamic light scattering (DLS) was used to determine the average hydrodynamic diameter of PEI in dilute solution, which is found to be 2.19 nm. The sample is found to have a narrow size distribution. Thermal gravimetric analysis (TGA) was used for confirmation of ionic bond formation. Concise details of the synthesis procedure are available in Refs. [67, 94].

2.2 *Material Characterization*

2.2.1 Basic Characterization

Gel permeation chromatography (GPC) (Aqueous GPC System, Agilent) measurements were conducted at 298 K in THF with a 1 mL/min flow rate as the mobile liquid phase. The polydispersity index (PID) and number–average molecular weight, M_n, of PMMA graft chains and PMMA matrices were determined using conventional calibration analysis using polystyrene standards. A TA Instruments Q-2000 Differential Scanning Calorimeter was used to determine the glass transition temperatures T_g of the polymer nanocomposites. The procedure involved placing ~5 mg of the polymer nanocomposite in an aluminum pan that was subsequently subjected to a heat/cool/heat cycle from 323–433 K at a rate of 10 K/min using nitrogen as the purge gas. The T_g of polymer nanocomposites is reported as the temperature corresponding to the maximum of the derivative of specific heat capacity in the second heating curve. A TA Discovery Series Thermogravimetric Analyzer (TA Instruments) was used to determine the amount of silane and PMMA graft chains attached to the nanoparticle surface after each surface modification step described above. The average grafting density of PMMA chains on the nanoparticle surface, σ, was calculated using Eq. 1:

$$\sigma = \left(\frac{W_{PMMA}}{100 - W_{PMMA}} - \frac{W_{amine}}{100 - W_{amine}}\right)\frac{N_A}{M_n S_{NP}} \tag{1}$$

where N_A is Avogadro's number, M_n is the molecular weight of the PMMA-grafted chains, $S_N P$ is the specific surface area of the 14 nm silica nanoparticles (= 2.07×10^{20} nm^2/g), W_{PMMA} and W_{amine} represent the percentage mass loss of PMMA-grafted nanoparticles and the mass loss of the amine-functionalized nanoparticles, respectively (obtained from TGA). Further characterization was performed using ^{1}H NMR.

2.2.2 Rheology

Rheology measurements were performed using a TA Instruments Discovery Hybrid Rheometer (HR2) in small-amplitude oscillatory shear (SAOS) mode using 3 mm stainless steel parallel plates. The real and imaginary parts of the complex shear modulus and the complex viscosity were recorded at various temperatures between 213 K and 243 K controlled within 0.1 K of the set point using liquid nitrogen coolant. The data was processed using TA Instruments TRIOS software to produce master curves and shift factors based on time–temperature superposition.

2.2.3 X-Ray Scattering

Small- and wide-angle X-ray scattering measurements were performed using a SAXSLAB Ganesha instrument equipped with a Cu Kα 50 kV Xenocs Genix ULD SL X-ray source (beam energy of 8 keV) and a 170 μm pixel-spaced, single-photon counting Dectris Pilatus 300 k 20 Hz detector. Quartz capillary tubes 1 mm in diameter were filled with the samples and sealed prior to loading into the instrument. Additional samples were prepared by placing a drop of the polymer or NOHM between strips of Kapton tape. A blank measurement was performed on the Kapton tape without the sample included. SAXS and WAXS data were collected at room temperature under vacuum in the q-range from 0.002 to 2.6 Å^{-1}.

2.2.4 Nuclear Magnetic Resonance (NMR)

PEI and NOHM-I-PEI samples were dried under high vacuum conditions at 323 K for 3 h in order to remove residual water before being transferred in a standard 5 mm NMR tube. A Bruker Avance III 400 WB spectrometer—operated at 100.61 MHz, coupled with a temperature controller—was employed to record ^{13}C NMR spectra. The ^{13}C spin–lattice relaxation time (T_1) spectra were measured using the inversion-recovery method with a set of 16 recovery (τ) periods. The number of repetitive acumulated scans was set to 256 for each τ value. The relaxation delay was at least 5 T_1 plus the acquisition time. The τ values were varied up to 1.5 T_1. The spectra were phase-corrected, and T_1 values were calculated from the signal intensity data using MestreNova. All the carbon atoms present in the molecule have at least one directly bonded hydrogen.

2.2.5 Broadband Dielectric Spectroscopy (BDS)

Dielectric measurements were executed on a High-Resolution Alpha Analyzer equipped with a Quatro system with the capability to control temperature within an error margin of 0.1 K (Novocontrol Technologies). PEI and the NOHM-I-PEI samples were placed in a parallel-plate capacitor arrangement comprising polished brass electrodes of diameter 15 mm. 100 μm thick silica spacers were employed to maintain the sample thickness in the course of the measurements. In order to remove any adsorbed water, the sample films were annealed at 420 K in inert conditions ensured by dry nitrogen flow for ~8 h prior to substantive dielectric measurements. BDS experiments were done in the frequency range 1 mHz–10 MHz over cooling and heating cycles in the temperature range 290–430 K (for PMMA-based systems) and 140–400 K (for PEI and NOHM-I-PEI).

3 Results and Discussion

3.1 PNCs and PGNCs

Broadband dielectric spectroscopy probes molecular dynamics and charge transport in broad temperature and frequency ranges making it a versatile tool for studying composite systems. The technique measures the complex dielectric function, ε^*, which is related to the complex conductivity function, σ^*. The two functions are related by the relation $\sigma^*(\omega, T) = i\omega\varepsilon_0\varepsilon^*(\omega, T)$, where $i = \sqrt{-1}$, ε_0 is the permittivity of free space, $\omega = 2\pi f$ is the angular frequency of the applied electric field, and f is the measuring frequency; effectively, $\varepsilon' = \sigma''/\omega\varepsilon_0$ and $\varepsilon'' = \sigma'/\omega\varepsilon_0$ [96]. In Fig. 2, the imaginary part, ε'' of the complex permittivity, ε^*, is presented as measured during heating and cooling runs for neat PMMA, 120 kg/mol and PGNC 120 kg/mol. The reproducibility of the data over the cycles indicates that the samples are thermally stable in the temperature range of interest. The flank on the low-frequency side is due to contributions from the DC ionic conductivity. There are two peaks in the spectra: on the low- frequency side, we observe a weak intensity peak assigned to the α-process due to fluctuations at the segmental level, on a nanometer length scale [97], which determine the glass transition. On the higher frequency side is the β-relaxation arising from the localized motion of the carboxylate ester groups about their bonds to segments on the chain backbone. The β-process dominates the dielectric spectra due to the strong dipole associated with the $-COOCH_3$ groups (see the structure of PMMA in the inset of Fig. 2). Only a small portion of the total dipole moment contributes to segmental dynamics reflected in the α-process [98, 99]. In order to better visualize the peak due to the α-relaxation, we apply the derivative approach popularized by Wübbenhorst and Turnhout [100], where the derived loss function is given by

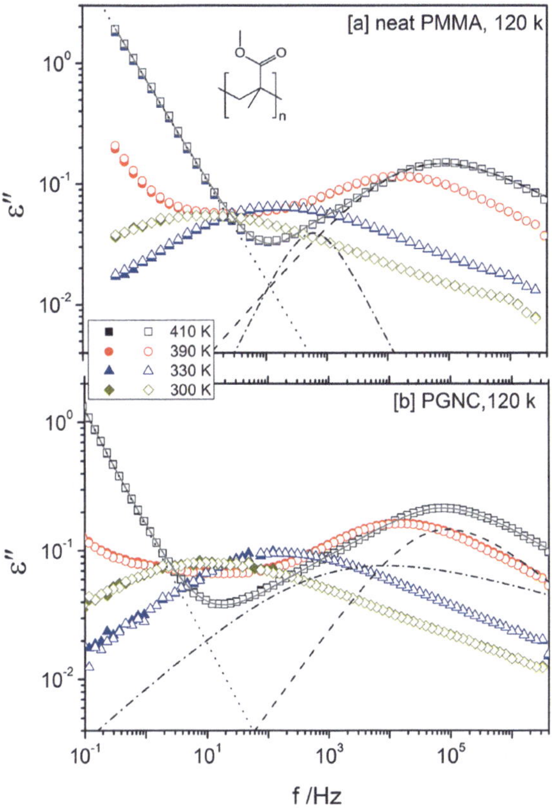

Fig. 2 Imaginary part of the complex dielectric function, $\varepsilon^* = \varepsilon' - i\varepsilon''$, presented as a function of frequency for **a** neat PMMA with molecular weight 120 kg/mol, and **b** PGNC (5 wt% of silica nanoparticles, and PMMA matrix of 120 kg/mol). Full and empty symbols represent data obtained in cooling and heating runs, respectively. Inset: chemical structure of PMMA. The dashed dotted and dashed lines are HN fits to the α- and β-processes, respectively, while the dotted line is the conductivity contribution at 410 K. Solid lines are the total fit functions. See text for details about the fitting procedure. The error bars are comparable to the size of the symbols unless otherwise specified. Reprinted with permission from Ref. [95]. Copyright (2020) American Chemical Society

$$\varepsilon''_{der}(\omega) = -\frac{\pi}{2}\frac{\partial\varepsilon'(\omega)}{\partial ln\omega} \tag{2}$$

where $\varepsilon'(\omega)$ is the real part of the complex dielectric function. Figure 3 shows both ε'' (panel [a]) and ε''_{der} (panel [b]) representation of data measured for systems with a molecular weight of 15 kg/mol. The α-peak shows up, in [b] as compared to [a], as a slightly more pronounced shoulder on the low-frequency side of the peak due to the β-relaxation. Figure 3 also shows subtle changes in the spectra as the sample is changed from neat PMMA to PNCs to PGNCs. These changes, with respect to mean relaxation times, and the shapes of the peaks, are considered further in the analysis and discussion. Because of the presence of two peaks and the strong upturn at low

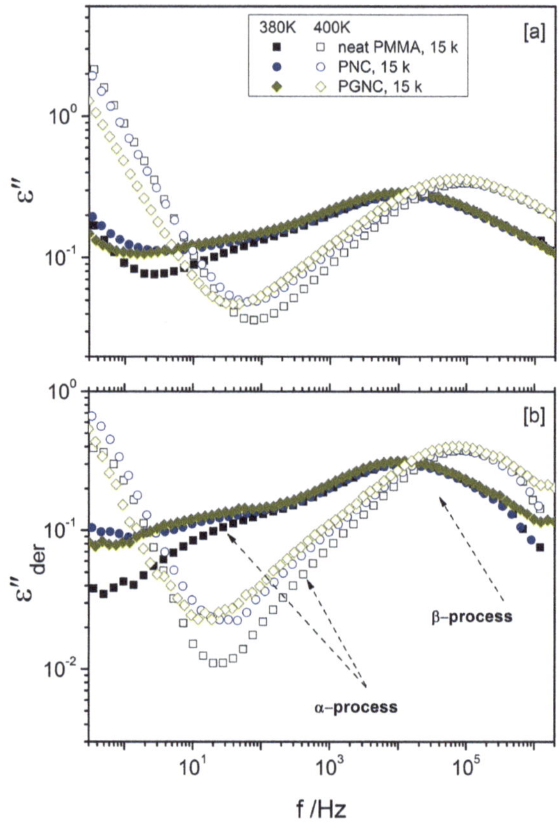

Fig. 3 **a** Dielectric loss, ε'', *versus* frequency as measured at 380 and 400 K for three different systems: (i) neat PMMA, 15 kg/mol; (ii) a PNC comprising 14 nm bare silica nanoparticles dispersed in 15 kg/mol PMMA at 5 wt% loading; and (iii) a PGNC with 5 wt% loading of PMMA-g-silica in 15 kg/mol PMMA matrix. **b** Derivative representation of the same data calculated according to Eq. 2 which makes it easier to visualize the two relaxations present in the spectra. The error bars are smaller than or comparable to the size of the symbols unless otherwise specified. Reprinted with permission from Ref. [95]. Copyright (2020) American Chemical Society

frequency, a linear combination of two Havriliak–Negami (HN) functions [101] and a term accounting for the DC ionic conductivity is applied to describe the data:

$$\varepsilon^* = \varepsilon_\infty + \frac{\Delta\varepsilon_1}{(1 + (i\omega\tau_{HN_1})^{\beta_1})^{\gamma_1}} + \frac{\Delta\varepsilon_2}{1 + (i\omega\tau_{HN_2})^{\beta_2})^{\gamma_2}} - i\frac{\sigma_0}{\varepsilon_0\omega} \tag{3}$$

where $\varepsilon^* = \varepsilon' - i\varepsilon''$ is the complex dielectric function, ε_0 is the dielectric permittivity of free space, ε_∞ is the permittivity of the relaxed medium, σ_0 is the DC ionic conductivity of the sample, $\Delta\varepsilon_i$ is the dielectric relaxation strength, τ_{HN_i} is the characteristic relaxation, and the parameters β_i and γ_i describe, respectively, the symmetric and asymmetric broadening of the spectra.

In order to uniquely describe the α-process given the strong β-relaxation, a systematic fitting procedure was followed. Starting with the neat PMMA, Eq. 3 was used to fit spectra obtained from the derivative representation (Eq. 2). τ_{HN} values thus obtained were used as fixed values in the next step, thereby reducing the number of fit parameters. Equation 3 was then fit to the dielectric loss (ε''), with β, γ, and $\Delta\varepsilon$ set as free parameters. It is important to note here that the ε''_{der} approach narrows the shape of the peak compared to ε''—see illustration in Fig. 4—and hence it was crucial to analyze the latter rather than the former peak [102]. We made an

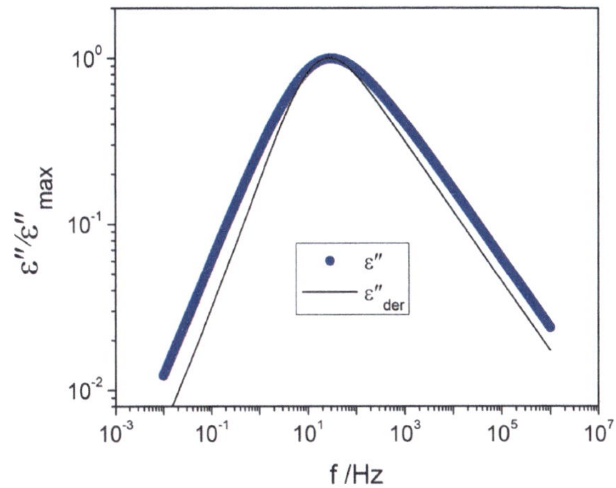

Fig. 4 A demonstration of the effect of the derivative (of the real part of the complex permittivity) approach on the shape of the peak. The *thick blue curve* is the imaginary part of the complex permittivity calculated using HN shape parameters, $\beta = 0.70$ and $\gamma = 0.60$ (see Eq. 3). The *thin black curve* is the first derivative of the corresponding real part of the complex permittivity with the said HN parameters. A fit of the derived spectrum by the HN function is altered in shape ($\beta = 0.83$ and $\gamma = 0.54$). For ease of comparison, the spectra have been normalized by the maximum loss value

assumption in fitting the data for the nanocomposites; at any given temperature, the shape of the β-relaxation was assumed to be that of neat PMMA. This requirement is however not imposed on the characteristic relaxation rate. The aforesaid supposition is well-founded given the very localized nature of the relaxation. In an investigation of several methacrylate polymers, it was found that the shape, position, and activation energy of the β-relaxation remain unchanged across the systems [103]. Furthermore, in studies of confined PMMA, it has been shown that the mean relaxation rates remain bulk-like over a wide temperature range [104–106].

Figure 5 shows several sets of data and the corresponding fit functions at 350, 390, and 405 K for systems bearing the 15 kg/mol PMMA matrix. Demonstrably, from the coincidence of the measured data points and the solid lines (which represent the total fit function according to Eq. 3), it is clear that the fitting procedure described above achieves an excellent description of the raw measured data. Now, from the HN parameters, the mean relaxation times, τ_m, were calculated using Eq. 4:

$$\tau_m = \tau_{HN} \sin\left(\frac{\beta\gamma\pi}{2+2\gamma}\right)^{\frac{1}{\beta}} \sin\left(\frac{\beta\pi}{2+2\gamma}\right)^{-\frac{1}{\beta}} \quad (4)$$

Figure 6 displays the mean relaxation times for the two processes present in the systems under study. The full symbols represent data for the α-relaxation while the empty ones represent times for the β-relaxation. Several observations can be made from this figure. First, concerning the β-process, we note that the mean relax-

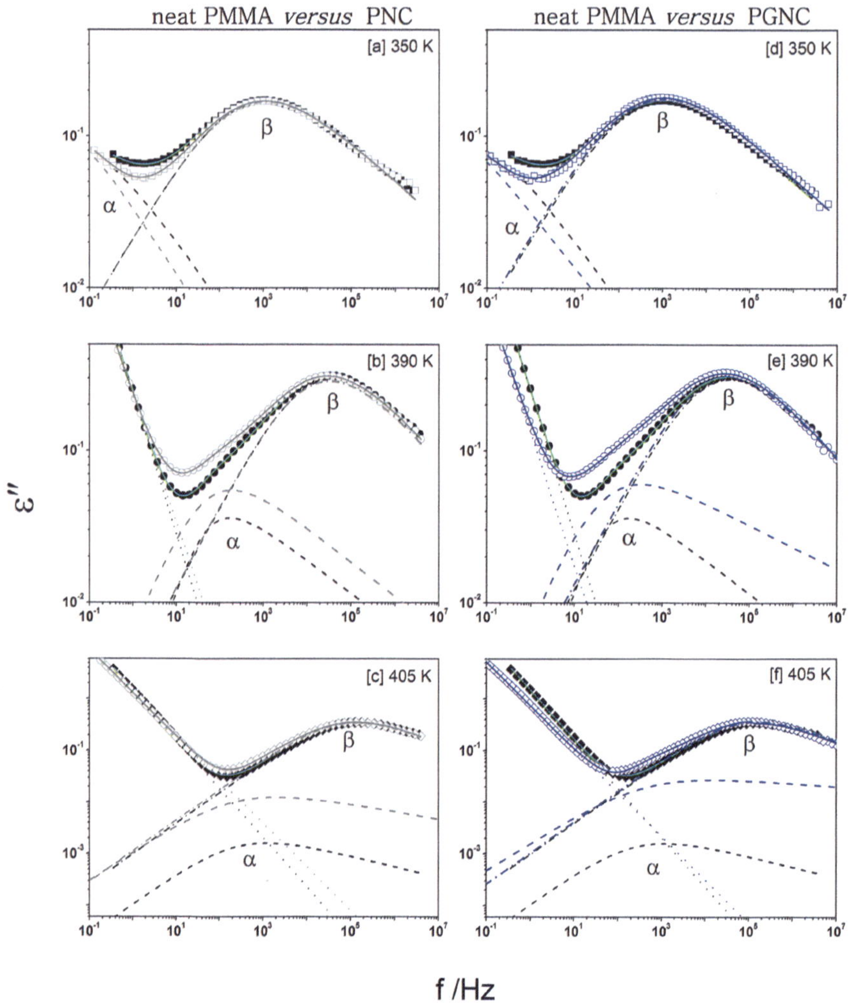

Fig. 5 Dielectric loss, ε'', *versus* frequency at three selected temperatures, as indicated, for neat PMMA, 15 kg/mol (*closed black symbols*) compared to PNC, 15 kg/mol (open gray symbols; [(**a**), (**b**), and (**c**)], and PGNC, 15 kg/mol [(*open blue symbols*; (**d**), (**e**), and (**f**)]. Depending on the temperature, and hence the relaxations observable in the accessible frequency window, data are described by using either one or two Havriliak–Negami functions plus a DC ionic conductivity contribution (Eq. 3) as shown by using dash-dotted lines (β-process), *dashed lines* (α-process), and *dotted lines* (DC conductivity contribution). The solid lines are sums of the individual functions used to model the entire spectrum. PNCs and PGNCs are constituted at 5 wt% based on the mass of silica. The error bars are smaller than or comparable to the size of the symbols unless otherwise specified. Reprinted with permission from Ref. [95]. Copyright (2020) American Chemical Society

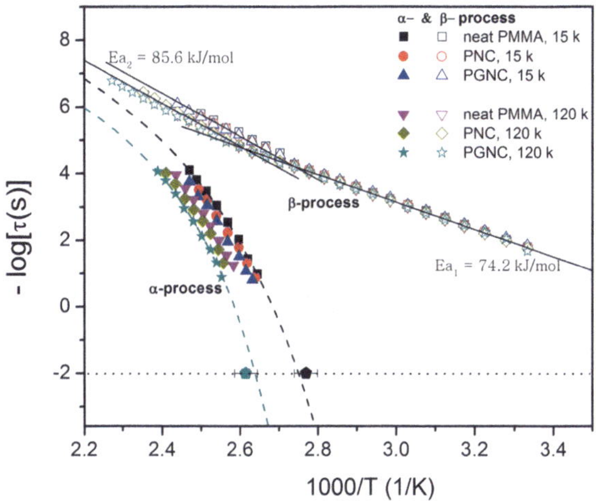

Fig. 6 Mean relaxation times of the α- and β-processes for neat PMMA, PNCs, and PGNCs as functions of inverse temperature. For graphical clarity, data are shown only for neat polymers and polymer nanocomposites with 5 wt% (based on silica nanoparticles) loading. The *solid* and *dashed* lines are fits to Eqs. 5 and 6, respectively. Representative values of the calorimetric T_g determined from DSC measurements for neat PMMA 15 k (*black pentagon symbols*) and PGNC 120 k (*gray pentagon symbols*) are also shown (by invoking the convention that T_g takes place at a segmental relaxation time of 100 s). VFT fits are displayed only for two samples for graphical clarity, and agreement between BDS and DSC data is evident. 15 k and 120 k stand for the respective molecular weight of the system in kg/mol. The error bars are smaller than or comparable to the size of the symbols unless otherwise indicated. Reprinted with permission from Ref. [95]. Copyright (2020) American Chemical Society

ation times for PNCs and PGNCs are identical—within the limits of experimental accuracy—to those calculated for neat PMMA. This trend is in tandem with the understanding that the β-process being a local relaxation is not affected in its rate, by polymer/nanoparticle interactions. Similar observations have been reported in the literature [105, 106]. Second, we describe the temperature dependence of the mean relaxation times of the β-process using the Arrhenius equation

$$\tau(T) = \tau_{\infty}\exp\left(\frac{-E_A}{\kappa_B T}\right) \tag{5}$$

where τ_{∞} is the relaxation time in the high-temperature limit, E_A the activation energy, and κ_B the Boltzmann constant. It is found that at the calorimetric glass transition temperature, its activation energy changes from ~74 to ~86 kJ/mol. This trend has been observed in several methacrylate polymers [99, 103, 107–110], and is attributed by one school of thought to a change in the mechanism of the relaxation, i.e., the onset of cooperativity of the α-relaxation [109, 110]. Third, the mean relaxation time of the α-process slows down as the system is changed from neat PMMA to PNCs to PGNCs. This corresponds to a systemic increase in the glass transition temperature

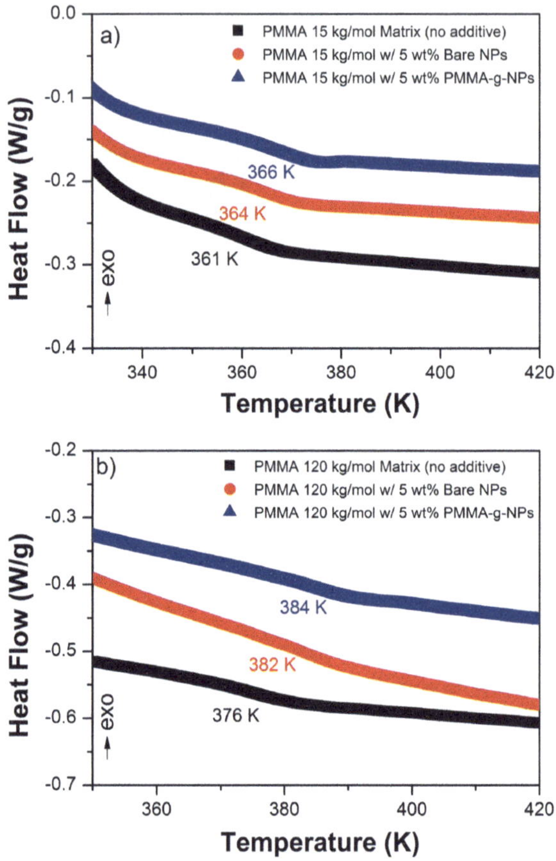

Fig. 7 DSC traces show that the addition of either bare or polymer-grafted (PMMA–g–NPs) nanoparticles in **a** 15 kg/mol and **b** 120 kg/mol PMMA matrices increases the T_g of the nanocomposite, with the systems containing PMMA–g–NPs being the most drastically affected. Adapted with permission from Ref. [95]. Copyright (2020) American Chemical Society

as obtained from DSC experiments—see Fig. 7. The temperature dependence of the characteristic mean relaxation times of the α-process was described using Eq. 6, commonly known as Vogel–Fulcher–Tammann (VFT) equation:

$$\tau(T) = \tau_0 \exp\left(\frac{-DT_0}{T - T_0}\right) \tag{6}$$

where τ_0 and D are constants, and T_0 is the Vogel temperature [111–113]. Extrapolating the fits down to a relaxation time of 100 s as displayed in Fig. 5 reveals a coincidence with the calorimetrically determined T_g, a conventionally accepted procedure for identifying the structural relaxation [96]. Although the narrow temperature range in which we can analyze the α-process introduces higher uncertainty in the extrapolation, the coincidence nonetheless lends credence to our assignment of the relaxation to structural dynamics.

Table 2 Values the relaxation dielectric strength as obtained from HN fits of the segmental relaxation at representative temperatures for samples studied in this work

NP loading (wt%)	PNC, 15 kg/mol		PGNC, 15 kg/mol	
	$\Delta\varepsilon$ (390 K)	$\Delta\varepsilon$ (405 K)	$\Delta\varepsilon$ (390 K)	$\Delta\varepsilon$ (405 K)
0	0.43	0.38	0.56	0.49
1	0.41	0.39	0.55	0.51
3	0.46	0.43	0.60	0.58
5	0.56	0.48	0.66	0.61
NP loading (wt%)	PNC, 120 kg/mol		PGNC, 120 kg/mol	
	$\Delta\varepsilon$ (390 K)	$\Delta\varepsilon$ (405 K)	$\Delta\varepsilon$ (390 K)	$\Delta\varepsilon$ (405 K)
0	0.52	0.40	0.76	0.58
1	0.53	0.39	0.81	0.57
3	0.61	0.42	1.12	0.68
5	0.68	0.44	1.09	0.70

Interactions at the interface of the nanoparticles and the host matrix are known to lead to changes in the glass transition temperature of the system. Specifically, negative and positive deviations from bulk values are attributed, respectively, to neutral or repulsive nanofiller/matrix interactions [8, 89, 114], and attractive interactions [114–117]. In the composites under study, the hydroxyl groups on the surfaces of silica nanoparticles participate in hydrogen-bonding interactions with the carbonyl groups present in monomeric repeat unit of PMMA [117, 118]. These nonbonded attractive interactions lead to a reduction in the number of degrees of freedom for the motion of the PMMA segments in the vicinity of the surface, hence slowing down segmental relaxation—resulting in an interphase characterized by reduced mobility relative to the bulk [88]. For PGNCs, an additional interaction is introduced: segments of the matrix interact with those of the grafted polymer. These interactions up the number density of entanglements and interdiffusion of chains within the interfacial regions, leading to relatively higher T_g in PGNCs than PNCs [2, 3]. It is additionally observed that the dielectric relaxation strength, $\Delta\varepsilon$, associated with the structural relaxation increases with the loading level of nanoparticles and the molecular weight of the host matrix (see Table 2). We understand this to arise from the increased contributions of interfacial polarization given the growth of the total interfacial volume in the sample, and the fact that $-OH$ groups on the surface of the nanoparticles likely contribute to that polarization.

We now turn our attention to the shape of the α-peak as characterized by the shape parameters introduced in Eq. 3. While the mean relaxation times can be adequately obtained from the derivative representation, dynamic heterogeneity—which is quantifiable in terms of the distribution of the relaxation times—can only be deduced

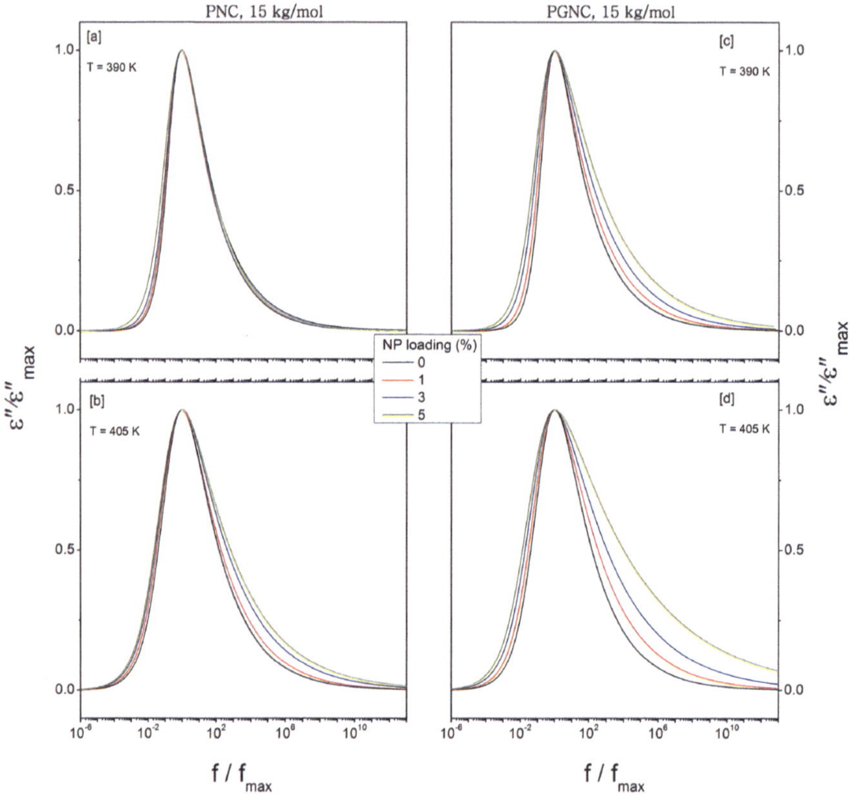

Fig. 8 Havriliak–Negami functions that describe the α-relaxation for systems comprising 15 kg/mol PMMA matrix at different loading levels of either bare nanoparticles [(**a**) and (**b**)] or PMMA–g–NPs [(**c**) and (**d**)] at 390 and 405 K. To facilitate comparisons of the peak shapes, each function is normalized with respect to the maximum loss, ε''_{max}, as well as the corresponding frequency value, f_{max}. Reprinted with permission from Ref. [95]. Copyright (2020) American Chemical Society

from the original dielectric loss spectrum. Figure 8 displays the Havriliak–Negami functions for PNC and PGNC systems having a 15 kg/mol PMMA matrix at two representative temperatures for varying nanoparticle loading levels. The functions are normalized with respect to the frequency position of maximum loss as well as the corresponding maximal dielectric loss value. Compared to the neat PMMA, the spectral shape for both the PNCs and PGNCs is broadened in dependence on the molecular weight of the matrix, nanoparticle loading, and temperature. At low temperatures, the α-peak for PNCs is largely comparable to that of neat PMMA, while that for the PGNCs is broadened asymmetrically, with a skew toward the high-frequency side. For the 120 kg/mol matrix, the broadening in the PGNCs is rather symmetrical (data not shown). At higher temperatures, the spectrum is broadened significantly on the high-frequency side for both systems. These changes become

Table 3 Percentage increase in the width of the α-relaxation peak with respect to that of neat PMMA at four temperatures for two nanoparticle loading levels

		±2 % increase on the low-frequency side				±2 % increase on the high-frequency side			
Temp (K)		390	395	400	405	390	395	400	405
PNC, 15 kg/mol	1 wt%	3	3	3	3	4	4	5	6
	5 wt%	6	6	6	5	4	14	20	27
PNC, 120 kg/mol	1 wt%	4	27	76	101	5	26	74	96
	5 wt%	9	36	83	105	9	38	91	167
PGNC, 15 kg/mol	1 wt%	5	5	5	5	7	7	9	9
	5 wt%	18	18	19	20	32	35	38	41
PGNC, 120 kg/mol	1 wt%	25	25	78	105	25	50	81	110
	5 wt%	52	62	103	207	51	67	110	196

more pronounced with nanoparticle loading. Table 3 shows the degree of broadening estimated by calculating the change in the area under the peak with respect to neat PMMA.

The widening of the peak of the α-relaxation implies an broadening in the distribution of the underlying microscopic relaxation times which is evidence of the dynamic heterogeneity in the system. We ascribe the slower relaxation times to interactions at the polymer/nanoparticle interface. As earlier pointed out, for the case of PNCs, hydrogen-bonding interactions between carbonyl and hydroxyl groups lead to the formation of an interphase within the dynamics are slowed down close to the nanoparticles. In the case of PGNCs, polymer matrix chains, or at least some part of the chains for the longer ones, wet and penetrate the graft. This leads to a reduced degree of freedom for the motion of segments in the interacting zone. Importantly, for the systems in our study, a separate slow process—analogous to that reported in poly(vinyl acetate)- and poly(2-vinyl pyridine)-based nanocomposites [33, 34]—is not observed. The implication of this is that the interphase with modified dynamics should be visualized, not as one with uniform mobility, but rather as having a gradient of dynamics that gradually become bulk-like as an imagined molecular probe moves away from the nanofiller.

The emergence of faster relaxation times, evident from the broadening of the α-peak on the short-time side, is hypothesized to have its origin in an increased free volume due to chain entanglements and interdiffusion. As long as the asymmetry between matrix and graft chains is not so high that matrix chains start to dewet, entanglements and chain interdiffusion become more pronounced with molecular weight. As chains begin to experience spatial restriction in the interphase zone between the hard nanofillers and the bulk-like matrix, packing frustrations arise. Consequently, density fluctuations and more free volume result from nonuniform segment density. Indeed, previous studies have demonstrated that the free volume is affected by adding nanoparticles into a polymeric material [119–121]. With this picture in mind,

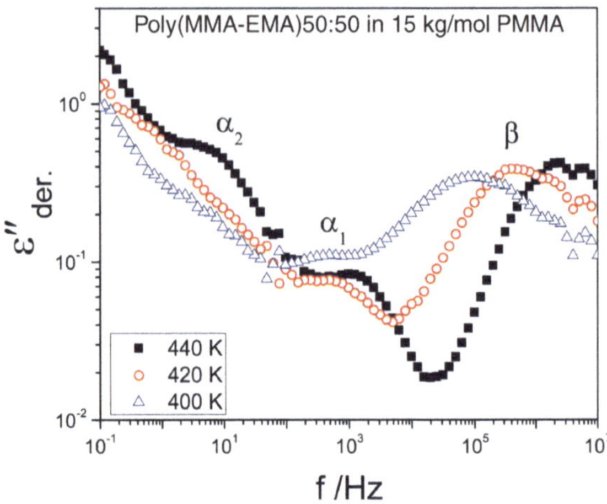

Fig. 9 Derivative loss, ε''_{der}, plotted as a function of frequency at three select temperatures for a PGNC sample whose matrix is neat PMMA (15 kg/mol), and graft is made of poly(methyl methacrylate-r-ethyl methacrylate) [poly(MMA-r-EMA)] random copolymers

this effect should be more pronounced in PGNCs compared to PNCs given the fact that the nonuniform interphase region in PGNCs is characterized by local changes that influence the local balance between mixing (wetting) of graft chains and matrix chains and dewetting [93]. Indeed as observed in Fig. 8, for a given loading level, there is less broadening in PNCs than in PGNCs. The role of the loading level can be explained as determining the amount of interphase region per unit volume between nanoparticles and matrix chains.

These findings open new pathways for tuning the properties of polymer-grafted nanocomposites. For instance, for the PMMA matrices studied in this work, the entropy of mixing between graft and host chains may be altered by grafting a random methacrylate-based copolymer onto the nanoparticles, and probing the changes that arise due to modified interfacial interactions. Figure 9 shows preliminary results obtained for the case of a PGNC system comprising silica nanoparticles grafted with poly(methyl methacrylate-r-ethyl methacrylate) (poly(MMA-r-EMA)) random copolymers with a 50:50 molar composition of the co-monomers. In addition to the structural relaxation (here identified as α_1) and the β-process, an additional relaxation, α_2, slower than α_1, is observed. Evidently, the dynamics of this system are markedly different from those of the PGNCs discussed earlier, and warrant detailed analysis to unravel the underlying molecular origin.

3.2 Nanoparticle Organic Hybrid Materials

Figure 10 shows X-ray scattering profiles for both PEI and NOHM-I-PEI. The correlation distances, d_1 ($\sim$10.4 nm) and d_2 ($\sim$5.2 nm), are the diameter and radius, respectively, of the silica nanoparticles as estimated assuming a simple form factor for monodisperse spheres. As expected, these values are higher than the lengths determined for the neat (nondecorated) nanoparticles ($\sim$ 7 nm, diameter) from trans-

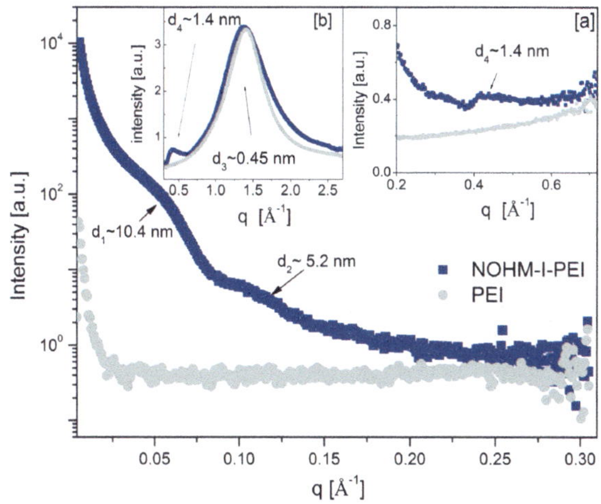

Fig. 10 Experimentally determined small-angle X-ray scattering (SAXS) data for NOHM-I-PEI (*blue symbols*) and PEI (*light gray symbols*), showing various correlation distances estimated from the peak position as $d_i = 2\pi/q$, where q is the scattering vector, and corroborated by calculations using a form factor for well-dispersed spheres. The insets **a** and **b** show the corresponding medium-angle (MAXS) and wide-angle X-ray scattering (WAXS) profiles. Adapted with permission from Ref. [123]. Copyright (2021) American Chemical Society

mission electron microscopy measurements. $d_3 \sim 0.45$ nm is present in both the neat polymer and the nanocomposite implying that this correlation distance is an intrinsic property of the polymer, leading to the conclusion that it is the backbone-to-backbone separation [122]. Finally, the correlation distance d_4 ($\sim$1.4 nm) is observed in the NOHM but not for the neat PEI. We conjecture that in the presence of nanoparticles, there is some level of organization in the polymeric regions of the hybrid, leading to mesoscale structures. This idea is revisited later in the discussion of the dynamics. Instructively, it is observed that the peak for the NOHM is slightly broader than that for PEI showing that the silica/PEI interactions lead to heterogeneities in the packing of chains.

Rheological measurements of PEI and NOHM-I-PEI samples were carried out to study their flow characteristics. Figure 11 shows the master curves of the real (G') and imaginary (G'') part of the complex shear modulus constructed at a reference temperature of 218 K (−55 °C), with the corresponding shift factors presented in the inset. The construction of the master curves is realized by invoking the time–temperature superposition principle where G' and G'' measured over a span of oscillatory frequencies and a number of temperatures are shifted to a reference temperature by multiplying the measured curves by a temperature-dependent factor, a_T. We observe from Fig. 11 that G' and G'' intersect at different oscillatory frequencies for the two samples. Among other things, this implies that the underlying fundamental relaxations responsible for mechanical energy dissipation in PEI and NOHM-I-PEI have time scales that are different by about 3 orders of magnitude at the reference

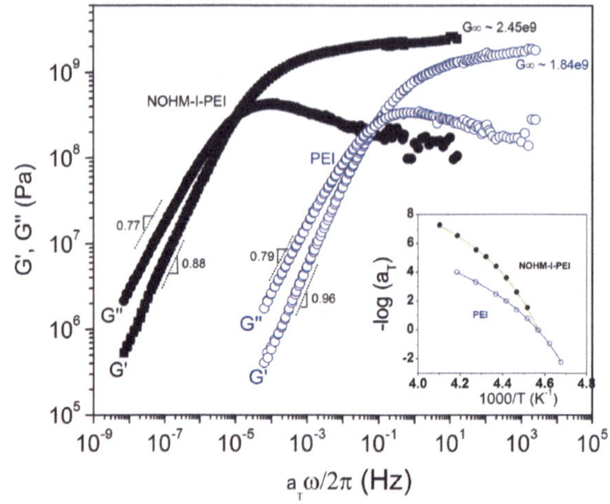

Fig. 11 Master curves (reference temperature, $T_{ref} = -55\,^{\circ}\mathrm{C}$) showing the real, G', and imaginary, G'', parts of the complex shear modulus, G^*, for PEI (*empty circles*) and NOHM-I-PEI (*filled squares*). The shift factors used to construct the master curves are shown in the inset. The solid lines are fits to Eq. 6. The error bars are comparable to the size of the symbols unless otherwise indicated. Reprinted with permission from Ref. [123]. Copyright (2021) American Chemical Society

temperature. Significantly, the temperature dependencies of a_T for the two systems are different, precluding the possibility that a similar relaxation process underlies mechanical energy dissipation in PEI and NOHM-I-PEI. For the neat polymer, the mean relaxation times from rheology coincide with those of the structural relaxation as probed by dielectric spectroscopy (see Fig. 14). This has the implication that the glass transition controls the mechanical response in PEI. However, the mean times from BDS for NOHM-I-PEI are considerably faster than those determined from rheology. Evidently, the length scale of the motions reported by rheology are larger than those of the α-relaxation. One may think of the fundamental times obtained from the shear–stress experiments for NOHM-I-PEI as being impacted by the characteristic times of the motion of silica nanoparticles. These nanoparticles are moving in a liquid medium whose viscosity is determined by the tethered PEI chains, and the characteristic length scale mainly set by the particle–particle interactions. Indeed, Cheng et al. have shown that nanoparticles contribute to the structural dynamics in nanocomposites [124].

With this in mind, the diffusivity, D_1, of the inorganic nanoparticles can be estimated as $D_1 = R_c^2/2\tau$, where $R_c = (3V_c/4\pi)^{\frac{1}{3}}$ is the particle jump length, $\tau \sim 1/\omega_c$ is the hopping time, $V_c \sim (1-\phi)\mathrm{KB}T/G'(\omega_c)$, ϕ is the particle volume fraction, ω_c is the angular frequency at which the dynamic loss and storage modulus intersect, KB is the Boltzmann constant, and T is the reference temperature at which the master curves are constructed [87]. Inserting values from Fig. 9 into this equation, we find $D_1 \sim 9.1 \times 10^{-26}\mathrm{m}^2/\mathrm{s}$. The diffusivity can also be calculated from the Stokes–Einstein equation, $D_2 = \mathrm{KB}T\omega_c/(6\pi R_p|G^*(\omega_c)|)$, where R_p is the radius of the nanoparticles (~3.5 nm), and $|G^*(\omega_c)| = \sqrt{2G'^2(\omega_c)}$ is the absolute value of the complex modulus at ω_c [125]. This delivers $D_2 \sim 5.8 \times 10^{-26} \mathrm{m}^2/\mathrm{s}$. The suggested meaning of the rheological result is plausible given the fact that $D_1 \approx D_2$. However, the calculated values of diffusivity are too low to be independently confirmed as they cannot be accessed by state-of-the-art techniques such as Pulsed Field Gradient NMR.

Concerning the slopes of the low-frequency side of G' and G'', we find values of 0.96 and 0.79, respectively, for the polymer, and 0.88 and 0.77 for the NOHM. The Zimm model estimates a slope of ~2/3 for the glass-to-rubber transition region of G' and G'' (plotted against ω), and this is found to hold for many polymers [126, 127]. On the other hand, simple molecular glass-forming liquids are known to have one mechanical relaxation corresponding to the glass transition with low-frequency slopes of 2 and 1 for G' and G'', respectively [128]. There is no doubt the slopes obtained for PEI and NOHM-I-PEI point to intricate phenomena not captured by these models, probably, due to the highly branched chains in PEI. Lastly, the high-frequency limiting shear modulus, G_∞, slightly increases from 1.8×10^9 for PEI to 2.5×10^9 Pa for the NOHM, signifying mechanical enhancement in the hybrid, which is also evident in the increased viscosity of NOHM-I-PEI (data not shown).

Figure 12 displays the imaginary part ε^* of the complex permittivity plotted as a function of frequency, measured by heating and cooling runs for PEI and NOHM-I-PEI; the thermal reproducibility of the measurements is assured. At low temperatures, one peak is observed, while at higher temperatures, two peaks appear in addition to a conductivity contribution to the spectrum. It is clear from this figure, as well as Fig. 13—which provides similar data for a wider temperature range, and also includes the real part of the complex conductivity function—that the spectral differences between the two data sets are rather subtle. Equation 3 was used to quantitatively analyze the data, e.g., as displayed in Fig. 10a, b by dotted and solid lines, respectively. Mean relaxation times for the two processes were calculated using Eq. 4, and are presented as functions of inverse temperature in Fig. 14. Also displayed in the same graph are the mean structural relaxation times calculated from Fig. 11 as $\tau_m(T) = \tau_{ref}(T_{ref})/a_T$, where τ_{ref} is the mean relaxation time at the reference temperature, T_{ref}, and a_T is the corresponding shift factor. It is observed that the mean times for the two relaxations in PEI coincide with the corresponding ones in NOHM-I-PEI as deduced from BDS data. The process at high temperatures is described by Eq. 6 which, when extrapolated to a relaxation time of 100 s, coincides with the calorimetrically determined glass transition temperature (215 K) for the two materials. Also, the mean times accessed from rheology for PEI agree with those obtained from BDS. This process is therefore assigned to the fluctuation of PEI segments. However, as already discussed—and ascribed to the contributions from the networks of silica cores—the mean times delivered from rheology for the hybrid material are significantly slower than those from dielectric spectroscopy.

The relaxation at low temperatures has an Arrhenius-type thermal activation and is therefore well-described by Eq. 5. This is characteristic of localized dipolar relaxations, and is identified as a β-process arising from the localized motion of the amine-terminated side groups (with an activation energy of 37.9 kJ/mol). The tem-

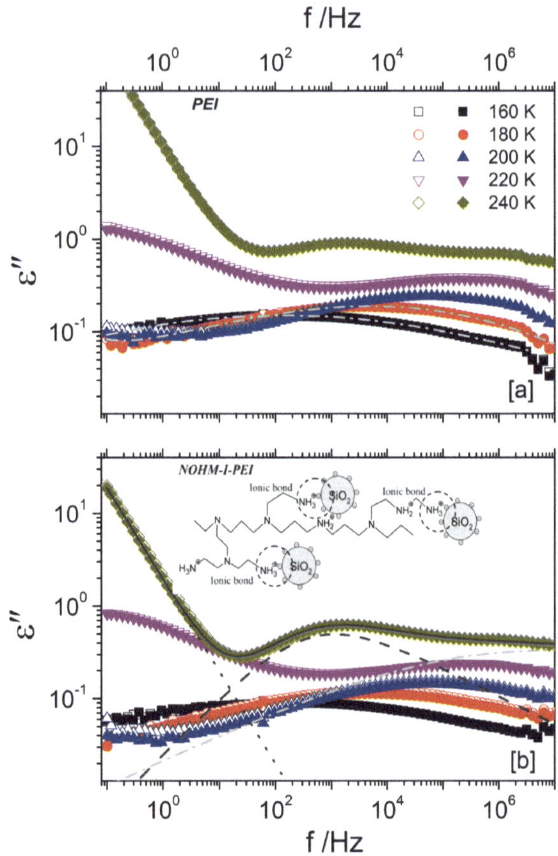

Fig. 12 Dielectric loss, ε'', as a function of frequency, f, at different temperatures as indicated for **a** neat PEI and **b** NOHM-I-PEI. At low temperatures, the data can be described by one Havriliak–Negami (HN) function as demonstrated in (**a**), while at higher temperatures, two HN functions are required. An example of the fitting is shown in panel (**b**) for data taken at 240 K. Filled and empty symbols represent data collected in cooling and heating runs, respectively. Inset: the chemical structure of NOHM-I-PEI obtained from Ref. [67]. The error bars are comparable to the size of the symbols unless otherwise indicated. Reprinted with permission from Ref. [123]. Copyright (2021) American Chemical Society

perature dependence of the relaxation strength, $\Delta\varepsilon_\beta$, of this process is given in the inset of Fig. 14, and is observed to undergo a change in slope at T_g. This phenomenon can be explained by the idea of "cage breaking" [129–133]: below T_g, the side group undergoes localized motions while attached to static segments of the backbone. The restrictions imposed by the "frozen" segments resemble those of a cage, so much so that as the temperature is increased, the cage becomes larger by virtue of their movement. The motion of the segments on the backbone gives the side group more degrees of freedom for its movements, and at some finite temperature above T_g, the two relaxations merge. Now, given the identical temperature dependence of $\Delta\varepsilon_\beta$ for PEI and NOHM-I-PEI, it is inferred that the underlying process has exactly the same origin and mechanism in both materials. Both the α- and β-relaxations take place on very short length scales; it is therefore not surprising that the introduction of nanoparticles does not affect these processes, at least in terms of their *mean* relaxation times. Processes occurring at longer length scales would be affected in dependence on molecular weight as well as the grafting density; using a *Type A* polymer [134]

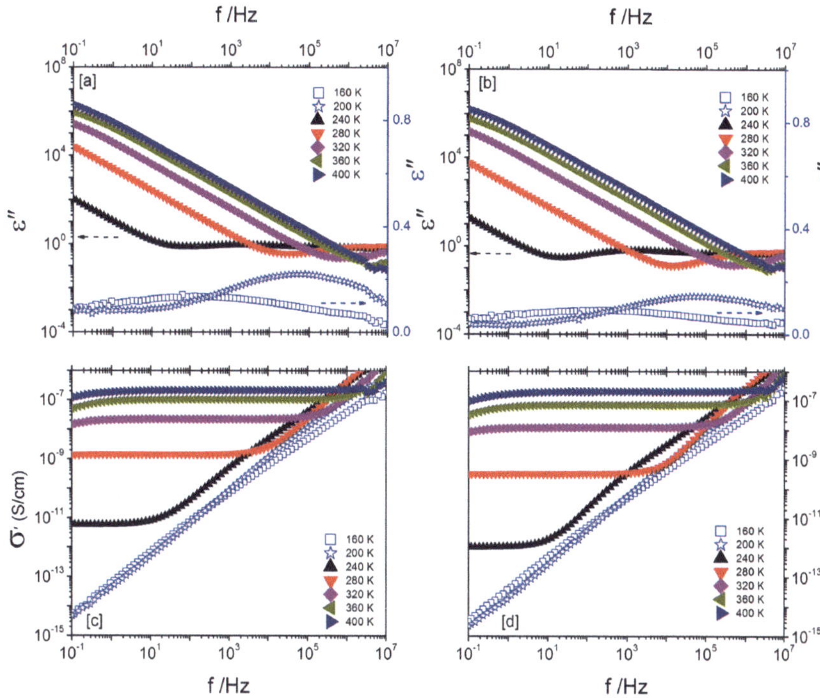

Fig. 13 The imaginary part of the complex dielectric function, ε^*, and the real part of the complex conductivity function, σ', for PEI **a**, **c** and NOHM-I-PEI **b**, **d**. The error bars are comparable to the size of the symbols unless otherwise indicated. Reprinted with permission from Ref. [123]. Copyright (2021) American Chemical Society

as canopy, Kim et al. have demonstrated this for a NOHM material comprising a poly(*cis*–1,4–isoprene) [135].

Concerning the shape of the α-process, representative HN functions obtained by fitting Eq. 3 to the dielectric loss data are shown in Fig. 15. For ease of comparison, the functions are normalized with respect to the maximal loss value. It is observed that the peaks are narrower for the neat polymer compared to the NOHM material, a finding that runs counter to the results discussed in Sect. 3.1 (PMMA-based composites), and to other reports about conventional nanocomposites [22, 34, 95]. This is nevertheless consistent with an observed broader glass transition in PEI than in the NOHM (see inset, Fig. 15). As discussed earlier for PMMA-based nanocomposites, the broadening of the distribution of the relaxation times for the primary α- relaxation arises from interfacial interactions, and from an increase in free volume due to chain packing frustrations in the wetting/dewetting zones of the graft/matrix. The polymer used for the preparation of NOHMs in this work is highly branched PEI, and therefore the neat system has more free volume than a simple linear polymer [136, 137]. The NOHM system by definition is self-suspended, i.e., no bulk-free polymer matrix exists; we posit that the silica nanoparticles take up some of the free volume that

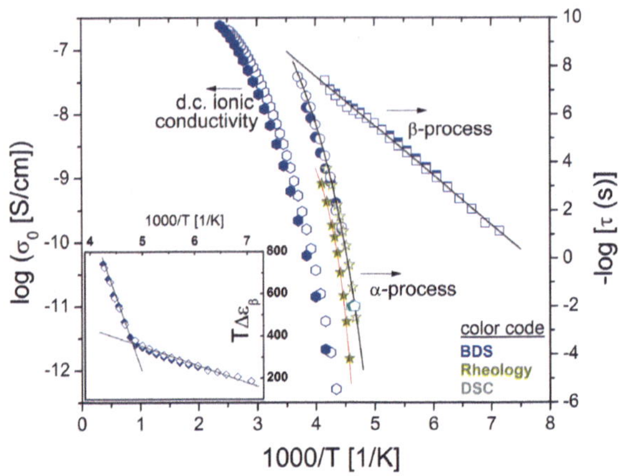

Fig. 14 Mean relaxation times of the α- and β-relaxations (*right y-axis*) for PEI and NOHM-I-PEI obtained from BDS (*blue symbols*) and rheology (*dark yellow*). The mean times, τ_m, from the rheology data are estimated as $\tau_m(T) = \tau_{ref}(T_{ref})/a_T$, where τ_{ref} is the mean structural relaxation time at the reference temperature T_{ref} and a_T is the corresponding shift factor. The DC ionic conductivity (*left y-axis*) *versus* inverse temperature for the two samples (*hexagons*) is also presented. The T_g values determined from DSC are shown by *pentagon symbols* and are plotted using the assumption that the calorimetric glass transition temperature coincides with a segmental relaxation time of 100 s. Inset: the temperature evolution of the dielectric relaxation strength associated with the β-process in the neat polymer and the NOHM. In all cases, *empty symbols* are for PEI, while *filled symbols* represent NOHM-I-PEI data. The error bars are comparable to the size of the symbols unless otherwise indicated. Reprinted with permission from Ref. [123]. Copyright (2021) American Chemical Society

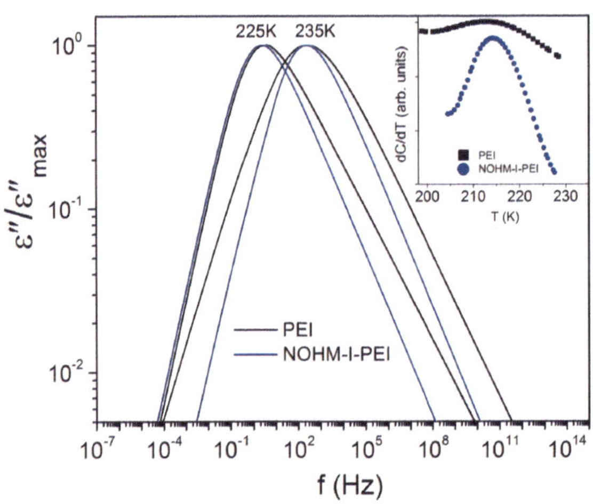

Fig. 15 HN functions used to describe the structural relaxation in the polymer (PEI) and the hybrid material (NOHM-I-PEI) at 225 and 235 K. Inset: plots of the temperature derivative of the heat capacity traces showing the T_g and breadth of the transition in PEI and NOHM-I-PEI. Interestingly, the hybrid material has a narrower peak than the neat polymer. Adapted with permission from Ref. [123]. Copyright (2021) American Chemical Society

would otherwise be present in neat PEI. This effectively "freezes" out some faster modes, a phenomenon that manifests itself in terms of a narrower peak in the hybrid. Furthermore, as deduced from the X-ray scattering data, the NOHM is hierarchically organized including mesoscale structures of about 1.4 nm in size. Cosby et al. have shown, for other liquids, that the dynamic signature for a mesoscale organization is a Debye-like relaxation that is slower than the structural process [138]. While we do not find a new sub-α-peak in the dielectric spectra of the NOHM material, the narrowing of the α-process may be indicative of the contribution of mesostructures to the total polarization in the hybrid material. More experimental and simulation efforts are required to shed more light on this aspect of dynamics in NOHMs.

Finally, we now consider the long-range ionic conductivity in these samples. As observed in Fig. 13, σ' exhibits a plateau at temperatures above 200 K in our frequency window. This plateau corresponds to the DC ionic conductivity, σ_0, of the material. The decrease observed upon further lowering of the frequency, see, for instance, data taken at 360 and 400 K, is due to electrode polarization which has its origin in the accumulation of charges at the electrodes, as well as the Maxwell–Wagner–Sillars polarization due to the presence of interfaces. These phenomena are outside the scope of the current study. The values of σ_0 are plotted as functions of inverse temperature in Fig. 14. The levels of ionic conductivity observed here are typical for many amorphous polymers that basically arise from minute amounts of impurities left over from polymerization. However, since σ_0 in the neat polymer is slightly higher than that of the hybrid material, this may be deciphered to be proton conductivity: hydrogen bonding in NOHM-I-PEI causes a reduction in the number of protons available to contribute to long-range charge transport compared to the case in neat PEI. This idea is worth substantive investigation using appropriate systems such as NOHMs bearing polymerized ionic liquids as canopies.

^{13}C NMR was used to gain a mechanistic understanding of the local motions of methylene groups in both PEI and NOHM-I-PEI by comparing their respective spin–lattice relaxation, thereby helping provide greater insight into the information derived from dielectric spectroscopy. Figure 16 displays the NMR spectra for the hybrid material as measured in the temperature range, 280–400 K. There are eight distinguishable peaks corresponding to different methylene groups identified (see inset) according to data available in the literature [139–141]. The degree of branching of the nitrogen atoms in the α and β positions of the carbon atom can be associated with the chemical shift of ^{13}C. Peaks *a* and *b* represent methylene groups located between a secondary and tertiary amino group, with the former being adjacent to secondary and the latter to tertiary groups. Peaks *c* and *d* are associated with a pair of equivalent methylene groups between two tertiary groups and between two secondary groups, respectively. The peaks *e*–*f* and *g*–*h* are found between a tertiary and primary group, and a secondary and primary group, respectively. The relaxation of the interior methylene groups of the PEI chain would therefore be represented by peaks *a*, *b*, *c*, and *d*, while that of the exterior is reflected in *e*, *f*, *g*, and *h*. As the temperature is lowered from 320 to 280 K, it is observed that the peaks undergo appreciable broadening so much so that the temperature-dependent data exhibit a T_1 minimum. This kind of minimum takes place at the temperature at which the

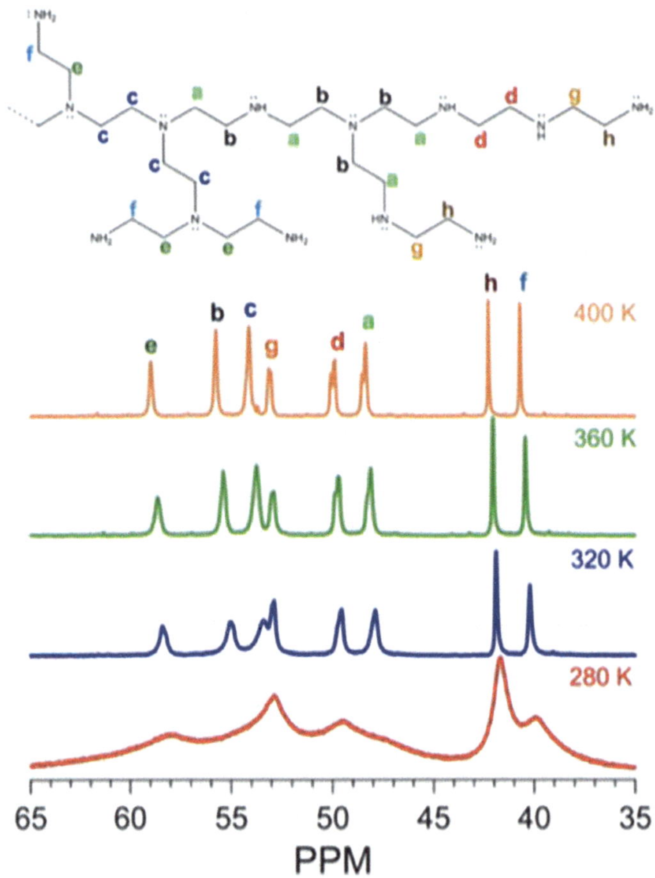

Fig. 16 Scheme of the partial structure of the branched PEI (top) with alphabetic labels representing different methylene groups in the polymer chain. The ^{13}C NMR spectra (bottom) of neat, dry NOHM-I-PEI at different temperatures with the respective assignments. Reprinted with permission from Ref. [123]. Copyright (2021) American Chemical Society

molecular motions are close to the observatory frequency, (i.e., 100.1 MHz in this case), satisfying the requirement $\omega\tau_c \sim 6.16$ [142].

Figure 17 shows T_1 values plotted as a function of inverse temperature for the relaxation of both the interior and exterior methylene groups in the hybrid material. Given the fact that the minimum occurs at lower temperatures for the methylene groups at the ends of the chains compared to those in the interior, it follows that there are longer correlation times for the latter groups. For both PEI and NOHM-I-PEI, therefore, the local motion is reduced for the carbons located in the interior parts of the polymer chain. For each methylene group, there is remarkable similarity of the findings between PEI and NOHM-I-PEI in terms of shape and temperature of T_1 minimum (see insets, Fig. 17) implying that the presence of silica nanoparticles

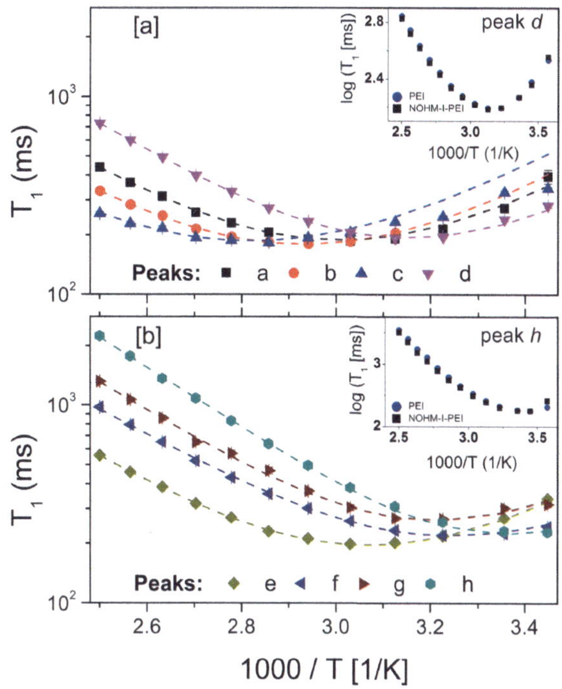

Fig. 17 A plot of the relaxation times for the interior methylene groups **a** and exterior methylene groups **b** in NOHM-I-PEI as a function of inverse temperature. The dashed lines are calculated from the estimated correlation times based on Eq. 7. Insets: T_1 relaxation times for peak *d* **a** and peak *h* **b** in the neat PEI and NOHM-I-PEI canopy as a function of inverse temperature. Adapted with permission from Ref. [123]. Copyright (2021) American Chemical Society

does not hinder the local polymer dynamics in the canopy. This is consistent with the findings from BDS about the α- and β-processes, and with other related studies [70].

As described by Gouverneur et al. [142], using the Bloembergen–Purcell–Pound (BPP) model, the correlation times were extracted from the relaxation times as

$$\frac{1}{T_1} = C\left(\frac{\tau_C}{1+\omega_0^2\tau_C^2} + \frac{4\tau_C}{1+4\omega_0^2\tau_C^2}\right) \tag{7}$$

where the relaxation rate is related to the correlation time τ_C with C and ω_C constant [143]. The temperature trend of the correlation time has Arrhenius-type relationship in this analysis, allowing the estimation of the activation energies. The dashed lines in Fig. 17 represent the relaxation time calculated from the correlation times for the NOHM-I-PEI. Specifically, there is good agreement between the model and the acquired data for peaks associated with methylene groups close to primary amino groups, which would correspond to the side and end groups of the polymer (i.e., *e*, *f*, *g*, and *h*) and those situated between two secondary amino groups (i.e., peak *d*). This means that the local dynamics can be explained by one mode of motion. The relaxation of the exterior methylene groups is in tandem with that of the β-process as identified from BDS in terms of the activation energy (~25 kJ/mol). Methylene groups close to tertiary amino groups (i.e., *a*, *b*, and *c*) show the most significant

Table 4 Activation energies estimated from the Bloembergen–Purcell–Pound (BPP) and Arrhenius models for the neat PEI and NOHM-I-PEI samples. Also included in this table are estimated temperatures at the T_1 minimum for the two samples

Peak	PEI			NOHM-I-PEI		
	E_a [±0.05 kJ/mol]	T ($T_{1,min}$) [±0.1 K]	T^{-1} ($T_{1,min}$) [±0.1 K^{-1}]	E_a [±0.05 kJ/mol]	T ($T_{1,min}$) [±0.1 K]	T^{-1} ($T_{1,min}$) [±0.1 K^{-1}]
a	25.81	330.5	3.0	25.52	329.9	3.0
b	25.39	341.9	2.9	25.16	340.1	2.9
c	22.26	353.3	2.8	23.72	350.2	2.9
d	26.51	316.6	3.1	26.50	316.8	3.2
e	26.46	324.4	3.1	26.15	326.4	3.1
f	25.68	304.6	3.3	24.93	304.7	3.1
g	28.71	312.1	3.2	27.59	310.6	3.2
g	29.39	293.7	3.4	29.50	297.2	3.4

deviations from the rest. This could be explained in terms of either a deviation from Arrhenius character (as is the case for the α-process identified from BDS studies) or the significant broadening of the corresponding peaks at low temperatures. As seen from the activation energies recorded in Table 4, although neat PEI has slightly higher values, the absolute numbers are very similar for all studied peaks.

4 Conclusions

A series of conventional PMMA-based nanocomposites, with and without graft chains, have been studied by varying the molecular weight of the matrix and the loading levels (0, 1, 3, and 5 wt%, based on the mass of the silica nanoparticles) of nanoparticles. The main difference between PNCs and PGNCs is found to show up in terms of the shape of their respective primary structural relaxation. PGNCs show pronounced broadening on the high-frequency side as a function of the molecular weight of the matrix and the nanoparticle loading level. Interestingly, in spite of the presence of these faster modes, calorimetric studies show that for a given molecular weight of the host matrix, the T_g of the PGNCs is higher than that of PNCs. Instead, the trend in values of T_g scales with the mean relaxation times of the α-process. These studies suggest that faster modes for the α-relaxation arise from density fluctuations driven by chain packing frustrations due to entanglements and chain interdiffusion. The character of this nonuniform interphase region is then to be considered as an essential attribute that affects the dynamic behavior and dispersion of PNCs and PGNCs. Figure 18 shows a schematic representation of the changes that happen to the segmental mobility due to matrix–graft and matrix–nanoparticle interactions. The picture drawn with the PNC and PGNC systems in mind is unified: spatial confinement of chains to the interface hinders the motion of segments, while the very

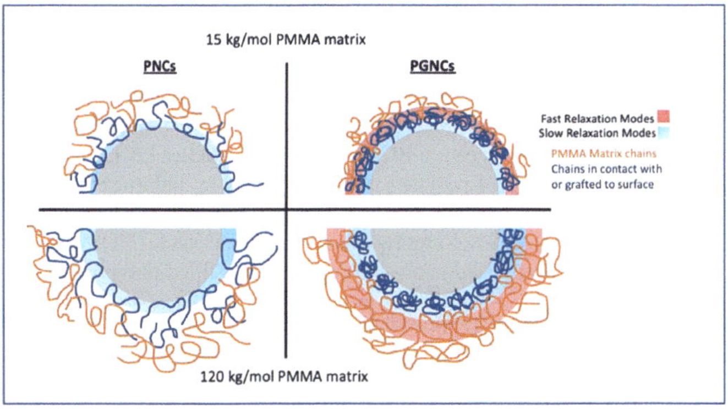

Fig. 18 Schematic representation of polymer-based nanocomposites, depicting the changes in segmental mobility that are envisioned in the vicinity of silica nanoparticles in PNCs and PGNCs. Reprinted with permission from Ref. [95]. Copyright (2021) American Chemical Society

presence of an interfacial region generated between the surface (whether bare or decorated) and the matrix gives rise to faster modes due to an increase in free volume in the interfacial zone. For the self-suspended nanocomposite system based on highly branched polyethylenimine, intricate dynamics are observed: the structural relaxation has a narrower peak in the hybrid than the neat polymer. Additionally, while BDS results show identical structural dynamics in the NOHM and PEI in terms of their mean relaxation times, rheological experiments reveal a slowed-down structural process in the hybrid. By combining insights from structural and dynamics studies, it is shown that the network of silica nanoparticles in NOHMs contribute to their mechanical energy dissipation, and that hierarchical organization in the system influences both dipolar relaxations and charge transport.

Acknowledgements This work was supported as part of the Breakthrough Electrolytes for Energy Storage (BEES), an Energy Frontier Research Center funded by the U.S. Department of Energy, Office of Science, Basic Energy Sciences, under Award No.: DE- SC0019409. E. U. M. and J. S. acknowledge partial financial support from the National Science Foundation, Division of Materials Research, Polymers Program, through DMR-1905597. M. A. H. is grateful for financial support from the National Science Foundation, the Division of Chemistry, through no. CHE-1753282. S. M. K. and D. P. S. are grateful for partial support from The Department of Energy's Kansas City National Security Campus, which is operated and managed by Honeywell Federal Manufacturing & Technologies, LLC, under Contract DE-NA-0002839. In addition, S. M. K. acknowledges partial support from the National Science Foundation (CBET-1512221). Rheology measurements were conducted at the Oak Ridge National Laboratory's Center for Nanophase Materials Sciences, which is sponsored by the Division of Scientific User Facilities, Office of Basic Energy Sciences, U.S. Department of Energy. SAXS measurements were performed at the Shared Materials Instrumentation Facility (SMIF) at Duke University.

References

1. Bansal A, Yang H, Li C, Cho K, Benicewicz BC, Kumar SK et al (2005) Quantitative equivalence between polymer nanocomposites and thin polymer films. Nat Mater 4:693–698
2. Bansal A, Yang H, Li C, Benicewicz BC, Kumar SK, Schadler LS (2006) Controlling the thermomechanical properties of polymer nanocomposites by tailoring the polymer-particle interface. J Polym Sci Part B: Polym Phys 44:2944–2950
3. Rong MZ, Zhang MQ, Zheng YX, Zeng HM, Walter R, Friedrich K (2001) Structure-property relationships of irradiation grafted nano-inorganic particle filled polypropylene composites. Polymer 42:167–183
4. Rong MZ, Zhang MQ, Ruan WH (2006) Surface modification of nanoscale fillers for improving properties of polymer nanocomposites: a review. Mater Sci Technol 22:787–796
5. Agrawal A, Choudhury S, Archer LA (2015) A highly conductive, non-flammable polymer-nanoparticle hybrid electrolyte. RSC Adv 5:20800–20809
6. Wang H, Zeng C, Elkovitch M, Lee LJ, Koelling KW (2001) Processing and properties of polymeric nano-composites. Polym Eng Sci 41:2036–2046
7. Balazs AC, Emrick T, Russell TP (2006) Nanoparticle polymer composites: where two small worlds meet. Science 314:1107–1110
8. Kim D, Srivastava S, Narayanan S, Archer LA (2012) Polymer nanocomposites: polymer and particle dynamics. Soft Matter 8:10813–10818
9. Askar S, Li L, Torkelson JM (2017) Polystyrene-grafted silica nanoparticles: investigating the molecular weight dependence of glass transition and fragility behavior. Macromolecules 50:1589–1598
10. Krishnamoorti R, Vaia RA (2007) Polymer nanocomposites. J Polym Sci, Part B: Polym Phys 45:3252–3256
11. Kumar SK, Benicewicz BC, Vaia RA, Winey KI (2017) 50th anniversary perspective: are polymer nanocomposites practical for applications? Macromolecules 50:714–731
12. Hong RY, Fu HP, Zhang YJ, Liu L, Wang J, Li HZ et al (2007) Surface-modified silica nanoparticles for reinforcement of PMMA. J Appl Polym Sci 105:2176–2184
13. Brown D, Mele P, Marceau S, Alberola ND (2003) A molecular dynamics study of a model nanoparticle embedded in a polymer matrix. Macromolecules 36:1395–1406
14. Senses E, Akcora P (2013) An interface-driven stiffening mechanism in polymer nanocomposites. Macromolecules 46:1868–1874
15. Chevigny C, Jouault N, Dalmas F, Boué F, Jestin J (2011) Tuning the mechanical properties in model nanocomposites: influence of the polymer-filler interfacial interactions. J Polym Sci, Part B: Polym Phys 49:781–791
16. Ueno K, Inaba A, Kondoh M, Watanabe M (2008) Colloidal stability of bare and polymer-grafted silica nanoparticles in ionic liquids. Langmuir 24:5253–5259
17. Cheng S, Mirigian S, Carrillo JM, Bocharova V, Sumpter BG, Schweizer KS et al (2015) Revealing spatially heterogeneous relaxation in a model nanocomposite. J Chem Phys 143:194704
18. de Gennes PG (1987) Polymers at an interface: a simplified view. Adv Colloid Interface Sci 27:189–209
19. Glomann T, Schneider GJ, Allgaier J, Radulescu A, Lohstroh W, Farago B et al (2013) Microscopic dynamics of polyethylene glycol chains interacting with silica nanoparticles. Phys Rev Lett 110:178001
20. Gin P, Jiang N, Liang C, Taniguchi T, Akgun B, Satija SK et al (2012) Revealed architectures of adsorbed polymer chains at solid-polymer melt interfaces. Phys Rev Lett 109:265501
21. Ghanbari A, Rahimi M, Dehghany J (2013) Influence of surface grafted polymers on the polymer dynamics in a silica-polystyrene nanocomposite: a coarse-grained molecular dynamics investigation. J Phys Chem C 117:25069–25076
22. Gong S, Chen Q, Moll JF, Kumar SK, Colby RH (2014) Segmental dynamics of polymer melts with spherical nanoparticles. ACS Macro Lett 3:773–777

23. Cheng S, Holt AP, Wang H, Fan F, Bocharova V, Martin H et al (2016) Unexpected molecular weight effect in polymer nanocomposites. Phys Rev Lett 116:038302
24. Cheng S, Carroll B, Lu W, Fan F, Carrillo JMY, Martin H et al (2017) Interfacial properties of polymer nanocomposites: role of chain rigidity and dynamic heterogeneity length scale. Macromolecules 50:2397–2406
25. Jouault N, Moll JF, Meng D, Windsor K, Ramcharan S, Kearney C et al (2013) Bound polymer layer in nanocomposites. ACS Macro Lett 2:371–374
26. Jouault N, Jestin J (2015) Intra- and interchain correlations in polymer nanocomposites: a small-angle neutron scattering extrapolation method. ACS Macro Lett 5:1095–1099
27. Carroll B, Cheng S, Sokolov AP (2017) Analyzing the interfacial layer properties in polymer nanocomposites by broadband dielectric spectroscopy. Macromolecules 50:6149–6163
28. Koga T, Jiang N, Gin P, Endoh MK, Narayanan S, Lurio LB et al (2011) Impact of an irreversibly adsorbed layer on local viscosity of nanoconfined polymer melts. Phys Rev Lett 107:225901
29. Bogoslovov RB, Roland CM, Ellis AR, Randall AM, Robertson CG (2008) Effect of silica nanoparticles on the local segmental dynamics in poly(vinyl acetate). Macromolecules 41:1289–1296
30. Holt AP, Bocharova V, Cheng S, Kisliuk AM, White BT, Saito T et al (2016) Controlling interfacial dynamics: covalent bonding versus physical adsorption in polymer nanocomposites. ACS Nano 10:6843–6852
31. Carrillo JMY, Cheng S, Kumar R, Goswami M, Sokolov AP, Sumpter BG (2015) Untangling the effects of chain rigidity on the structure and dynamics of strongly adsorbed polymer melts. Macromolecules 48:4207–4219
32. Sargsyan A, Tonoyan A, Davtyan S, Schick C (2007) The amount of immobilized polymer in PMMA SiO_2 nanocomposites determined from calorimetric data. Eur Polym J 43:3113–3127
33. Holt AP, Griffin PJ, Bocharova V, Agapov AL, Imel AE, Dadmun MD et al (2014) Dynamics at the polymer/nanoparticle interface in poly(2-vinylpyridine)/silica nanocomposites. Macromolecules 47:1837–1843
34. Füllbrandt M, Purohit PJ, Schönhals A (2013) Combined FTIR and dielectric investigation of poly(vinyl acetate) adsorbed on silica particles. Macromolecules 46:4626–4632
35. Harton SE, Kumar SK, Yang H, Koga T, Hicks K, Lee H et al (2010) Immobilized polymer layers on spherical nanoparticles. Macromolecules 43:3415–3421
36. Kim SY, Meyer HW, Saalwächter K, Zukoski CF (2012) Polymer dynamics in PEG-silica nanocomposites: effects of polymer molecular weight, temperature and solvent dilution. Macromolecules 45:4225–4237
37. Betancourt BA, Douglas JF, Starr FW (2013) Fragility and cooperative motion in a glass-forming polymer-nanoparticle composite. Soft Matter 9:241–254
38. Cheng S, Xie SJ, Carrillo JY, Carroll B, Martin H, Cao PF et al (2017) Big effect of small nanoparticles: a shift in paradigm for polymer nanocomposites. ACS Nano 11:752–759
39. Zhang Q, Archer LA (2002) Poly(ethylene oxide)/silica nano-composites: structure and rheology. Langmuir 18:10435–10442
40. Berriot J, Lequeux F, Monnerie L, Montes H, Long D, Sotta P (2002) Filler-elastomer interaction in model filled rubbers, a 1H NMR study. J Non-Cryst Solids 307–310:719–724
41. Kaufman S, Slichter WP, Davis DD (1971) Nuclear magnetic resonance study of rubber–carbon black interactions. J Polym Sci: Part A–2, 9:829–839
42. Krutyeva M, Wischnewski A, Monkenbusch M, Willner L, Maiz J, Mijangos C et al (2013) Effect of nanoconfinement on polymer dynamics: surface layers and interphases. Phys Rev Lett 110:108303
43. Mapesa EU, Shahidi N, Kremer F, Doxastakis M, Sangoro J (2021) Interfacial dynamics in supported ultrathin polymer films-from the solid to the free interface. J Phys Chem Lett 12:117–125
44. Bagwe RP, Hilliard LR, Tan W (2006) Surface modification of silica nanoparticles to reduce aggregation and nonspecific binding. Langmuir 22:4357–4362

45. Ganesan V, Jayaraman A (2014) Theory and simulation studies of effective interactions, phase behavior and morphology in polymer nanocomposites. Soft Matter 10:13–38
46. Kumar SK, Jouault N, Benicewicz B, Neely T (2013) Nanocomposites with polymer grafted nanoparticles. Macromolecules 46:3199–3214
47. Pandey YN, Papakonstantopoulos GJ, Doxastakis M (2013) Polymer/nanoparticle interactions: bridging the gap. Macromolecules 46:5097–5106
48. Kim D, Archer LA (2011) Nanoscale organic-inorganic hybrid lubricants. Langmuir 27:3083–3094
49. Moganty SS, Jayaprakash N, Nugent JL, Shen J, Archer LA (2010) Ionic-liquid-tethered nanoparticles: hybrid electrolytes. Angew Chem Int Ed Engl 49:9158–9161
50. Warren SC, Banholzer MJ, Slaughter LS, Giannelis EP, DiSalvo FJ, Wiesner UB (2006) Generalized route to metal nanoparticles with liquid behavior. J Am Chem Soc 128:12074–12075
51. Voevodin AA, Vaia RA, Patton ST, Diamanti S, Pender M, Yoonessi M et al (2007) Nanoparticle-wetted surfaces for relays and energy transmission contacts. Small 3:1957–1963
52. Patton ST, Voevodin AA, Vaia RA, Pender M, Diamanti SJ, Phillips B (2008) Nanoparticle liquids for surface modification and lubrication of mems switch contacts. J Microelectromech Syst 17:741–746
53. Park SI, Lim JH, Kim JH, Yun HI, Roh JS, Kim CG et al (2004) Effects of surfactant on properties of magnetic fluids for biomedical application. Phys Status Solidi B 241:1662–1664
54. Morais PC, Santos JG, Silveira LB, Gansau C, Buske N, Nunes WC et al (2004) Susceptibility investigation of the nanoparticle coating-layer effect on the particle interaction in biocompatible magnetic fluids. J Magn Magn Mater 272–276:2328–2329
55. Yang Y, Zhang ZG, Grulke EA, Anderson WB, Wu G (2005) Heat transfer properties of nanoparticle-in-fluid dispersions (nanofluids) in laminar flow. Int J Heat Mass Transfer 48:1107–1116
56. Murshed SMS, Leong KC, Yang C (2006) A model for predicting the effective thermal conductivity of nanoparticle-fluid suspensions. Int J Nanosci 05:23–33
57. Jang SP, Choi SUS (2004) Role of brownian motion in the enhanced thermal conductivity of nanofluids. Appl Phys Lett 84:4316–4318
58. Jwo CS, Teng TP, Chang H (2007) A simple model to estimate thermal conductivity of fluid with acicular nanoparticles. J Alloys Compd 434–435:569–571
59. Wang X, Xu X, Choi SUS (1999) Thermal conductivity of nanoparticle-fluid mixture. J Thermophys Heat Transfer 13:474–480
60. Eastman JA, Choi SUS, Li S, Yu W, Thompson LJ (2001) Anomalously increased effective thermal conductivities of ethylene glycol-based nanofluids containing copper nanoparticles. Appl Phys Lett 78:718–720
61. Pil Jang S, Choi SUS (2007) Effects of various parameters on nanofluid thermal conductivity. J Heat Transfer 129:617–623
62. Kim I, Svendsen HF (2011) Comparative study of the heats of absorption of post-combustion CO_2 absorbents. Int J Greenhouse Gas Control 5:390–395
63. Shin JY, Lee BS, Jung Y, Kim SJ, Lee S-g (2007) Palladium nanoparticles captured onto spherical silica particles using a urea cross–linked imidazolium molecular band. Chem Commun 5238–5240
64. Ma Z, Yu J, Dai S (2010) Preparation of inorganic materials using ionic liquids. Adv Mater 22:261–285
65. Park Y, Shin D, Jang YN, Park A-HA (2011) CO_2 capture capacity and swelling measurements of liquid-like nanoparticle organic hybrid materials via attenuated total reflectance fourier transform infrared spectroscopy. J Chem Eng Data 57:40–45
66. Park Y, Petit C, Han P, Alissa Park A-H (2014) Effect of canopy structures and their steric interactions on CO_2 sorption behavior of liquid-like nanoparticle organic hybrid materials. RSC Adv 4:8723

67. Lin K-YA, Park A-HA (2011) Effects of bonding types and functional groups on CO_2 capture using novel multiphase systems of liquid-like nanoparticle organic hybrid materials. Environ Sci Technol 45:6633–6639
68. Bourlinos AB, Ray Chowdhury S, Herrera R, Jiang DD, Zhang Q, Archer LA et al (2005) Functionalized nanostructures with liquid-like behavior: expanding the gallery of available nanostructures. Adv Funct Mater 15:1285–1290
69. Bourlinos AB, Herrera R, Chalkias N, Jiang DD, Zhang Q, Archer LA et al (2005) Surface-functionalized nanoparticles with liquid-like behavior. Adv Mater 17:234–237
70. Jespersen ML, Mirau PA, Ev M, Vaia RA, Rodriguez R, Giannelis EP (2010) Canopy dynamics in nanoscale ionic materials. ACS Nano 4:3735–3742
71. Nugent JL, Moganty SS, Archer LA (2010) Nanoscale organic hybrid electrolytes. Adv Mater 22:3677–3680
72. Agarwal P, Qi H, Archer LA (2010) The Ages in a self-suspended nanoparticle liquid. Nano Lett 10:111–115
73. Liu X, Abel BA, Zhao Q, Li S, Choudhury S, Zheng J et al (2019) Microscopic origins of caging and equilibration of self-Suspended hairy nanoparticles. Macromolecules 52:8187–8196
74. Andrew Lin K-Y, Park Y, Petit C, Park A-HA (2014) Thermal stability, swelling behavior and CO_2 absorption properties of nanoscale ionic materials (NIMs). RSC Adv 4:65195–65204
75. Lin K-YA, Petit C, Park A-HA (2013) Effect of SO_2 on CO_2 capture using liquid-like nanoparticle organic hybrid materials. Energy Fuels 27:4167–4174
76. Petit C, Lin K-YA, Park A-HA (2013) Design and characterization of liquidlike poss-based hybrid nanomaterials synthesized via ionic bonding and their interactions with CO_2. Langmuir 29:12234–12242
77. Park Y, Decatur J, Lin K-YA, Park A-HA (2011) Investigation of CO_2 capture mechanisms of liquid-like nanoparticle organic hybrid materials via structural characterization. Phys Chem Chem Phys 13:18115–18122
78. Yu W, Wang T, Park A-HA, Fang M (2020) Toward sustainable energy and materials: CO_2 capture using microencapsulated sorbents. Ind Eng Chem Res 59:9746–9759
79. Wang J, Fang W, Luo J, Gao M, Wan Y, Zhang S et al (2019) Selective separation of CO_2 using novel mixed matrix membranes based on pebax and liquid-like nanoparticle organic hybrid materials. J Membr Sci 584:79–88
80. Yu W, Wang T, Park A-HA, Fang M (2019) Review of liquid nano-absorbents for enhanced CO_2 capture. Nanoscale 11:17137–17156
81. George M, Weiss RG (2001) Chemically reversible organogels: aliphatic amines as "latent" gelators with carbon dioxide. J Am Chem Soc 123:10393–10394
82. George M, Weiss RG (2002) Chemically reversible organogels via "latent" gelators. Aliphatic amines with carbon dioxide and their ammonium carbamates. Langmuir 18:7124–7135
83. da Silva EF, Svendsen HF (2004) Ab initio study of the reaction of carbamate formation from CO_2 and alkanolamines. Ind Eng Chem Res 43:3413–3418
84. Gabrielsen J, Michelsen ML, Stenby EH, Kontogeorgis GM (2005) A Model for estimating CO_2 solubility in aqueous alkanolamines. Ind Eng Chem Res 44:3348–3354
85. Liu YD, Zhang LZ, Watanasiri S (1999) Representing vapor-liquid equilibrium for an aqueous MEA-CO2 system using the electrolyte non-random-two-liquid model. Ind Eng Chem Res 38:2080–2090
86. Zhang J, Zhang S, Dong K, Zhang Y, Shen Y, Lv X (2006) Supported absorption of CO2 by tetrabutylphosphonium amino acid ionic liquids. Chem Eur J 12:4021–4026
87. Bourlinos AB, Giannelis EP, Zhang Q, Archer LA, Floudas G, Fytas G (2006) Surface-functionalized nanoparticles with liquid-like behavior: the role of the constituent components. Eur Phys J E Soft Matter 20:109–117
88. Street DP, Mah AH, Patterson S, Pickel DL, Bergman JA, Stein GE et al (2018) Interfacial interactions in PMMA/silica nanocomposites enhance the performance of parts created by fused filament fabrication. Polymer 157:87–94

89. Natarajan B, Li Y, Deng H, Brinson LC, Schadler LS (2013) Effect of interfacial energetics on dispersion and glass transition temperature in polymer nanocomposites. Macromolecules 46:2833–2841
90. Li Y, Benicewicz BC (2008) Functionalization of silica nano-particles via the combination of surface-initiated RAFT polymerization and click reactions. Macromolecules 41:7986–7992
91. Li C, Han J, Ryu CY, Benicewicz BCA (2006) Versatile method to prepare RAFT agent anchored substrates and the preparation of PMMA grafted nanoparticles. Macromolecules 39:3175–3183
92. Jiao Y, Akcora P (2012) Assembly of polymer-grafted magnetic nanoparticles in polymer melts. Macromolecules 45:3463–3470
93. Martin TB, Mongcopa KI, Ashkar R, Butler P, Krishnamoorti R, Jayaraman A (2015) Wetting-dewetting and dispersion-aggregation transitions are distinct for polymer grafted nano-particles in chemically dissimilar polymer matrix. J Am Chem Soc 137:10624–10631
94. Cantillo NM, Bruce M, Hamilton ST, Feric TG, Park A-HA, Zawodzinski TA (2020) Electrochemical behavior of copper ion complexed with nanoparticle organic hybrid materials. J Electrochem Soc 167:116508
95. Mapesa EU, Street DP, Heres MF, Kilbey SM, Sangoro J (2020) Wetting and chain Packing across interfacial zones affect distribution of relaxations in polymer and polymer-grafted nanocomposites. Macromolecules 53:5315–5325
96. Kremer F, Schönhals A (eds) (2003) Broadband dielectric spectroscopy. Springer, Berlin
97. Bahar I, Erman B, Kremer F, Fischer EW (1992) Segmental motions of cis-polyisoprene in the bulk state: interpretation of dielectric relaxation data. Macromolecules 25:816–825
98. Schmidt-Rohr K, Kulik AS, Beckham HW, Ohlemacher A, Pawelzik U, Boeffel C et al (1994) Molecular nature of the beta-relaxation in poly(methyl methacrylate) investigated by multidimensional NMR. Macromolecules 27:4733–4745
99. Bergman R, Alvarez F, Alegría A, Colmenero J (1998) The merging of the dielectric α- and β-relaxations in poly(methyl methacrylate). J Chem Phys 109:7546–7555
100. Wübbenhorst M, van Turnhout J (2002) Analysis of complex dielectric spectra. I. one-dimensional derivative techniques and three- dimensional modelling. J Non-Cryst Solids 305:40–49
101. Havriliak S, Negami S (1967) A complex plane representation of dielectric and mechanical relaxation processes in some polymers. Polymer 8:161–210
102. Heres M, Cosby T, Mapesa EU, Sangoro J (2016) Probing nanoscale ion dynamics in ultrathin films of polymerized ionic liquids by broadband dielectric spectroscopy. ACS Macro Lett 5:1065–1069
103. Ribelles JLG, Calleja RD (1985) The beta dielectric relaxation in some methacrylate polymers. J Polym Sci, Polym Phys Ed 23:1297–1307
104. Serghei A, Mikhailova Y, Huth H, Schick C, Eichhorn KJ, Voit B et al (2005) Molecular dynamics of hyperbranched polyesters in the confinement of thin films. Eur Phys J E: Soft Matter Biol Phys 17:199–202
105. Boucher VM, Cangialosi D, Alegría A, Colmenero J, Gonzaléz-Irun J, Liz-Marzan LM, (2011) Physical aging in PMMA/silica nanocomposites: enthalpy and dielectric relaxation. J Non-Cryst Solids 357:605–609
106. Cangialosi D, Boucher VM, Alegría A, Colmenero J (2012) Enhanced physical aging of polymer nanocomposites: the key role of the area to volume ratio. Polymer 53:1362–1372
107. Bergman R, Alvarez F, Alegria A, Colmenero J (1998) Dielectric relaxation in PMMA revisited. J Non-Cryst Solids 235–237:580–583
108. Sasabe H, Saito S (1968) Dielectric relaxations and electrical conductivities of poly(alkyl methacrylates) under high pressure. J Polym Sci 6:1401–1418
109. Garwe F, Schönhals A, Lockwenz H, Beiner M, Schröter K, Donth E (1996) Influence of cooperative α dynamics on local β relaxation during the development of the dynamic glass transition in poly(n-alkyl methacrylate)s. Macromolecules 29:247–253
110. Garwe F, Schönhals A, Beiner M, Schröter K, Donth E (1994) Molecular cooperativity against locality at glass transition onset in poly(n–butyl methacrylate). J Phys (1994) Condens Matter 6:6941–6945

111. Fulcher GS (1992) Analysis of recent measurements of the viscosity of glasses - reprint. J Am Ceram Soc. 75:1043–1059
112. Tammann G, Hesse W (1926) Die Abhängigkeit der Viscosität von der Temperatur bei unterkühlten Flüssigkeiten. Z anorg allg Chem 156:245–257
113. Vogel H (1921) The law of the relationship between viscosity of liquids and the temperature. Phys Z 22:645–646
114. Ash BJ, Siegel RW, Schadler LS (2004) Mechanical behavior of alumina/poly(methyl methacrylate) nanocomposites. Macromolecules 37:1358–1369
115. Salavagione HJ, Martínez G, Gómez MA (2009) Synthesis of poly(vinyl alcohol)/reduced graphite oxide nanocomposites with improved thermal and electrical properties. J Mater Chem 19:5027
116. Ramanathan T, Stankovich S, Dikin DA, Liu H, Shen H, Nguyen ST (2007) Graphitic nanofillers in PMMA nanocomposites - an investigation of particle size and dispersion and their influence on nanocomposite properties. J Polym Sci, Part B: Polym Phys 45:2097–2112
117. Rittigstein P, Priestley RD, Broadbelt LJ, Torkelson JM (2007) Model polymer nanocomposites provide an understanding of confinement effects in real nanocomposites. Nat Mater 6:278–82
118. Akcora P, Kumar SK, Garcïa Sakai V, Li Y, Benicewicz BC, Schadler LS (2010) Segmental dynamics in PMMA-grafted nanoparticle composites. Macromolecules 43:8275–8281
119. Sharma SK, Sudarshan K, Sahu M, Pujari PK (2016) Investigation of free volume characteristics of the interfacial layer in poly(methyl methacrylate)-alumina nanocomposite and its role in thermal behaviour. RSC Adv 6:67997–68004
120. Harms S, Rätzke K, Faupel F, Schneider GJ, Willner L, Richter D (2010) Free volume of interphases in model nanocomposites studied by positron annihilation lifetime spectroscopy. Macromolecules 43:10505–10511
121. Sharma SK, Prakash J, Sudarshan K, Sen D, Mazumder S, Pujari PK (2015) Structure at interphase of poly(vinyl alcohol)-SiC nanofiber composite and its impact on mechanical properties: positron annihilation and small-angle x-ray scattering studies. Macromolecules 48:5706–5713
122. Sharma KP, Choudhury CK, Srivastava S, Davis H, Rajamohanan PR, Roy S, Kumaraswamy G (2011) Assembly of polyethyleneimine in the hexagonal mesophase of nonionic surfactant: effect of ph and temperature. J Phys Chem B 115:9059–9069
123. Mapesa EU, Cantillo NM, Hamilton ST, Harris MA, Zawodzinski TA Jr, Park A-HA et al (2021) Localized and collective dynamics in liquid-like polyethylenimine-based nanoparticle organic hybrid materials. Macromolecules 54:2296–2305
124. Cheng S, Xie S-J, Carrillo J-MY, Carroll B, Martin H, Cao P F et al (2017) Big effect of small nanoparticles: a shift in paradigm for polymer nanocomposites. ACS Nano 11:752–759
125. Mason TG (2000) Estimating the viscoelastic moduli of complex fluids using the generalized Stokes-Einstein equation. Rheol Acta 39:371–378
126. Rubinstein M, Colby RH (2003) Polymer physics. Oxford University Press, London
127. Choi UH, Ye Y, Salas de la Cruz D, Liu W, Winey KI, Elabd YA et al (2014) Dielectric and viscoelastic responses of imidazolium-based ionomers with different counterions and side chain lengths. Macromolecules 47:777–790
128. Wyss HM (2016) Fluids, colloids and soft materials: an introduction to soft matter physics. Wiley, Hoboken, NJ
129. Caporaletti F, Capaccioli S, Valenti S, Mikolasek M, Chumakov AI, Monaco G (2019) A microscopic look at the Johari-Goldstein relaxation in a hydrogen-bonded glass-former. Sci Rep 9:14319
130. Khodadadi S, Sokolov AP (2015) Protein dynamics: from rattling in a cage to structural relaxation. Soft Matter 11:4984–4998
131. Lappala A, Sefton L, Fenimore PW, Terentjev EM (2019) Connectivity and free-surface effects in polymer glasses. Sci Rep 9:3830
132. Smith GD, Bedrov D (2007) Relationship between the α and β-relaxation processes in amorphous polymers: insight from atomistic molecular dynamics simulations of 1,4-polybutadiene melts and blends. J Polym Sci, Part B: Polym Phys 45:627–643

133. Yu H-B, Yang M-H, Sun Y, Zhang F, Liu J-B, Wang CZ et al (2018) Fundamental link between beta relaxation, excess wings, and cage-breaking in metallic glasses. J Phys Chem Lett 9:5877–5883
134. Stockmayer WH (1967) Dielectric dispersion in solutions of flexible polymers. Pure Appl Chem 15:539–554
135. Kim SA, Mangal R, Archer LA (2015) Relaxation dynamics of nanoparticle-tethered polymer chains. Macromolecules 48:6280–6293
136. Xie H, Wang D, Tao T, Wang L (2014) Synthesis of highly branched sulfonated polymers and the effects of degree of branching on properties of branched sulfonated polymers as proton exchange membranes. J Power Sources 15:328–337
137. Neelakandan S, Liu D, Wang L, Hu M, Wang L (2019) Highly branched poly(arylene ether)/surface functionalized fullerene-based composite membrane electrolyte for DMFC applications. Int J Energy Res 43:3756–3767
138. Cosby T, Vicars Z, Wang Y, Sangoro J (2017) Dynamic-mechanical and dielectric evidence of long-lived mesoscale organization in ionic liquids. J Phys Chem Lett 8:3544–3548
139. Lukovkin GM, Pshezhetsky VS, Murtazaeva GA (1973) NMR ^{13}C study of the structure of polyethyleneimine. Eur Polym J 9:559–565
140. Pierre TS, Geckle M (1985) $^{13-}$NMR analysis of branched polyethyleneimine. J Macromol Sci Chem 22:877–887
141. Holycross DR, Chai M (2013) Comprehensive NMR studies of the structures and properties of PEI polymers. Macromolecules 46:6891–6897
142. Gouverneur M, Jeremias S, Schönhoff M (2015) ^{7}Li nuclear magnetic resonance studies of dynamics in a ternary gel polymer electrolyte based on polymeric ionic liquids. Electrochim Acta 175:35–41
143. Bloembergen N, Purcell EM, Pound RV (1948) Relaxation effects in nuclear magnetic resonance absorption. Phys Rev 73:679–712

Special Systems

Polymer Composites with Molecular Fillers: Microscopic Views into Supramolecular Reinforcement

Kay Saalwächter

Abstract This chapter provides an overview of the insights that dielectric spectroscopy can provide into mechanisms of supramolecular reinforcement of polymer chains equipped with associating polar "sticky" groups. This situation features analogies to the pinning of polymer chains to attractive nanoparticles, but the dielectric response of the polar groups provides direct access to the dynamics of the sticky sites. This is of potential use in elucidating the molecular origin of the viscoelastic properties of such materials. Analogies and differences between systems composed of defined binary associates versus large aggregates are discussed, and ambiguities as well as open challenges in interpreting the experimental results are highlighted.

Keywords Associative networks · Transient networks · Telechelic polymers · Supramolecular polymers · Ionomers · Aggregation · Hydrogen bonds · Self-healing

Symbols and Abbreviations

BIIR	Brominated isobutylene isoprene rubber (also: butyl rubber)
BLrN	Bond lifetime renormalization
ΔH_{st}	Reaction enthalpy for sticky bond formation
c_x	Number/mole concentration of species x
$\Delta\varepsilon$	Dielectric strength of a relaxtion
BDS	Broadband dielectric spectroscopy
DP	Degree of polymerization
DSC	Differential scanning calorimetry
ε', ε''	Real and imaginary part of the complex dielectric function
ϵ_b	Microscopic energy barrier for sticker association/dissociation

K. Saalwächter (✉)
Institut für Physik – NMR, Martin-Luther-Universität Halle-Wittenberg,
Betty-Heimann-Str. 7, 06120 Halle (Saale), Germany
e-mail: kay.saalwaechter@physik.uni-halle.de
URL: https://www.physik.uni-halle.de/nmr

A. Schönhals and P. Szymoniak (eds.), *Dynamics of Composite Materials*,
Advances in Dielectrics, https://doi.org/10.1007/978-3-030-89723-9_6

E_a^{st}	Activation energy for sticker dissociation
HN	Havriliak-Negami (fitting function)
ILM	Interfacial layer model
IR	Infrared (spectroscopy)
$J(\tau_{open})$	Number of returns of a free sticker
$K_{a/d}$	Association/dissociation equilibrium constant
k_+	Association rate
k_-	Dissociation rate
μ	Electric dipole moment, mostly in Debye (1 D = 3.336×10^{-30} C·m)
MWS	Maxwell-Wagner-Sillars (interfacial polarization process)
$\bar{n}$	(mole) number density
τ_α^* or τ_{α_2}	Dielectric relaxation time associated with sticky groups
τ_d^*	Timescale for sticker dissociation ($= 1/k_-$)
τ_{CR}	Dielectric relaxtion time related to sticker clusters; unknown origin
τ_{open}	Timescale of free-sticker diffusion
τ_{rot} and τ'_{rot}	Rotational correlation time of a free sticker and a sticker pair, resp.
τ_{st}^*	Rheologically relevant relaxation time of a sticky chain segment
T_g	Glass transition temperature
PB	Polybutadiene
PDMS	Polydimethylsiloxane
PI	Polyisoprene
PIB	Polyisobutylene
ω	Angular frequency in rad/s
SAXS	Small-angle X-ray scattering

1 Introduction

In this chapter, we turn to "molecular nanocomposites", where the reinforcement is provided by associating supramolecular groups, henceforth referred to as sticky groups or stickers. These may either be distributed sparsely along the polymer backbone or be located at the chain ends (the case of telechelic polymers). We restrict ourselves to the case of unpolar polymer backbones, which provide an enhanced tendency of association of the more polar or even ionic stickers. While the topic is somewhat at the edge of the scope of this book, a glimpse into this field is worthwhile, as Broadband Dielectric Spectroscopy (BDS) as well as other spectroscopic techniques such as Nuclear Magnetic Resonance (NMR) enable a rather unobstructed view into the interactions and dynamics of the stickers. This provides data that can potentially help elucidating reinforcement mechanisms related to transiently pinned chains, which are also relevant for conventional nanocomposites.

Figure 1 provides an overview of the scope of this chapter and the relevant dynamic processes related to the stickers and the chains they are attached to. Our main focus is the case of unspecific aggregation into large clusters, which are often assumed to be (but may actually not be) spherical. This is the more common case given

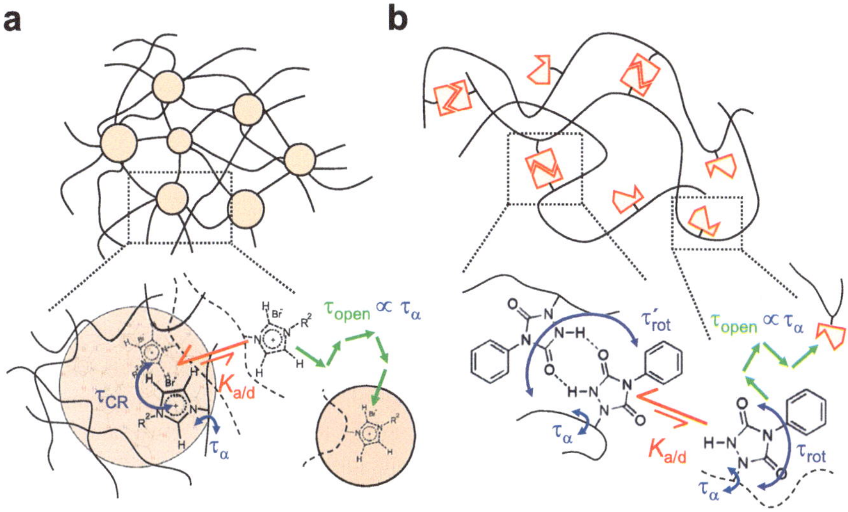

Fig. 1 Two different scenarios for supramolecular reinforcement provided by polar or ionic sticky groups: **a** Unspecific aggregation into clusters versus **b** defined pairwise association as assumed for phenylurazole groups [2]. Different relevant dynamic processes are indicated, such as the segmental α relaxation of the polymer backbone (τ_α), a cluster-related relaxation (τ_{CR}), the tumbling time of a free sticker (τ_{rot}) or a binary associate (τ'_{rot}), association/dissociation with equilibrium constant $K_{d/a} = k_+/k_-$ ($k_{+,-}$ being the respective rates), and the timescale of open sticker diffusion τ_{open}

a large mismatch in polarity or, more generally, the relevant molecular solubility parameters. In special cases, stickers with defined hydrogen-bonding motifs may favor a pairwise sticky bonding. This is mostly realized in solvent-containing systems given the right solubility balance, but in the bulk such a scenario cannot always be proven unambiguously. It is attractive because it lends itself to more quantitative treatments in terms of data interpretation and the application of microscopic models such as the sticky reptation model [1].

This contribution tries to give an up-to-date critical account of what BDS can teach us in this class of materials. We will first provide a short summary of insights obtained for classical ionomers, where the clusters always involve a significant fraction of backbone segments with largely reduced mobility. We then turn to elastomeric materials where sticker association does not change backbone segmental dynamics (i.e., the timescale of the α relaxation) of more distant Kuhn segments to a significant degree, which comprise supramolecular groups interacting *via* hydrogen bonds or π-π interactions. Such transient networks hold promise for applications as smart materials, allowing for complex functions such as self-healing, reprocessing or shape memory, with application perspectives in actuation, 3D printing or biomedicine [3–6]. We will first focus on the mentioned model cases with an assumed defined binary association, and then address different cluster-forming systems showing rather similar phenomenology, comprising uncharged as well as charged systems (the latter based upon ionic-liquid-like moieties that associate without strong mobility reduction of the backbone). It will be shown that "supramolecular reinforcement" in cluster-forming systems can share strong similarities to the reinforcement arising from nanoparticles, in particular with regards to the relevance of interface-related motional restrictions to the polymer dynamics.

2 Classical Ionomers

The case of high-polarity aggregates in ionomers [7], i.e., polymers equipped with sparse ionic sites, has significant relevance in polymer technology. In these classical cases comprising unpolar backbones such as polystyrene functionalized with carboxylic or sulfonic acid groups and then neutralized by metal ions, a picture like the one in Fig. 1a is widely accepted. According to Eisenberg, Hird, and Moore [8], the charged moieties form small ($\sim$1 nm) multiplets surrounded by a corona of strongly immobilized backbone segments of a few nm diameter, depending on backbone stiffness. If such structures overlap to form larger percolated aggregates, a separate T_g may be detectable. Ionomers are thus thermoplastic materials that are self-reinforced with nanometric inclusions, and are mostly in use for high-modulus performance applications, DuPont's polyethylene-based Surlyn^{TM} and its use for golf balls being a good example. But the concept also extends to elastomers [9], where NMR experiments from our group have revealed that the rigid clusters act as a reinforcing nanoscopic filler [10].

Detailed BDS studies on such ionomers have been performed by the Winey and Runt groups [11, 12]. Apart from features related to the weak conductivity, including electrode polarization (EP) and Maxwell-Wagner-Sillars (MWS) relaxations, one can usually identify one or two segmental relaxations. It is noted that dielectric spectra $\varepsilon''(\omega)$ of the materials described in this chapter can rarely be analyzed directly, due to an overwhelming conductivity contribution at low frequencies. This can be circumvented by calculating the derivative quantity

$$\varepsilon''_{\text{deriv}}(\omega) = -(\pi/2)(\partial\varepsilon'(\omega)/\partial \ln \omega) \tag{1}$$

as provided by the Kramers-Kronig relations [13]. This quantity can then be subjected to the usual decomposition into several contributions, using fits to combinations of common model functions such as Havriliak Negami (HN) in a suitably adapted derivative form.

In this way, one finds in ionomers either a single broadened α relaxation or another separate, 1–2 decades slower α_2 or α^* relaxation, whose molecular origin is not fully clear. Details depend on the degrees of acid functionalization and metal hydroxide neutralization as well as the type of metal ion. Molecular relaxations in such systems are generally associated with dipole fluctuations of a small fraction of free acid groups, while the charge-related di- and multipoles within the aggregates are thought to be screened and not contribute to the dielectric response. This is corroborated by trends of decreasing ionic strength of both processes with increasing

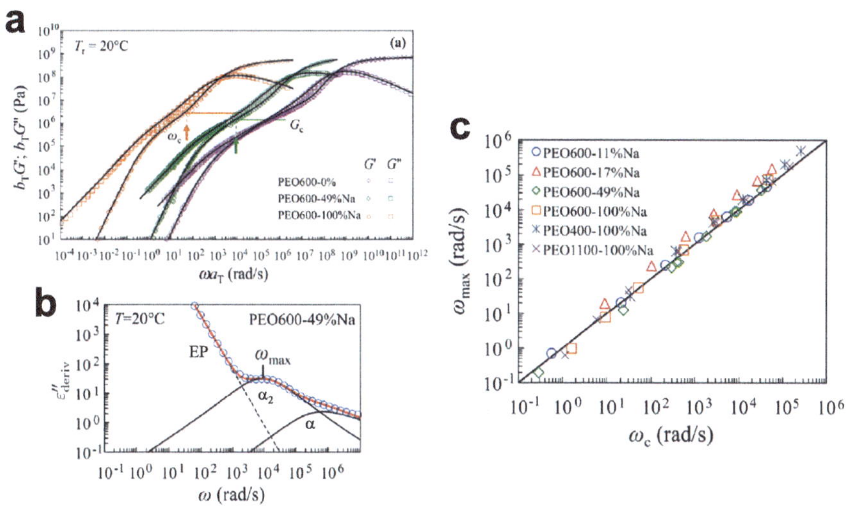

Fig. 2 **a** Rheological master curves of the storage and loss moduli of different PEO-based ionomers with a given fraction of sodium sulfonate isophtalate moieties, identifying the characteristic frequency (ω_c) of sticky Rouse unit; **b** dielectric loss derived from ϵ' via derivative analysis for one sample, with a fit using a power law for EP and two HN functions, and **c** dielectric peak frequency for the α_2 process (ω_{max}) versus the characteristic rheological frequency ω_c associated with ionic dissociation. Reproduced from Ref. [14] with permission from the Society of Rheology

metal content [12]. A final picture is, however, not available, and it is well possible that groups transiently leaving the aggregates or even relaxations within them could play a role.

Due to their main application as high-modulus materials, flow above T_g and chain dynamics in these classical ionomers have been rarely studied. In an important contribution from the Colby group [14], it could be shown that the maximum frequency of the dielectric α_2 process in a non-crystallizing PEO-based ionomer is nearly identical to a characteristic frequency ω_c derived from the onset of flow in rheological master curves, as illustrated in Fig. 2. Note that the correlation presented in Fig. 2c comprises several samples with different levels of functionalization and chain length, measured at different temperatures. The frequency ω_c corresponds to the inverse timescale of the relaxation of a part of the chain between two stickers, which was associated with the "sticky Rouse" unit, constituting a limiting case of sticky reptation theory [1].

In the given samples, where stickers are closely spaced and where the backbone is sufficiently polar to facilitate diffusion of a free ionic sticker, one can thus directly associate the dielectric α_2 process with the rearrangement of ionic groups, i.e., the process of an ion pair leaving a given cluster and finding a new one. It could thus share some similarities with the process of conductivity relaxation, which is related to local rearrangements of the ions and discussed as the elementary process of charge transport in molecular ion conductors [15]. It turns out, however, that the observation of matching microscopic and macroscopic timescales is a rare exception rather than the rule. There are many reasons for a given rheological timescale to be different from a local sticker fluctuation, an obvious one being the possible return of a dissociated sticker to its original binding site [16]. In the studied system, this probability was considered to be low due to the high concentration and close spatial proximity of the many clusters.

3 Transient Elastomers Formed by Sticker Pairs

We now turn to a special and sufficiently simple model system that has enabled very detailed analyses of BDS results and will serve as a starting point of our discussion of more complex materials. This is the case of stickers that form defined binary interactions as illustrated in Fig. 1b, thus turning long chains with multiple stickers into a transient elastomer (in case of telechelics, only supramolecular linear chains would be formed).

Pioneering work is due to Stadler and coworkers [2, 17, 18], who studied well-entangled polybutadiene (PB) randomly grafted with phenylurazole groups (their structure is shown in Fig. 1b), with spacings roughly equal to or higher than the entanglement molecular weight. While the presence of larger aggregates can probably not be fully excluded, this system was concluded to be governed by pairwise interactions on the basis of results from infrared (IR) spectroscopy [2], which enables the determination of the sticker pair and free sticker concentrations (c_{S_2} and c_S, respectively). The data shown in Fig. 3 demonstrates that the system has a rather featured dielectric response and enables the separation and quantification of the α and β relaxations that are also found in bulk PB, and one additional process (α^*) attributed to the stickers (see Fig. 3a). A consistent analysis of this system was presented by Müller et al. [18], and shall be summarized here due to its exemplary character.

A relaxation map, i.e., an Arrhenius plot of characteristic frequencies (inverse correlation times) on a log scale versus $1/T$, is shown in Fig. 3b (for better comparison, all following relaxation maps in this chapter will follow this convention). It demonstrates that the β relaxation of PB is unaffected, while the segmental α relaxation slows down upon modification with stickers. The α^* process also slows down upon increasing the sticker concentration and broadens on cooling, which we take up below. The timescales of both the α and the α^* relaxations exhibit a Vogel-Fulcher (VF) behavior. Their divergence at a common Vogel temperature is a common feature of relaxations in such sticky polymer systems, and is usually attributed to a coupling of sticker-related and segmental dynamics, where the timescale of the latter is thought to represent the microscopic attempt time of the former [14, 16],

$$\tau_\alpha^*(T) = \tau_\alpha(T)\,\exp\{E_\mathrm{a}^\mathrm{st}/RT\}\,. \tag{2}$$

Here, the activation energy E_a^st may be associated with the opening of the sticky bonds in cases where τ_α^* is related to the bond-opening process. This is the case in the given system, as corroborated by the fact that a timescale of macroscopic mechanical

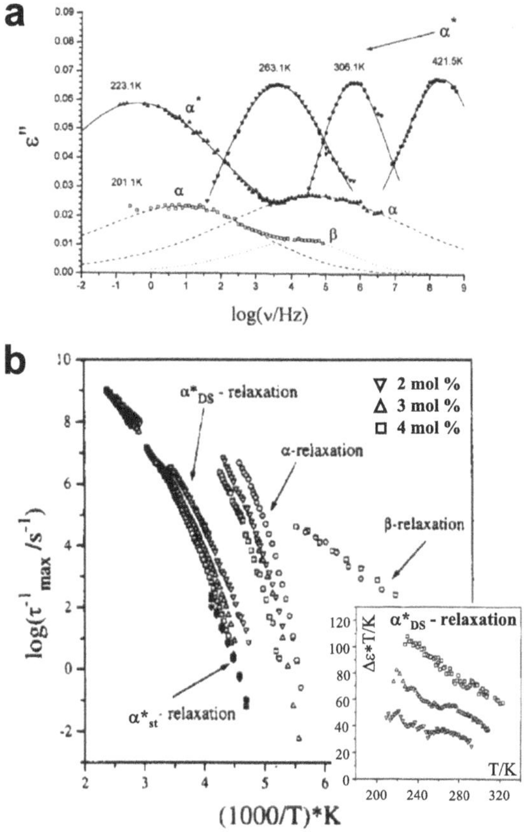

Fig. 3 **a** Dielectric loss curves for polybutadiene (PB) modified with 4 mol% of phenylurazole groups, where 3 processes can be separated via HN fits. Reproduced from Ref. [17] with permission from Springer Nature. **b** Relaxation map for samples with different degrees of modification, including also the relaxation of a sticky chain segment from rheological measurements (α^*_{st}). The inset shows the temperature dependence of the temperature-normalized relaxation strength $\Delta\varepsilon\, T$ of the sticker-related process. Reproduced from Ref. [18] with permission from the American Chemical Society

relaxation, also shown in Fig. 3b and henceforth referred to as sticky chain relaxation process α^*_{st} (with the sticky (sub)chain lifetime τ^*_{st}), has the same temperature dependence. Note that this process is detected in addition to the terminal flow time, which happens on much longer timescales as captured by the sticky reptation model [1]. Interestingly, that timescale shows no significant dependence on the sticker fraction, meaning that it is associated with subchains of variable length between two stickers as constraints dominating the relaxation timescale.

Müller et al. [18] worked out a near-quantitative understanding of the dielectric response, based upon the assumed pairwise binding motif. The bonding equilibrium is given by

$$S + S \underset{k_-}{\overset{k_+}{\rightleftharpoons}} S_2 \ , \tag{3}$$

where k_+ and k_- are the rates of association and dissociation, respectively. These are related to the equilibrium constant *via* $K_{\text{a/d}} = c_{S_2}/c_S^2 = k_+/k_-$. Its enthalpic and entropic contributions can be obtained from the temperature-dependent c_{S_2} and c_S as taken from the IR spectroscopy results [2].

With regards to BDS, we have to consider two different entities, S and S_2, with their respective dipole moments. These fluctuate in time through a number of different processes, which comprise rotations of the different entities with different correlation times $\tau_{\rm rot}$ and $\tau'_{\rm rot}$, respectively (see Fig. 1b), but also *chemical relaxation* [19]. The latter arises from the change in dipole moment upon association or dissociation with the respective rate constants, see the reaction scheme (3). In the system with 4 mol % stickers, the dielectic α^* process and the corresponding mechanical sticky subchain relaxation are very close, suggesting a direct relation to sticker dissociation, $\tau_{\rm d}^* = 1/k_-$. The decorrelation of the relevant dipole moment is, however, also affected by the overall rotation of the dimer with $\tau'_{\rm rot}$,

$$\tau_\alpha^* = \left(1/\tau_{\rm d}^* + 1/\tau'_{\rm rot}\right)^{-1}. \tag{4}$$

Thus, the faster of the two constituent correlation times will dominate the measured τ_α^*, and distributions of these correlation times would dominate either the high- or the low-frequency tail of the observed process.

The above considerations can only be correct when a contribution of the dissociated unimers to τ_α^* can be excluded. This was achieved by quantitative analyses of measured relaxation strength $\Delta\varepsilon = \varepsilon_{\rm s} - \varepsilon_\infty$, where $\varepsilon_{\rm s}$ and ε_∞ are the permittivities in the low- and high-frequency limits of the investigated process. $\Delta\varepsilon$ is taken from the HN fits, and can be related to the effective fluctuating dipole moment $\mu_{\rm eff}$ via the Onsager equation (neglecting a Kirkwood-Fröhlich correlation factor),

$$\mu_{\rm eff}^2 = \frac{9kT}{4\pi\bar{n}}\varepsilon_0\frac{\Delta\varepsilon(2\varepsilon_{\rm s}+\varepsilon_\infty)}{\varepsilon_{\rm s}(\varepsilon_\infty+2)^2}\,, \tag{5}$$

where $\bar{n}$ is the number density of fluctuating units.

Two pieces of evidence were combined. First, Eq. (5) suggests that the product $\Delta\varepsilon\, T$ should be proportional to $\bar{n}$, which can be either c_{S_2} or c_S. The inset of Fig. 3b shows that $\Delta\varepsilon\, T$ increases from sample to sample, proportionally to the overall sticky group content, and that it *decreases as temperature increases*. The concentration of unimers can only increase with temperature; thus it was concluded that dimers must dominate the α^* process. Second, associating $\bar{n}$ with the known c_{S_2} of all samples, Eq. (5) provides a $\mu_{\rm eff}$ of about 0.6 D for all samples. This value was discussed to be in agreement with modeling calculations of the dimer structure, and much lower than the 2.6 D associated with the dissociated urazole groups. Note that the coplanar centrosymmetric structure of the dimer sketched in Fig. 1b implies a zero dipole moment, however, force-field calculations support a twisted structure with finite dipole moment.

Finally, Müller et al. argued that the unimer contribution to the dielectric spectra is dominated by their fast rotation within the PB matrix on a timescale close to the segmental τ_α; the observed asymmetric broadening of the α relaxation was explained in this way. The association reaction was estimated to be too slow to make a contribution here. Similarly, the considerable broadening of the α^* relaxation on cooling could be attributed to a change in relevance of $\tau_{\rm d}^*$ versus $\tau'_{\rm rot}$, see Eq. (4).

This observation was, however, not analyzed further, and one may suspect that the occurrence of larger aggregates could also play a role.

The missing piece of the puzzle concerns the variable timescale separation of the localized mechanical process (being independent of sticker concentration) and the sample-dependent τ_α^*. It was argued that at lower sticker contents, τ_α^* should be increasingly dominated by $\tau'_{\rm rot}$, which decreases due to a lower $T_{\rm g}$ of the samples with less stickers ($\Delta T_{\rm g}$ amounts to about 7.5 K from pure PB to the 4 mol % sample). But a discussion of actual timescales was not provided, and additional factors could play a role.

3.1 Sticky Bond Lifetime Renormalization

At this point, more recent insights from theory come into play. In the original sticky reptation model [1], the local stress relaxation process of a given chain segment was associated with the dissocation rate k_- of the equilibrium (3). However, an important feature of an equilibrium is the possibility of a backwards reaction, which for a given pair does not change the overall connectivity and should not affect the mechanical response. Thus, Rubinstein and Semenov [16] suggested that a dissociated sticker must find another bonding partner for stress relaxation to occur and that the sticky bond lifetime should be renormalized. The relevant quantity is the average number of returns to the original parter $J(\tau_{\rm open})$. In later work [20], Rubinstein and coworkers refined the bond lifetime renormalization (BLrN) model and provided a prediction for the rheologically active stick bond lifetime and the dissociation time,

$$\tau_{\rm st}^* = J(\tau_{\rm open})\tau_{\rm d}^* + \tau_{\rm open} \;, \tag{6}$$

where $\tau_{\rm open}$ is introduced as the time it takes a sticker to find a new partner by diffusive exploration (see Fig. 1). Both $\tau_{\rm open}$ and $J(\tau_{\rm open})$ were suggested to reflect the restrictions to diffusion by the local chain connectivity (leading to Rouse-like subdiffusion with the mean-square displacement scaling as $t^{1/2}$) as well as the typical distance to the next bonding partner, from which $\tau_{\rm open}$ and $J(\tau_{\rm open})$ arise from the solution of a first-passage problem. Importantly, the theory assumes strictly binary interactions and calculates the average next-free-neighbor distance from their concentration provided by the association/dissociation equilibrium.

These factors could well play a role in the system of Stadler and coworkers. Only recently, an analogous system of phenylurazole stickers attached to polyisoprene (PI) was prepared and studied in detail by the Richter group [21, 22]. The aim of Gold et al. was to provide evidence for the validity of the BLrN model. They allowed the subdiffusive exponent to vary, and found a value of 0.36, which is very reasonable for a system with additional entanglement constraints.

However, cautionary remarks are on order. Gold et al. assumed their τ_α^* to coincide with $\tau_{\rm d}^*$, thus neglecting the possible dipole decorrelation due to $\tau'_{\rm rot}$. They just pointed out the alleged centrosymmtric symmetry of the sticker pair and its consequently vanishing dipole moment. No proof was given, and it seems unlikely that the situation in the PI matrix is different from the one in PB. Further, no assessment of the relaxation strength was presented, which was shown above to provide very relevant insights into this question.

A second aspect of concern is the interpretation and validity of Eq. (2) of this chapter. The microscopic view behind this equation has been fundamentally challenged very recently by Ghosh and Schweizer [23], but we here restrict ourselves to even more obvious shortcomings. The equation is often used in the given context, and the activation energy is nearly always associated with the bond energy, i.e., the enthalpic contribution to the equilibrium (3). This means that the recombination reaction is barrier-free ($\epsilon_b = 0$) and is assumed to happen instantaneously. The relevance of an additional barrier $\epsilon_b > 0$ has actually been pointed out in Rubinstein's and Semenov's earlier work [16], explicitly stating that its presence would lead to significant slowdown, as two open sticky sites have to spend considerable time close to each other to associate. This would provide an additional contribution to BLrN that has been neglected in the later elaboration [20]. A high barrier would in fact switch the scenario to "reaction control" rather than diffusion control [24], for which the BLrN model does not hold.

It is, in fact, very difficult to decide on these aspects, as the energy gain upon association on the one hand, and the height of a potential barrier on the other hand, are rarely accessible in an objective way. Notably, relying on a characterization of the association–dissociation equilibrium on the basis of IR spectroscopy, Gold et al. did provide a comparison of E_a^{st} obtained from Eq. (2) and the reaction enthalpy ΔH_{st} derived from IR spectroscopy [21, 22], and concluded on a significant barrier ($\epsilon_b =$ 13 kJ/mol for lowly functionalized PI and even 46 kJ/mol for 4 mol% of stickers, to be compared to a ΔH_{st} of -16 kJ/mol). They consequently used the latter value for their estimation of the next-free-neighbor concentration. But this does not remedy the neglect of the slowdown of sticker recombination arising from the barrier.

A potential barrier was completely neglected in more recent, allegedly successful tests of the BLrN model published by Sokolov and coworkers [25, 26]. They relied on the estimation of ΔH_{st} based on Eq. (2), which may thus impart two potentially large errors, the additional one being an underestimation of free stickers available for recombination (because $|\Delta H_{st}| < E_a^{st}$). Also, the BDS-based τ_α^* was again equated with the dissociation timescale without further critical tests. The systems of study were linear telechelic polymers of different kinds equipped with comparably simple hydrogen-bonding groups such as –COOH and –NH_2. The dominance of binary sticky bonding was claimed, but not furnished with a clear proof. It, in fact, seems unlikely that such simple endgroups could not also form larger multiplets or even clusters. The existence of such structures would require a significant modification of the BLrN theory, as such less specific aggregates would always be open to receive a free sticker, thus facilitating sticky chain relaxation considerably [27].

3.2 Complex Cases

A considerably higher level of complexity, without the possibility to unambiguously separate and/or assign different relaxation processes in both the dielectric as well as the rheological response, was reported for other systems. Early on, Stadler and coworkers published on a carboxy-substituted phenylurazole-modified polyisobutylene (PIB), in which quasi crystalline clusters appear that even show a melting endotherm in DSC [28]. Here, an additional slower dielectric relaxation could be associated with the clusters, and was discussed to originate from the interior of the clusters rather than in/out fluctuations. In more recent work, even three (!) α^* processes were found in an acrylate polymer randomly equipped with sticky hydrolyzed acrylic acid sidegroups, along with a featureless and much delayed rheological response for higher functionalization degrees [29]. For telechelic polydimethylsiloxanes (PDMS) with chemically more complex sticky endgroups [30, 31], two sticker-related processes were found in BDS, both with a qualitative correlation to relaxation processes identified in shear rheology data, yet without clear systematic trends. In all these cases, the authors postulated the coexistence of low-valency (possibly binary) sticky bonds and larger clusters. The latter are addressed in the following section.

4 Transient Elastomers with Sticker Clusters

As already pointed out early on by Stadler [18], stickers forming clusters may actually be the rule rather than the exception. This holds in particular for comparably polar stickers attached to unpolar polymer backbones. Thus, we now switch to systems that are entirely dominated by cluster-forming sticky groups, which are again dielectrically and rheologically simple enough to discuss the relevant relationship between macroscopic relaxations seen in rheology and the microscopic processes observed by BDS. We also include additional insights into the chain relaxation on the molecular level and into the clusters by different solid-state NMR spectroscopy techniques. Finally, similarities to nanoparticle-based reinforcement related to the potential relevance of interface-related mobility reduction of polymer segments are discussed.

4.1 Theoretical Considerations

The polymer-physics implications of polymers equipped with sticky groups forming large clusters have been worked theoretically by Halperin [32] addressing diblock copolymers, and Semenov and colleagues [33, 34]. These works focus on micellar solutions, but are readily generalized to the bulk by adjusting the concentration. They emphasize the important role of chain-level entropic effects, which concern crowding and local stretching effects in the brush-like micellar corona and the spatial separation of the stickers. These factors, together with the energy gain on cluster formation and surface-related effects, determine the size (and also shape) of the aggregates, where it is reasonable to assume that there is a favorable cluster size with a given aggregation number AN characterized by a rather flat free-energy minimum. Cooperative relaxation mechanisms involving the contact of two clusters are considered unlikely due to the repulsive coronas. The association–dissociation equilibrium governing the sticky chain dynamics is thus qualitatively different from (3):

$$S + S_{AN} \underset{k_-}{\overset{k_+}{\rightleftharpoons}} S_{AN+1} \, . \tag{7}$$

The physically relevant difference is that stickers do not have to diffuse for a long time to find another free sticker to pair up with, but will always find another cluster closeby that can accept another sticker. To a good approximation, since $AN \approx AN+1$, the equilibrium constant is given by $K_{a/d} = k_+/k_- = c_S^{-1}$, where the concentration of free stickers c_S is expected to be rather small. In analogy to Eq. (2), the dissociation rate should follow

$$k_- = k_0 \exp\{-E_a^{st}/RT\} \, , \tag{8}$$

where the activation energy $E_a^{st} = |\Delta H_{st}| + \epsilon_b$ may include a relevant barrier term ϵ_b, and where the attempt rate k_0 may (or may not) be associated with the inverse segmental relaxation time $\tau_\alpha(T)$. As we cannot be sure about the latter point, in particular in a system forming large clusters that may be dynamically decoupled from the surrounding polymer matrix, we henceforth assume a T-independent pre-exponential factor and discuss only apparent activation energies E_a^{app} taken directly from Arrhenius plots of the respective dynamic quantities. For details, we refer to our recent publications [35–37].

4.2 *Ionic Stickers*

We now move to a first example illustrated in Fig. 4. The basis of the studied systems is a modified industrial product, namely, brominated isobutylene isoprene rubber (BIIR) with high molecular weight, where the bromine atoms are attached to the rare, randomly distributed isoprene units (one in ~100 monomers) and aid in crosslinking of the product. The backbone is thus essentially polyisobutylene (PIB). It can be modified in a one-step reaction with different alkyl imidazole derivatives to form a self-healing rubber material with high potential for actual applications [38]. The resulting ionic stickers (alkyl imidazolium bromide) form clusters with $AN \approx 20$, as taken from small-angle X-ray scattering (SAXS) experiments [35].

The BDS results in Fig. 4 are rather comparable to the ones of Fig. 3, except that the charge-related conductivity requires the derivative analysis using Eq. (1). This

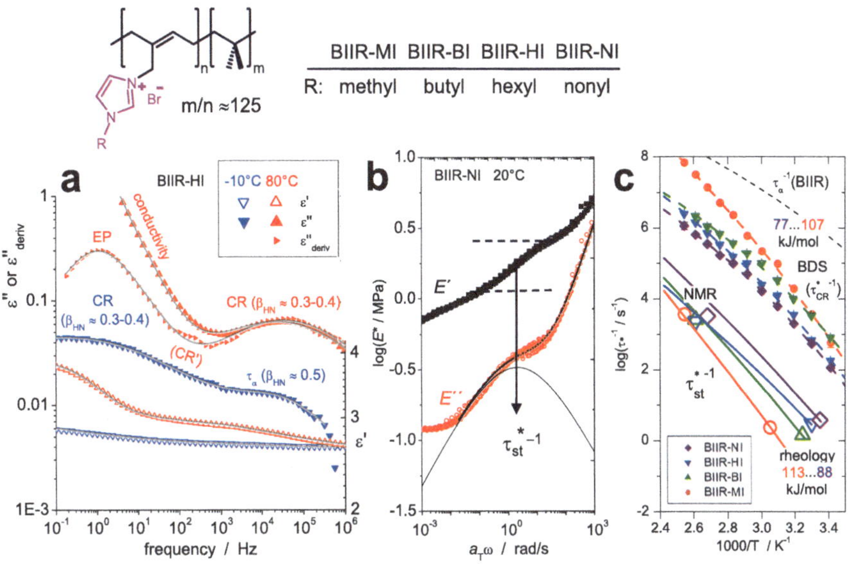

Fig. 4 Dynamics in alkylimidazole-modified BIIR: **a** Dielectric function of BIIR-HI measured at two different temperatures, with symmetric broadening parameters β of the HN fits for the different processes indicated. **b** Master curve of the complex modulus of BIIR-NI showing the drop in storage modulus (dashed lines) related to sticky chain relaxation, and the corresponding relaxation peak in the loss modulus fitted to an HN function (solid line). **c** Relaxation map comparing the BDS-based cluster relaxation (solid symbols) with the segmental relaxation of pure BIIR (dashed line) and the chain-level relaxation measured independently by rheology and NMR (open symbols). The solid lines reflect the T-dependence from the rheological shift factors. The ranges of apparent activation energies E_a^{app} are indicated. A sketch of the chemical structures is given at the top. Reproduced in parts from Ref. [35] with permission from the American Chemical Society

reveals an additional process attributed to electrode polarization (EP). The α^* process is here referred to as cluster relaxation (CR), which is at high frequencies followed by the segmental relaxation (we neglect a minor relaxation between EP and CR, which may well be due to impurities). Pure PIB has an extremely low dipole moment and thus a low relaxation strength [39]. The α process observed in the BIIR is actually tied to the polar groups acting as dilute probes. It is difficult to separate it from the CR, which is why the relaxation map in Fig. 4c just shows the segmental relaxation for pure BIIR. The key questions are now whether the CR is a measure of the sticky bond dissociation and whether the α process may provide its attempt frequency; see Eq. (8).

In our previous work [35], we have thus conducted dynamic-mechanical analyses and proton time-domain multiple quantum (MQ) NMR experiments to study the relaxation of the sticky chain sections characterized by τ_{st}^* on the macroscopic as well as microscopic levels. A sample of the mechanical results is shown in Fig. 4b. The terminal relaxation is actually beyond the frequency and temperature ranges of the experiments, owing to the large molecular weight, but τ_{st}^{*-1} can readily be read off from the intermediate maximum of the loss modulus as separated by an HN fit. We refrain from discussing details concerning the MQ NMR method [35], and just mention that this method provides a sensitive means to detect the very weak connectivity-related anisotropy of segmental motion. This readily reflects the motions of the chains and allows one to extract a segmental orientation autocorrelation function, from which τ^*_{st} can be estimated on a microscopic level.

Since the materials are thermorheologically complex, meaning that the temperature dependence of the measured dynamics switches from being segment-controlled at low temperatures to sticker-controlled at high temperatures, absolute-scale information for τ^*_{st} is only available in isothermal experiments, given the limited frequency or time ranges of the techniques. The relaxation map in Fig. 4c demonstrates the near-quantitative correspondence of ${\tau^*_{st}}^{-1}$ measured on the macro- and micro-scales. It is seen that in the samples with longer alkyl chains the sticky subchains relax faster, in line with their improved self-healing ability.

But in comparison to the chain dynamics, the BDS-based results for the CR constitute a puzzle. Here, the methyl-modified stickers relax fastest, i.e., the timescale separation between the chain-level relaxation and the potential sticker relaxation (tentatively associating ${\tau^*_{CR}}^{-1}$ with ${\tau^*_{d}}^{-1} = k_-$) increases from somewhat less than 2 decades to more than 4 decades. This is completely incompatible with any BLrN argument, since the theory behind Eq. (6) is only based upon geometric arguments (average sticker distances) and the segmental timescale [20], which are both nearly constant across the sample series.

Interestingly, as is also apparent from Fig. 4c, the range of apparent activation energies E_a^{app} is nearly the same for ${\tau^*_{st}}^{-1}$ and ${\tau^*_{CR}}^{-1}$, with a significant variation across the sample series. This indicates that τ^*_{CR} does provide the microscopic "clock" for sticker detachment. It is worth noting that the EP process, presumably related to the large-scale motion of charge-bearing stickers through the PIB matrix, also follows the order of the rheology/NMR-detected chain-based relaxation across the sample series.

Some clues to the potential solution of this puzzle are collected in Fig. 5. We start with the relaxation strength $\Delta\varepsilon_{CR}$, which increases somewhat on heating across the sample series, and is larger for the stickers with longer alkyl chains. This may indeed suggest a relation to the dissociation reaction, with a detachment of the more hydrophobic stickers being increasingly favored. In Fig. 5a, we attempt to fit $\Delta\varepsilon_{CR}(T)$ with a Van t' Hoff relation ($\bar{n}(T) \propto \exp\{\Delta H/RT\}$), testing whether this quantity may be proportional to the free-sticker density $\bar{n}(T)$ with fixed dipole moment μ_{eff} according to Eq. (5): $\Delta\varepsilon \propto \bar{n}(T)\mu^2_{eff}/T$. The result provides a constant *apparent* sticky bond energy of order 3.6 kJ/mol for all samples. This is hardly a physically reasonable result, as it is of the same order as RT and would imply a large free-sticker fraction, considering also that the latter is entropically favored. However, the SAXS results shown in Fig. 5b do not suggest any significant change in AN (which is encoded in the peak position and shape, while the relative intensity reflecting scattering contrast may also change due to thermal expansion effects). In addition, also the elasticity moduli of the samples (upper dashed line in Fig. 4b) are not compatible with a significant free-sticker fraction. Thus, $\Delta\varepsilon_{CR}$ is not likely associated with the

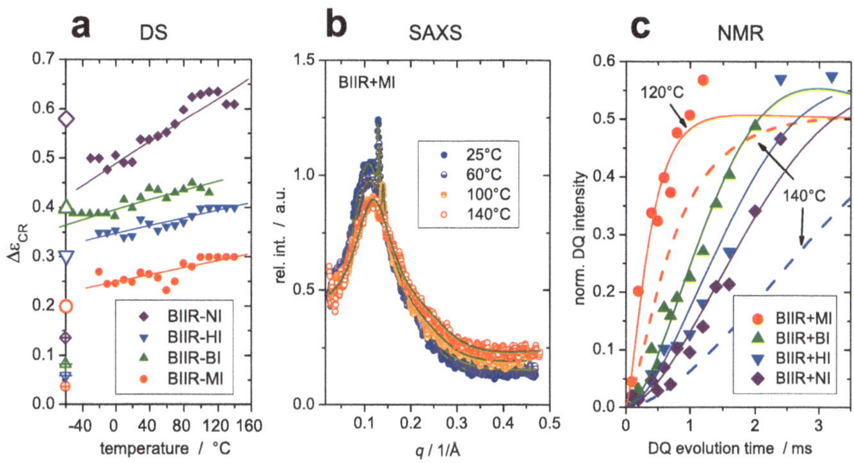

Fig. 5 **a** Dielectric strength of the cluster relaxation of functionalized BIIR. The lines are Van t' Hoff fits to $\Delta\varepsilon$ (see Eq. (5) and main text), and the symbols on the y axis are estimates for an MWS process (see text). **b** SAXS results for BIIR-MI at different temperatures along with fits to determine the cluster size. **c** Proton double-quantum (DQ) NMR build-up curves detected at the position of the imidazolium resonances of functionalized BIIR, reflecting the degree of motional anisotropy of all stickers. The fitting lines (solid for 120 °C) also include results for samples BIIR-MI and BIIR-HI taken at 140°C (dashed). Reproduced in parts from Ref. [35] with permission from the American Chemical Society

fixed dipole moment of free stickers. The E_a^{app} provided in Fig. 4c appear to be much more reasonable estimates of the association energy.

An alternative interpretation of the CR could be that of an MWS interface polarization process arising from motions of the charges within the clusters. Although the scale-invariant mean-field theory description provided by Eqs. (13.8)–(13.11) of Ref. [40] is probably at its limit for the given nanometer size range, we attempt a consistency check. Reasonable estimates of the required quantities, namely, the volume fraction of the clusters and the permittivities of the polymer and the associated endgroups can be taken from the literature, thanks to the alkylimidazolum moiety being a popular component of ionic liquids. The relevant control parameter is the "shape factor" of the clusters, which is $n = 1/3$ for spheres, and significantly smaller for elongated structures. The crossed small symbols on the y axis of Fig. 5a are the predictions for $n = 1/3$, while the large open symbols are obtained for an adjusted n of 1/9. Recent computer simulations actually support a stringlike shape of aggregates in ionomers with randomly spaced charged pendant groups [41]. Notably, the order of increasing $\Delta\varepsilon$ across the sample series is predicted correctly; it is rooted mainly in the increasing cluster volume fraction and the decrease in permittivity, both related to the increase in alkyl chain length.

The timescale of an MWS process is inversely related to the conductivity of the clusters. Also, in this regard, we thus have a consistent explanation for the observation in Fig. 4c, where the methylimidazolium bromide having the potentially highest conductivity has the shortest τ^*_{CR}. However, the predicted timescale is a few orders of magnitude shorter than observed one. The tethering of the moiety to the polymer backbone could of course play a role here, but definite conclusions are hardly possible on the basis of the available theory and data.

Figure 5c shows results from high-resolution magic-angle spinning (MAS) double-quantum (DQ) NMR experiments. The shown build-up curves are detected at the spectral positions of the imidazolium resonances. From the spectra (not shown), we take that the intensity of these resonances quantitatively reflects their total proton fraction in the sample. This means that these curves tell us about the dynamics of *all* stickers present, not just an unknown fraction. The rise time of these curves reflects the degree of motional anisotropy. Without going into detail, we can conclude that at the given temperatures, all stickers must perform rotational dynamics on a timescale much faster than about 25 μs, with order parameters $S = \langle P_2(\cos\theta)\rangle$ ranging from about 0.1 for methyl imidazolium to 0.02 for the nonyl derivative (with θ specifying the orientation of the 2 closest protons in the imidazolium group). Heating from 120 °C to 140 °C decreases these values further by a factor of about 2 (see the dashed lines, representing the fits to data points not shown). Thus, the motional amplitude of the stickers increases both upon heating and for longer alkyl residues. These trends are thus the same as for $\Delta\varepsilon_{\mathrm{CR}}$. Therefore, we have concluded [35] that the BDS-detected τ^*_{CR} most likely reflects sticker motions *within* the clusters. While more work, possibly molecular simulation, is clearly needed to resolve the open questions, it must be concluded that BDS does in the given case not provide an access to the relevant timescale of sticky bond dissociation.

4.3 Hydrogen-Bonding Stickers

As a second example, we summarize more recent results [36, 37] on telechelic PIB functionalized with uncharged hydrogen-bonding endgroups, specifically, barbituric acid (BA) and thymine (Th); see Fig. 6. In this case, the clusters are considerably larger, ranging from AN in the order of 50 for linear chains to more than 100 for three- and four-arm star molecules. An earlier study of similar micelle-forming monofunctional linear PIBs showed that their high viscosity is best described by models assuming a colloidal suspension of micelles [42]. This system should thus comply with the only available dedicated micellar model of Semenov and colleagues [33, 34], which states that terminal flow is governed by collective micellar rearrangements.

In order to check the relevance of micellar packing and collective processes for the terminal flow of the samples based upon telechelic chains (which are capable of forming bridges and lead to a high elasticity modulus), we have applied the same combination of rheology and MQ NMR as described above [36]. The data summarized in Fig. 6 speak a clear language. For all studied cases, the macroscopic and the microscopic probes provide again fully consistent results for the relaxation rate of the sticky chains τ_{st}^{*-1}. Since all samples are essentially unentangled, the sticky chain relaxation coincides with the onset of terminal flow in all samples. This proves that

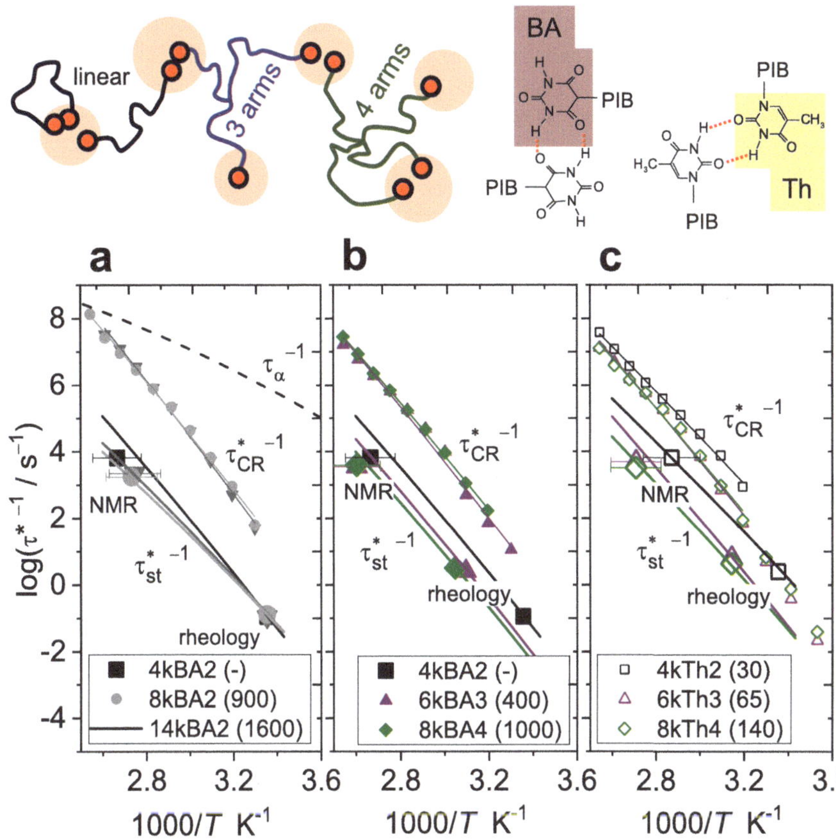

Fig. 6 Relaxation maps for telechelic PIB polymers with barbituric acid (BA) or thymine (Th) endgroups: **a** linear BA-PIB with different chain length, **b** linear and star-shaped PIB-BA and **c** linear and star-shaped PIB-Th. All panels include the BDS-based cluster relaxations and the chain-level relaxations measured independently by rheology and NMR. Panel **a** also shows the segmental relaxation of pure PIB. The parenthesized numbers in the legends are the approximate "slowdown ratios" of the chain-based relaxations compared with the cluster relaxation. Data from Ref. [37]

flow is governed by single-chain relaxation in the spirit of the sticky Rouse model [14], and that micellar rearrangements are not required to enable flow. If the latter were relevant, then the timescale of macroscopic flow would have to be exponentially longer than the NMR-detected single-chain relaxation. The effect of topology, resulting from the comparison of linear and star polymers with equal subchain length, was found to be moderate, with stars relaxing more slowly by somewhat less than a decade.

Aside from these findings of general polymer-physical relevance, we were, of course, also interested in the relation of the sticky chain relaxation to fluctuations of the endgroups [37]. Evaluation of the slopes of the activation plots in Fig. 6 reveals nearly the same E_a^{app} of about 140–150 kJ/mol for all samples, where the weaker Th-based stickers exhibit somewhat lower values. The magnitude of E_a^{app} reflects the strong aggregation tendency also in this system, but the role of a potential barrier remains again uncertain. The BDS data for these systems look very similar to those shown in Fig. 4a, with the main difference being that the conductivity-related processes have variable magnitude and are related to impurities (mostly residual water). The dominant cluster relaxation with t rate ${\tau_{CR}^*}^{-1}$ shows in this case a temperature-independent $\Delta\varepsilon_{CR}\, T$, which may indicate a constant number of fluctuating dipoles. Beyond the uncertainties of such arguments relying on the validity of Eq. (5), the fact that the measured $\Delta\varepsilon_{CR}$ decreases on heating is strongly suggestive of CR not being connected to an association–dissociation equilibrium.

Yet again, it is observed that ${\tau_{CR}^*}^{-1}$ has nearly the same apparent activation energy (slope in the relaxation maps) as the sticky chain relaxation rate τ_{st}^{*-1}, which is about twice as high as the segmental activation in the given temperature interval. The CR must thus play the role of an attempt process for sticky chain relaxation. The numbers collected in the legends of Fig. 6 are the approximate "slowdown factors" by which the chain-based relaxation is slower than the CR. Apart from a comparably weaker effect of molecular weight (Fig. 6a) and topology (Fig. 6b, c), the main variation of about a decade is seen to be governed by the type of the endgroup (Fig. 6b, c). The Th-based systems with a significantly faster terminal relaxation show a CR of comparable rate as the BA-based systems, but they exhibit a considerably faster terminal flow. Also, in this system, the two endgroup series are virtually isostructural, with similar molecular weights and AN or cluster-cluster separation. Thus again, BLrN arguments cannot be invoked to explain the slowdown, meaning that there is no direct relation between the timescales of the BDS-based CR and the sticky bond dissociation. It is again possible to argue that the CR is an intra-aggregate process, based upon the consistency of the observed $\Delta\varepsilon_{CR}$ with the dielectric properties of neat melts of heterocyclic molecules forming hydrogen bonds [43], just considering the dilution of the nanometric domains in the unpolar bulk PIB. A similar conclusion was drawn earlier by the Sokolov group for a telechelic polydimethylsiloxane (PDMS) featuring small hydrogen-bonding and clustering endgroups [31]. This is the system we turn to discussing next.

4.4 Supramolecular Reinforcement by Sticker Clusters

So far, the function of the supramolecular bonds was mainly to provide crosslinks and thus rubber-like elasticity. In case of sticker clusters, which can be considered as giant crosslinks with large functionality, the elasticity is enhanced by the resulting suppression of crosslink fluctuations, as captured by the phantom model of rubber elasticity. We now turn to a final example represented by comparably short telechelic PDMS chains equipped with hydrogen-bonding endgroups (see Fig. 7a), where the clusters provide additional, actual reinforcement by acting as nanometric fillers [31, 44, 45]. This system is special, in that, its calorimetric response features a second

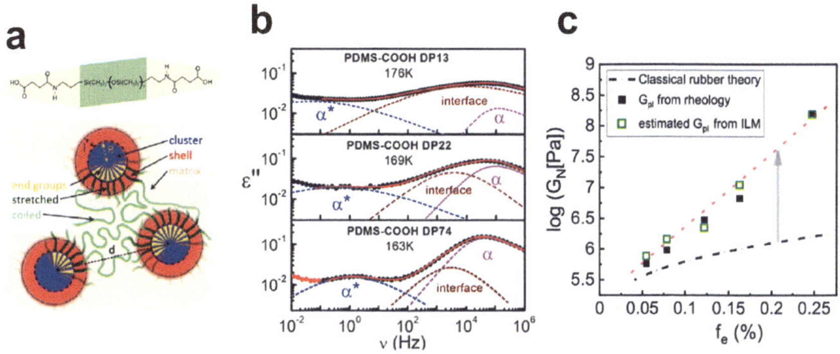

Fig. 7 **a** Chemical structure of cluster-forming telechelic PDMS and sketch of the cluster-related contributions to reinforcement, **b** frequency-dependent dielectric loss for three samples with different degree of polymerization (DP) revealing three contributions including an interface-related shifted α process of PDMS and **c** plateau modulus (G_N or G_{pl}) of samples with increasing end-group volume fraction f_e (decreasing DP) along with theoretical predictions including an interfacial layer model (ILM). Reproduced from Refs. [44, 45] with permission from the American Chemical Society

glass transition, which could be assigned to the phase-separated stickers [31]. In fact, the stronger one of the two cluster-related relaxations seen in BDS (denoted α_2, not visible in Fig. 7b) has a timescale of roughly 100 s at this second calorimetric T_g. This strongly suggests that the process is originating from motions within the 2–3 nm large clusters. It has consequently no simple relation to the timescale of debonding, nicely supporting the conclusions from the last sections. The very broad α^* process seen in Fig. 7b is close in frequency to a shallow peak in the rheological loss modulus, but it is not clear if it is simply related to the relevant lifetime of the sticky bonds.

The most interesting feature of this system is, however, the increasing asymmetric broadening of the well visible segmental α relaxation of the PDMS backbone upon decreasing the chain length, as illustrated in Fig. 7b. This phenomenology is reminiscent of the similar signatures observed when mixing polymers with rigid nanoparticles offering favorable adsorptive interactions with the polymer. In such cases, the so-called hydrodynamic reinforcing effect of the particles depends, in the absence of filler-network effects, on the volume fraction of the added undeformable material. It is typical that attractive nanoparticles exhibit significantly larger reinforcement than predicted by their bare volume fraction, because they are nearly always covered by a layer of interfacial polymer with significantly reduced mobility, which effectively increases the size of the high-modulus filler. This is exactly the phenomenology observed for the telechelic PDMS system [44, 45].

Sokolov and his coworkers analyzed the dielectric response shown in Fig. 7b in the same way as in attractive nanocomposites and we refer to his contribution to the book in hand, i.e., they applied a simple additive approach to separate the interfacial contribution, which is delayed by about one decade. They then used the reduction of the dielectric signal from the bulk-like polymer to estimate the interfacial volume fraction (the micellar corona), where the chains must be strongly stretched and possibly also more tightly packed than in the bulk. Assuming that the PDMS segments with reduced mobility form a layer around the particles, its thickness is estimated to about 0.8 nm independent of chain length. Based on this information, and estimates of the mechanical properties of the cluster core and this corona, an interfacial layer model [46] could be applied to predict the level of reinforcement of the plateau modulus. This result is compared in Fig. 7c with the experimental results. The striking agreement speaks for itself, thus concludes this section with a perfect model case of supramolecular reinforcement.

4.5 Common Aspects for Cluster-Forming Stickers

In summary, for three separate systems, we could highlight that BDS data of cluster-forming supramolecular self-reinforced elastomers clearly reflect aspects of the relevant sticky bond energies and activation barriers. In special cases, even polymer backbone segments with strongly reduced mobility at the cluster interface, which play a crucial role in reinforcement, can be detected. Unfortunately, BDS cannot readily be used to learn about the timescale of sticky bond dissociation. The interpretation of the absolute timescale of the observed cluster relaxations remains an open problem; in many cases, strong arguments point toward an origin from the inside of the clusters. The key challenge for all characterization methods is certainly the high degree of structural and dynamic disorder in the given systems, as the observed relaxations are all far from Debye-like.

5 Conclusions

In this chapter, I have provided a "historical" account of works using BDS to unravel the microscopic underpinnings of transient crosslinking and reinforcement of polymer melts by sticky sidegroups or endgroups. While the case of cluster formation seems to be the prevailing one for polar stickers embedded in unpolar polymer matrices, a truly quantitative understanding of the BDS response of such supramolecularly reinforced systems has so far only been claimed for a model case of binary sticker contacts mediated by defined hydrogen-bonding motifs.

But even in this case, open questions remain, such as direct evidence for the sole relevance of binary association and the separation of the two contributions to the relevant dielectric process, i.e., sticky bond opening and overall tumbling of the molecular complex. A third one is the presence of an energy barrier in the association–dissociation equilibrium. Theoretical approaches treating the non-trivial problem of sticky bond lifetime renormalization, taking into account the rheologically inactive immediate reversal of a sticky bond opening event (as opposed to sticky bond exchange), critically rely on such information. It seems that, given the current state of the field, the problem should be considered unsolved.

The situation seems even less clear for systems forming large clusters. The rheological behavior and its relation to microscopic relaxation events on the chain level seem to be well understood in some cases, but relating the latter with the timescale of directly detected sticker dissociation events remains a challenge. The relevant dielectric process observed in different systems of systematically varied sticky groups is clearly related to sticker dissociation, in that it often (but not always) shows nearly the same apparent activation energy. However, diametrically opposing trends in the measured data clearly show that its absolute timescale cannot be associated with sticker dissociation. This implies unknown contributions of possibly entropic origin, and an important role of energy barriers for sticky bond formation, requiring likely many attempts before a new sticky bond is formed.

It appears that many of the open question could be tackled by molecular simulations with a sufficient level of chemical detail, allowing for a well-defined association mode (either binary or clustering) and controlled association energy and barrier. Moreover, for the case of large aggregates, it seems necessary to have more direct access to the actual structure of the associates, possibly benefiting from new developments in electron microscopy. I hope that this chapter may inspire new work along these lines, further advancing our understanding of this fascinating class of soft materials.

Acknowledgements I am indebted to all members of the groups of Ralph Colby and Jim Runt for their hospitality during my sabbatical at Penn State University in spring 2018, teaching an old NMR dog new dielectric tricks. I particularly thank Ciprian Iacob for guiding my first steps into hands-on dielectric spectroscopy, as well as Aijie Han and Josh Bostwick for sustained support. Ralph's critical views on ambiguous data and numerous discussions are highly appreciated. I also thank Thomas Thurn-Albrecht and Klaus Schröter for valuable comments and suggestions.

References

1. Leibler L, Rubinstein M, Colby R (1991) Dynamics of reversible networks. Macromolecules 24:4701–4707
2. de Lucca Freitas L, Auschra C, Abetz V, Stadler R (1991) Hydrogen bonds in unpolar matrix—Comparison of complexation in polymeric and low molecular-weight systems. Colloid Polym Sci 269:566–575
3. Seiffert S (ed) (2015) Supramolecular polymer networks and gels. Springer, Cham
4. Hager MD, Zwaag SVD, Schubert US (eds) (2016) Self-healing materials. Springer, Cham
5. Campanella A, Döhler D, Binder WH (2018) Self-Healing in Supramolecular Polymers. Macromol Rapid Commun 39:1700739
6. Rupp H, Döhler D, Hilgeroth P, Mahmood N, Beiner M, Binder WH (2019) 3D Printing of Supramolecular polymers: impact of nanoparticles and phase separation on printability. Macromol Rapid Commun 40:1900467
7. Eisenberg A, Kim J-S (1998) Introduction to Ionomers. Wiley, New York
8. Eisenberg A, Hird B, Moore RB (1990) A new multiplet-cluster model for the morphology of random Ionomers. Macromolecules 23:4098–4107

9. Macknight WJ, Lundberg RD (1984) Elastomeric Ionomers. Rubber Chem Technol 57:652–663
10. Malmierca MA, González-Jiménez A, Mora-Barrantes I, Posadas P, Rodríguez A, Ibarra L, Nogales A, Saalwächter K, Valentín JL (2014) Characterization of network structure and chain dynamics of elastomeric Ionomers by means of ^{1}H low-field NMR. Macromolecules 47:5655–5667
11. Castagna AM, Wang W, Winey KI, Runt J (2011) Structure and dynamics of Zinc-neutralized Sulfonated Polystyrene Ionomers. Macromolecules 44:2791–2798
12. Castagna AM, Wang W, Winey KI, Runt J (2011) Influence of Cation Type on structure and dynamics in Sulfonated Polystyrene Ionomers. Macromolecules 44:5420–5426
13. Wübbenhorst M, van Turnhout J (2002) Analysis of complex dielectric spectra. I. One-dimensional derivative techniques and three-dimensional modelling. J Noncryst Solids 305:40–49
14. Chen Q, Tudryn GJ, Colby RH (2013) Ionomer dynamics and the sticky Rouse model. J Rheol 57:1441–1461
15. Gainaru C, Stacy EW, Bocharova V, Gobet M, Holt AP, Saito T, Greenbaum S, Sokolov AP (2016) Mechanism of conductivity relaxation in liquid and polymeric electrolytes: direct link between conductivity and diffusivity. J Phys Chem B 120:11074–11083
16. Rubinstein M, Semenov N (1998) Thermoreversible gelation in solutions of associating polymers. 2. Linear dynamics. Macromolecules 31:1386–1397
17. Müller M, Kremer F, Stadler R, Fischer EW, Seidel U (1995) The molecular dynamics of thermoreversible networks as studied by broadband dielectric spectroscopy. Colloid Polym Sci 273:38–46
18. Müller M, Stadler R, Kremer F, Williams G (1995) On the motional coupling between chain and junction dynamics in Thermoreversible networks. Macromolecules 28:6942–6949
19. Williams G (1972) The use of the dipole correlation function in dielectric relaxation. Chem Rev 72:55–69
20. Stukalin EB, Cai L-H, Kumar NA, Leibler L, Rubinstein M (2013) Self-healing of Unentangled polymer networks with reversible bonds. Macromolecules 46:7525–7541
21. Gold BJ, Hövelmann CH, Lühmann N, Székely NK, Pyckhout-Hintzen W, Wischnewski A, Richter D (2017) Importance of compact random walks for the rheology of transient networks. ACS Macro Lett 6:73–77
22. Gold BJ, Hövelmann CH, Lühmann N, Pyckhout-Hintzen W, Wischnewski A, Richter D (2017) The microscopic origin of the rheology in supramolecular entangled polymer networks. J Rheol 61:1211–1226
23. Ghosh A, Schweizer KS (2021) Physical bond breaking in associating copolymer liquids. ACS Macro Lett 10:122–128
24. Hoy RS, Fredrickson GH (2009) Thermoreversible associating polymer networks. I. Interplay of thermodynamics, chemical kinetics, and polymer physics. J Chem Phys 131:224902
25. Tress M, Xing K, Ge S, Cao P, Saito T, Sokolov AP (2019) What dielectric spectroscopy can tell us about supramolecular networks. Eur Phys J E 42:133
26. Ge S, Tress M, Xing K, Cao P, Saito T, Sokolov AP (2020) Viscoelasticity in associating oligomers and polymers: experimental test of the bond lifetime renormalization model. Soft Matter 16:390–401
27. Amin D, Likhtman AE, Wang Z (2016) Dynamics in Supramolecular polymer networks formed by associating Telechelic chains. Macromolecules 49:7510–7524
28. Müller M, Dardin A, Seidel U, Balsamo V, Iván L, Spiess HW, Stadler R (1996) Junction dynamics in Telechelic hydrogen bonded Polyisobutylene Networksn Macromolecules 29:2577–2583
29. Goldansaz H, Fustin C-A, Wübbenhorst M, van Ruymbeke E (2015) How Supramolecular Asssemblies control dynamics of associative polymers: toward a general picture. Macromolecules 49:1890–1902
30. Xing K, Tress M, Cao P-F, Fan F, Cheng S, Saito T, Sokolov AP (2018) The role of chain-end association lifetime in segmental and chain dynamics of telechelic polymers. Macromolecules 51:8561–8573

31. Xing K, Tress M, Cao P-F, Cheng S, Saito T, Novikov VN, Sokolov AP (2018) Hydrogen-bond strength changes network dynamics in associating telechelic PDMS. Soft Matter 14:1235–1246
32. Halperin A, Alexander S (1989) Polymeric micelles: their relaxation kinetics. Macromolecules 22:2403–2412
33. Semenov AN, Joanny J-F, Khokhlov AR (1995) Associating polymers: equilibrium and linear viscoelasticity. Macromolecules 28:1066–1075
34. Rubinstein M, Panyukov S (2002) Elasticity of polymer networks. Macromolecules 35:6670–6686
35. Mordvinkin A, Suckow M, Böhme F, Colby RH, Creton C, Saalwächter K (2019) Hierarchical sticker and sticky chain dynamics in self-healing butyl rubber ionomers. Macromolecules 52:4169–4184
36. Mordvinkin A, Döhler D, Binder WH, Colby RH, Saalwächter K (2020) Terminal flow of cluster-forming supramolecular polymer networks: single-chain relaxation or micelle reorganization? Phys Rev Lett 125:127801
37. Mordvinkin A, Döhler D, Binder WH, Colby RH, Saalwächter K (2021) Rheology, sticky chain and sticker dynamics of supramolecular elastomers based upon cluster-forming telechelic linear and star polymers. Macromolecules 54:5065–5076. https://doi.org/10.1021/acs.macromol.1c00655
38. Das A, Sallat A, Böhme F, Suckow M, Basu D, Wießner S, Stöckelhuber KW, Voit B, Heinrich G (2015) Ionic modification turns commercial rubber into a self-healing material. ACS Appl Mater Interfaces 7:20623–20630
39. Richter D, Arbe A, Colmenero J, Monkenbusch M, Farago B, Faust R (1998) Molecular motions in Polyisobutylene: a neutron spin echo and dielectric investigation. Macromolecules 31:1133–1143
40. Kremer F, Schönhals A (eds) (2003) Broadband dielectric spectroscopy. Springer, Berlin
41. Hall LM, Seitz ME, Winey KI, Opper KL, Wagener KB, Stevens MJ, Frischknecht AL (2012) Ionic aggregate structure in Ionomer melts: effect of molecular architecture on aggregates and the Ionomer peak. J Am Chem Soc 134:574–587
42. Yan T, Schröter K, Herbst F, Binder WH, Thurn-Albrecht T (2014) Nanostructure and rheology of hydrogen-bonding Telechelic polymers in the melt: from Micellar liquids and solids to Supramolecular gels. Macromolecules 47:2122–2130
43. Pulst M, Balko J, Golitsyn Y, Reichert D, Busse K, Kressler J (2016) Proton conductivity and phase transitions in 1,2,3-triazole. Phys Chem Chem Phys 18:6153–6163
44. Tress M, Ge S, Xing K, Cao P-F, Saito T, Genix A-C, Sokolov AP (2021) Turning rubber into a glass: mechanical reinforcement by Microphase separation. ACS Macro Lett 10:197–202
45. Ge S, Samanta S, Tress M, Li B, Xing K, Dieudonné-George P, Genix AC, Cao P-F, Dadmun M, Sokolov AP (2021) Critical role of the interfacial layer in associating polymers with Microphase separation. Macromolecules 54:4246–4256
46. Maurer FHJ (1990) An interlayer model to describe the physical properties of particulate composites. In: Ishida H (ed) Controlled interphases in composite materials, pp. 491–504, Cleveland. Springer, Dordrecht

Dynamics of Hyperbranched Polymers Under Severe Confinement in Intercalated Nanocomposites

Kiriaki Chrissopoulou and Spiros H. Anastasiadis

Abstract The dynamic response of different hyperbranched polymers in the bulk and under severe confinement within the ~1 nm spacing of the inorganic galleries in intercalated nanocomposites is discussed. Three generations of a polyester polyol hyperbranched polymer of the Boltorn family as well as the poly(ester amide) Hybrane are mixed with sodium montmorillonite, Na^+-MMT and graphite oxide, GO, in compositions for which all polymers are intercalated and there are no free chains outside the completely filled galleries. Thus, we aim at investigating the effect of the severe confinement and of the different polymer/surface interactions on the hyperbranched polymer dynamics. In both cases, the relaxation processes identified for the neat polymers are found in the nanohybrids, however, with different temperature dependences. Moreover, the different polymer/surface interactions result in a different manifestation of the relaxation processes in the different confining environments.

Keywords Polymer dynamics · Hyperbranched polymers · Nanocomposites · Intercalated structure · Confinement · Dielectric spectroscopy · Segmental relaxation

Abbreviations

DSC	Differential scanning calorimetry
GO	Graphite oxide
HBPs	Hyperbranched polymers
HN	Havriliak-Negami
MWS	Maxwell–Wagner-Sillars

K. Chrissopoulou · S. H. Anastasiadis (✉)
Institute of Electronic Structure and Laser, Foundation for Research and Technology—Hellas, N. Plastira 100, 70013 Heraklion Crete, Greece
e-mail: spiros@iesl.forth.gr

S. H. Anastasiadis
Department of Chemistry, University of Crete, P.O. Box 2208, Heraklion Crete, Greece

A. Schönhals and P. Szymoniak (eds.), *Dynamics of Composite Materials*, Advances in Dielectrics, https://doi.org/10.1007/978-3-030-89723-9_7

Na^+-MMT	Sodium montmorillonite
QENS	Quasi-elastic neutron scattering
T_g	Glass transition temperature
VFT	Vogel-Fulcher-Tammann
XRD	X-ray diffraction

1 Introduction

The investigation of the dynamics of polymeric systems has attracted scientific interest because of its inherent complexity over many length- and time-scales, which influences the structure–property relation and affects significantly many of the macroscopic properties of the materials [1–9]. Polymer dynamics, which includes vibrational motions, rotations of side groups, the segmental α-process as well as the overall chain dynamics, covers a very broad time regime of more than ten decades from the pico-second (ps) to the second (s) regime [10]. Each dynamical mode may exhibit a different length-scale dependence that needs to be investigated; therefore, a number of experimental techniques (dielectric relaxation spectroscopy, neutron and light scattering, nuclear magnetic resonance among others) with complementary spatial and temporal capabilities need to be utilized for its investigation [11–14]. Among the different techniques, quasi-elastic neutron scattering and dielectric spectroscopy have been utilized to investigate the very local rotation and reorientation of side groups, which are usually observed at temperatures below the polymer glass transition temperature, T_g. These processes exhibit Arrhenius temperature dependencies, $\tau = \tau_0 exp[E/RT]$, where R is the gas constant. At temperatures above the polymer T_g, these techniques can probe the segmental dynamics or alpha (α-) relaxation process due to the cooperative motion of (a few) chain segments. Characteristic features of this process are the non-exponentiality of the respective relaxation function and the non-Arrhenius temperature dependence of its relaxation times when approaching the T_g, which is empirically described by the Vogel-Fulcher-Tammann equation $\tau = \tau_0 exp[B/(T - T_0)]$ [8, 12–17].

Further than the investigation of polymer dynamics in the bulk, a subject of great interest during the recent years is the study of the dynamic behavior close to interfaces and/or in very thin films since significant differences can emerge when the molecules are confined over distances comparable to their sizes [18–21]. At the same time, a great number of investigations have appeared on the effect of confinement on the glass transition temperature and polymer dynamics in polymer nanocomposites [22–40]. Polymer nanohybrids consist of inorganic, metal/metal oxide or carbon-based nanoadditives (e.g., graphene [41–43], carbon nanotubes [44], silica or titania nanoparticles [45–48], clays [22, 25, 26, 28, 29, 33, 35, 39]) finely dispersed, in most cases, within a polymer matrix. These materials exhibit enhanced and often innovative physicochemical properties due to the nanometric size of the additives, which leads to a significant increase of the interfacial area, thus creating a large proportion of "interfacial" polymer chains [49–51]. The key parameter for developing a nanocomposite with promising properties and, at the same time, one of the most important experimental challenges is the fine dispersion of the additives within the polymeric matrix that depends on the polymer/surface interactions. An especially interesting case among the various nanoadditives is the case of layered additives. In that case, the interactions not only affect the dispersion of the additive within the polymer matrix, but they determine the final structure of the nanocomposite material as well. Three different structures can be found in polymer/layered materials nanohybrids which with increasing polymer/surface interactions which are [52, 53]: (i) the phase separated, when unfavorable interactions exist between the fillers and the polymer chains and the two ingredients rest in their own phase; (ii) the intercalated one [35, 37, 54–57], where the chains penetrate within the galleries of the additive forming a structure of alternating organic–inorganic phase; and (iii) the exfoliated one [58–65], where the interactions between the polymer and the surfaces are so favorable that the layered structure of the additive is destroyed and the inorganic platelets are homogeneously dispersed in the polymer. In addition to the optimized properties, for example offer the ability to investigate the structure and dynamics of the polymer under severe confinement. Among the various layered materials, graphite oxide and natural layered silicate are of special interest. Both materials are hydrophilic and thus they are favorite candidates for mixing with hydrophilic polar polymers, however, their surface structure and chemical composition are very unlike, and therefore, they are expected to manifest different interactions and different affinity with certain polymers.

A characteristic that influences many physical properties of polymeric materials, for example, critical transition temperatures, their solubility in different solvents, and thus their size in the solution, viscosity, etc., is their architecture. Dendritic polymers are an innovative class of materials that have been investigated and at the same time utilized in many applications during the last years due to the excellent properties they exhibit, like their low viscosity and/or their high density; the large number of different groups they may possess in the periphery of the molecules renders them as very useful functional materials [66, 67]. They can be divided into two broad categories: *dendrimers* and *hyperbranched polymers*, HBPs. The former possesses a well-defined molecular weight and has a well-ordered and totally symmetric structure that is arranged in perfect generations, whereas the latter has a larger degree of polydispersity, a varying number of arm points, and an asymmetry in their structure. Whereas both materials exhibit similar properties, HBPs have the additional advantage of a cost-effective synthesis due to the chemical methods utilized and the absence of any necessary purification [68]. HBPs already constitute basic key components in different high-added-value applications, like in nanolithography [69], as surface modifiers [70], as dispersion and processing agents [71–73] as well as in the pharmaceutical industry and nanomedicine for the encapsulation of substances, targeted delivery of drugs, as therapeutic agents [74–76], etc.

In this work, a short overview of the dynamics of hyperbranched polymers under severe confinement in intercalated nanocomposites will be presented. The dynamic response of different hyperbranched polymers mixed with two different layered materials, sodium montmorillonite, Na^+-MMT, and graphite oxide, GO, is investigated for temperatures both below and above the glass transition temperature of the polymers to investigate the effect of both the confinement and the different polymer/surface interactions on the local and the segmental relaxation; in each case, the dynamics are compared to that of the polymers in the bulk. In both systems, the concentration of the nanohybrids is such that all polymer is intercalated between the layers of the nanoadditives and that there is no free, bulk-like, polymer outside the completely filled galleries. The chapter is organized as follows: Initially, the dynamics of the neat polymers will be reported followed by the description of the relaxation behavior of the nanocomposites with Na^+-MMT and with GO. All measurements were performed utilizing Dielectric Relaxation Spectroscopy (DRS).

2 Hyperbranched Polymer Dynamics

Most of the experimental studies on hyperbranched polymers reported the procedure of their chemical synthesis and their basic (macro)molecular characterization [77–79]. At the same time, there are not many works that investigate their hydrogen bond forming ability and/or the relation between their structure, their measured properties, and their dynamics [80, 81]. The latter has been studied experimentally for temperatures below [35, 39, 40, 82–84] and above the HBPs glass transition temperature, T_g [35, 39, 40]. Two of the most frequently utilized experimental techniques for the investigation of their dynamics have been dielectric spectroscopy and quasi-elastic neutron scattering (QENS). QENS identified the relaxation processes that exist for HBPs and studied the way the formed network of hydrogen bonds in these systems affects the dynamics; it is noted that such an effect is not usually observed in the investigation of the linear polymer [33, 36]. Furthermore, computational methods have been employed to investigate the static properties and the dynamic behavior of HBPs in the melt over many length- and time-scales and to study both intra- and inter-molecular hydrogen bond formation as well as their lifetime duration [33, 75, 80, 85].

An interesting class among the different HBPs is that of hyperbranched polyesters like the three generations of the Boltorn™ polyester polyols (Polymer Factory, Sweden) which are shown in Scheme 1 and the hyperbranched poly(ester amide) Hybrane® S1200 (DSM, the Netherlands and Polymer Factory, Sweden) that is shown in Scheme 2. Boltorns are amorphous polymers with their branches consisting mainly of ester groups, whereas they possess an increasing number of hydroxyl end-groups with increasing generation (i.e., an average of 16 for H20, 32 for H30, and 64 for H40) as well as an increasing total average molecular weight of 1750, 3608, and 7323 g/mol. Differential Scanning Calorimetry (DSC) measurements provided the glass transition temperatures of the three polymers as $T_{g,H20} = 14$ °C, $T_{g,H30} = 35$ °C, and $T_{g,H40} = 46$ °C indicating a significant effect of the generation on the T_g

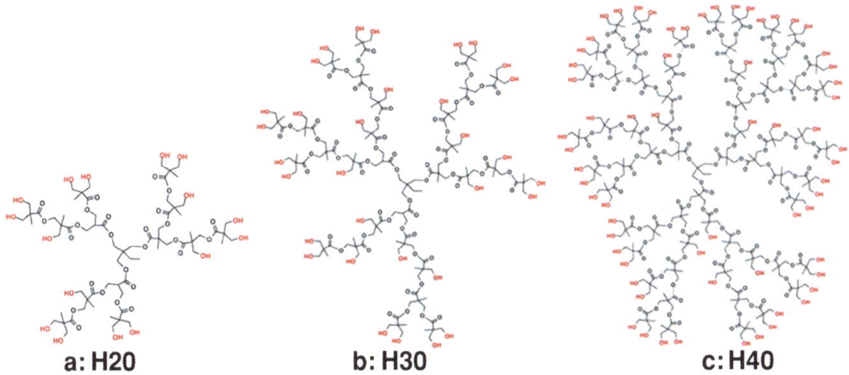

Scheme 1 The three generations of the hyperbranched polyester polyols Boltorn^{TN} (Images Reproduced with permission from Polymer Factory Sweden AB Copyright, https://www.polymerfactory.com/boltorn). a: H20 b: H30 c: H40

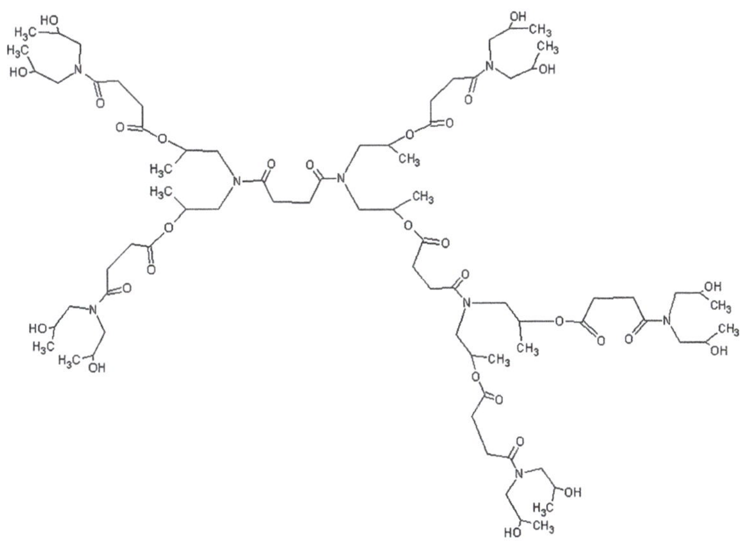

Scheme 2 Chemical formula of a probable structure of hyperbranched poly(ester amide) Hybrane® S1200. Adopted with permission from Polymer Factory Sweden AB Copyright

[35]. Hybrane is also amorphous with a $T_g = 43$ °C [39], having an average molecular weight of 1200 g/mol, whereas it possesses amide groups besides hydroxyl and carboxyl groups. The presence of these functional groups increases, on the one hand, the polarity and the hydrophilicity of the polymers, and on the other hand, their capability to form hydrogen bonds (both intra- and inter-molecular) between the oxygens of the ester groups and the hydroxyl groups.

Dielectric measurements were performed for all these materials over a broad range of frequencies and temperatures both above and below the respective glass transition temperatures. The complex dielectric permittivity, ε^*, given by Eq. 1, is measured as a function of the angular frequency $\omega = 2\pi\nu$, where ν is frequency, providing the real and the imaginary parts, ε' and ε'' of ε^*:

$$\varepsilon^*(\omega) = \varepsilon'(\omega) - i\varepsilon''(\omega) \quad (1)$$

The measurements of the dynamics were made with a dielectric spectrometer Alpha Analyzer from Novocontrol, in the frequency interval 10^{-2}–10^7 Hz. A film (hundreds of micrometers) of the material was positioned between the parallel plates of a stainless steel capacitor. In the case of the pure polymers, the samples were heated at 80 °C and pressed between two electrodes of 30 mm diameter. 100 μm diameter fibers of fused silica were used as spacers. The samples were annealed at 140 °C in vacuum for 24 h to remove any remaining water. In the case of the nanocomposites, the powder was pressed to form disks 12 mm in diameter and 0.3–0.6 mm in thickness. The pellets were annealed at 140 °C in vacuum for 24 h and placed between indium foils to improve the electrical contact with the electrodes. During the measurements, the samples were kept in a pure nitrogen atmosphere, whereas the temperature was controlled through a heated flow of nitrogen gas, by a Quatro Cryosystem.

Figure 1 shows the frequency dependence of the imaginary part of the complex permittivity, ε'', for the H20, as a representative of the Boltorn family, (Fig. 1a, b) and for the Hybrane (Fig. 1c, d) hyperbranched polymers. All materials were annealed at 100 °C for 30 min before the measurements and were thermally equilibrated at successively decreasing temperatures prior to the isothermal data collection. In all cases, the existence of multiple relaxation processes is evident [35, 36, 39]. In the low-temperature regime, below the glass transition temperatures of the polymers, a very broad peak of approximately seven orders of magnitude in frequency, that moves to higher frequencies with increasing temperature is observed. Both the poly(ester amide) Hybrane and all three generations of the hyperbranched polyester polyols Boltorn exhibit this broad peak at temperatures below 10 °C, which shows very analogous characteristics. This sub-T_g relaxation peak may reflect the motion of hydroxyl groups and/or the orientation fluctuations of the ester groups, the so-called γ- and β-process, respectively, according to the literature [82–84].

At temperatures above the calorimetric glass transition, another peak appears, which is related to the glass transition associated dynamics (α-process) as identified by the quantitative analysis, which is presented below. In the case of the Hybrane, however, an additional process can be observed between the broad peak seen at temperatures below T_g and the α-relaxation. This intermediate process can be qualitatively observed in the spectra of Fig. 1d from 0 °C (wine squares) to 50 °C (blue pentagons), whereas at higher temperatures it is partly obscured by the tail of the α-process peak that enters in the frequency window [39]. At even higher temperatures (depending on the material), the observation of any relaxation processes is partially obscured by the conductivity contribution in all polymers.

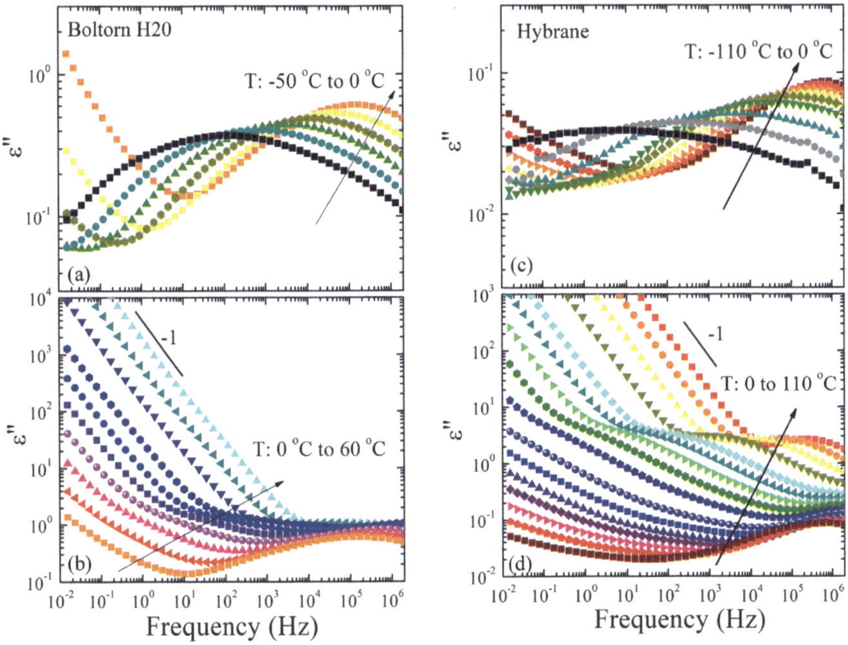

Fig. 1 Imaginary part of the complex dielectric permittivity, ε'', as a function of frequency, for the neat H20 (**a, b**) and the neat Hybrane (**c, d**) polymers at (**a, c**) the low temperature range and (**b, d**) the high temperature range. Reprinted with permission from Ref [35], copyright (2015) American Chemical Society (**a, b**) and from Ref [39], copyright (2021) Elsevier Ltd (**c, d**)

The dynamics of similar hyperbranched polymers have been studied in the past by dielectric spectroscopy. However, in many cases, the high contribution of dc conductivity obscured the observation and clear identification of various processes. For example, when H20, H30, H40 hyperbranched polymers were investigated [83], only one peak was observed, at low temperatures, identified as the γ-process. In a similar study of the same hyperbranched polymers, the γ- and the β-processes were resolved below the glass transition temperature, especially for the higher generations, possessing high activation energies, but no relaxation process was resolved above T_g [84]. Two distinct sub-T_g processes associated with the hydroxyl motion and the ester reorientation, respectively, with similar activation energies, E, were observed in the case of a hydroxyl-terminated hyperbranched polymer [82], whereas in hyperbranched polyesters with terminal acetate, benzoate or aliphatic groups, only one peak, attributed to the ester group motion, was observed but with much larger activation energy together with the segmental relaxation [82]. Furthermore, only the γ- and faintly the β-processes could be identified in a poly(ester amine) [86] and a poly(amido amine) [87] with the segmental relaxation not being detected unambiguously due to the dc conductivity contribution; the segmental dynamics of the poly(ester amine) [86] could be identified with the complementary use of specific heat spectroscopy.

The quantitative analysis of the dielectric spectra was performed by fitting a superposition of empirical Havriliak-Negami (HN) functions to the $\varepsilon^*(\omega)$ data:

$$\varepsilon^*(\omega) = \varepsilon_\infty + \sum_k \left[\frac{\Delta\varepsilon_k}{\left\{1 + \left(i\omega\tau_{HN,k}\right)^{\alpha_k}\right\}^{\beta_k}} \right] - i\frac{\sigma_{dc}}{\varepsilon_0\omega} \tag{2}$$

where k is the index of each process,$\tau_{HN,k}$,$\Delta\varepsilon_k$, and α_k, β_k ($0 < \alpha_k$, $\alpha_k\beta_k \leq 1$) are the Havriliak-Negami relaxation time, the relaxation strength, and the exponents describing the symmetric and asymmetric broadening of the distribution of relaxation times of each process, respectively, ε_0 is the dielectric constant of vacuum and ε_∞ the dielectric permittivity at frequencies much higher than the relaxation frequencies of all considered processes. An ionic conductivity contribution to be added at low frequencies and high temperatures is accounted for by the last term of Eq. (2), where σ_{dc} is the dc conductivity [13]. It is noted that the Havriliak-Negami time, τ_{HN}, and the position of the maximal loss, ω_{peak}, are related via the expression $\omega_{peak} = \frac{1}{\tau_{HN}}\left[sin\frac{\alpha\pi}{2+2\beta}\right]^{1/\alpha}\left[sin\frac{\alpha\beta\pi}{2+2\beta}\right]^{-1/\alpha}$.

Figure 2 illustrates the analysis of the dielectric data based on the above equation for three temperatures of Boltorn H20 (Fig. 2a–c) and of Hybrane (Fig. 2d–f). It is evident from the analysis that, in all cases, multiple HN-functions are necessary to obtain a good fit to the data, two at temperatures below the glass transition and an additional one above T_g. From the spectra of H20 at low temperatures, the shape and relaxation strength parameters are determined for the two sub-T_g processes; for the faster process, the β exponent is fixed, whereas the shape exponent α as well as the dielectric strength, $\Delta\varepsilon$, increase with increasing temperature. At temperatures above T_g, the α-process emerges as shown in Figs. 2b and 2c (dash-dotted line); for this process, $\Delta\varepsilon$ decreases from 9.3–3.7 with increasing temperature as anticipated for the segmental relaxation. These values are very similar for all three generations of the Boltorn polymers.

In the case of Hybrane, two HN functions are also necessary to analyze the dielectric curves at low temperatures because the peak is too broad to be fitted by a single HN-function. The faster sub-T_g process is highly asymmetric, whereas the slower sub-T_g process can be identified only at few temperatures. At higher temperatures, an intermediate process can be detected above −10 °C, with $\Delta\varepsilon = 0.14$–0.97, increasing with increasing temperature; this process is partially obscured by the α-process. Above the Hybrane T_g, the segmental relaxation (α-process) is observed with $\Delta\varepsilon$ decreasing with increasing temperature from 19.4 to 8.9.

Figure 3a shows the results of the above described analysis concerning the relaxation times in an Arrhenius representation for the three Boltorn polymers. The relaxation times of the two sub-T_g processes coincide, follow an Arrhenius temperature dependence, $\tau = \tau_0 exp[E/RT]$ and can be identified as the γ- and the β-processes

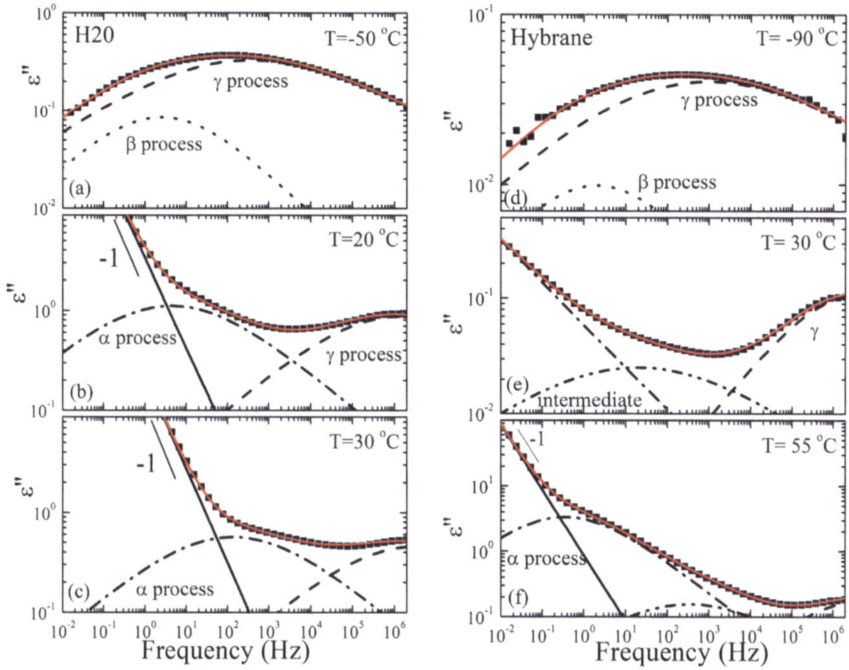

Fig. 2 Imaginary part of the dielectric permittivity, ε'', for the H20 (**a–c**) and for Hybrane (**d–f**) at three temperatures far below (**a, d**), around (**b, e**), and above (**c, f**) each polymer glass transition temperature. The processes needed for the analysis of the spectra are shown with dashed (γ-process), dotted (β-process), dash-dotted (α-process), and dash-dot-dot (intermediate process) lines, whereas the red solid lines are the total fits (together with the conductivity at high temperatures). Reprinted with permission from Ref. [35], copyright (2015) American Chemical Society (**a–c**) and from Ref. [39], copyright (2021) Elsevier Ltd (**d–f**)

related to the rotation of the hydroxyl and the reorientation of the ester groups, respectively. The activation energies of the γ-processes are $E_{\gamma,\mathrm{H20}} = 65.0 \pm 1.5$ kJ/mol, $E_{\gamma,\mathrm{H30}} = 69.5 \pm 1.0$ kJ/mol, and $E_{\gamma,\mathrm{H40}} = 66.5 \pm 1.5$ kJ/mol, whereas $\tau_0 =$ is in the order of 10^{-18} s. It is noted that the values we report for the activation energies of the γ-process of these polymers are lower than the ones reported in the literature for similar hyperbranched polymers [82, 84] but still high in comparison to local processes of other linear polymers [25] or of a different poly(ester amine) [86]. As far as the β-process is concerned, there are few temperatures from which we can derive a relaxation time since, as temperature increases, it is in close proximity to the other two processes. The temperature dependence of its relaxation times appears very similar to the one of the γ-relaxation process. The activation energies derived for the three polymers are $E_{\beta,\mathrm{H20}} = 70 \pm 3$ kJ/mol, $E_{\beta,\mathrm{H30}} = 81 \pm 9$ kJ/mol, and $E_{\beta,\mathrm{H40}} = 86 \pm 2$ kJ/mol, respectively. For this process, an apparent weak dependence of the activation energy on the generation is observed probably indicating that the reorientation of the carbonyl groups becomes more difficult as the molecules become

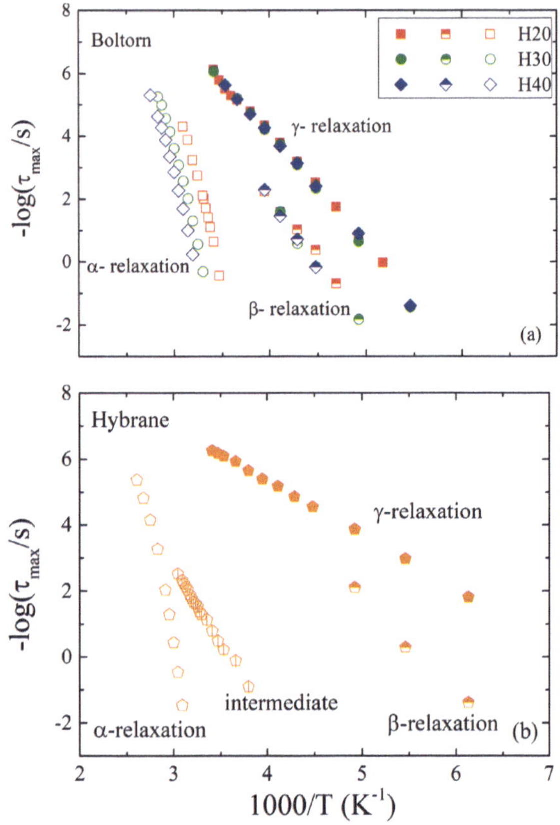

Fig. 3 Arrhenius relaxation map of **a** H20, H30, and H40 Boltorn polyester polyols and **b** Hybrane poly(ester amide). Adapted with permission from Ref. [35], copyright (2015) American Chemical Society **a** and from Ref. [39], copyright (2021) Elsevier Ltd **b**

denser. The large value of the activation energies for both hydroxyl rotation and the ester reorientation is attributed to intra- and inter-molecular hydrogen bonds that are formed in these systems due to the large number of functional groups that exist in the periphery of the molecules. Furthermore, the existence of strong hydrogen bonds can be corroborated by their X-ray diffraction measurements [35], in which a sharp peak is observed on top of the amorphous halo at $2\theta = 17.3°$ that corresponds to a distance of ~0.5 nm indicating a chain packing of the polymers due to an extensive hydrogen bond network [41, 75, 80, 81].

At higher temperatures and as the glass transition of each polymer is approached (according to DSC measurements) [35], the segmental process becomes evident. However, the temperature that this process appears is different for the three polymers in accordance with the increase of T_g with increasing generation. On the other hand, for all three polymers, its temperature dependence follows the Vogel-Fulcher-Tammann (VFT) equation $\tau = \tau_0 exp[B/(T - T_0)]$. Fitting the experimental data provides all relevant parameters; it is noted that in all cases $\tau_0 = 10^{-13}$ s was kept fixed to decrease the error of the analysis. The resulting parameters are: $B = 1939 \pm 29$ K and $T_0 = 225.6 \pm 1.0$ K for H20, $B = 2113 \pm 24$ K and $T_0 = 233.5 \pm 1.0$ K for

H30, and $B = 2227 \pm 29$ K and $T_0 = 237.3 \pm 1.0$ K for H40. The Vogel temperature T_0 increases with generation and shows a similar dependence with that of the calorimetric glass transition temperature, as anticipated. Moreover, the fragility parameter that is $D = B/T_0 = 8.6$, 9.0, and 9.4 for H20, H30, and H40, respectively, denoting that the increase of the generation results in slightly stronger glasses.

For the Hybrane, the three sub-T_g processes follow an Arrhenius temperature dependence, with activation energies $E = 33.0 \pm 2$ kJ/ mol (faster secondary process, γ), 60 ± 4 kJ/mol (slower secondary process, β), and 83 ± 3 kJ/mol (intermediate process). Clear differences can be observed compared to the results for the Boltorn hyperbranched polymers described above. The relaxation times for the fastest process are shorter than the respective ones of the Boltorns. Moreover, its activation energy is significantly smaller, resembling the values found for linear polymers. This means that this process corresponds to a hydroxyl rotation that is not restricted by a strong hydrogen bond network. On the other hand, both the relaxation times and the activation energy of the second secondary process resemble the constrained hydroxyl rotation. It is noted that when the dynamics of a similar cross-linked poly(ester amide) was investigated, comparable activation energies were found [88]. Moreover, molecular dynamics simulations showed that the existence of hydrogen bonds leads to a slower secondary sub-T_g process with high activation energy [85]. The existence of hydroxyl groups where their motion is not affected by hydrogen bonds cannot be identified for Boltorns, as discussed above, indicating that the network formed is much stronger. This could be anticipated by the fact that the X-ray diffractograms of the three Boltorns had a characteristic peak attributed to a hydrogen bond network which cannot be observed in the respective measurement of Hybrane [33, 39]. At the same time, it is clear from the spectra and the analysis that no process related to ester reorientation can be resolved; this can be understood on the one hand, by the smaller number of ester groups of Hybrane when compared to the Boltorns and on the other hand, by the presence of the intermediate process. It is noted that, although we again utilize the Greek letters γ and β to name the two secondary processes for the Hybrane, as is usually done in reports for polymer dynamics, the assignment of the sub-T_g processes of the Boltorn and the Hybrane polymers are different.

The intermediate process in Hybrane appears much slower than the other two sub-T_g relaxations and possesses a significantly high activation energy (83 ± 3 kJ/mol, as mentioned earlier in the text). This relaxation mode can be related to the motion of a larger part of Hybrane polymer. A similar relaxation was reported when molecular dynamics simulations were employed to study the same polymer; its characteristics correlated it with a motion of the branches of the molecule, whereas a significant effect of the intra- and inter-molecular hydrogen bond network was evident [33]. Moreover, in the case of an H-shaped copolymer of poly(ethylene oxide) and polyethylene, the presence of a similar process was related to the relaxation of parts of the PEO arms situated close to its free ends [38].

The segmental relaxation can be identified at temperatures higher than the glass transition temperature of the neat Hybrane [33]. As anticipated, its temperature dependence follows the VFT equation. The resulting parameters are $B = 2127 \pm 25$ K, the Vogel temperature $T_0 = 259 \pm 1.0$ K, and the fragility parameter is $D = 8.2$, which are very close to the values obtained before for the three different generations of the hyperbranched Boltorn polyols (H20, H30, and H40), whereas $\tau_0 = 10^{-13}$ s was kept fixed in this case as well.

3 Polymer Dynamics in Na$^+$-MMT Nanocomposites

A nanoadditive that has been widely utilized for the development of polymer nanocomposites is sodium montmorillonite. It is a material consisting of two-dimensional layers that are stacked together forming a multilayered structure due to weak van der Waals forces; each layer structure comprises two tetrahedral sheets of silica that envelop an octahedral sheet of alumina. The layers possess negative charges due to isomorphic substitution that are counterbalanced by Na$^+$ cations in the interlayer galleries. It is the presence of these charges that make the material polar and favors the interactions with hydrophilic polymers.

Polymer nanocomposites with Na$^+$-MMT exhibit unique properties such as enhanced strength and thermal resistance, reduced gas permeability, reduced flammability, etc. [53, 89]. Moreover, intercalated systems provide the opportunity to study the properties of macromolecules under strong confinement utilizing, however, macroscopic samples and conventional analytical techniques. Such studies on the dynamics of intercalated polymers reveal that the relaxation processes of the bulk polymer are strongly affected by the interaction with the inorganic layers. A number of investigations showed [22, 24, 25, 28, 90] that when polymer chains exist under strong confinement, their segmental process (α-relaxation) shows a different behavior than the respective process of an unconfined polymer. It exhibits a significantly different temperature dependence that deviates from the usual VFT and resembles an Arrhenius behavior, it can be resolved even at much lower temperatures than the T_g of the bulk polymer, whereas it possesses much faster relaxation times. In contrast to the segmental relaxation, the more localized β-relaxation does not show important differences between a polymer in the bulk and under confinement [25]. In systems, however, where the polymer-inorganic interactions are more attractive, the α-relaxation observed is much slower than that of the bulk polymer [30, 91, 92]. Recently, the structure of nanohybrids of hyperbranched polymers with Na$^+$-MMT (supplied by Southern Clay) was studied and the dynamics of the confined polymers in intercalated nanohybrids have been investigated [33, 35, 39].

X-ray diffraction, XRD, was utilized to characterize the structure of both the pure polymer and the nanocomposites. The measurements were performed using a RINT-2000 Rigaku Diffractometer with a 12 kW rotating anode generator, a copper anode as well as a secondary pyrolytic graphite monochromator. The Kα radiation of copper was utilized with a wavelength $\lambda_{CuK\alpha} = 1.54$ Å and diffraction angles 2θ from 1.5° to 30° were scanned with a step of 0.02°. For layered materials that exhibit a periodic structure, like the ones used in the current work, the measured XRD patterns show the characteristic (*00l*) diffraction peaks; the resulting spacing is calculated using Bragg's relationship, $n\lambda = 2d_{00l}\sin\theta$, where d_{00l} is the interlayer distance, 2θ the diffraction angle, λ the wavelength of the radiation and n the order of diffraction. The existence of the respective peaks in the nanocomposites, together with the diffraction angles at which these peaks are observed, will provide information on the structure of the nanohybrid materials.

Figure 4 shows XRD measurements of nanohybrids composed of the H20, H30, and H40 hyperbranched polymers (Fig. 4a) and of the Hybrane (Fig. 4b) and Na^+-MMT. The main (001) peak of pure Na^+-MMT is found at $2\theta = 8.9 \pm 0.1°$, corresponding to an interlayer spacing of $d_{001} = 0.99 \pm 0.05$ nm for the empty galleries. For both Bolton and Hybrane nanocomposites, a clear shift of the diffraction peak of the inorganic material toward lower angles is observed, indicating that an intercalated

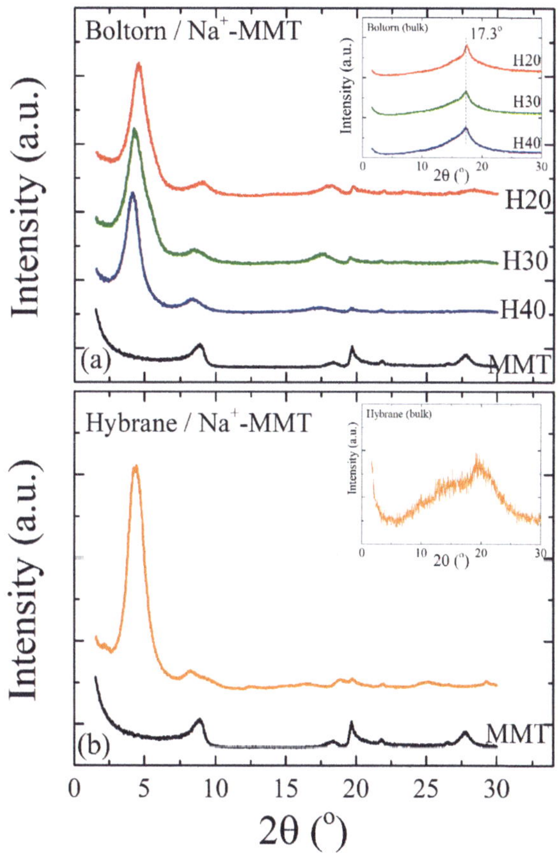

Fig. 4 X-ray diffractograms of **a** Na^+-MMT and nanocomposites of Boltorns (H20, H30, and H40) with Na^+-MMT with 50 wt% polymer content and **b** Na^+-MMT and Hybrane/Na^+-MMT nanocomposite with 30 wt% polymer content after annealing at 200 °C under vacuum. The inset in both shows the diffractograms of the respective neat polymers. The curves have been shifted vertically for clarity. Reprinted with permission from Ref. [35], copyright (2015) American Chemical Society **a** and from Ref. [39], copyright (2021) Elsevier Ltd **b**

structure has been formed due to favorable interactions between the polymers and the inorganic surfaces. More specifically, for H20/Na$^+$-MMT the main diffraction peak is found at $2\theta = 4.5 \pm 0.1^\circ$ corresponding to an interlayer spacing $d_{001} = 1.95 \pm 0.05$ nm, for H30 / Na$^+$-MMT at $2\theta = 4.3 \pm 0.1^\circ$ and $d_{001} = 2.05 \pm 0.05$ nm and for H40 / Na$^+$-MMT at $2\theta = 4.1 \pm 0.1^\circ$ and $d_{001} = 2.15 \pm 0.05$ nm; in all nanohybrids, the polymer content is 50 wt%. It is noted that, when nanohybrids with different polymer content are synthesized, discrete interlayer distances are obtained indicating that the intercalation proceeds in such a way that the galleries are filled initially with monolayers and then, progressively, with bilayers of polymer chains until they are completely filled and the remaining of the polymer resides outside of the completely filled galleries [35]. It is estimated that at a composition of 50 wt% polymer, there are no polymer chains residing outside the completely filled galleries. The situation is very similar with Hybrane/Na$^+$MMT nanohybrids. In this case, it is estimated that the galleries are full for 30 wt% polymer content with the main diffraction peak (Fig. 4b) observed at $2\theta = 4.35 \pm 0.1^\circ$, which corresponds to an interlayer distance of $d_{001} = 2.03 \pm 0.05$ nm [33, 39].

Dielectric measurements were performed for all these nanohybrids in an attempt to investigate the effect of the severe confinement within the galleries of Na$^+$-MMT on polymer dynamics. Figure 5 shows the frequency dependence of the imaginary part, ε'', of the complex permittivity of the H20 / Na$^+$-MMT, the H30 / Na$^+$-MMT, and the H40 / Na$^+$-MMT nanocomposites for different temperatures both above

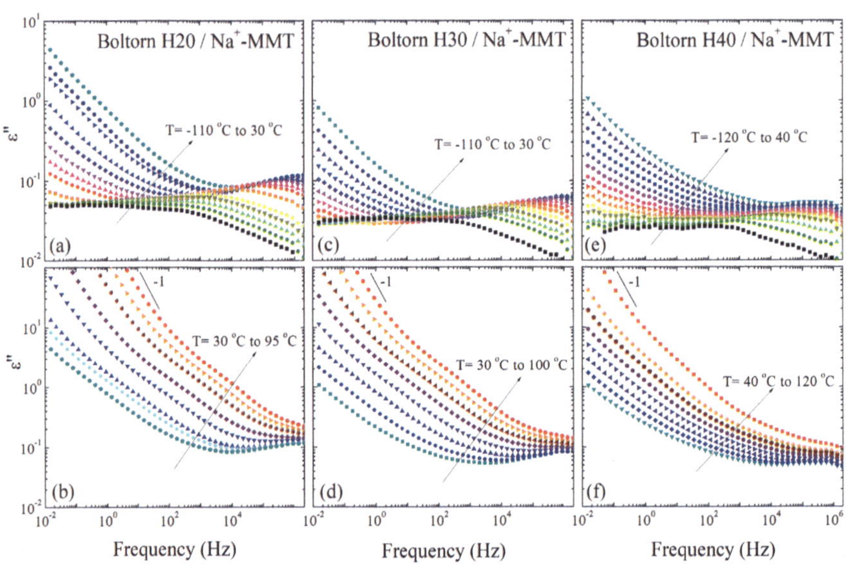

Fig. 5 Frequency dependence of the imaginary part of the dielectric permittivity, ε'', for the H20 / Na$^+$-MMT **a**, **b**, the H30 / Na$^+$-MMT **c**, **d** and the H40 /Na$^+$-MMT **e**, **f**, with 50 wt% polymer content for low (**a, c, e**) and high (**b, d, f**) temperatures. Reprinted with permission from Ref. [35], copyright (2015) American Chemical Society (**a**, **b**)

and below the glass transition temperatures of the neat polymers [35]. Right away, one should point to the fact that relaxation processes can be identified for a much broader range of temperatures than for the respective polymers both in the low and high-temperature regime, whereas clear differences can be observed in their spectra compared with the ones of the neat polymers.

At low temperatures, the spectra seem to have multiple relaxation processes like in those of the pure polymers; however, the peaks have a very different shape with lower dielectric strength and significantly faster relaxation times than those for the bulk (when compared at the same temperature). For higher temperatures, but still below the glass transition temperature of the neat polymers (for the intercalated hybrids, no calorimetric T_g was observed by DSC [35]), the presence of an additional process becomes evident. In studies of the dynamics of linear polymers confined within the galleries of layered silicates, a process with an Arrhenius temperature dependence, which was observed below the polymer glass transition temperature, was identified as the segmental relaxation, which was significantly modified due to confinement [22, 25], whereas a similar behavior was identified for glass formers confined within pores [93–95].

The dielectric data for the nanocomposites were analyzed utilizing a superposition of Havriliak-Negami functions, similarly with the analysis for the neat polymers discussed above. Three relaxation processes were identified together with an additional one observed at very high temperatures. The three faster processes will be mentioned from here on with the same Greek letters like in the neat polymers followed by a prime apex, i.e., γ', β', and α' from the fastest to the slowest, whereas the slowest one can be related to the presence of the inorganic surfaces since it is not observed in the polymers. For all three generations, the β exponent of the γ' process is fixed to 1.0 whereas the values for the α exponents and the dielectric strengths, $\Delta\varepsilon$, were derived based on the dependencies from the low-temperature spectra. At higher temperatures, but still below the bulk polymer T_g, another process appears and, at even higher temperatures, a slow process becomes evident; for the former, $\Delta\varepsilon$ decreases by increasing temperature, whereas for the latter, $\Delta\varepsilon$ increases. At low frequencies and high temperatures, the addition of a contribution to the fit due to dc conductivity is necessary.

Multiple relaxation processes are evident in the dielectric spectra of the Hybrane/Na^+-MMT nanocomposite with 30 wt% polymer content, with the imaginary part of the permittivity, ε'', shown in Fig. 6 as a function of frequency [39]. Two relaxation processes are observed at low temperatures similar to the case of the pure Hybrane; nevertheless, all processes have a much narrower shape and lower dielectric strength than the process of the bulk polymer. In this case, however, no relaxation processes are observed at temperatures around the bulk polymer T_g; it is only at even higher temperatures, 90 °C (dark yellow diamonds) to 150 °C (red circles), where another process apparently enters the measurement window, as evidenced by the peculiar increase of the dielectric strength in the respective data. Superposition of many relaxation processes is necessary, according to the performed analysis, to fit the data of Hybrane/Na^+-MMT nanohybrid in a satisfactory way. The values of the characteristic relaxation strength, $\Delta\varepsilon$, as well as the shape parameters for the

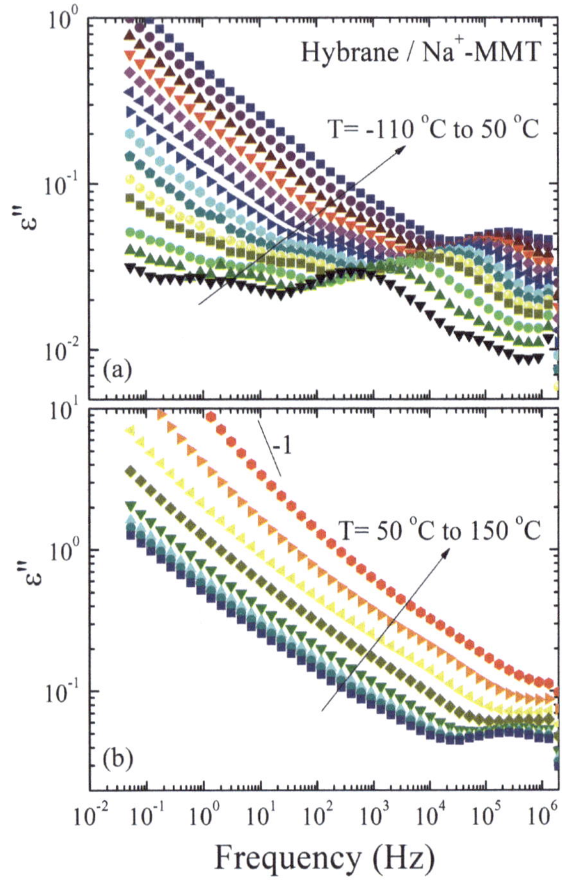

Fig. 6 Imaginary part of the complex dielectric permittivity, ε'', for the Hybrane / Na^+-MMT nanocomposite with 30 wt% polymer as a function of frequency **a** at low temperatures and **b** at high temperatures. Reprinted with permission from Ref. [39], copyright (2021) Elsevier Ltd

sub-T_g processes, are obtained from the low-temperature measurements. The fast sub-T_g relaxation has Δε ~0 .15 and is relatively asymmetric and narrow, whereas the slower one is very weak and its dielectric strength decreases with the increase of the temperature. When the temperature is higher than the glass transition temperature of the neat polymer, another process appears, called α', with Δε that decreases with increasing temperature. At the highest temperatures measured, a conductivity contribution is evident together with a process attributed to Maxwell–Wagner-Sillars (MWS) polarization.

The relaxation times obtained from the analysis for all nanocomposites are shown in Fig. 7 in an Arrhenius representation together with the ones of the bulk polymers (shown as lines). Figure 7a demonstrates that, on the one hand, the relaxation times of the two sub-T_g processes obtained for the Boltorn nanohybrids are very similar between the three polymers, and on the other hand, they are significantly faster (especially when compared at low temperatures) for the systems under confinement in relation to the unconfined ones; moreover, their temperature dependence is much

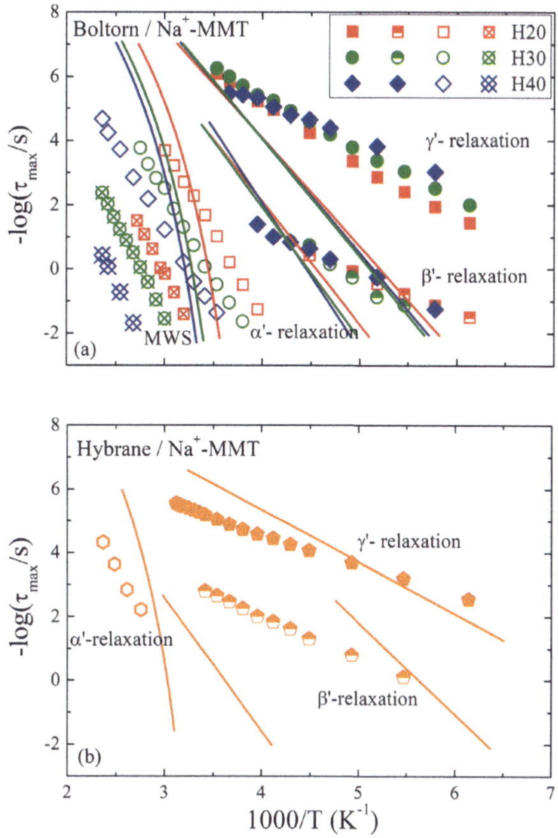

Fig. 7 Arrhenius relaxation map of **a** the H20/Na$^+$-MMT, H30/Na$^+$-MMT, and H40/Na$^+$-MMT with 50 wt% polymer and **b** the Hybrane/Na$^+$-MMT with 30 wt% polymer, nanocomposites. The solid lines are the respective relaxation times of the neat polymers. The error is assumed smaller than the size of the points. Adapted with permission from Ref. [35], copyright (2015) American Chemical Society **a** and from Ref. [39], copyright (2021) Elsevier Ltd **b**

weaker than the respective of the bulk polymers. The latter results in lower activation energies for the fastest process which is called γ' and can be correlated with the motion of the polar hydroxyl groups of the confined polymers, $E_{\gamma',\mathrm{H20}} = 32.5 \pm 2.5$ kJ/mol, $E_{\gamma',\mathrm{H30}} = 31 \pm 2$ kJ/mol, and $E_{\gamma',\mathrm{H40}} = 23 \pm 1$ kJ/mol. Similar activation energies have been found in the past when the dynamics of a linear poly(ethylene oxide) was investigated below the polymer glass transition temperature [25]. As discussed earlier, the Boltorn hyperbranched polymers are able to form in the bulk a network due to hydrogen bonds that leads to high activation energies of the local processes because of the difficulties it raises in the motion. When the polymers experience a strong confinement inside the inorganic galleries, the chains are forced to get more flatten conformations, and therefore, the formation of this hydrogen bond network is inhibited; therefore, lower activation energy, similar to the one obtained for systems that do not have the ability to form hydrogen bonds, is anticipated. Similarly weak temperature dependence is obtained for the second sub-T_g process as well, resulting in activation energies $E_{\beta',\mathrm{H20}} = 23.5 \pm 1$ kJ/mol, $E_{\beta',\mathrm{H30}} = 32.5 \pm 2.5$ kJ/mol, and $E_{\beta',\mathrm{H40}} = 26 \pm 1$ kJ/mol. This process that is called β' can be related to the reorientation of the ester groups under confinement which becomes easier as well due to the lower number of hydrogen bonds that can be formed.

Equally important differences between the dynamic response of the bulk and the confined polymers can be observed in the higher temperature regime, as well. Another process with Arrhenius temperature dependence appears, below the glass transition temperature of the bulk polymer, that has similar behavior with the α'-process reported in previous studies of PMPS [22] and PEO [25] under confinement. The Arrhenius temperature dependence of the α'-process in contrast to the VFT dependence of the usual segmental relaxation results in faster relaxation times for the confined Boltorns at low temperatures or close to the glass transition temperature and in significantly slower ones at higher temperatures in comparison to the ones of the neat polymers. However, the dielectric strength derived for the α' process shows a similar temperature dependence with the usual one of the segmental relaxation, i.e., it decreases with increasing temperature. At the same time, its activation energy indicates that it is a process that is not related to the motion of a single sub-unit. Its values are especially high ($E_{\alpha',H20} = 99.4 \pm 2$ kJ/mol, $E_{\alpha',H30} = 100.7 \pm 2$ kJ/mol, and $E_{\alpha',H40} = 99.2 \pm 2$ kJ/mol) and within the error are the same for the three polymers under confinement. On the other hand, this process shows a specific dependence on the generation since it enters the frequency window at −30 °C for the H20 / Na^+-MMT, at −10 °C for the H30 / Na^+-MMT, and at 10 °C for the H40 / Na^+-MMT nanohybrid. The difference between these temperatures seems to follow the difference in the glass transition temperatures of the three polymers; in all three cases, the α'-process can be resolved at ~40 °C below the respective T_g of each polymer. Thus, the observed dynamic behavior indicates enhanced mobility at temperatures even below the T_g. It is noted, however, that DSC did not show any thermal transitions for the specific nanohybrids; it is only for nanocomposites with higher polymer content that a transition can be observed at temperatures even higher than the one of the pure polymers [35].

Two sub-T_g processes are observed, as well, in the Hybrane/Na^+-MMT nanohybrid, with weaker Arrhenius temperature dependencies when compared to the neat polymer; the activation energies are obtained as 19.0 ± 1 kJ/mol for the faster (γ') process and 25.5 ± 0.5 kJ/mol for the slower one (β'); the primes are again utilized to facilitate comparison with the respective relaxation processes of the bulk Hybrane polymer. Both processes are attributed to the less and more constrained hydroxyl motion with the restrictions becoming weaker in the nanocomposite [39]. A similar behavior was reported when quasi-elastic neutron scattering (QENS) measurements and computer simulations [33] were employed to investigate the sub-T_g dynamic behavior of a Hybrane / Na^+-MMT nanohybrid. At higher temperatures, in the regime where the intermediate and the segmental relaxation were resolved for neat Hybrane, no relaxation process can be observed in the case of the Hybrane / Na^+-MMT nanocomposite, whereas it is only at much higher temperatures and for a narrow temperature regime that a slow α'-relaxation appears. The relaxation times of the α'-process are longer than the respective ones of the bulk polymer and they follow Arrhenius temperature dependence with a significantly high activation energy, 101

$\pm$ 5 kJ/mol [39]. It is noted that a similar process was found in the high-temperature regime for the Boltorn / Na^+-MMT nanohybrids as well.

The results of the slower segmental relaxation under confinement for the Hybrane/Na^+-MMT nanohybrid are in agreement with quasi-elastic neutron scattering measurements which were performed on the same specimen at the IN10 backscattering spectrometer of the Institut Laue Langevin (ILL) in Grenoble, France [33]. In this technique, the sum of the coherent, $S_{\mathrm{coh}}(q,\omega)$, and the incoherent, $S_{\mathrm{inc}}(q,\omega)$, structure factors are measured as a function of scattering wavevector, q, and frequency, ω. The energy variation was performed by moving the monochromator and exploiting the Doppler effect; the range of energy transfer was set to $-13 < \Delta E = \hbar\omega < 13$ μeV. The incident wavelength was $\lambda = 6.271$ Å, and the scattering vector, q, range was set to $0.5 \leq q \leq 2.0$ Å^{-1}. The samples were contained in flat aluminum holders with sample thicknesses $0.2 \leq d \leq 1$ mm. The experimental resolution function, $R(q,\omega)$ was obtained by performing measurements at 1.5 K, when the scattering of the samples within the observation window of the instrument is entirely elastic.

Initially, the energy-resolved elastically scattered intensity from the samples was measured as a function of temperature at various wavevectors. In general, this intensity has a maximum at the very low (liquid Helium) temperatures, when the sample consists of elastic scatterers, whereas increasing temperature causes an enhancement of the molecular dynamics, which leads to a significant decrease of the elastically scattered intensity. The wavevector dependence of the elastic intensity can be utilized to evaluate the mean squared displacement of the scatterers using the equation $I_{el}(q,\omega=0) = I_{el}(q=0,\omega=0)exp\left(-\langle u^2\rangle q^2/6\right)$. The effective mean squared displacements$\langle u^2\rangle$, msd, of the neat Hybrane, the neat Na^+-MMT and the nanohybrid with 30 wt% polymer are shown in Fig. 8. The msd of Hybrane is almost insensitive to temperature between T ~ 2-100 K, it then weakly increases as the sub-T_g dynamic motions set in, it reaches a plateau, and then it increases abruptly above the glass transition temperature T_g, when the segmental motion becomes active. The corresponding msd of the hybrid shows an initial increase that is very similar to that

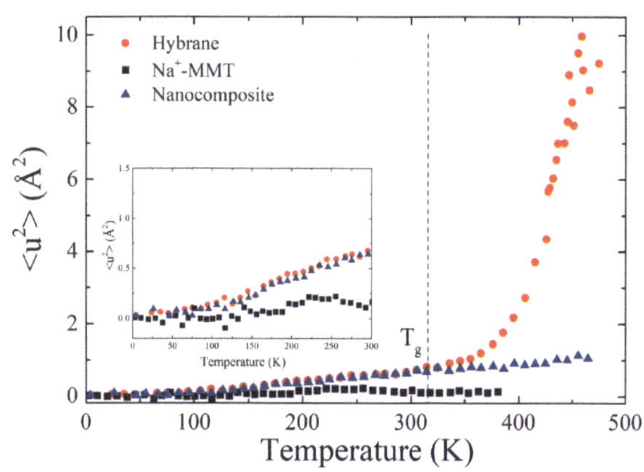

Fig. 8 Temperature dependence of the mean squared displacement, $< u^2 >$, for the Hybrane (●), the Na^+-MMT (■), and the 30 % Hybrane—70 % Na^+-MMT nanocomposite (▲). Reprinted with permission from Ref. [33] copyright (2013) American Chemical Society

of the polymer showing the same temperature dependence at the low temperatures ($T < T_g$). However, for temperatures above the bulk polymer T_g, the msd of the hybrid does not show any further increase but remains almost constant indicating the absence of any dynamics within the time window of the experimental setup. This leads to the conclusion that the majority of the polymer segments must be immobilized onto the inorganic surfaces of the silicate and they do not participate in the segmental motion within the time window of the experimental setup. It is noted that the msd of the inorganic material is quite small and almost constant throughout the whole temperature range.

The incoherent structure factor $S(q,\omega)$ was also measured for both the polymer and the intercalated nanocomposite at selected temperatures both below and above the glass transition temperature of Hybrane to study the sub-T_g and the segmental motions. The measurements performed at low temperatures (150–200 K) for Hybrane showed only a weak temperature effect on the quasi-elastic broadening; however, no scattering vector dependence was observed as is expected for a very local motion of a small functional group of the molecule. For the nanohybrid, the weak changes in the broadening of the quasi-elastic wings indicated even weaker temperature dependence than the one of the pure polymer at the corresponding temperatures. Moreover, the dynamic structure factor of the nanocomposite is almost q-independent. This behavior is reflected on the relaxation times that are shown as a function of the scattering vector in Fig. 9 [33]. It is clear that no scattering vector dependence of the average relaxation times can be seen, which, for the specific temperature, are found to be somewhat faster for the confined system than for the bulk polymer. The absence of any scattering vector dependence in conjunction with the similarity of the relaxation times at temperatures below the polymer glass transition temperature confirms the assumption that this is a local relaxation process that is not affected to any significant extent by the confinement in the order of 1 nm between the inorganic layers. If anything, the specific motion becomes somewhat easier; this can again be attributed

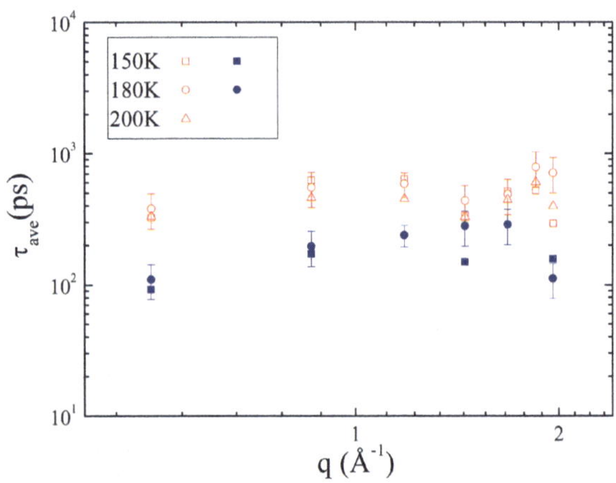

Fig. 9 Scattering vector dependence of the average relaxation time, τ_{ave}, of Hybrane (open symbols) and of the nanocomposite with 30 wt% Hybrane (solid symbols) for the temperatures indicated, which are lower than the bulk polymer T_g. Reprinted with permission from Ref. [33] copyright (2013) American Chemical Society

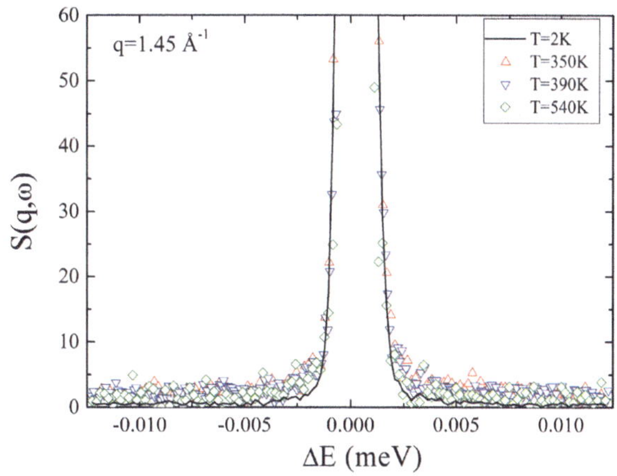

Fig. 10 Incoherent structure factor, $S(q,\omega)$, of the nanocomposite with 30 % Hybrane for various temperatures above the bulk polymer T_g. The line is the instrumental resolution measured at a temperature of 2 K. The y-axis is shown at 10 % of the intensity of the data for the temperature of 350 K. Reprinted with permission from Ref. [33] copyright (2013) American Chemical Society

to a reduced number of hydrogen bonds under confinement, which potentially leads to a reduction of the effects of the constraints on the motion [33].

The incoherent structure factor for Hybrane measured for temperatures above its T_g showed a peculiar scattering vector and temperature dependence. It was only at the highest temperature measured that the scattering vector dependence of the relaxation times resembled a behavior anticipated for the segmental relaxation, i.e., the data showed a q^{-2} dependence and they were in qualitative and quantitative agreement with simulation results. At lower temperatures (but still above T_g), a non-trivial temperature dependence indicated the existence of another relaxation process that cannot be distinguished from the segmental one due to the small frequency window of the technique. Indeed, molecular dynamic simulations predicted the existence of a relaxation process attributed to a kind of branch motion in agreement with the results obtained from the analysis of the dielectric measurements described above.

The incoherent structure factor measurements of the nanohybrid showed a significantly different behavior; the data are shown in Fig. 10 together with the measurement of the instrumental resolution function measured at 1.5 K where no molecular motion exists. It is clear that the data do not show any temperature dependence even at the highest temperature measured and that all spectra almost coincide with the instrumental resolution, indicating frozen dynamics at least for the experimental time range that is accessible by the IN10 backscattering spectrometer of ILL.

The absence of any dynamics for the confined polymer above the glass transition temperature of Hybrane is in agreement with the results obtained by dielectric spectroscopy and discussed above that denote the absence of both the intermediate and the segmental motions at least within the temperature regime they are observed for neat Hybrane; this illustrates the effect of both the confinement and the Hybrane/Na^+-MMT interactions on the polymer dynamics.

4 Polymer Dynamics in Graphite Oxide (GO) Nanocomposites

Graphite oxide, GO, has a multilayered structure similarly with other graphitic materials and like sodium montmorillonite that was discussed above. Recently, graphitic materials have attracted the scientific interest because of their exceptional morphological and structural features, their potential for chemical modification as well as their excellent mechanical, electrical and thermal properties. Polymer/graphitic material nanocomposites comprise an interesting category of nanohybrids that have been already used in a great number of applications as energy materials, in solar cells, in water splitting, in environmental and catalytic technologies, for electromagnetic shielding as well as in various biomedical applications [96–100]. Among the various graphitic materials, graphite oxide has the advantage of a hydrophilic character due to the various polar functional groups that result from the oxidation of graphite, and thus it can be easily mixed with hydrophilic polymers, producing nanocomposites with either exfoliated or intercalated structures. Although it is mainly graphene or reduced graphene oxide that have been reported in the literature for the preparation of polymer nanocomposites, there are a few works that utilized graphite oxide [101–106]; in most of these, just a few percent of additive was used, which precluded the investigation of the polymer structure and dynamics under confinement [31, 32, 39, 40, 107, 108].

When the hydrophilic hyperbranched polymers Boltorns and Hybrane are mixed with graphite oxide, favorable polymer/surface interactions are anticipated. Figure 11 shows the X-ray diffraction measurements of such nanohybrids with 50 wt% polymer and 50 wt% GO (ACS Materials) content. For all four polymers, the intercalation of the polymer is evident by the shift of the main diffraction peak of the GO toward lower angles; this peak for GO (after annealing at 120 °C under vacuum for 12 h to remove humidity) is observed (Fig. 11d) at $2\theta = 12.7 \pm 0.1°$, which corresponds to an interlayer spacing of $d_{001} = 0.70 \pm 0.05$ nm. The main diffraction peak for the Boltorn nanocomposites are found at $2\theta = 6.4 \pm 0.1°$ for H20 / GO and at $2\theta = 5.4 \pm 0.1°$ for H30/GO and H40/GO, which correspond to interlayer distances of $d_{001} = 1.40 \pm 0.05$ nm for the former and $d_{001} = 1.60 \pm 0.05$ nm for the latter two. The intercalated structures formed in the nanocomposites can be attributed to the favorable interactions between the polyester polyols and the surface of graphite oxide that may explain also the shorter interlayer distance for the second generation H20 polymer, which possesses a smaller number of hydroxyl and ester groups [40]. Along the same lines, Hybrane, which possesses the smallest number of functional groups and, thus, should have the weakest interactions with the GO surfaces, exhibits a diffraction peak at $2\theta = 8.8°$, which corresponds to $d_{001} = 1.0$ nm and, thus, a thin film of only 0.3 nm is formed within the galleries of GO (Fig. 11d) [39].

The dynamics of the hyperbranched polymers when they reside between the platelets of graphite oxide was measured with dielectric spectroscopy as well [39, 40]. Figure 12 a, b shows the imaginary part of the complex permittivity, ε'', as a function of frequency for H20/GO nanocomposite for many temperatures below (Fig. 12a)

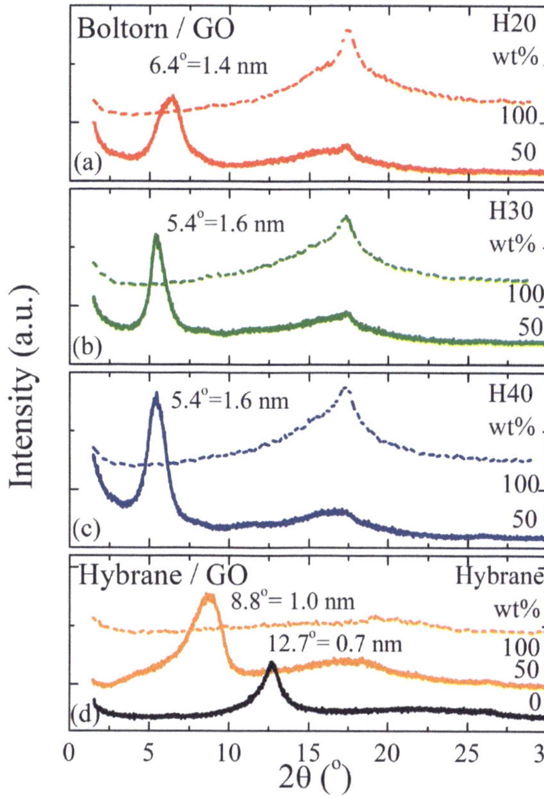

Fig. 11 X-ray diffractograms of **a** H20 / GO, **b** H30 / GO, **c** H40 / GO, and **d** Hybrane / GO with 50 wt% polymer content. In all plots, the diffractograms of the respective polymers are included (dash lines), whereas in **d** the measurement of GO (black line) is shown as well. The curves have been shifted vertically for clarity. Adapted with permission from Ref. [39], copyright (2021) Elsevier Ltd **d**

and above (Fig. 12b) the glass transition temperature. At low temperatures, multiple relaxation processes exist, evident by the width of the dielectric curves, similarly to the case of the neat H20 (Fig. 1a); however, the data have a very different profile and are of low dielectric strength. At temperatures around −20 to − 10 °C another relaxation process becomes evident in the spectra; it is noted that these temperatures are below the glass transition temperature of the neat polymers. At even higher temperatures, a MWS polarization mode, present due to the existence of surfaces as well as a conductivity contribution can be observed. The behavior is very similar for the nanocomposites of higher generations of the Boltorn family.

Figure 12c, d, e demonstrate the representative analysis of the spectra of H20/GO for three temperatures, two below,−50 °C (Fig. 12c) and −10 °C (Fig. 12d), and one above (20 °C, Fig. 12e) the T_g of the neat H20 polymer. A superposition of Havriliak-Negami functions was necessary to successfully fit the data, similarly to the case of the neat polymers. All relaxation processes identified for the polymers should be included in the analysis for the nanocomposites as well. Moreover, the behavior is qualitatively very similar between the three nanohybrids. In this case, as in the case of the Boltorn / Na^+-MMT nanocomposites, the same Greek letters used for the neat polymers will be utilized followed by a prime apex (α′, β′,γ′). From the spectra at

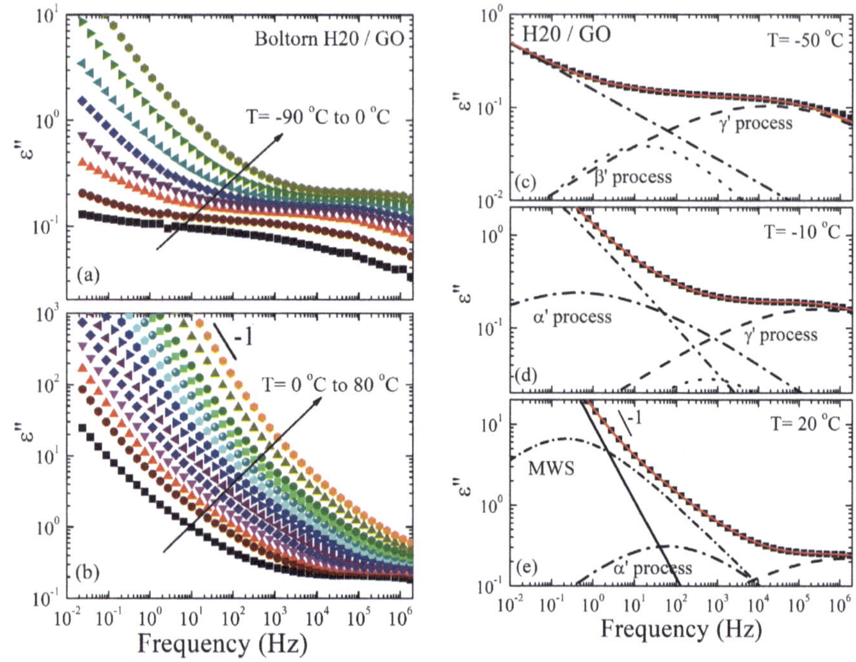

Fig. 12 **a**,**b** Frequency dependence of the imaginary part of the complex permittivity, ε'', for 50 wt% H20 / 50 wt% GO nanohybrid in the temperature range of **a** – 90–0 °C and **b** 0–80 °C. **c, d, e** Analysis of the imaginary part of the dielectric permittivity, ε'', for the H20/GO with 50 wt% polymer at – 50 °C (**c**), –10 °C (**d**), and 20 °C (**e**). The HN-functions needed for the deconvolution of the spectra are shown with dashed (γ'), dotted (β'), dash-dotted (α'), and short dash-dot (MWS) lines, whereas the solid lines are the summation of the processes (best fit)

low temperatures, the shape parameters and the relaxation strength are determined for the γ' and β' processes of the nanocomposites. For all three generations, the β exponent of all processes in the nanocomposites was fixed to 1.0. For both the γ' and the β' processes, the dielectric strength increases with temperature. At higher temperatures, but still below the polymer glass transition temperature, the α' process appears (dash-dotted line) and, at even higher temperatures, the slower MWS process becomes evident. For the α' process, Δε decreases with temperature similarly for all three nanocomposites; such dependence is anticipated for segmental relaxation. At low frequencies and high temperatures, the addition of an ionic conductivity contribution is necessary with a ω^{-1} dependence.

The frequency dependence of the imaginary part ε'' of the complex permittivity for a Hybrane/GO nanocomposite with 50 wt% polymer content is shown in Fig. 13a, b [39] for low and high temperatures (relative to the polymer glass transition) together with the analysis at selected temperatures (Fig. 13c–e). For the Hybrane/GO nanohybrid, one rather strong relaxation process can be clearly observed in the low-temperature regime, while at a few very low temperatures, a shoulder in the data

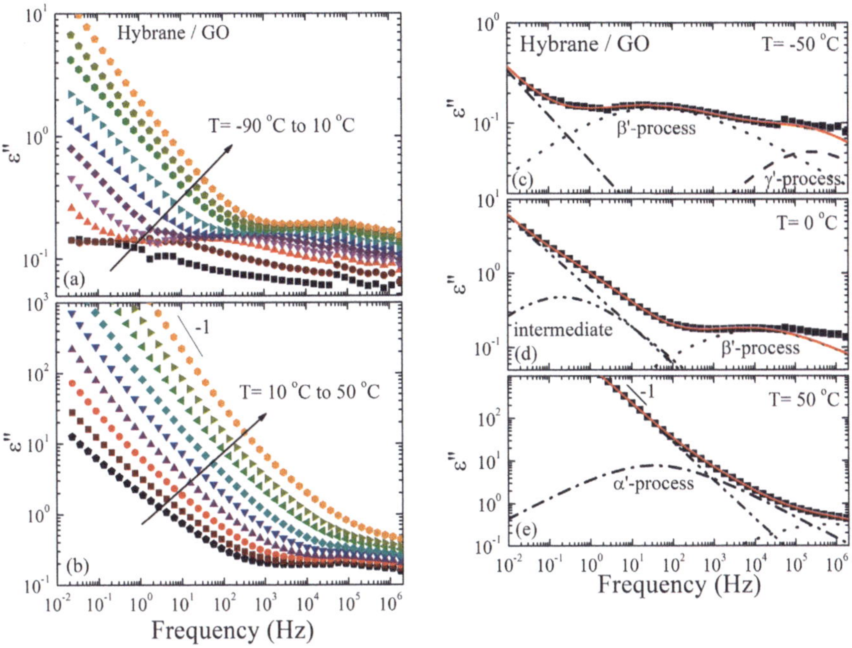

Fig. 13 **a, b** Imaginary part of the complex dielectric permittivity, ε'', for the Hybrane / GO with 50 wt% polymer as a function of frequency in (**a**) the low temperature regime and (**b**) the high-temperature regime. Imaginary part of the dielectric permittivity, ε'', of Hybrane/GO (**c–e**) at (**c**): $T = -50$ °C far below the bulk polymer T_g, (**d**): $T = 0$ °C below the T_g and (**e**): $T = 50$ °C above the T_g. In each case, the individual processes necessary for the deconvolution of the spectra (γ', β', intermediate and α' relaxation) are shown with black lines, whereas the red lines are the total fits. Adapted with permission from Ref. [39], copyright (2021) Elsevier Ltd

could be attributed to a faster secondary relaxation process, with very low dielectric strength. By increasing temperature, but still below the bulk polymer T_g (−43 °C), an additional intermediate relaxation process is observed, partly obscured by the emergence of the segmental process as well as of the conductivity. The analysis showed two sub-T_g processes at very low temperatures similar to the case of the neat Hybrane and provided the values of their characteristic parameters; the fast process can be resolved only at a few low temperatures and possesses low dielectric strength, whereas the slow one is much stronger and prevails in the whole low-temperature region. Two more relaxation processes can be observed in the spectra at higher temperatures, however, still below the neat Hybrane T_g; an “intermediate” relaxation observable at few temperatures only and a slower one; the dielectric strength $\Delta\varepsilon$ of both relaxation processes decrease with increasing temperature. An ionic conductivity contribution is introduced as well to account for the increase of the data at low frequencies.

Figure 14 shows the temperature dependence of the relaxation times for all processes obtained from the analysis of all neat (lines) and confined (symbols)

Fig. 14 Arrhenius relaxation map for **a** Boltorn / GO [40] and **b** Hybrane / GO nanocomposites with 50 wt% polymer. The lines correspond to the fitting of the data of the neat polymers with a VFT equation for the α-relaxation and with an Arrhenius equation for the other sub-T_g processes. The errors in the times are smaller than the size of the points. Adapted with permission from Ref. [39], copyright (2021) Elsevier Ltd (**b**)

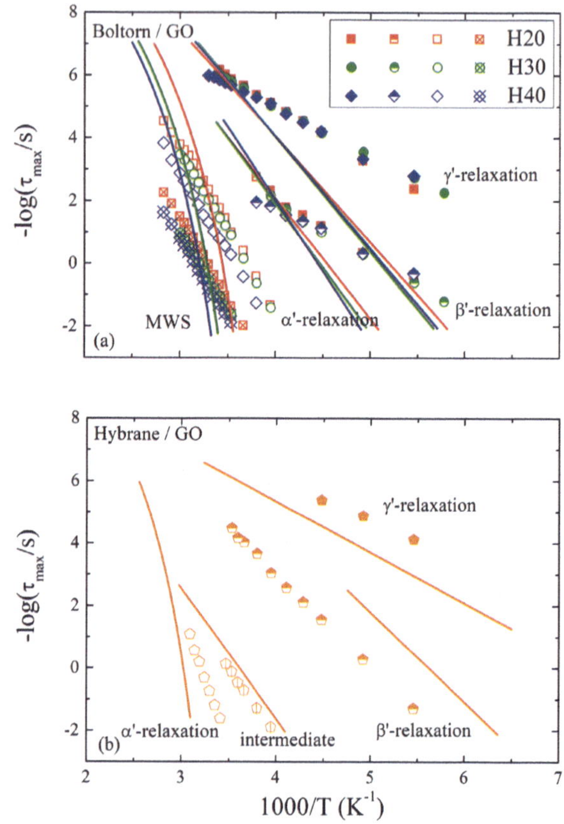

hyperbranched polymers plotted in an Arrhenius representation. For the Boltorns/GO nanocomposites (Fig. 14a), there are three main relaxation processes which all appear in the experimental spectra at temperatures below the glass transition temperatures of the neat polymers and all follow Arrhenius temperature dependence. Starting from the ones that are seen at the lowest temperatures, it is evident that there is no specific generation dependence. Both the temperature and generation dependencies of these sub-T_g relaxations resemble the respective ones of the Boltorns, however, the values of the relaxation times and of the activation energies are lower than the ones obtained for the neat polymers. The two relaxations can be identified as the hydroxyl rotation and the ester reorientation that can become unconstrained under confinement due to the smaller number of hydrogen bonds that can be created within the GO galleries. Indeed, a more quantitative analysis provides the values of the activation energies as $E_{\gamma',H20} = 36 \pm 1.5$ kJ/mol, $E_{\gamma',H30} = 31 \pm 1.0$ kJ/mol, and $E_{\gamma',H40} = 30 \pm 1.0$ kJ/mol for the γ'-process and $E_{\beta',H20} = 33.5 \pm 2.0$ kJ/mol, $E_{\beta',H30} = 35 \pm 2$ kJ/ mol, and $E_{\beta',H40} = 28 \pm 2$ kJ/mol for the β'-process, i.e., lower than what was found for the three polymers and similar to cases where the formation of hydrogen bonds is not anticipated. The third process that appears in the spectra can be identified as the segmental dynamics under confinement based on the temperature dependence of its dielectric strength and the values of the relaxation times that are in the proximity of the relaxation times of the pure polymer segmental relaxation and despite their different temperature dependence that deviate from the usual VFT dependence. No dependence on the generation can be observed, in this case as well, with the respective activation energies for the nanohybrids being $E_{\alpha',H20} = 99.5 \pm 5.0$ kJ/mol, $E_{\alpha',H30} = 98 \pm 3.0$ kJ/mol, and $E_{\alpha',H40} = 97 \pm 2.0$ kJ/mol, whereas even the relaxation times show only a slight trend with somehow slower segmental dynamics with increasing generation; however, it is noted that this dependence cannot be related to the different T_gs of the polymers which show larger differences, as discussed above. Finally, at even higher temperatures, the slow MWS process appears in the nanocomposites due to the presence of a large number of interfaces that may trap ion species in their proximity. The dynamic response of the three Boltorn polymers, when they are confined within the GO galleries, is very similar to the one obtained in the Na^+-MMT nanocomposites with respect to both the number and the temperature dependence of the various relaxation processes; the only observed difference is that the generation dependence of the segmental process appears stronger for the nanocomposites with Na^+-MMT than for those with the GO.

Figure 14b shows the relaxation map for the Hybrane/GO nanocomposite. At very low temperatures, two relaxation processes exist in this case as well, with relaxation times faster and slower, respectively, than the fast and the slow sub-T_g relaxations of Hybrane. Another difference between the polymer and the nanohybrid is in the dielectric strength of the two processes; the dielectric strength $\Delta\varepsilon$ of the fast process is larger than that of the slow for for the neat Hybrane polymer whereas the oppsite is observed for the GO nanocomposite. Their activation energies can be calculated as $E_{\gamma'} = 25 \pm 2$ kJ/mol for the faster process and $E_{\beta'} = 58 \pm 1$ kJ/mol for the slower one; these values indicate that the two processes can be related to the free and the constrained motion of the hydroxyl functional groups that exist in the polymer. As temperature increases, there is an additional process that resembles the intermediate relaxation that is found in Hybrane with similar relaxation times and an Arrhenius temperature dependence ($E = 80 \pm 5$ kJ/mol). At even higher temperatures, the α' relaxation, that is usually identified under confinement is resolved in this case as well, with activation energy, $E = 189 \pm 2$ kJ/mol, i.e., exhibiting an Arrhenius temperature dependence. At the highest temperatures that measurements were performed, the presence of conductivity dominates the dielectric spectra.

When one compares the dynamics in the two Hybrane nanohybrids, i.e., with sodium montmorillonite, Na^+-MMT, and graphite oxide, GO, significant differences can be found, demonstrating that the specific polymer/surface interactions influence both the number of relaxation processes as well as their characteristics. For example, in the case of Hybrane/GO nanocomposite, three sub-T_g processes exist similarly with the neat polymer, whereas in the Hybrane / Na^+-MMT hybrid, the intermediate process cannot be resolved. Moreover, the relaxation times as well as the values and the temperature dependencies of the dielectric strength of the two faster processes indicate the different effects of the hydrogen bond network on the dynamics. For the GO nanohybrid, the fast γ'-process becomes weaker, whereas the slow β'-relaxation becomes slower than the respective processes of the bulk polymer; this can be potentially attributed to hydrogen bonds formed between the hydroxyl groups of the polymer and the functional groups of the GO surface, that causes a reduction in the number of freely rotating hydroxyls. On the contrary, for Hybrane/Na^+-MMT system, both the fast and the slow sub-T_g motions become easier, indicating less hydrogen bonds either between the functional groups of the polymer or between the polymer and the surface. However, in both cases, the segmental relaxation exhibits a change from a VFT to an Arrhenius behavior; this is attributed to the strong confinement that the polymer is subjected to between the layers of both additives, whereas the overall Hybrane/montmorillonite interactions that seem to be more favorable result in slower segmental dynamics as well.

5 Conclusions

The dynamic response of different hyperbranched polymers in the bulk and in nanocomposites is investigated utilizing dielectric spectroscopy. Three generations of a polyester polyol of the Boltorn family, as well as the poly(ester amide), Hybrane, are mixed with sodium montmorillonite, Na^+-MMT, and graphite oxide, GO X-ray diffraction shows that the synthesized nanocomposites exhibit intercalated structure with the interlayer distance that depends on the number of functional groups of the polymers, and thus on the strength of the interactions between the polymers and the surfaces. The dynamics are investigated for nanohybrids with compositions for which all polymers are intercalated and that there are no free chains outside the completely filled galleries; thus the effects of the severe confinement and of the different polymer/surface interactions of the dynamics are investigated.

The relaxation processes identified for the neat polymers are observed in the nanohybrids as well with different temperature dependencies; however, differences in their identification are observed between the different polymers. In all cases, the sub-T_g processes follow Arrhenius temperature dependencies but with a weaker temperature dependencies, and thus smaller activation energies than the respective processes of the neat polymers due to the smaller number of hydrogen bonds that can be formed inside the galleries. Additionally only one intermediate relaxation is observed in Hybrane and attributed to a kind of branch motion that is observed only in Hybrane/GO nanohybrid, whereas it cannot be identified in the Hybrane/Na^+-MMT nanocomposite. The segmental dynamics is either faster or slower under confinement depending on the polymer/surface interactions; however, in all cases, the segmental processes exhibit Arrhenius temperature dependencies in confinement in contrast to the VFT temperature dependence of the neat polymers. In conclusion, the number and the kind of functional groups of the polymers that control their ability for hydrogen bond formation between the chains themselves and between the polymer and the surface as well as the severe confinement define the dynamics in intercalated polymer nanocomposites.

Acknowledgements The authors thank all the co-workers that contributed in the works reviewed in this chapter: Dr. K. Androulaki, Dr. S. Fotiadou, Ms. C. Karageorgaki, Prof. K. Karatasos, Dr. I. Tanis, Dr. D. Tragoudaras, Prof. M. Labardi, Dr. D. Prevosto and Dr. B. Frick. The support of COST Actions MP0902-COINAPO (STSM-MP0902-14971), MP1202-HINT (STSM-MP1202-020615-059356), and CA15107-MultiComp for Short Time Scientific Missions in the University of Pisa as well as the funding from the AENAO research project, Action KRIPIS, project MIS-5002556, funded by the General Secretariat for Research and Technology, Ministry of Education, Greece and the European Regional Development Fund (Sectoral Operational Programme: Competitiveness and Entrepreneurship, NSRF 2007-2013) are greatly acknowledged.

References

1. Sakai VG, Arbe A (2009) Quasielastic neutron scattering in soft matter. Curr Opin Colloid Interface Sci 14:381–390. https://doi.org/10.1016/j.cocis.2009.04.002
2. Paul W, Smith GD (2004) Structure and dynamics of amorphous polymers: computer simulations compared to experiment and theory. Rep Prog Phys 67:1117–1185. https://doi.org/10.1088/0034-4885/67/7/R03
3. Baschnagel J, Varnik F (2005) Computer simulations of supercooled polymer melts in the bulk and in confined geometry. J Phys Condens Matter 17:R851–R953. https://doi.org/10.1088/0953-8984/17/32/R02
4. Lee HN, Paeng K, Swallen SF, Ediger MD (2009) Direct measurement of molecular mobility in actively deformed polymer glasses. Science 323:231–234. https://doi.org/10.1126/science.1165995
5. Chen K, Saltzman EJ, Schweizer KS (2009) Segmental dynamics in polymers: from cold melts to ageing and stressed glasses. J Phys Cond Matt 21:503101. https://doi.org/10.1088/0953-8984/21/50/503101
6. Roland CM (2010) Relaxation phenomena in vitrifying polymers and molecular liquids. Macromolecules 43:7875–7890. https://doi.org/10.1021/ma101649u
7. Freed KF (2011) The descent into glass formation in polymer fluids. Acc Chem Res 44:194–203. https://doi.org/10.1021/ar100122w
8. Colmenero J, Arbe A (2013) Recent progress on polymer dynamics by neutron scattering: from simple polymers to complex materials. J Polym Sci Part B Polym Phys 51:87–113. https://doi.org/10.1002/polb.23178
9. Alegria A, Colmenero J (2016) Dielectric relaxation of polymers: segmental dynamics under structural constraints. Soft Matter 12:7709–7725. https://doi.org/10.1039/c6sm01298a
10. Frick B, Richter D (1995) The microscopic basis of the glass-transition in polymers from neutron-scattering studies. Science 267:1939–1945. https://doi.org/10.1126/science.267.5206.1939
11. Peter C, Kremer K (2009) Multiscale simulation of soft matter systems—from the atomistic to the coarse-grained level and back. Soft Matter 5:4357–4366. https://doi.org/10.1039/b912027k
12. Runt JP, Fitzgerald JJ (1997) Dielectric spectroscopy of polymeric materials: fundamentals and applications. American Chemical Society, Washington, DC
13. Kremer F, Schönhals A (eds) (2003) Broadband dielectric spectroscopy. Springer-Verlag, Berlin
14. Arbe A, Alvarez F, Colmenero J (2020) Insight into the structure and dynamics of polymers by neutron scattering combined with atomistic molecular dynamics simulations. Polymers 12:3067. https://doi.org/10.3390/polym12123067
15. Vogel H (1921) The temperature-dependent viscosity law for liquids. 22:645

16. Fulcher GS (1925) Analysis of recent measurements of the viscosity of glasses. J Am Cerm Soc 8:339–355. https://doi.org/10.1111/j.1151-2916.tb16731.x
17. Tammann G, Hesse W (1926) The dependence of viscosity on temperature in supercooled liquids. Z Anorg All Chem 156:245–257. https://doi.org/10.1002/zaac.19261560121
18. O'Connel PA, McKenna GB (2005) Rheological measurements of the thermoviscoelastic response of ultrathin polymer films. Science 307:1760–1763. https://doi.org/10.1126/science.1105658
19. Napolitano S, Wübbenhorst M (2011) The lifetime of the deviations from bulk behaviour in polymers confined at the nanoscale. Nature Comm 2:260. https://doi.org/10.1038/ncomms1259
20. Zhang W, Douglas JF, Starr FW (2018) Why we need to look beyond the glass transition temperature to characterize the dynamics of thin supported polymer films. PNAS 115:5641–5646. https://doi.org/10.1073/pnas.1722024115
21. Panagopoulou A, Rodríguez-Tinoco C, White RP, Lipson JEG, Napolitano S (2020) Substrate roughness speeds up segmental dynamics of thin polymer films. Phys Rev Lett 124:027802. https://doi.org/10.1103/PhysRevLett.124.027802
22. Anastasiadis SH, Karatasos K, Vlachos G, Manias E, Giannelis EP (2000) Nanoscopic-confinement effects on local dynamics. Phys Rev Lett 84:915–918. https://doi.org/10.1103/PhysRevLett.84.915
23. Ash BJ, Siegel RW, Schadler LS (2004) Glass-transition Temperature Behavior of Alumina/PMMA Nanocomposites. J Polym Sci Part B: Polym Phys 42:4371–4383. https://doi.org/10.1002/polb.20297
24. Frick B, Alba-Simionesco C, Dosseh G, Le Quellec C, Moreno AJ, Colmenero J, Schönhals A, Zorn R, Chrissopoulou K, Anastasiadis SH, Dalnoki-Veress K (2005) Inelastic neutron scattering for investigating the dynamics of confined glass-forming liquids. J Non-Cryst Sol 351:2657–2667. https://doi.org/10.1016/j.jnoncrysol.2005.03.061
25. Elmahdy MM, Chrissopoulou K, Afratis A, Floudas G, Anastasiadis SH (2006) Effect of confinement on polymer segmental motion and ion mobility in PEO/layered silicate nanocomposites. Macromolecules 39:5170–5173. https://doi.org/10.1021/ma0608368
26. Chrissopoulou K, Afratis A, Anastasiadis SH, Elmahdy MM, Floudas G (2007) Structure and dynamics in PEO nanocomposites. Europ Phys J Sp Top 141:267–271. https://doi.org/10.1140/epjst/e2007-00051-9
27. Rittigstein P, Priestley RD, Broadbelt LJ, Torkelson JM (2007) Model polymer nanocomposites provide an understanding of confinement effects in real nanocomposites. Nat Mater 6:278–282. https://doi.org/10.1038/nmat1870
28. Chrissopoulou K, Anastasiadis SH, Giannelis EP, Frick B (2007) Quasielastic neutron scattering of poly(methyl phenyl siloxane) in bulk and under severe confinement. J Chem Phys 127:144910. https://doi.org/10.1063/1.2775449
29. Anastasiadis SH, Chrissopoulou K, Frick B (2008) Structure and dynamics in polymer/layered silicate nanocomposites. Mater Sci Eng B 152:33–39. https://doi.org/10.1016/j.mseb.2008.06.008
30. Vo LT, Anastasiadis SH, Giannelis EP (2011) Dielectric study of poly(styrene-co-butadiene) composites with carbon black, silica and nanoclay. Macromolecules 46:6162–6171. https://doi.org/10.1021/ma200044c
31. Barroso-Bujans F, Fernandez-Alfonso F, Cerveny S, Parker SF, Alegria A, Colmenero J (2011) Polymers under extreme two-dimensional confinement: poly(ethylene oxide) in graphite oxide. Soft Matter 16:7173–7176. https://doi.org/10.1039/c1sm05661a
32. Barroso-Bujans F, Cerveny S, Alegria A, Colmenero J (2013) Chain length effects on the dynamics of poly(ethylene oxide) confined in graphite oxide: a broadband dielectric spectroscopy study. Macromolecules 46:7932–7939. https://doi.org/10.1021/ma401373p
33. Fotiadou S, Karageorgaki C, Chrissopoulou K, Karatasos K, Tanis I, Tragoudaras D, Frick B, Anastasiadis SH (2013) Structure and dynamics of hyperbranched polymer/layered silicate nanocomposites. Macromolecules 46:2842–2855. https://doi.org/10.1021/ma302405q

34. Alexandris S, Sakellariou G, Steinhart M, Floudas G (2014) Dynamics of unentangled cis-1,4-polyisoprene confined to nanoporous alumina. Macromolecules 47:3895–3900. https://doi.org/10.1021/ma5006638
35. Androulaki K, Chrissopoulou K, Prevosto D, Labardi M, Anastasiadis SH (2015) Dynamics of hyperbranched polymer under confinement: a dielectric relaxation study. ACS Appl Mater Interf 7:12387–12398. https://doi.org/10.1021/am507571y
36. Chrissopoulou K, Anastasiadis SH (2015) Effects of nanoscopic-confinement on polymer dynamics. Soft Matter 11:3746–3766. https://doi.org/10.1039/c5sm00554j
37. Androulaki K, Chrissopoulou K, Prevosto D, Labardi M, Anastasiadis SH (2019) Structure and dynamics of biobased polyester nanocomposites. Biomacrocules 20:164–176. https://doi.org/10.1021/acs.biomac.8b01231
38. Elmahdy MM, Gournis D, Ladavos A, Spanos C, Floudas G (2020) H-shaped copolymer of polyethylene and poly(ethylene oxide) under severe confinement: phase state and dynamics. Langmuir 36:4261–4671. https://doi.org/10.1021/acs.langmuir.0c00127
39. Androulaki K, Chrissopoulou K, Labardi M, Anastasiadis SH (2021) Effect of interfacial interactions on static and dynamic behavior of hyperbranced polymers: comparison between different layered nanoadditives. Polymer 222:123646. https://doi.org/10.1016/j.polymer.2021.123646
40. Chrissopoulou K, Androulaki K, Labardi M, Anastasiadis SH (2021) Static and dynamic behavior of polymer/graphite oxide nanocomposites before and after thermal reduction. Polymers 13:2008. https://doi.org/10.3390/polym13071008
41. Karatasos K (2014) Graphene/hyperbranched polymer nanocomposites: insight from molecular dynamics simulations. Macromolecules 47:8833–8845. https://doi.org/10.1021/ma502123a
42. Sun XX, Huang CJ, Wang LD, Liang L, Cheng YJ, Fei WD, Li YB (2020) Recent progress in graphene / polymer nanocomposites. Adv Mater 33:2001105. https://doi.org/10.1002/adma.202001105
43. Papageorgiou DG, Li ZL, Liu MF, Kinloch IA, Young RJ (2020) Mechanisms of mechanical reinforcement by graphene and carbon nanotubes in polymer nanocomposites. Nanoscale 12:2228–2267. https://doi.org/10.1039/c9nr06952f
44. Chen JJ, Liu BF, Gao XH, Xu DG (2018) A review of the interfacial characteristics of polymer nanocomposites containing carbon nanotubes. RSC Adv 8:28048–28085. https://doi.org/10.1039/c8ra04205e
45. Rissanou AN, Papananou H, Petrakis VS, Doxastakis M, Andrikopoulos KS, Voyiatzis GA, Chrissopoulou K, Harmandaris V, Anastasiadis SH (2017) Structural and conformational properties of poly(ethylene oxide) / silica nanocomposites: effect of confinement. Macromolecules 50:6273–6284. https://doi.org/10.1021/acs.macromol.7b00811
46. Papananou H, Perivolari E, Chrissopoulou K, Anastasiadis SH (2018) Tuning polymer crystallinity via the appropriate selection of inorganic nanoadditives. Polymer 157:111–121. https://doi.org/10.1016/j.polymer.2018.10.018
47. Kunzo P, Lobotka P, Kovacova E, Chrissopoulou K, Papoutsakis L, Anastasiadis SH, Krizanova Z, Vavra I (2013) Nanocomposites of polyaniline and titania nanoparticles for gas sensors. Phys Status Solidi A 210:2341–2347. https://doi.org/10.1002/pssa.201329289
48. Alkhodairi H, Russell ST, Pribyl J, Benicewicz BC, Kumar SK (2020) Compatibilizing immiscible polymer blends with sparsely grafted nanoparticles. Macromolecules 53:10330–10338. https://doi.org/10.1021/acs.macromol.0c02108
49. Balazs AC, Emrick T, Russell TP (2006) Nanoparticle polymer composites: where two small worlds meet. Science 314:1107–1110. https://doi.org/10.1126/science.1130557
50. Winey KI, Vaia RA (2007) Polymer nanocomposites. MRS Bull 32:314–358. https://doi.org/10.1557/mrs2007.229
51. Pourrahimi AM, Hoang TA, Liu DM, Pallon LKH, Gubanski S, Olsson RT, Gedde UW, Hedenqvist MS (2016) Highly efficient interfaces in nanocomposites based on polyethylene and ZnO nano/hierarchical particles: a novel approach toward ultralow electrical conductivity insulations. Adv Mater 28:8651–8657. https://doi.org/10.1002/adma.201603291

52. Giannelis EP (1996) Polymer layered silicate nanocomposites. Adv Mater 8:29–35
53. Ray SS, Okamoto M (2003) Polymer/layered silicate nanocomposites: a review from preparation to processing. Prog Polym Sci 28:1539–1641. https://doi.org/10.1016/j.progpolymsci.2003.08.002
54. Chrissopoulou K, Andrikopoulos KS, Fotiadou S, Bollas S, Karageorgaki C, Christofilos D, Voyiatzis GA, Anastasiadis SH (2011) Crystallinity and chain conformation in PEO/layered silicate nanocomposites. Macromolecules 44:9710–9722. https://doi.org/10.1021/ma201711r
55. Wang J, Hu KH, Xu YF, Hu XG (2008) Structural, thermal and tribological properties of intercalated polyoxymethylene / molybdenum disulfide nanocomposites. J Appl Polym Sci 110:91–96. https://doi.org/10.1002/app.28519
56. La Mantia FP, Scaffaro R, Ceraulo M, Mistretta MC, Dintcheva NT, Bota L (2016) A simple method to interpret the rheological behaviour of intercalated polymer nanocomposites. Comp Part B-Eng 98:382–388. https://doi.org/10.1016/j.compositesb.2016.05.045
57. Habel C, Maiz J, Olmedo-Martínez JL, López JV, Breu J, Müller AJ (2020) Competition between nucleation and confinement in the crystallization of poly(ethylene glycol)/large aspect ratio hectorite nanocomposites. Polymer 202:122734. https://doi.org/10.1016/j.polymer.2020.122734
58. Kawasumi M, Hasegawa N, Kato M, Usuki A, Okada A (1997) Preparation and mechanical properties of polypropylene-clay hybrids. Macromolecules 30:6333–6338. https://doi.org/10.1021/ma961786h
59. Chrissopoulou K, Altintzi I, Anastasiadis SH, Giannelis EP, Pitsikalis M, Hadjichristidis N, Theophilou N (2005) Controlling the miscibility of polyethylene/layered silicate nanocomposites by altering the polymer/surface interactions. Polymer 46:12440–12451. https://doi.org/10.1016/j.polymer.2005.10.106
60. Kim DH, Fasulo PD, Rodgers WR, Paul DR (2007) Structure and properties of polypropylene-based nanocomposites: effect of PP-g-MA to organoclay ratio. Polymer 48:5308–5323. https://doi.org/10.1016/j.polymer.2007.07.011
61. Chrissopoulou K, Altintzi I, Andrianaki I, Shemesh R, Retsos H, Giannelis EP, Anastasiadis SH (2008) Understanding and controlling the structure of polypropylene/layered silicate nanocomposites. J Polym Sci Part B Polym Phys 46:2683–2695. https://doi.org/10.1002/polb.21594
62. Chrissopoulou K, Anastasiadis SH (2011) Polyolefin/layered silicate nanocomposites with functional compatibilizers. Eur Polym J 47:600–613. https://doi.org/10.1016/j.eurpolymj.2010.09.028
63. Biswas S, Fukushima H, Drzal LT (2011) Mechanical and electrical property enhancement in exfoliated graphene nanoplatelet /liquid crystalline polymer nanocomposites. Compos Part A-Appl Sci Manufact 42:371–375. https://doi.org/10.1016/j.compositesa.2010.12.006
64. Karevan M, Kalaitzidou K (2013) Understanding the property enhancement mechanism in exfoliated graphite nanoplatelets reinforced polymer nanocomposites. Compos Int 20:255–268. https://doi.org/10.1080/15685543.2013.795752
65. Farahanchi A, Malloy RA, Sobkowicz MJ (2018) Extreme shear processing for exfoliating organoclay in nanocomposites with incompatible polymers. Polymer 145:117–126. https://doi.org/10.1016/j.polymer.2018.04.056
66. Tomalia DA (2005) The dendritic state. Mater Today 8:34–46. https://doi.org/10.1016/S1369-7021(05)00746-7
67. Yan D, Gao C, Frey H (2011) Hyperbranched polymers: synthesis, properties and applications. Wiley Series on Polymer Engineering and Technology, Wiley, New York
68. Hult A, Johansson M, Malmstrom E (1999) Hyperbranched Polymers in Adv. Polym. Sci. 143:1–34. Springer, Berlin, Heidelberg. https://doi.org/10.1007/3-540-49780-3_1
69. Lebib A, Chen Y, Cambril E, Youinou P, Studer V, Natali M, Pepin A, Janssen HM, Sijbesma RP (2002) Room-temperature and low-pressure nanoimprint lithography. Microelectron Eng 61–62:371–377. https://doi.org/10.1016/S0167-9317(02)00485-9
70. Ondaral S, Wågbergb L, Enarsson L-E (2006) The adsorption of hyperbranched polymers on silicon oxide surfaces. J Colloid Interface Sci 301:32–39. https://doi.org/10.1016/j.jcis.2006.04.052

71. Wågberg L, Ondaral S, Enarsson L-E (2007) Hyperbranched polymers as a fixing agent for dissolved and colloidal substances on fiber and SiO_2 surfaces. Ind Eng Chem Res 46:2212–2219. https://doi.org/10.1021/ie061108b
72. Wang JL, Kontopoulou M, Ye ZB, Subramanian R, Zhu SP (2008) Chain-topology-controlled hyperbranched polyethylene as effectove polymer processing ais (PPA) for extrusion of a metallocene linear-low-density polyethylene (mLLDPE). J Rheol 52:243. https://doi.org/10.1122/1.2807445
73. Anastasiadis SH, Hatzikiriakos SG (1998) The work of adhesion of polymer/wall interfaces and its association with the onset of wall slip. J Rheol 42:795–812. https://doi.org/10.1122/1.550909
74. Suttiruengwong S, Rolker J, Smirnova I, Arlt W, Seiler M, Luederitz L, de Diego YP, Jansens PJ (2006) Hyperbranched polymers as drug carriers: microencapsulation and release kinetics. Pharm Dev Technol 11:55–70. https://doi.org/10.1080/10837450500463919
75. Tanis I, Karatasos K, Assimopoulou AN, Papageorgiou VP (2011) Modeling of hyperbranched polyesters as hosts for the multifunctional bioactive agent shikonin. Phys Chem Chem Phys 13:10808–10817. https://doi.org/10.1039/c1cp20271e
76. Arsinidis P, Karatasos K (2019) Computational study of the interaction of a PEGylated hyperbranched polymer/doxorubicin complex with a bilipid membrane. Fluids 4:17. https://doi.org/10.3390/fluids4010017
77. Gelade ETF, Goderis B, de Koster CG, Meijerink N, van Benthem R, Fokkens R, Nibbering NMM, Mortensen K (2001) Molecular structure characterization of hyperbranched polyesteramides. Macromolecules 34:3552–3558. https://doi.org/10.1021/ma001266t
78. Zagar E, Zigon M, Podzimek S (2006) Characterization of commercial hyperbranched aliphatic polyesters. Polymer 47:166–175. https://doi.org/10.1016/j.polymer.2005.10.142
79. Dritsas GS, Karatasos K, Panayiotou C (2009) Investigation of thermodynamic properties of hyperbranched aliphatic polyesters by inverse gas chromatography. J Chromatogr A 1216:8979–8985. https://doi.org/10.1016/j.chroma.2009.10.050
80. Tanis I, Karatasos K (2009) Local dynamics and hydrogen bonding in hyperbranched aliphatic polyesters. Macromolecules 42:9581–9591. https://doi.org/10.1021/ma901890f
81. Zagar E, Huskic M, Zigon M (2007) Structure-to-properties relation of aliphatic hyperbranched polyesters. Macrom Chem Phys 208:1379–1387. https://doi.org/10.1002/macp.200600672
82. Malmstrom E, Hult A, Gedde UW, Liu F, Boyd RH (1997) Relaxation processes in hyperbranched polyesters: effect of terminal groups. Polymer 38:4873–4879. https://doi.org/10.1016/S0032-3861(97)00019-0
83. Zhu PW, Zheng S, Simon G (2001) Dielectric relaxation in a hyperbranched polyester with terminal hydroxyl groups: effects of generation number. Macromol Chem Phys 202:3008–3017. https://doi.org/10.1002/1521-3935(20011001)202:15%3c3008::AID-MACP3008%3e3.0.CO;2-9
84. Turky G, Shaaban SS, Schönhals A (2009) Broadband dielectric spectroscopy on the molecular dynamics in different generations of hyperbranched polyester. J Appl Polym Sci 113:2477–2484. https://doi.org/10.1002/app.30046
85. Tanis I, Tragoudaras D, Karatasos K, Anastasiadis SH (2009) Molecular dynamics simulations of a hyperbranched poly(ester amide): statics, dynamics and hydrogen bonding. J Phys Chem B 113:5356–5368. https://doi.org/10.1021/jp8097999
86. Omara SS, Abdel Rehim MH, Ghoneim A, Madkour S, Thünemann AF, Turky G, Schönhals A (2015) Structure-property relationships of hyperbranched polymer/kaolinite nanocomposites. Macromolecules 48:6562–6573. https://doi.org/10.1021/acs.macromol.5b01693
87. Omara SS, Turky G, Ghoneim A, Thünemann AF, Abdel Rehim MH, Schönhals A (2017) Hyperbranched poly(amidoamine)/kaolinite nanocomposites: Structure and charge carrier dynamics. Polymer 121:64–74. https://doi.org/10.1016/j.polymer.2017.06.017
88. Fernández-Francos X, Rybak A, Sekula R, Ramis X, Ferrando F, Okrasa L, Serra A (2013) Modification of epoxy-anhydride thermosets with a hyperbranched poly(ester amide). II. Thermal, dynamic mechanical, and dielectric properties and thermal reworkability. J Appl Polym Sci 128:4001–4013. https://doi.org/10.1002/app.38453

89. Tjong SC (2006) Structural and mechanical properties of polymer nanocomposites. Mat Sci Eng R 53:73–197. https://doi.org/10.1016/j.mser.2006.06.001
90. Fotiadou S, Chrissopoulou K, Frick B, Anastasiadis SH (2010) Structure and dynamics of polymer chains in hydrophilic nanocomposites. J Polym Sci Part B Polym Phys 48:1658–1667. https://doi.org/10.1002/polb.21974
91. Lorthioir C, Laupretre F, Soulestin J, Lefebvre JM (2009) Segmental dynamics of poly(ethylene oxide) chains in a model polymer/clay intercalated phase: solid state NMR investigation. Macromolecules 42:218–230. https://doi.org/10.1021/ma801909s
92. Prevosto D, Lucchesi M, Bertoldo M, Passaglia E, Ciardelli F, Rolla P (2010) Interfacial effects on the dynamics of ethylene-propylene copolymer nanocomposite with inorganic clays. J Non-Cryst Sol 356:568–573. https://doi.org/10.1016/j.jnoncrysol.2009.09.035
93. Schönhals A, Goering H, Schick C, Frick B, Zorn R (2005) Polymers in nanoconfinement: what can be learned from relaxation and scattering experiments? J Non-Cryst Sol 351:2668–2677. https://doi.org/10.1016/j.jnoncrysol.2005.03.062
94. Schönhals A, Goering H, Schick C, Frick B, Mayorova M, Zorn R (2007) Segmental dynamics of poly(methyl phenyl siloxane) confined to nanoporous glasses. Eur Phys J Sp Top 141:255–259. https://doi.org/10.1140/epjst/e2007-00049-3
95. Bras AR, Fonseca IM, Dionisio M, Schönhals A, Affouard F, Correia NT (2014) Influence of nanoscale confinement on the molecular mobility of ibuprofen. J Phys Chem C 118:13857–13868. https://doi.org/10.1021/jp500630m
96. Kim H, Abdala AA, Macosko CW (2010) Graphene/polymer nanocomposites. Macromolecules 43:6515–6530. https://doi.org/10.1021/ma100572e
97. Hu KS, Kulkarni DD, Choi I, Tsukruk VV (2014) Graphene-polymer nanocomposites for structural and functional applications. Prog Polym Sci 39:1934–1972. https://doi.org/10.1016/j.progpolymsci.2014.03.001
98. Georgakilas V, Tiwari JN, Kemp KC, Perman JA, Bourlinos AB, Kim KS, Zboril R (2016) Noncovalent functionalization of graphene and graphene oxide for energy materials, biosensing, catalytic, and biomedical applications. Chem Rev 116:5464–5519. https://doi.org/10.1021/acs.chemrev.5b00620
99. Papageorgiou DG, Kinloch IA, Young RJ (2017) Mechanical properties of graphene and graphene-based nanocomposites. Prog Mater Sci 90:75–127. https://doi.org/10.1016/j.pmatsci.2017.07.004
100. Kenanakis G, Vasilopoulos KC, Viskadourakis Z, Barkoula NM, Anastasiadis SH, Kafesaki M, Economou EN, Soukoulis CM (2016) Electromagnetic shielding effectiveness and mechanical properties of graphite-based polymeric films. Appl Phys A-Mater Sci Proc 122:802. https://doi.org/10.1007/s00339-016-0338-7
101. Li Z, Young RJ, Wang R, Yang F, Hao L, Jiao W, Liu W (2013) The role of functional groups on graphene oxide in epoxy nanocomposites. Polymer 54:5821–5829. https://doi.org/10.1016/j.polymer.2013.08.026
102. Chang Y-W, Lee K-S, Lee Y-W, Bang JH (2015) Poly(ethylene oxide)/graphene oxide nanocomposites: structure, properties and shape memory behavior. Polym Bull 72:1937–1948. https://doi.org/10.1007/s00289-015-1381-9
103. Larsen RM, Jensen EA (2016) Epoxy–graphite oxide nanocomposites: mechanical properties. J Appl Polym Sci 43591 doi: https://doi.org/10.1002/app.43591
104. Liu D, Bian Q, Li Y, Wang Y, Xiang A, Tian H (2016) Effect of oxidation degrees of graphene oxide on the structure and properties of poly(vinyl alcohol) composite films. Comp Sci Technol 129:146–152. https://doi.org/10.1016/j.compscitech.2016.04.004
105. Botlhoko OJ, Ramontja J, Ray SS (2017) Thermal, mechanical, and rheological properties of graphite and graphene oxide-filled biodegradable polylactide/poly(E-caprolactone) blend composites. J Appl Polym Sci 45373. doi: https://doi.org/10.1002/app.45373
106. Panova TV, Efimova AA, Efimov AV, Berkovich AK (2019) Physico-mechanical properties of graphene oxide / poly(vinyl alcohol) composites. Coll Polym Sci 297:485–491. https://doi.org/10.1007/s00396-018-04465-3

107. Barroso-Bujans F, Fernandez-Alonso F, Pomposo JA, Enciso E, Fierro JLG, Colmenero J (2012) Tunable uptake of poly(ethylene oxide) by graphite-oxide based materials. Carbon 50:5232–5241. https://doi.org/10.1016/j.carbon.2012.07.008
108. Barroso-Bujans F, Alegría A, Pomposo JA, Colmenero J (2013) Thermal stability of polymers confined in graphite oxide. Macromolecules 46:1890–1898. https://doi.org/10.1021/ma302407v

Industrial Applications

Interfacial Effects on the Dielectric Properties of Elastomer Composites and Nanocomposites

Bo Li, Georgios Polizos, and Evangelos Manias

Abstract Elastomer-based composites and nanocomposites are discussed, with the emphasis on the filler and interfacial effects on their dynamics, and the resulting manifestation in their macroscopic dielectric or mechanical response. Specifically, selected polydimethylsiloxane (PDMS)/barium-titanate and styrene-butadiene rubber (SBR)/graphene oxide nanocomposites are discussed, as examples where controlled spatial distribution of filler (structured composites) gives rise to significantly different dielectric and thermomechanical behaviors. Also, ethylene–propylene–diene (EPDM)/carbon black (CB) composites are presented, as examples of systems that exhibit an abnormal temperature dependence of dielectric relaxation; and this response is discussed in the context of CB cluster polarization and quantified through a simple scaling model of the same. Finally, multi-filler EPDM/CB/ceramic elastomer nanocomposites are discussed, as examples of systems with antagonistic interfacial effects between conductive and dielectric fillers, as well as with strong interphase responses that can overwhelm the dielectric contributions from the fillers; in these systems, interphasial responses result in counter-intuitive dielectric behaviors and in strong deviations from standard design principles typically employed in the design of dielectric composite systems.

Keywords Dielectric properties · Interfacial effects · Polymer matrix composites · Polymer nanocomposites · Synthetic elastomers · Synthetic rubber · Barium titanate · Graphene oxide · Carbon black · Ceramic nanoparticles · Multi-filler composites · Composites with clustered nanofillers

---

B. Li
Piezoelectric Technology, PolyK Technologies, LLC, State College, PA 16803, USA

G. Polizos
Electrification and Energy Infrastructure Division, Oak Ridge National Laboratory, Oak Ridge, TN 37831, USA
e-mail: polyzosg@ornl.gov

E. Manias (✉)
Department of Materials Science and Engineering, and Center for Dielectrics and Piezoelectrics, CDP, Materials Research Institute, The Pennsylvania State University, University Park, PA 16802, USA
e-mail: manias@psu.edu

A. Schönhals and P. Szymoniak (eds.), *Dynamics of Composite Materials*, Advances in Dielectrics, https://doi.org/10.1007/978-3-030-89723-9_8

Abbreviations

CB	Carbon black
DRS	Dielectric relaxation spectroscopy
EPDM	Random copolymer from ethylene, propylene, and diene monomers
GO	Graphene oxide
HN	Havriliak–Negami relaxation
MW	Maxwell–Wagner polarization
MWS	Maxwell–Wagner–Sillars polarization
PDMS	Polydimethylsiloxane
ppb	Pixels-per-box (viz. TEM cluster analysis)
rGO	Reduced graphene oxide
SBR	Styrene-butadiene rubber

α,β	Peak shape parameters (e.g., Havriliak–Negami equation)
d_f	Fractal dimension
d_w	Diffusion exponent
D	Diffusion coefficient
e	Electron charge
ε^*	Complex permittivity [function, $\varepsilon^*(\omega, T, \ldots)$, with $\varepsilon^* = \varepsilon' - i\varepsilon''$]
ε_0	Vacuum permittivity
ε'	Permittivity, real part of complex permittivity (function, $\varepsilon' \equiv \mathrm{Re}[\varepsilon^*]$)
ε''	Imaginary part of complex permittivity (function, $\varepsilon'' \equiv \mathrm{Im}[\varepsilon^*]$)
$\Delta\varepsilon$	Dielectric relaxation strength [permittivity change between the onset and the end of a relaxation, $\Delta\varepsilon \equiv \varepsilon'(\omega_{\text{onset}}) - \varepsilon'(\omega_{\text{end}})$]
f	Frequency ($f = \omega/2\pi$)
f_{max}	Characteristic frequency of a relaxation (f at ε'' peak maximum/inflection-point)
k_B	Boltzmann's constant
k	Dielectric constant [empirical; typically, the permittivity ε' value at the application relevant frequency range and temperature; often a weak function (of frequency, temperature) rather than a mathematical constant]
ξ	Cluster size
σ	Conductivity (function, can be complex/real/imaginary, σ^*, σ', σ'', *See* permittivity)
s, σ_0	Conductivity parameters (frequency power and amplitude as $\omega \to 0$, e.g., HN equation)
T	Temperature
τ	Relaxation time
$tan\,\delta$	Dissipation factor, loss tangent [function, for dielectric $\tan\delta(\omega) = \varepsilon''(\omega)/\varepsilon'(\omega)$; for mechanical: lossmodulus over storage modulus]
ω	Frequency ($\omega = 2\pi f$)

1 Introduction

Compared to other material classes, such as ceramics, polymers typically possess higher electrical breakdown strengths but lower values of dielectric permittivity. At the same time, materials with high dielectric constants are vital to electrical and electronic applications—such as power systems, electronic equipment, and energy storage devices—and oftentimes flexible geometries are required, or high mechanical strains are expected to occur during the usual operation. In such cases, polymer elastomers are typically the only viable material choice that can combine high mechanical flexibility and toughness with sufficiently high dielectric permittivity. However, elastomeric polymers are typically characterized by relatively low dielectric permittivities, and proper composite design is required to substantially improve the dielectric performance (e.g., increased dielectric permittivity, at low dielectric loss and high breakdown strength), so as to meet application requirements, ranging from simple electrical insulation, to advanced capacitors and high-k dielectrics, to efficient electromagnetic interference shielding. Currently, the predominant approach to obtain elastomer-type materials with such property sets is the addition of high-permittivity inorganic fillers and/or conductive carbon black and graphitic fillers, i.e., through elastomer composites. The design of these multi-component materials relies heavily on established models that predict composite dielectric responses based on the constituents' permittivities, i.e., of the polymer matrix and filler(s): In such design approaches, models (e.g., micromechanical mixing rules, Lichtenecker's logarithmic mixing rule, Maxwell–Wagner equation, Bruggeman model, etc.) are employed for selecting the type and concentration of dielectric fillers in composites, so as to achieve the desired composite properties [1, 2]. However, such models result in increasingly inaccurate estimations of the composite dielectric properties when interfacial effects begin to dominate (e.g., for high aspect-ratio fillers, sub-micron and nanosized particles, multi-filler composites, or whenever extensive polymer/filler interphases with complex nature and an a priori unknown dielectric behavior exist) [1–5]. Furthermore, composites with anisotropic distribution of fillers can also have significant differences in their response compared to conventional composite models, since the dielectric response can vary substantially between different loci in a composite: e.g., composites with aligned or clustered fillers versus composites with random filler dispersions, or multi-filler composites versus single-type filler composites.

Here, we discuss selected examples where synthetic elastomer polymer composites, reinforced by ceramic-dielectric particles and/or carbon-based conductive fillers, exhibit behaviors that depart from the typical dielectric responses. Hence, the science underlying such non-typical behaviors can provide ideas, opportunities, approaches, and paradigms, for elastomer composites with fundamentally new, emergent, or highly improved dielectric responses.

2 Anisotropic Composite Elastomers

Polydimethylsiloxane with Aligned Barium Titanate Nanoparticles
Elastomer composites with anisotropic distribution of the fillers can have significant advantages over conventional composites with random dispersion of the fillers. Filler alignment can be induced using a shear force or by applying an external electric field with high strength. The tailored filler anisotropy can result in marked improvements in energy storage systems, as well as in flexible membranes with enhanced separation or barrier properties [6–13].

Anisotropic composite silicone elastomers based on polydimethylsiloxane (PDMS, Sylgard184®) filled with $BaTiO_3$ (median particle size 190 nm) were synthesized using dielectrophoretic alignment of the fillers [9]. An external electric field (1.6 kV/mm at 100 Hz) was applied to the elastomer/particle suspension during the curing process. The spatial distribution of the aligned fillers in the elastomer matrix is shown in the scanning electron microscopy (SEM) image in Fig. 1. Dielectric relaxation spectroscopy (DRS [14]) was used to measure the dielectric properties

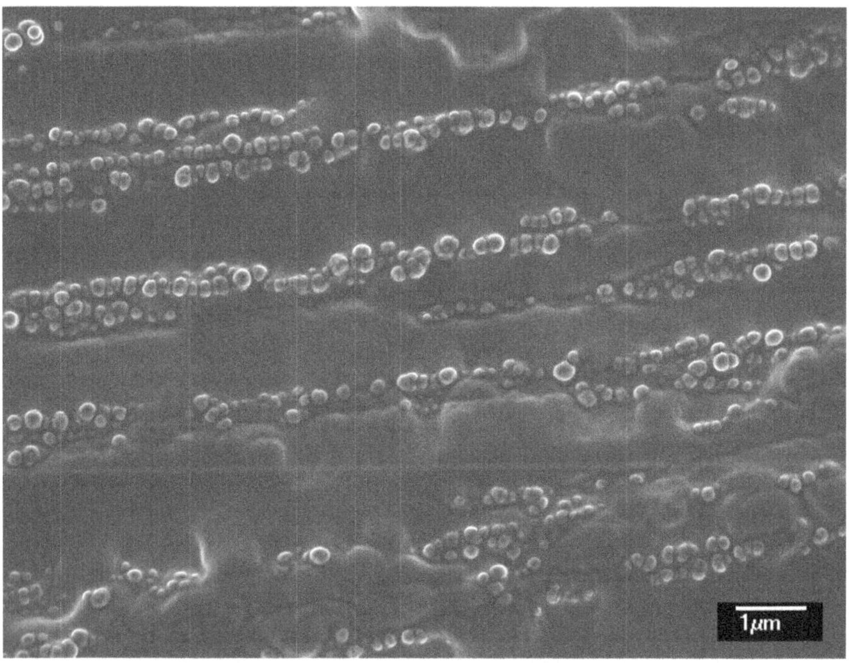

Fig. 1 SEM image of the PDMS elastomer with dielectrophoretically aligned $BaTiO_3$ fillers (10 vol%). Figure reproduced with permission from Ref. [9]. Copyright 2008 Amer. Inst. Physics

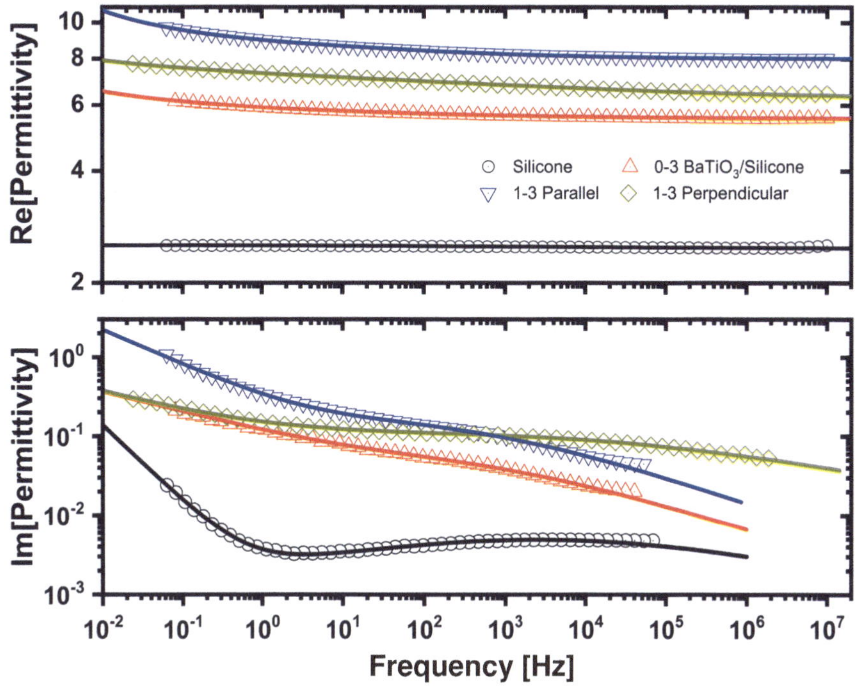

Fig. 2 Real and imaginary parts of the complex permittivity at 293 K (20 °C) of PDMS composites filled with 22.5 vol% $BaTiO_3$. The lines are the best fit of the HN expression and a conductivity power law contribution to the experimental data. Figure reproduced with permission from Ref. [9]. Copyright 2008 Amer. Inst. Physics

of the structured composites at directions parallel (1–3 parallel[1]) and perpendicular (1–3 perpendicular. See Footnote 1) to the direction of the applied field. The pristine silicone elastomer and an elastomer with random filler dispersion (0–3 composite. See Footnote 1) were also measured. The pristine elastomer was subjected to the same thermal cycling under the same external field as the structured elastomers. The real and imaginary parts of the complex permittivity of the elastomers filled with 22.5 vol% $BaTiO_3$ are shown in Fig. 2 Least-square fitting to the experimental data was performed using the Havriliak-Negami (HN) expression and a conductivity contribution [14, 15]:

$$\varepsilon^*(\omega) = \varepsilon_\infty + \frac{\Delta\varepsilon}{\left[1 + (i\omega\tau_0)^{1-\alpha}\right]^\beta} - i\frac{\sigma_0}{\varepsilon_0}\omega^{-s} \tag{1}$$

[1] 0–3 stands for a composite with the fillers not being aligned in any direction. 1–3 parallel and 1–3 perpendicular denote composites with the fillers being aligned parallel and perpendicular to the direction of the electric field, respectively.

Table 1 Fitting parameters of Eq. (1) to the experimental data in Fig. 2. Table was adapted with permission from Ref. [9]. Copyright 2008 Amer. Inst. Physics

Sample	α	Β	Δε	f_{max} (Hz)
Silicone	0.76 ± 0.01	1	0.05 ± 0.01	2518 ± 641
0–3 $BaTiO_3$/silicone	0.65 ± 0.03	1	0.15 ± 0.02	194 ± 47
1–3 parallel	0.66 ± 0.01	1	0.76 ± 0.01	68 ± 8
1–3 perpendicular	0.78 ± 0.02	0.75 ± 0.08	1.14 ± 0.06	649 ± 143

The values of the shape parameters (α, β), dielectric strength ($\Delta\varepsilon$), and maximum frequency f_{max} of the relaxation mechanism are summarized in Table 1. A weak mechanism process is shown in the dielectric losses of the pristine polydimethylsiloxane elastomer due to interfacial Maxwell–Wagner–Sillars (MWS) polarization effects that are associated with ionic (catalyst, space charge) residues at the interfaces of loci with different crosslinking densities [16]. The dielectric strength of this relaxation mechanism is significantly higher in the elastomer composites and particularly in the structured composites (which possess a preferred filler orientation, parallel or perpendicular to the electric field direction). The increase in the $\Delta\varepsilon$ values originates from the interactions between the elastomer and the high permittivity $BaTiO_3$ fillers. The variation in the values depends on the filler spatial configuration. A shift toward lower frequencies of the f_{max} is observed for the composites due to the slowdown of the dynamics at the elastomer/filler interface. The effect of the filler spatial distribution is also evident in the real part of the permittivity. The $\varepsilon'(\omega)$ values increase substantially in the $BaTiO_3$ composites and the highest values were measured at direction parallel to the filler alignment. The increase of the permittivity due to the filler alignment becomes even more prominent at high electric fields (Fig. 3) and can result in important improvements in the energy storage of capacitors operating at high voltage [9].

3 Styrene-Butadiene Elastomers Filled with Functionalized Graphene Oxide

Styrene-butadiene rubber (SBR) is one of the most widespread and used elastomers with numerous applications. Commercially available reinforced SBR composites are typically based on metal oxides (i.e., silica) or carbon black fillers. The dispersion of metal oxide fillers is thermodynamically favorable when sulfur-based silanes are used during the mechanical mixing and crosslinking of the SBR matrix. The dispersion of carbon black can also be easily accomplished due to the hydrophobic nature of the SBR matrix. Graphene oxide (GO) is a high performance, high surface area carbon filler. Its planar geometry and the strong interaction with the styrene groups of the SBR are ideal for achieving mechanically robust interfaces with the elastomer matrix. Moreover, the oxygen-based functional groups (i.e., C = O, C–O, O = C–OH) on

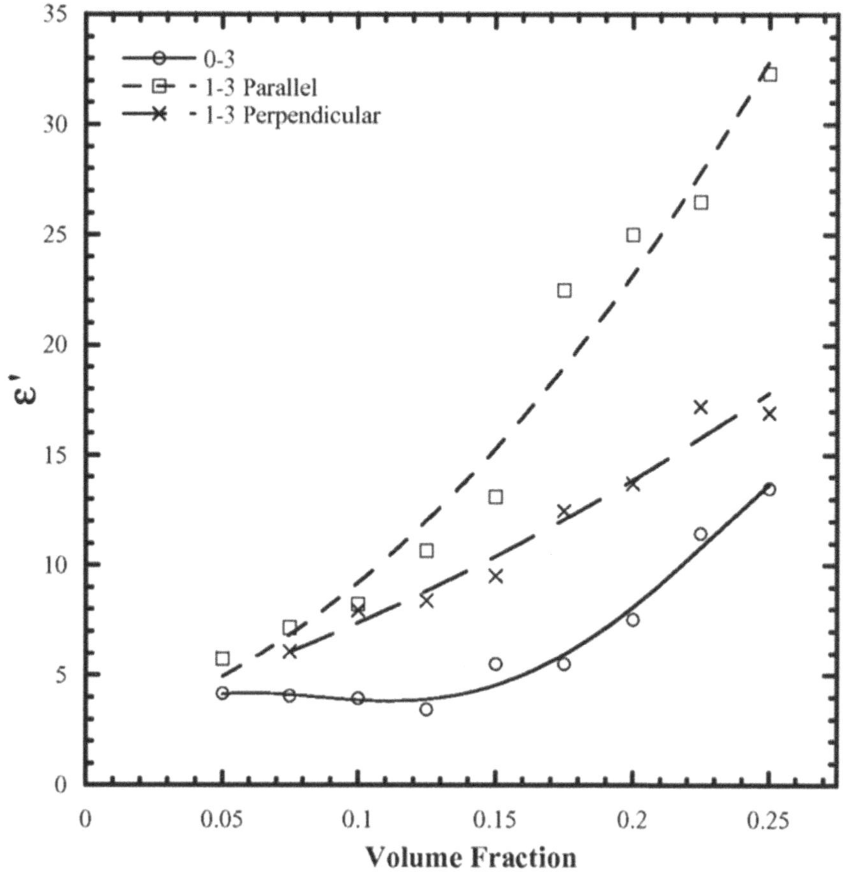

Fig. 3 Permittivity values of the structured composites at different $BaTiO_3$ volume fractions. Figure reproduced with permission from Ref. [9] Copyright 2008 Amer. Inst. Physics

the surface of the GO provide sites that can be functionalized, increase the d-spacing of the GO planes and improve their dispersion.

GO fillers with thiol and dodecylamine (GO–DA) functionalities were synthesized [17]. The thiol-modified GO fillers have reactive interfaces and can be crosslinked with the SBR matrix during the mechanical mixing. During the functionalization process, GO was partially reduced (rGO) and the oxygen content on the surface of the fillers was decreased [18, 19]. The rGO–SH fillers were dispersed in the SBR using a Brabender mixer that was operated at 110 °C. The dodecylamine functionality resulted in a hydrophobic interface on the GO surface and allowed the solution mixing of the GO–DA fillers with the SBR in toluene. The composite matrix was coagulated in methanol and was dried in vacuum at 80 °C. The high-resolution XPS spectra of the GO and the rGO–SH for the C1s and O1s are shown in Fig. 4. The reduction of the GO fillers during the functionalization is clearly shown by the significantly

Table 2 Filler properties: particle size; Filler permittivity ε_r (particle or bulk, the *expected* dielectric ε value of the fillers in the composites); and *measured effective* filler permittivity $\varepsilon_r^{effective}$ in composites (filler permittivity as quantified from the EPDM composite permittivity: the $\varepsilon_r^{effective}$ combines the filler ε_r plus any interfacial contributions, synergistic or antagonistic). Table adapted with permission from Ref. [32]. Copyright 2021 by the authors (Springer)

Filler[a]	Filler particle size (pm)	Filler		Filler $\varepsilon_r^{effectlve}$ in the composite				
		ε_r		20 Hz	60 Hz	140 Hz	1 kHz	10 kHz
SiO_2	1–2	&0	4	42	30	26	21	18
TiO_2	1–2	&0	260	158	138	155	148	142
Sakai-$BaTiO_3$	0.4	&1	3060	537	506	507	495	462
lonized-BaTi0_3	0.4	&1	3420	765	780	787	747	740
Ferro–BaTi O_3	(fine) 0.6–0.9	&0	9000	328	375	333	310	302
Ferro-BaTi O_3	(coarse) 0.9–1.7	&0	9000	601	605	570	565	550
(Nb, In)TiO_2	1–2	&2	60000	141	142	142	130	116

&0 Measured by the supplier; &1 Measured in suspension; &2 Measured at 100 Hz after sintering
[a]The ceramic fillers are: **(i)** *As-received* particulates: SiO_2 (S5631, Sigma-Aldrich); Sakai-$BaTiO_3$ (BT-04, Sakai Chemical Industries); Ferro-$BaTiO_3$ (ZL9000, Ferro Electronic Materials, sieved to separate coarse and fine particulates/agglomerates), and TiO_2 (Rutile 43047, Alfa Aesar). **(ii)** *Surface-modified* particulates: ionized-$BaTiO_3$ (by washing the Sakai-$BaTiO_3$ with distilled water, stirring overnight, and drying at 80 °C, increasing the mobile ions concentration at the $BaTiO_3$ external surface); and "colossal dielectric permittivity" (Nb + In) co-doped TiO_2 (termed as (Nb, In)TiO_2, measured to have ε' (100 Hz) = 60,000 after sintering, prepared as per [45])

lower intensity of the C−O peak in the C1s spectra. The surface composition of the C−O groups was 22.3 and 8.6 at % for the GO and rGO–SH fillers, respectively. The composition of the C (sp^2) was 7.9 and 43.3 at % for the GO and rGO–SH fillers, respectively. Similar trends were found for the composition of the O = C−OH groups that was 6.0 and 1.3 at % for the GO and rGO–SH fillers, respectively. TEM images of the non-crosslinked SBR/GO and SBR/rGO–SH composites are shown in Fig. 5. The mechanical properties of the composites were measured using dynamic mechanical analysis (DMA) at 1 Hz over a broad temperature range. The heating rate was 5 K/min. Comparison plots of the storage modulus and tanδ are shown in Fig. 6. The GO and GO-modified filler contents were 4 wt %. The properties of the GO-based composites were compared to the SBR filled with 9 wt% silica nanoparticles (SBR/SnP). Silica is the most frequently used filler material [20, 21].

The comparison with the SBR/SnP was performed to establish performance improvements with respect to baseline composites. All composites were non-crosslinked. The fillers in the SBR/SnP, SBR/GO, and SBR/rGO–SH composites were dispersed by mechanical mixing. The SBR/GO and SBR/rGO–SH composites have comparable modulus values. In the glassy state ($T < -40$ °C), their modulus plateau values are 44% higher than that of the SBR/SnP even though their filler weight content is less than half compared to that of the silica composite. The higher performance of the GO fillers can be associated with their higher surface area and planar geometry that resulted in improved interfacial adhesion with the SBR matrix.

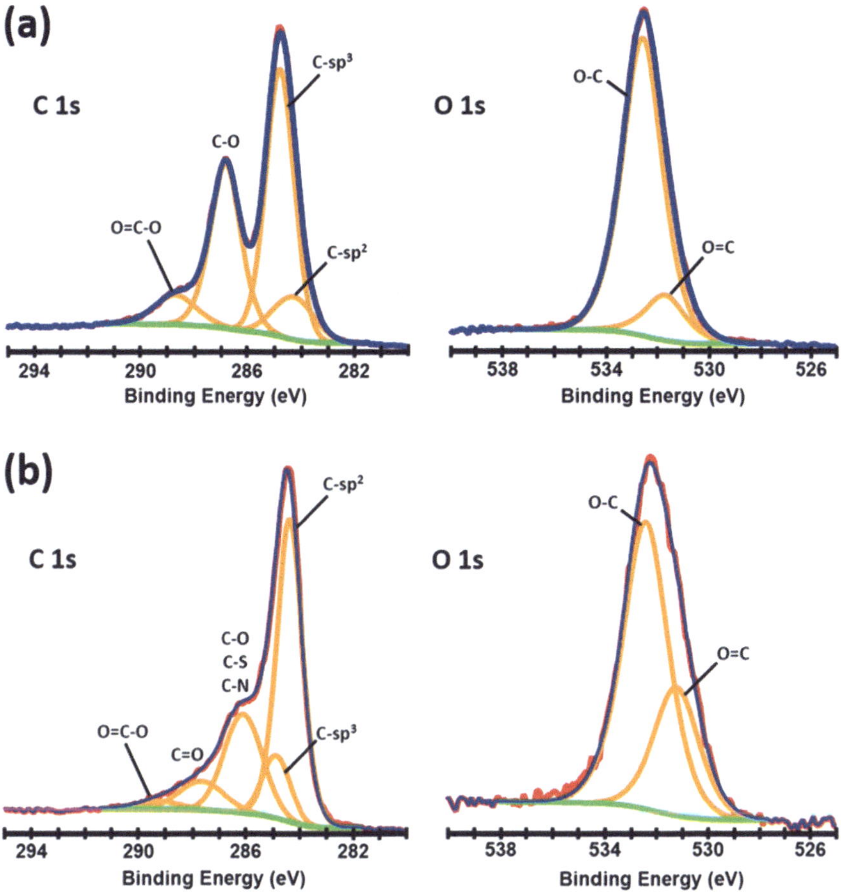

Fig. 4 High-resolution XPS spectra of the **a** GO and **b** rGO–SH. Figure reproduced from Ref. [17]. Copyright 2020 by the authors (CC attribution license)

The GO–DA fillers were dispersed in the SBR using solution mixing. The mixing was promoted by the hydrophobic functionality of the GO–DA fillers and the SBR/GO–DA composite showed more notable improvements. The storage modulus increased by 60 % in the glassy state and the rubbery plateau values at high temperatures (T > 20 °C) increased almost 200%. The tanδ values in Fig. 6b show a peak at −32 °C that is associated with the glass transition temperature of the elastomer matrix. The shoulder at higher temperatures is attributed to an interfacial relaxation mechanism with slower dynamics due to an interfacial SBR phase that is adsorbed on the surface of the fillers [22]. This interfacial mechanism is more pronounced in the SBR/GO–DA composite due to the better filler–elastomer interfacial adhesion. The peak intensity (and area) of the tanδ peak is also significantly lower for the SBR/GO–DA

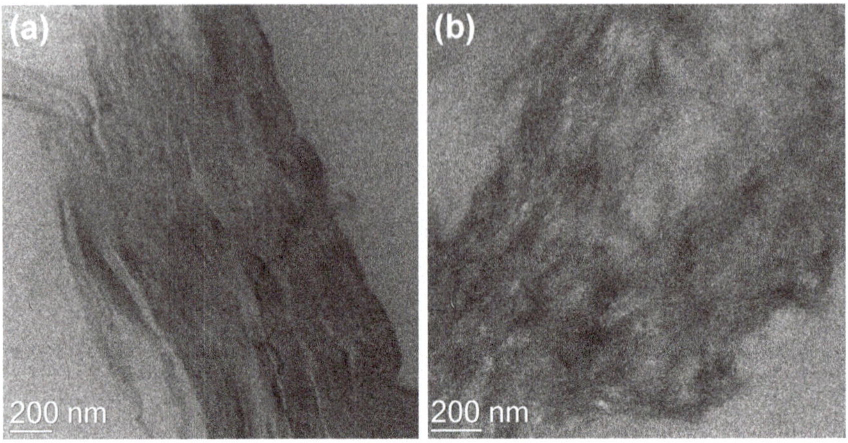

Fig. 5 TEM images of the **a** SBR/GO and **b** SBR/rGO–SH. Figure reproduced from Ref. [17]. Copyright 2020 by the authors (CC attribution license)

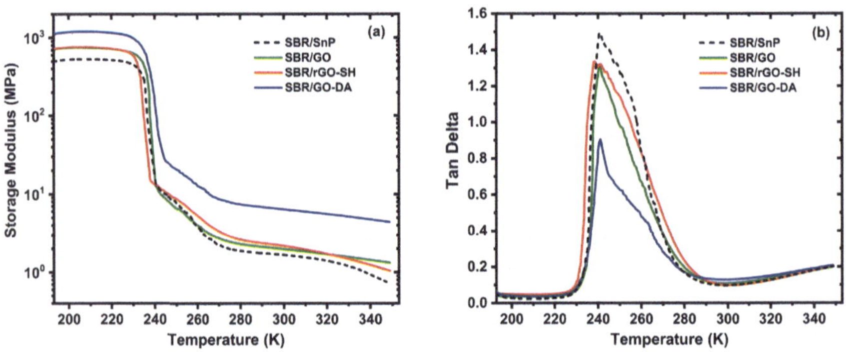

Fig. 6 Comparison plots of the mechanical properties of the SBR composites according to DMA measurements. **a** storage modulus and **b** tanδ. Figure reproduced from Ref. [17]. Copyright 2020 by the authors (CC attribution license)

composite due to less mobile SBR/GO–DA interface. Tailoring the interfacial chemistry of the GO fillers can make their mixing with hydrophobic elastomer matrices thermodynamically favorable and result in highly reinforced elastomer composites.

4 Elastomer Composites with Clustered Nanoparticulate Fillers

EPDM Elastomer Composites with Conductive Carbon Black

For many applications, the increased permittivity in high-k ceramic polymer composites cannot meet the desired requirements. Conductive fillers are therefore employed, as they can offer a dramatic increase in dielectric permittivity near percolation transition (ε' up to 10^3–10^4) [23], when appropriate measures are taken to reduce or prevent percolative electrical conduction. Such composites have attracted considerable attention for applications as dielectrics, including charge storage capacitors, high-k gate dielectrics, electromagnetic interference shielding, electroactive materials [24–28].

Typically, it is assumed that the substantial increase in dielectric permittivity is associated with the formation of microcapacitor networks, within which charge carriers can migrate too and accumulate at the filler–polymer interfaces (essentially an interfacial polarization, typically modeled as Maxwell–Wagner–Sillars relaxation [14, 23, 30, 31]). Despite the success of this approach in quantifying the behavior of multiple materials, there are many instances where there exit discrepancies [32]. Here, we review a recent study of a near-percolated polymer composite, based on an ethylene propylene diene polymer (EPDM, Vistalon 5420, ExxonMobil Chemical) with 32 wt% carbon black (CB, Spheron 6000, Cabot [33]). These nanocomposite materials deviate in dielectric response from the behavior expected under the usual assumptions of microcapacitor networks (*vide infra*).

Figure 7a shows representative TEM images of the composite structure. Individual CB particles with an average size of 100 nm are clustered into CB aggregates, which are well distributed across the entire composite sample. CB particles in the aggregates are strongly connected by nanometer-thin immobilized bound polymer layers [34], and the arrangement of CB can be described by a fractal dimension (d_f) [35, 36]:

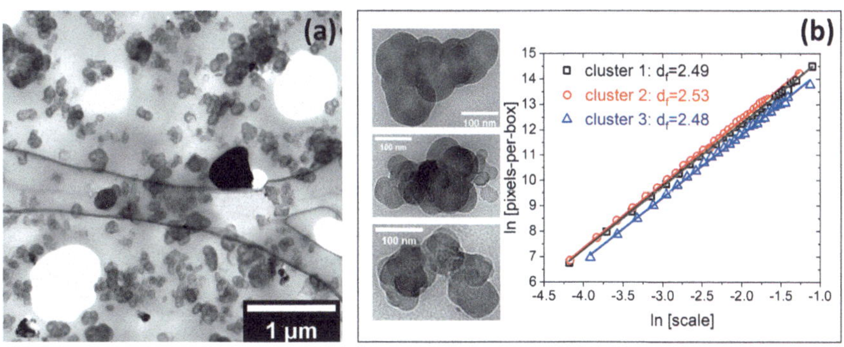

Fig. 7 **a** TEM image of CB polymer composite. The larger dark-colored particles are a second type of ceramic fillers. **b** Derivation of the fractal dimension for the clusters using the pixel-counting algorithm: TEM images of three CB clusters and the corresponding analyses. Figure and data adapted with permission from Ref. [29]. Copyright 2021 by the authors

$$N(r) \sim r^{d_f} \tag{2}$$

where r is the size of the cluster containing $N(r)$ number of CB particles. For the EPDM/CB composites in question, the fractal dimension of the CB clusters could not be directly measured by SAXS, due to the low angle limit of our instruments, and instead, we used the fractal box method (pixel-counting algorithm) to estimate its value [37, 38]. Specifically, grids with varying numbers of boxes are overlaid on a TEM image containing each CB cluster, and the average number of pixels in each box is counted as pixels-per-box (ppb). The fractal dimension of this CB cluster can be calculated from the pixel-counting plot (i.e., ln(ppb) vs. ln(box size)), as shown in Fig. 7b, and as expected different CB clusters in the same composite show a very similar fractal dimension, namely $d_f \cong 2.5$.

For these EPDM/CB composites, Fig. 8 provides the temperature- and frequency-dependent dielectric properties. At each temperature, a strong dielectric relaxation can be identified, manifested by a strong step change in dielectric permittivity and an associated peak in ε''. To better understand the polarization mechanism, the dielectric

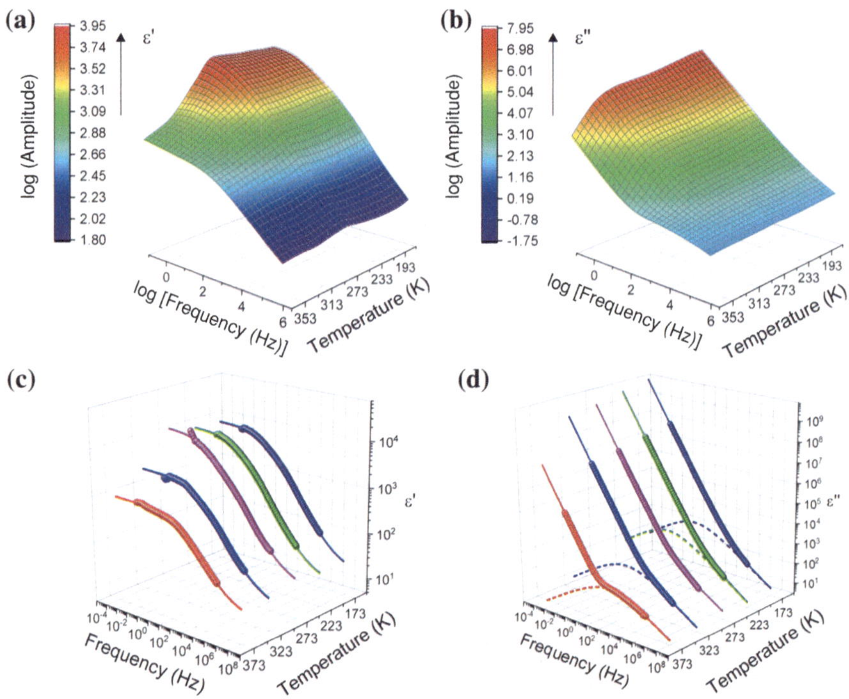

Fig. 8 **a, b** Dielectric spectra of a near-percolated EPDM/CB composite as a function of temperature and frequency, as well as **c, d** the best fitting Cole–Cole curves. Figure and data adapted with permission from Ref. [29]. Copyright 2021 by the authors

spectra of the polymer composite are analyzed within the framework of the Cole–Cole equation plus a conductivity contribution (Eq. 1 with β and s values set equal to 1.0, for a detailed discussion on the choice of fitting equation, we refer to [39]). The fitting curves are also shown in Fig. 7, indicating a good fitting of the experimental data to the Cole–Cole model.

The results from the best fit are plotted in Fig. 9a–d. Interestingly, the evolution of polarization does not follow the trends as generally expected from MWS polarization. Specifically, as shown in Fig. 9a, b, the relaxation strength and relaxation time increase with temperature in the lower-temperature region, whereas in the higher temperature region, both parameters decrease with temperature. This non-monotonic temperature evolution indicates a rather complex nature for the dielectric relaxation in the polymer composite, which cannot be described by a simple thermal activation model. The same is also observed in Fig. 9c, where the conductivity keeps decreasing with increasing temperature, again manifesting a departure from simple thermally activated mechanism. The shape parameter demonstrates a nearly constant value of 0.6 (Fig. 9d), and in fact, this value provides strong evidence for the relevant mechanism (*vide infra*, Eq. 7).

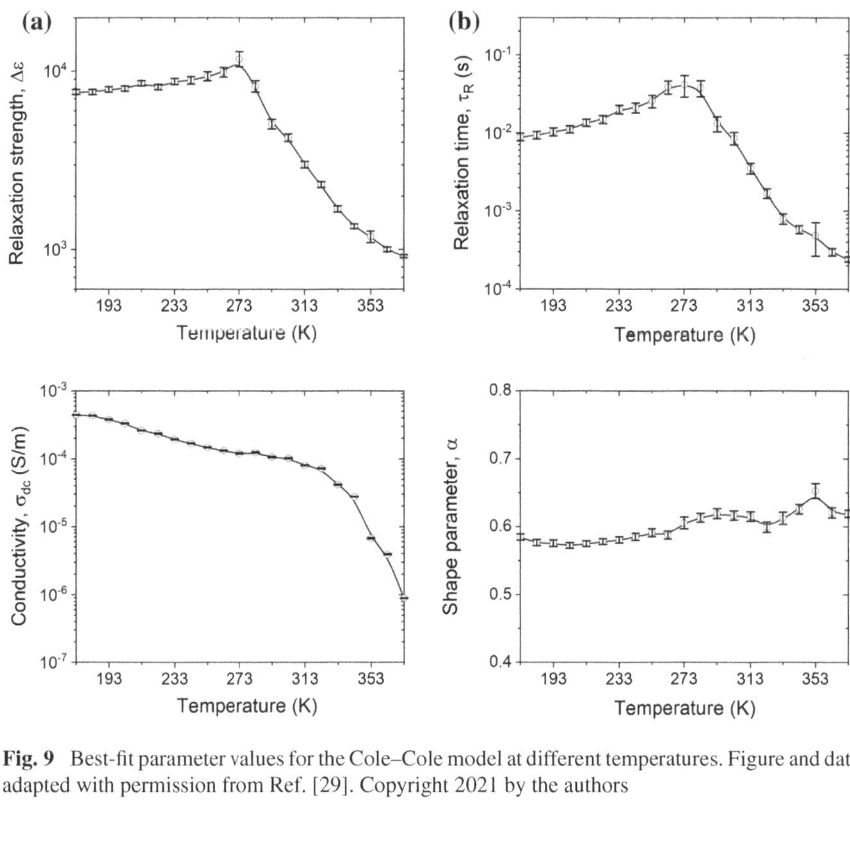

Fig. 9 Best-fit parameter values for the Cole–Cole model at different temperatures. Figure and data adapted with permission from Ref. [29]. Copyright 2021 by the authors

The evolution of the various dielectric variables indicates the significance of the morphology of the CB filler clusters, as defined through electrical connectivity, rather than simple topological/geometric definition. Reducing temperature evidently corresponds to better electrical connectivity between the CB particles within the cluster, which would be equivalent to shrinkage of the EPDM matrix and closer proximity of the CB particles, or also equivalent to increasing filler concentration, accounting for the increased conductivity with reducing temperature. Along the same lines, the T-dependence of relaxation strength and relaxation time is strongly reminiscent to the divergence of filler cluster size typically observed across the percolation threshold [36, 40], as long as both parameters are positively correlated with cluster size. From this perspective, the measured dielectric and electrical responses can be quantified by considering the electron transport along the electrically connected CB fillers (within clusters or across networks), and one can provide scaling relations between the various dielectric variables and the CB cluster morphology.

For these EPDM/CB polymer composites, since the CB network is self-similar over different length scales (fractal nature), one can define (Eq. 3) the fractal dimension (d_f) of the CB aggregates at different length scales, as

$$N(r) \sim r^{d_f} \quad \text{where} \quad \begin{cases} d_f = 2.5, & r < \xi \\ d_f = 3, & r > \xi \end{cases} \tag{3}$$

r is the size of a CB aggregate containing $N(r)$ number of CB particles (similar to Eq. 2). Within a CB cluster (of size ξ), the fractal dimension (d_f) of CB arrangement is 2.5, whereas macroscopically, beyond the single-cluster size ($r > \xi$), the CB fillers can be regarded as a homogenous three-dimensional dispersion with $d_f = 3$ [36, 41, 42]. Such a fractal CB particle arrangement, would result in electron transport that scales accordingly over the same length scales, i.e., with the mean square displacement of electrons $< r^2(t) >$ scaling with time (t) as [35, 36])

$$< r^2(t) > \sim t^{2/d_w} \quad \text{where} \quad \begin{cases} d_w = 3.8, & r < \xi \\ d_w = 2, & r > \xi \end{cases} \tag{4}$$

That is, electron transport within one CB cluster follows an "anomalous diffusion" with a diffusion exponent $d_w = 3.8$, whereas electrons travel between clusters is governed by normal (charges in external electric field, or even Fickian) diffusion$< r^2(t) > \sim t$.

5 Frequency Dependence of Dielectric Properties

For the frequency dependence of dielectric properties, one can first consider the frequency-dependent AC conductivity: At each temperature, the conductivity ($\sigma(w)$) is proportional to the number density of electrons participating in the conduction process (n) and the diffusion coefficient (D) via $\sigma(w) \sim nD$. The number density of conduction electrons (n) is proportional to the solid concentration of CB in a blob (as defined by a collection of CB particles in electrical contact to each other, i.e., a collection of CB particles that the electrons can explore during one period of measurement, t). This blob size (r) can be estimated from Eq. (4) as $r \sim t^{1/d_w}$ and using Eq. (3), one can get

$$n \sim \frac{N(r)}{r^3} \sim r^{d_f - 3} \sim t^{(d_f - 3)/d_w} \tag{5}$$

The diffusion coefficient (D) will simply be the mean square displacement of the electrons over the same period of measurement (t), and thus Eq. (4), yields

$$D \sim \frac{\langle r^2(t) \rangle}{t} \sim t^{(2-d_w)/d_w} \tag{6}$$

and combing Eqs. (5) and (6), would result in the frequency dependence of conductivity:

$$\sigma(w) \sim nD \sim t^{(d_f - d_w - 1)/d_w} \sim w^{(d_w - d_f + 1)/d_w} \sim \begin{cases} w^{0.605}, r < \xi \\ w^0, r > \xi \end{cases} \tag{7}$$

The conductivity of the polymer composite will exhibit different power law dependences on frequency, when electrons explore a single cluster versus when they explore larger length scales. The crossover frequency corresponds to the critical time scale when electrons can explore one entire cluster, and identifies the transition from anomalous diffusion to normal diffusion for electron transport.

It is interesting to note that the scaling exponent of the high-frequency conductivity (0.605), is very close to the shape parameter (α) measured from the experimental data (Fig. 9d). As defined in Eq. (1), α characterizes the slope of the loss (ε'') curve of dielectric relaxation and should have the same value as the slope of conductivity (σ_w) at the high-frequency end, as $\sigma_w = 2\pi f \varepsilon_0 \varepsilon''$. A consistent α value of ~0.6 measured across the whole temperature range confirms our assumption of electron displacement as the dominant relaxation mechanism in the CB polymer composite.

6 Cluster Size Dependence of Dielectric Properties

In order to confirm the validity of the above scaling relations, one can use these scaling relations to determine the cluster size dependence of the various dielectric variables measured by dielectric spectroscopy. Thus, by using the TEM-determined cluster size and fractal dimension and the DRS-determined dielectric quantities, one can directly compare the scaling laws versus the experimental trends. The dielectric variables that can be investigated are the DC conductivity (σ_{dc}), the relaxation strength ($\Delta\varepsilon$), and the relaxation time (τ_R), as quantified by the fitted parameter values of the Cole–Cole function.

7 Scaling of Conductivity (σ_{dc}) with Custer Size

At low-frequency limit, the DC conduction (σ_{dc}) can be calculated by

$$\sigma_{dc} = \frac{e^2 n}{k_B T} D \tag{8}$$

where e denotes the electron charge; n and D are the number density of conduction electrons and the diffusion coefficient, respectively, as defined before; k_B and T are the Boltzmann's constant and temperature, respectively [35, 41, 43]. For long-range electron transport, n and D are independent of time, but they are strong functions of cluster size (ξ), since the number density of conduction electrons is proportional to the solid concentration of CB in the cluster, that is

$$n \sim \frac{N(\xi)}{\xi^3} \sim \xi^{d_f - 3} \tag{9}$$

and from the definition of $N(\xi)$ and ξ, for a CB cluster was experimentally determined to be $d_f \cong 2.5$.

The diffusion coefficient can be calculated by simply considering that the electrons explore the cluster size ξ, i.e., a mean square displacement of ξ^2, over a time interval t_ξ (as described in Eq. 4, with $t_\xi \sim \xi^{d_w}$ and $d_w = 3.8$, as it applies to electron transport within a cluster):

$$D \sim \frac{\xi^2}{t_\xi} \sim \xi^{2-d_w} \tag{10}$$

combining Eqs. (9) and (10) with Eq. (8), the DC conductivity and cluster size ξ, scale as

$$\sigma_{dc} = \frac{e^2 n}{k_B T} D \sim \frac{nD}{T} \sim \frac{\xi^{d_f - d_w - 1}}{T} = \frac{\xi^{-2.3}}{T} \tag{11}$$

8 Scaling of Relaxation Time (τ_R) and Relaxation Strength ($\Delta\varepsilon$) with Cluster Size

The polarization of the conductive CB filler clusters originates from the electron transport within a cluster (over longer time scales, electrons can travel beyond the cluster, follow normal diffusion, result in a DC conductivity, and do not contribute to the polarization). Consequently, the relaxation time measured from the dielectric spectrum (τ_R) should scale with the time needed for electrons to explore a cluster (t_ξ, as defined in Eq. 4).

$$\tau_R \sim t_\xi \sim \xi^{d_w} = \xi^{3.8} \tag{12}$$

The dielectric relaxation strength that corresponds to the conductive filler cluster polarization, can be considered as the relaxation of a number of individual dipoles with a total dipole moment (P), for which [44]:

$$\Delta\varepsilon = \frac{NP^2}{3\varepsilon_0 k_B T} \tag{13}$$

where N is the number density of dipoles; ε_0 is the vacuum permittivity; k_B and T are the Boltzmann's constant and temperature, respectively. In analogy, for the EPDM/CB polymer composites, we can have

$$\Delta\varepsilon = \frac{NP^2}{3\varepsilon_0 k_B T} \sim \frac{Nq^2\xi^2}{T} \tag{14}$$

with N now being the number density of clusters; P being the equivalent dipole moment of polarized clusters, which is proportional to the number density of electrons in the cluster (q) and the cluster size (ξ). Since the electron density in a cluster (q) also scales with the concentration of CB in the cluster, i.e., $N(\xi)/\xi^3 \sim \xi^{d_f-3}$ (Eq. 9): Nq should be proportional to the CB concentration in the composite and, therefore, is a constant in this case. Thus, the relaxation strength can be rewritten as

$$\Delta\varepsilon \sim \frac{q\xi^2}{T} \sim \frac{\xi^{d_f-1}}{T} = \frac{\xi^{1.5}}{T} \tag{15}$$

Equations (11), (12), and (15) provide the dependences of measured/fitted dielectric parameters on the cluster size (ξ), which can be replaced (through Eq. 12) by its scaling with the relaxation time (τ_R). In this way, by using the scaling between ξ and τ_R, we can rewrite the scaling relations Eqs. (11), (12), and (15), as dependencies on the relaxation time, rather than on the cluster size:

$$\Delta\varepsilon \cdot T \sim \tau_R^{0.40};\ \sigma_{dc} \cdot T \sim \tau_R^{-0.61};\ \sigma_{dc} \cdot T \sim (\Delta\varepsilon \cdot T)^{-1.53} \tag{16}$$

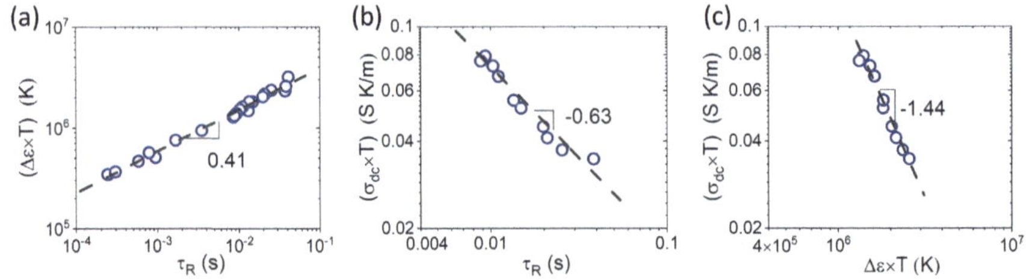

Fig. 10 Scaling relations between various dielectric parameters. The number on the graphs represents the slope of the linear regression on experimental data, which is in good agreement with the scaling prediction. Figure and data adapted with permission from Ref. [29]. Copyright 2021 by the authors

In this way, the derived scaling relations are expressed through only measured/fitted values, and one can use experimental data only, as plotted in Fig. 10a–c, to test the validity of the scaling relations. A good agreement is achieved—maybe surprisingly given some of the simplifications outlined above—but the scaling of the experimental data clearly confirm the scaling relations of the model, making a strong case that the electron transport within the CB clusters is in fact the dominant polarization mechanism.

Finally, since $\Delta\varepsilon$, τ_R, σ_{dc} and α are the fitted variables in HN or in Cole–Cole, once these parameters are experimentally determined, it is possible to quantify the details of the composite/filler morphology and of the filler cluster size based on dielectric data (as illustrated in Scheme 1).

9 Dielectric Response of Multi-filler Elastomer Composites

EPDM Elastomer Composites with Carbon Black and Ceramic Nanoparticles
In order to identify and highlight the interfacial contributions to the composites' dielectric properties, we present some results from elastomer/carbon black/ceramic composites[2]: In these multi-filler composite materials, we added a series of micron-sized ceramic fillers to the EPDM/CB composites discussed in the previous section.

In order to distinguish the relative contributions of the fillers and of the filler interfaces to the macroscopic composite permittivity, a series of ceramic particulates—that span orders of magnitude in permittivity values and have varied surface chemistries—were compared [32]. Specifically, a series of low-loss ceramic fillers were used, with 0.4–2 μm in size, differing in nature (i.e., dielectric/paraelectric SiO_2 and TiO_2 versus ferroelectric $BaTiO_3$), and spanning a wide range of dielectric permittivities (particulate or bulk ε', ranging from 4 to 60,000). For each filler, a series of EPDM composites was prepared, varying the ceramic filler loading from

[2] The text and figures of this section were adapted with permission from a conference proceedings paper published in Ref. [32], Copyright 2021 by the authors (B. Li, E. Manias, et al., MRS Advances, Springer).

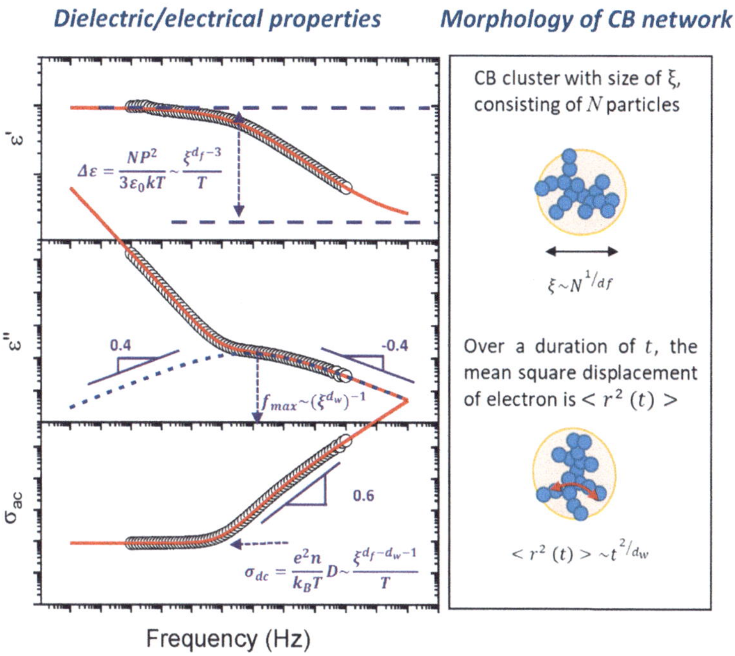

Scheme 1 Schematic illustrating the correlation between dielectric/electrical properties and the carbon black filler-network morphology. Figure and data adapted with permission from Ref. [29]. Copyright 2021 by the authors

5 to 25 wt%, while keeping the carbon black (CB) content constant at 15 wt% and the organic/elastomer matrix the same, i.e., all elastomers had constant monomer: plasticizer and monomer:curing agent ratios (30 phr and 5 phr, respectively). The dielectric properties of the composites were measured as a function of frequency (20–10 kHz), at 1 V, using a charge–voltage converter (GADD, integrated with a Stanford Research SR830 lock-in amplifier [32]) on rectangular specimens (4 × 1 cm wide, by 3 mm thickness) coated by colloidal silver electrodes, Results were collected by measuring the real part of permittivity, $\varepsilon'(\omega)$, Fig. 11a, and the "dissipation factor", $\tan\delta(\omega)$, Fig. 11b, as per [45].

In a first approach, keeping the carbon black (CB) content constant at 15 wt% across all systems, allows for simplifying the comparative discussion across the various ceramic fillers, by considering the matrix to be the combination of all organics (EPDM, plasticizer, and crosslinker) *plus* the carbon black [1]. In this case, the composite dielectric permittivity, ε, can be quantified by Lichtenecker's logarithmic mixing rule applied to the permittivity of the "matrix" and the ceramic fillers:

$$log\varepsilon = v_f log\varepsilon_r + (1 - v_f) log\varepsilon_m \tag{17}$$

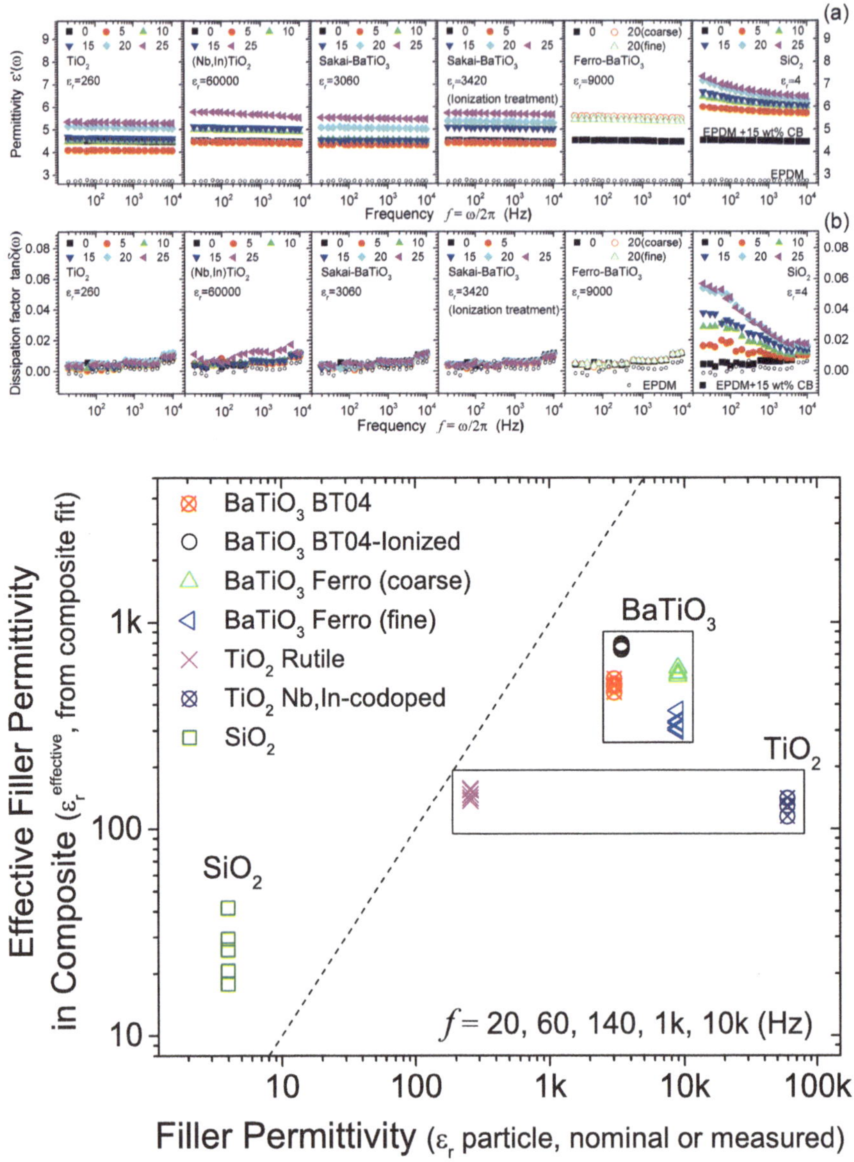

Fig. 11 **a** Dielectric permittivity, $\varepsilon'(\omega)$ and **b** dielectric dissipation factor, tan $delta(\omega)$, as a function of frequency (ω) at room temperature, for EPDM composites (with 0–25 wt% ceramic fillers). **c** The filler's effective dielectric permittivity in the composites (including interfacial contributions, as determined by ε in Lichtenecker's mixing rule) is compared against the particle/filler permittivity ε_r Fittings at five frequencies provide five values for each filler's effective ε (see Table 2). The dashed-line corresponds to a 1:1 relation, i.e., no interfacial contributions. Figure and data adapted with permission from Ref. [32]. Copyright 2021 by the authors (Springer)

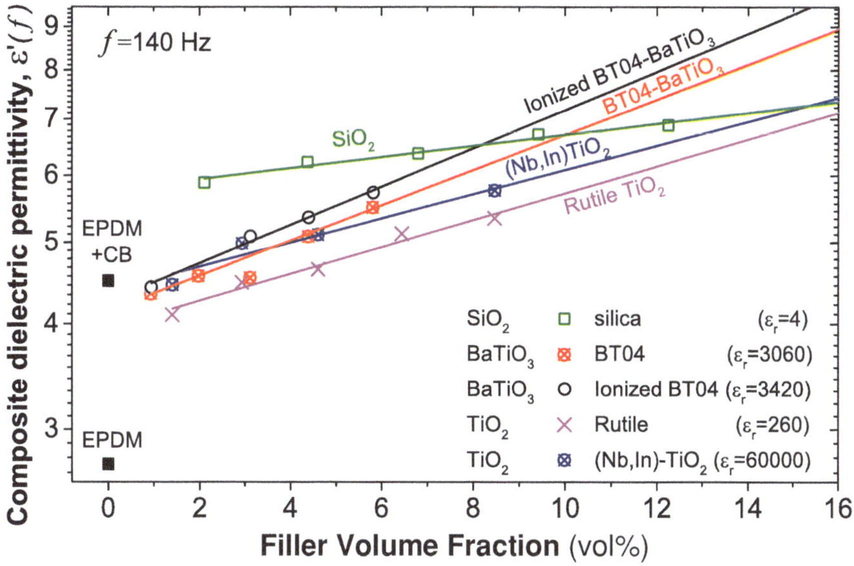

Fig. 12 **a** Measured composite permittivity at 140 Hz and fittings of the permittivities versus filler loading by Lichtenecker's logarithmic mixing rule. Figure and data adapted with permission from Ref. [32]. Copyright 2021 by the authors (Springer)

with ε_r and ε_m being the permittivities of the ceramic filler and the matrix, respectively, and v_f the volume fraction of the ceramic filler (Fig. 12). Subsequent measurement of the macroscopic composite permittivity (ε) as a function of ceramic filler loading (v_f), can thus yield an "effective dielectric permittivity" (ε_r) for each ceramic filler in the composite; i.e., a permittivity value that quantifies the *combined ceramic filler particle plus their interfacial contributions* to the measured composite permittivity (Table 2, Fig. 11c).

10 Effect of the Interphases on the Dielectric Properties of the Composites

The values of $\varepsilon_r{}^{effective}$ provide a measure of each filler's realistic contribution to the measured composite permittivity, and are found to be markedly different from the values expected (e.g., markedly different from the usual filler "dielectric constants" ε_r used in design considerations, or those used for the mixing rule above: which typically are bulk permittivity values, or particle permittivity values, often averaged over the application relevant frequency and temperature ranges, or measured particle-ε' values). This is a clear indication that the dielectric contributions from the interphases and the interfaces dominate the overall composite permittivity, rather than the ceramic fillers' ε_r dielectric values. Although addition of ceramic fillers does increase the permittivity of the composites (compared to the unfilled EPDM elastomer, Fig. 12), this increase is qualitatively and quantitatively different from what is expected.

Intuition and typical design models (e.g., mixing rules) predict a larger dielectric permittivity composite, when fillers with higher ε_r are incorporated. However, in Fig. 12, the silica composites exhibit the greatest permittivity (ε-value) among all the composites, despite SiO_2 being the lowest ε_r filler used, when comparing the various ceramic particulates. Also, the composites with the Sakai-$BaTiO_3$ ($\varepsilon_r = 3060$) fillers show higher permittivity values than the composites with Nb + In co-doped TiO_2 ($\varepsilon_r = 60{,}000$) fillers. These marked deviations of the measured composite permittivities from the natural expectation that "higher-*k* fillers should yield higher permittivity composites" [1], strongly indicate that the composite permittivities are dominated by interfacial contributions, which are not considered by most mixing rules [1, 2, 46]. Silica displays, maybe, the most straightforward interfacial behavior: A low-frequency relaxation process can be clearly identified in the EPDM/CB/SiO_2 composites (manifested both as a step in the permittivity and an associated peak in the loss, Fig. 11). This is a typical behavior of macroscopic-scale/domain polarization—i.e., Maxwell–Wagner-Sillars (MWS) polarization—arising from the low-frequency subdiffusive transport of the charge carriers (e.g., ions, surfactants, impurities) within interphases or other domain boundaries, which manifests as an additional increase in permittivity.

For the other fillers, interfacial effects also exist but, rather than a MWS polarization, they seem to influence the composite dielectric performance in a qualitatively different manner: As seen in Fig. 11, there exist no pronounced peaks in tan δ as would be expected from a high-strength MWS mechanism; instead the composite permittivity is almost frequency-independent. Also, in contrast to any common expectation, at low contents (at low ceramic volume fractions 1–2 vol%) the composites exhibit permittivity values lower than the EPDM + CB elastomer: Such composite permittivities, with values below the "matrix" (EPDM + CB) permittivity, cannot be explained by any weighted mixing of matrix and filler permittivities (as it would necessitate a negative filler ε_r value for the ceramics); rather, this behavior reflects contributions from the ceramic particle interphases that are antagonistic to the EPDM/CB dielectric response (e.g., local field effects, restricted mobility for filler-sorbed polymer chains, etc., yielding a reduced permittivity in the interphases compared to the matrix [32, 46]). One way to quantify these interfacial effects, within the Lichtenecker's approximation, would be to use an "effective permittivity" value for each filler (Fig. 12), which combines both filler and interfacial contributions, toward a value that can be substantially different than the filler's permittivity (Table 2). In Fig. 11c, when these "effective permittivities" [of the fillers + interphases in the composites] are compared against the expected values from the filler "dielectric constants", an interesting trend is observed: Fillers are clustered in groups of same filler type (enclosed in boxes in Fig. 11c), where the fillers' dielectric contribution to the composites' permittivity is measured to relate more to the filler type (e.g., same filler chemistry and, consequently, similar interphase characteristics), rather than to each filler's permittivity value. This is completely different to the expected behavior, e.g., mixing rules, where the fillers' dielectric contribution to the composites ε is expected to relate directly to the permittivity ε_r of the filler particles. For instance, the $\varepsilon_r{}^{effective}$ of the rutile TiO_2 is nearly identical to that of the Nb, In-doped TiO_2, even though the two fillers differ in permittivity by two orders of magnitude (230 × difference in ε_r); also, all four different $BaTiO_3$ fillers show similar contributions in the composites (similar values of $\varepsilon_r{}^{effective}$), despite large variations in their filler/particle dielectric permittivities ε_r (Table 2).

The above comparisons clearly demonstrate that the fillers' effective permittivity and, therefore, how much each filler realistically contributes to the macroscopically observed composite's permittivity, are largely dominated by the interphases and the interfaces, i.e., the nature of the filler (e.g., chemistry, surface characteristics and modifications, etc.), rather than by the filler's permittivity. This conclusion is further manifested in: **(i)** the measurable differences recorded in the fillers' $\varepsilon_r{}^{effective}$ when considering fillers that only differ in size (e.g., for the Ferro $BaTiO_3$ fillers, the coarse particles have a higher $\varepsilon_r{}^{effective}$ than the fine identical-type particles, due to lower volume-fractions of the unfavorable interfaces). **(ii)** when comparing fillers that vary only in surface treatment (e.g., the ionized $BaTiO_3$ demonstrates the highest effective permittivity among all other $BaTiO_3$ fillers, because of the extra mobile ions introduced on the filler surfaces, which provide higher space-charge polarization within the interface); and **(iii)** the counter-intuitive antagonistic interphasial response of the ceramics to the carbon black dielectric response (as evidenced by the measured reduction in permittivity upon addition of low concentrations of ceramics to EPDM + CB, Fig. 12), where the ceramic fillers and their interphases limit the carbon black cluster polarization, e.g., by interrupting the CB connectivity (electron tunneling) when ceramic particles are located between adjacent carbon black particles.

Acknowledgements Financial support was provided by the National Science Foundation, through the "*Center for Dielectrics and Piezoelectrics*" under NSF Grants IIP-1361571 and IIP-1361503. Additional support by Eaton-Cooper is also acknowledged. GP acknowledges financial support by the U.S. Department of Energy, Office of Electricity Delivery and Energy Reliability—Power Electronics Program. Oak Ridge National Laboratory is operated for the U.S. Department of Energy by U.T.—Battelle under Contract No. DE-AC05-00OR22725.

Conflict of Interest Declaration The authors declare no conflicts of interest.

References

1. Sihvola AH (1999) Electromagnetic mixing formulas, applications. I.E.T. Books, London, UK

2. Dang ZM, Yuan JK, Zha JW, Zhou T, Li ST, Hu GH (2012) Fundamentals, processes and applications of high-permittivity polymer-matrix composites. Prog Mater Sci 57:660–723
3. Barber P, Balasubramanian S, Anguchamy Y, Gong S, Wibowo A, Gao H, Ploehn HJ, Zur Loye HC (2009) Polymer composite and nanocomposite dielectric materials for pulse power energy storage. Materials 2:1697–1733
4. Smith RC, Liang C, Landry M, Nelson JK, Schadler LS (2008) The mechanisms leading to the useful electrical properties of polymer nanodielectrics. IEEE Trans Dielectrics Electr Insulation 15:187–196
5. Lewis TJ (2005) Interfaces nanometric dielectrics. J Phys D: Appl Phys 38:202–212
6. Li B, Xidas PI, Manias E (2018) High breakdown strength polymer nanocomposites based on the synergy of nanofiller orientation and crystal orientation for insulation and dielectric applications. ACS Appl Nano Mater 1:3520–3530
7. Li B, Xidas PI, Triantafyllidis KS, Manias E (2017) Effect of crystal orientation and nanofiller alignment on dielectric breakdown of polyethylene/montmorillonite nanocomposites. Appl Phys Lett 111:082906
8. Tomer V, Polizos G, Randall CA, Manias E (2011) Polyethylene nanocomposite dielectrics: implications of nanofiller orientation on high field properties and energy storage. J Appl Phys 109:074113
9. Tomer V, Randall CA, Polizos G, Kostelnick J, Manias E (2008) High- and low-field dielectric characteristics of dielectrophoretically aligned ceramic/polymer nanocomposites. J Appl Phys 103:034115
10. Wan J, Xie J, Kong X, Liu Z, Liu K, Shi F, Pei A, Chen H, Chen W, Chen J, Zhang X, Zong L, Wang J, Chen LQ, Qin J, Cui Y (2019) Ultrathin, flexible, solid polymer composite electrolyte enabled with aligned nanoporous host for lithium batteries. Nat Nanotechnol 14:705−711
11. Wang X, Zhai H, Qie B, Cheng Q, Li A, Borovilas J, Xu B, Shi C, Jin T, Liao X, Li Y, He X, Du S, Fu Y, Dontigny M, Zaghib K, Yang Y (2019) Rechargeable solid-state lithium metal batteries with vertically aligned ceramic nanoparticle/polymer composite electrolyte. Nano Energy 60:205−212
12. Zhai H, Xu P, Ning M, Cheng Q, Mandal J, Yang Y (2017) A flexible solid composite electrolyte with vertically aligned and connected ion-conducting nanoparticles for lithium batteries. Nano Lett 17:3182−3187
13. Gao Y, Picot OT, Tu W, Bilotti E, Peijs T (2018) Multilayer coextrusion of graphene polymer nanocomposites with enhanced structural organization and properties. J Appl Polym Sci 135:46041
14. Kremer F, Schönhals A (2003) Broadband dielectric spectroscopy. Springer, Berlin, Heidelberg
15. Havriliak S, Negami S (1966) A complex plane analysis of α-dispersions in some polymer systems. J Polym Sci Polym Symp 14:99−117
16. Fragiadakis D, Pissis P (2007) Glass transition and segmental dynamics in poly(dimethylsiloxane)/silica nanocomposites studied by various techniques. J Non-Cryst Solids 353:4344−4352
17. Park J, Sharma J, Monaghan KW, Meyer III HM, Cullen DA, Rossy AM, Keum JK, Wood III DL, Polizos G (2020) Styrene-based elastomer composites with functionalized graphene oxide and silica nanofiber fillers: mechanical and thermal conductivity properties. Nanomaterials 10:1682
18. Kang SM, Park S, Kim D, Park SY, Ruoff RS, Lee H (2011) Simultaneous reduction and surface functionalization of graphene oxide by mussel-inspired chemistry. Adv Funct Mater 21:108−112
19. Wang J, Fei G, Pan Y, Zhang K, Hao S, Zheng Z, Xia H (2019) Simultaneous reduction and surface functionalization of graphene oxide by cystamine dihydrochloride for rubber composites.Compos Part A: Appl Sci 122:18−26
20. Zou H, Wu S, Shen J (2008) Polymer/silica nanocomposites: preparation, characterization, properties, and applications. Chem Rev 108:3893–3957
21. Lee DW, Yoo BR (2016) Advanced silica/polymer composites: materials and applications. J Ind Eng Chem 38:1–12

22. Vo LT, Anastasiadis SH, Giannelis EP (2011) Dielectric study of poly(styrene-co-butadiene) composites with carbon black, silica, and nanoclay. Macromolecules 44:6162–6171
23. Nan C-W, Shen Y, Ma J (2010) Physical properties of composites near percolation. Ann Rev Mater Res 40:131–151
24. Dang ZM, Yuan JK, Yao SH, Liao RJ (2013) Flexible nanodielectric materials with high permittivity for power energy storage. Adv Mater 25:6334–6365
25. Dang ZM, Yuan JK, Zha J-W, Zhou T, Li S-T, Hu G-H (2012) Fundamentals, processes and applications of high-permittivity polymer–matrix composites. Prog Mater Sci 57:660–723
26. Wang B, Huang W, Chi L, Al-Hashimi M, Marks TJ, Facchetti A (2018) High-k gate dielectrics for emerging flexible and stretchable electronics. Chem Rev 118:5690–5754
27. Zhang Y, Qiu M, Yu Y, Wen B, Cheng L (2017) A novel polyaniline-coated bagasse fiber composite with core–shell heterostructure provides effective electromagnetic shielding performance. ACS Appl Mater Interfaces 9:809–818
28. Zhang QM, Li H, Poh M, Xia F, Cheng ZY, Xu H, Huang C (2020) An all-organic composite actuator material with a high dielectric constant. Nat 419:284–287
29. Li B, Randall CA, Manias E (2021) Submitted Manuscript
30. He F, Lau S, Chan HL, Fan J (2009) High dielectric permittivity and low percolation threshold in nanocomposites based on poly(vinylidene fluoride) and exfoliated graphite nanoplates. Adv Mater 21:710–715
31. Shen Y, Lin Y, Li M, Nan C-W (2007) High dielectric performance of polymer composite films induced by a percolating interparticle barrier layer. Adv Mater 19:1418–1422
32. Li B, Sarkarat M, Baker A, Randall CA, Manias E (2021) Interfacial effects on the dielectric properties of elastomer/carbon-black/ceramic composites. MRS Adv 6:247–251
33. Ryan DE, Manias E, Li B (2019) Elastomer composites with high dielectric constant. U.S. Patent 10,438,717, 8 Oct 2019
34. Litvinov VM, Orza RA, Klüppel M, van Duin M, Magusin PCMM (2011) Rubber–filler interactions and network structure in relation to stress–strain behavior of vulcanized, carbon black filled EPDM. Macromolecules 44:4887–4900
35. Stauffer D, Aharony A (1991) Introduction to percolation theory. Taylor & Francis, London
36. Bunde A, Havlin S (2012) Fractals and disordered systems. Springer, New York
37. Vega GFH (2019) Quantifying matrix complexity with the calculation of fractal dimensions from GC×GC chromatograms—a potential oil fingerprinting technique. Master Dissertation, Saint Mary's University
38. Captur G, Muthurangu V, Cook C, Flett AS, Wilson R, Barison A, Sado DM, Anderson S, McKenna WJ, Mohun TJ, Elliott PM (2013) Quantification of left ventricular trabeculae using fractal analysis. J Cardiovasc Magn Reson 15:36
39. Kalmykov YP, Coffey WT, Crothers DSF, Titov SV (2004) Microscopic models for dielectric relaxation in disordered systems. Phys Rev E 70:041103
40. Pötschke P, Dudkin SM, Alig I (2003) Dielectric spectroscopy on melt processed polycarbonate-multiwalled carbon nanotube composites. Polymer 44:5023–5030
41. Meier JG, Klüppel M (2008) Carbon black networking in elastomers monitored by dynamic mechanical and dielectric spectroscopy. Macromol Mater Eng 293:12–38
42. Meier JG, Mani JW, Klüppel M (2007) Analysis of carbon black networking in elastomers by dielectric spectroscopy. Phys Rev B 75:054202
43. Gefen Y, Aharony A, Alexander S, Anomalous S (1983) Diffusion on percolating clusters. Phys Rev Lett 50:77–80
44. Sidebottom DL (2009) Colloquium: understanding ion motion in disordered solids from impedance spectroscopy scaling. Rev Mod Phys 81:999–1014
45. Okamoto T, Long J, Wilkie R, Stitt J, Maier R, Randall CA (2016) A charge-based deep level transient spectroscopy measurement system and characterization of a ZnO-based varistor and a Fe-doped $SrTiO_3$ dielectric. Jap J Appl Phys 55:026601
46. Jones TB (1995) Electromechanics of particles. Cambridge University Press, New York

Microstructure and Segmental Dynamics of Industrially Relevant Polymer Nanocomposites

Julian Oberdisse and Anne-Caroline Genix

Abstract Recent progress in the filler microstructure and the dynamical properties of polymer nanocomposites of industrial relevance is reviewed. We focus mainly on systems used in car tire treads made of styrene-butadiene rubber (SBR) matrices in which precipitated amorphous silica like Zeosil® with a complex multi-scale structure is dispersed—while occasionally comparing to other experimental systems, including model studies with well-defined colloidal silica nanoparticles. Electron microscopy and small-angle scattering—namely SAXS and SANS—are powerful methods of structural analysis, and some recent developments including the correlation hole analysis giving access to local aggregate properties or reverse Monte Carlo approaches providing aggregate distribution functions are discussed. The dynamical response of such complex systems is then studied by broadband dielectric spectroscopy (BDS), starting from the individual components. Maxwell–Wagner–Sillars interfacial polarization processes including charge and water migration are shown to be able to probe large-scale microstructure, evidencing filler percolation effects as observed in rheology. Then, BDS is shown to be highly suitable for studies of the evolution of the segmental dynamics of rubber, giving insight into vulcanization mechanisms under the effect of industrial additives, as well as in transitions from heterogeneous to homogeneous dynamics of polymer blends. Finally, the difficulties of investigating segmental dynamics in industrially relevant filled systems by BDS are critically discussed, and some recent neutron spin echo results are reported evidencing only a small impact of filler surfaces on segmental dynamics in weakly interacting SBR-silica systems.

Keywords Small-angle scattering · Reverse Monte Carlo · Filler structure · Percolation · Segmental dynamics · Interfacial polarization · Rubber additives

---

J. Oberdisse · A.-C. Genix (✉)
Laboratoire Charles Coulomb (L2C), Université de Montpellier, CNRS, 34095 Montpellier, France
e-mail: anne-caroline.genix@umontpellier.fr

A. Schönhals and P. Szymoniak (eds.), *Dynamics of Composite Materials*,
Advances in Dielectrics, https://doi.org/10.1007/978-3-030-89723-9_9

Abbreviations

BDS	Broadband dielectric spectroscopy
ILM	Interfacial layer model
MWS process	Maxwell–Wagner–Sillars process
NP	Nanoparticle
NSE	Neutron spin echo
PNC	Polymer nanocomposite
QENS	Quasi-elastic neutron scattering
RMC	Reverse Monte Carlo
SAXS	Small-angle X-ray scattering
SANS	Small-angle neutron scattering

1 Introduction

Polymer nanocomposites (PNCs) are fascinating materials that display unique macroscopic and microscopic dynamical properties, the secret of which lies in the details of their organization on the nanoscale. They are usually made of nanoparticles (NPs) dispersed in a polymer matrix. The most common filler particles are carbon black and silica, while industrially relevant polymers are usually rubbery, i.e., with a glass transition temperature T_g below room temperature. In general, the matrix is crosslinked in order to limit the flow of the material, but in some fundamental studies including ours, crosslinking is avoided. In industrial applications like tire treads, the original material was natural rubber (NR) made of Hevea Brasiliensis latex, but geopolitical evolutions during the twentieth century made industry turn to synthetic polymers, in particular petrol-based polyisoprene and styrene-butadiene rubber (SBR), before becoming now again interested in NR under the pressure of environmental issues.

The dynamical properties of PNCs extend over many orders of magnitude in the frequency domain, covering the entire field from rheology related to chain reptation and flow, to mechanical behavior controlled by rubber elasticity and segmental dynamics. Obviously, the relevant frequencies for industrial applications like car tires are related to the rotational speed of the tire used in the rubbery regime. Unwanted viscous losses on this scale transform the kinetic energy of the car into heat and, thus, cause the rolling resistance, which ultimately increases fuel consumption and CO_2 production. At higher frequency related to fast material deformations on the rugosity of the road, the tire needs to be dissipative, as only this property makes it stick to the road. Such wet-skid properties enable the driver to break and steer under unfavorable meteorological conditions, which are after all desired capabilities that we are willing to pay for in terms of higher fuel consumption. To see this, just imagine a tire entirely made of steel—its dissipation would be zero, and the driver would be happy as long as he does not have to turn… As a result of such considerations, the fundamental dynamical properties of rubber materials on the nano- and microscale need to be understood and tuned in order to optimize macroscopic rheological properties. As the pure polymer matrix would not be stiff enough for applications, it is blended with hard nanoparticle fillers—taking advantage of the so-called reinforcement effect [1, 2]. The magnitude of the latter depends on many parameters, the most important one being the amount of hard material, followed by its spatial organization, and connection to the matrix polymer. The choice of the filler material has been restricted for a century to different grades of the same substance, carbon black. The latter is hydrophobic and it has strong attractive interactions with most polymer molecules and is, thus, easily miscible with matrices. A lower rolling resistance of tire treads formulated with blends of SBR and BR (butadiene rubber) is, however, achieved using chemically coupled hydrophilic silica as reinforcing filler instead of the traditional physically coupled carbon black. This change in the formulation has found its way into the commercial application through the "Michelin green tire" some thirty years ago. Nanoscale dynamics depends, thus, on various parameters linked to the filler structure and chemistry and of course also to the details of the polymer matrix itself: its chemistry, possibly blending, crosslinking, plasticization, and the impact of any additive. Moreover, the mechanical response of the material depends also in a non-linear manner on the imposed strain. These effects, in particular strain softening called the Payne effect [3, 4], are thought to be due to the filler reorganization undergone by the material under deformation. Although they are not the heart of our review, we will point out some of these aspects when appropriate for the present chapter.

Before doing so, a few definitions will be necessary. First of all, while we will focus on systems of industrial relevance, the comparison to model systems will come in handy. The idea is that real industrial systems are highly complicated, which makes physical understanding hard to achieve [5]. We, therefore, concentrate either on simplified-industrial systems with reduced numbers of components, but employing industrial silica filler of complex multi-scale structure, or on model systems based on colloidal silica NPs of well-defined spherical shape and size. Due to its origin, the industrial filler is usually a powder that is dispersed in the polymer in an internal mixer, a process termed solid-state mixing. On the other hand, the colloidal particles are suspended in a solvent, and samples are made by solvent casting. In all cases of our studies reported here, the polymer is a well-defined, low-polydispersity styrene-butadiene, so the system remains a "model" in this respect. The interaction between the polymer and the particles may be tuned, e.g., by covalent grafting of SBR or by the amount of coating agent. In addition to this, changing the chemistry of the polymer molecules allows modifying the interactions from basically "neutral" van der Waals interactions to attractive interactions possibly based on hydrogen bonding, accompanied usually by an increased glass-transition temperature. Obviously, switching to high-T_g, glassy materials opens the road to entirely new studies, where the stronger interactions lead to well-defined interfacial regions of new physical properties. While we refer the interested reader to works of Sokolov et al., this domain remains outside the scope of this chapter focusing on rubber systems. It is only noted that standard physico-chemical methods like the determination of bound rubber, differential scanning calorimetry (DSC), nuclear magnetic resonance (NMR), and of course broadband dielectric spectroscopy (BDS) give experimental access to such properties.

Our rationale is to develop an understanding of model systems, propose innovative methodological advances, and then transpose this to the simplified-industrial system. It is, thus, mandatory to be able to design samples in a highly controlled manner and then study their structure and dynamics, in order to extract correlations between parameters and properties, and ultimately gain insight into the underlying processes. Concerning the structure, our group has a strong background in scattering techniques (light, X-rays, neutrons). Scattering gives access to correlations in space and time, and depending on the design of the contrast situation, one may study NPs, or the polymer, or cross-correlations between both. In standard polymer-NP contrast, small-angle X-ray scattering (SAXS) is the method of choice. It allows characterizing individual filler NP dispersions, the formation of aggregates, or of networks. It can be combined sometimes to transmission electron microscopy (TEM), which is sensitive to the same contrast between particles and polymer. Although it is a fascinating topic, chain structure will not be discussed in this chapter [6]. It is the realm of neutron scattering, in particular SANS, with isotopic substitution of the polymer in order to label certain polymer chains, i.e., provide contrast by chemical design. Whatever the technique, the resulting intensity functions in reciprocal space are not straightforward to interpret. In the past, our group has proposed a variety of approaches giving quantitative access to NP structure in PNCs, and some of them are briefly outlined in this chapter.

Knowing the statistical properties of the NP dispersion allows one to imagine on which spatial scale dynamical processes take place. Different methods of characterization of dynamics have been used in the literature, and here we focus mainly on BDS, with some excursions into NMR, neutron spin echo (NSE), and rheology. These techniques give access to various processes, like the segmental motion of molecules or migration of charges. Our proposition is to combine them with the static properties, for instance imagining surface motion of ions along silica structures spanning the sample by analysis of the interfacial polarization.

The organization of this chapter serves the following scope: by putting together in a step-by-step approach structural and dynamical information available on all compounds and their blends, we wish to reach the difficult discussion of segmental dynamics in rubber nanocomposites. We, therefore, start with Sect. 2 which is devoted to the formulation of industrially relevant PNCs. In Sect. 3, we introduce our own recent contributions on filler structure in both model and industrial nanocomposites. Filler dispersion as characterized by scattering techniques like SAXS sets the landscape in which dynamical processes can take place. It is emphasized that the difficulty here does not lie in performing the experiments themselves, but in the developments of new quantitative data analyses, namely the so-called correlation hole approach giving access to internal densities and filler aggregate masses and, thus, space-filling properties up to percolation. In Sect. 4, the dynamical features of industrially relevant PNCs are discussed, first in terms of their individual components—filler and polymer—followed by an analysis of interfacial polarization of embedded filler described by Maxwell–Wagner–Sillars (MWS) processes. The latter include large-scale charge migration along the filler surfaces and can, thus, be used as a probe of filler structure in the matrix, which evolves as a function of, e.g., filler content, or grafting. In the last part of Sect. 4, segmental dynamics of the polymer matrix are reviewed. Its sensitivity to additives allows fine-tuning vulcanization and the glass transition and, thus, macroscopic rheological properties relevant for applications. Moreover, blending of different polymer molecules with their individual relaxation processes may or may not result in blending at the molecular level leading to a single relaxation. One of the key questions of the community refers to the influence of the filler surfaces on the polymer dynamics [7, 8], and some recent progress—including quasi-elastic neutron scattering (QENS) measurements—will be discussed. Different applications of dielectric spectroscopy, like nanoscale resolution, and a surprising measurement of the adsorption isotherm of coating agents onto silica buried in the polymer matrix will also be mentioned.

2 Nanocomposite Formulation

Samples related to industrial applications are usually designed in a way to resemble as much as possible the material finally used. Experimental studies of the structure and dynamics of such samples start by carefully formulating samples of complex composition, with many ingredients [9]. Among them, the foremost ones are obviously the rubber and the filler. Rubbers refer to a class of polymers with specific elastic properties that are entropic in nature and the have ability to sustain reversibly large deformations. They are made of long and flexible chains without bulky side groups, which is the reason why their T_g is significantly below room temperature. The vast majority of rubbers is formed by the addition of conjugated dienes as it is the case for natural rubber or its synthetic counterpart polyisoprene, and butadiene rubber, or by statistical copolymerization of diene and non-diene monomers as for SBR. In this chapter, our main focus lies on SBR-based PNCs but some examples based on NR or BR will be discussed. Regarding the industrial filler, we restrict ourselves to precipitated amorphous silica forming breakable agglomerates of ca. 100 μm size in a dried powder. The typical radius of the elementary silica NPs is in the 10–20 nm range. The polymer–filler interaction is an important parameter in system design as it has a direct consequence on both structural and dynamical properties. It can be tuned with coating or coupling agents. Functionalized polymer chains that are able to graft onto the silica surface during the hot mixing process are also commonly used. In the following, the label F- is added to the polymer acronym to highlight the presence of such chains in PNC (e.g., F-SBR). Interaction between polymer chains and silica surface-silanols is, thus, either covalent grafting via the functionalized chain-ends or a coupling agent, or weak non-specific van der Waals interactions favoring aggregation of bare silica NPs [10]. Surface modification using a coating agent is an intermediate situation allowing to reduce the filler–filler interaction, without chemical linkage with the polymer.

Rubber PNCs are prepared by solid-phase mixing using an internal mixer and/or an open two-roll mill. The different ingredients including liquid additives are introduced in subsequent steps. To improve rubber properties, the polymer chains are vulcanized, i.e., crosslinked with permanent covalent bonds using sulfur. However, this chemical process is slow and several additives termed accelerators and activators have to be included during mixing to reduce cure time and increase the reaction efficiency. The most common activators are metal oxides and fatty acids, whereas thiazoles and sulfonamides are often used as accelerators in combination with a basic molecule like diphenyl guanidine (DPG) acting as secondary accelerator. In Table 1, the most common additives to SBR-silica PNCs as used in the literature are summarized.

Nowadays, it is worth noting that the usefulness of most additives is well-motivated as their role (e.g., curative or antioxidant) has been widely studied in the literature. However, the combination of certain molecules incorporated jointly

Table 1 List of typical tire tread additives that are important for rubber processing, their abbreviation, name, and function. In the last column, their presence in industrial-simplified systems is indicated

Additive	Full name	Function	Simplified system
IPPD	N-isopropyl-N'-phenyl-para-phenylenediamine	Antioxidant	
6PPD	N-(1,3-dimethylbutyl)-N'-phenyl-para-phenylenediamine	Antioxidant	X
AO2246	2,2'-methylenebis-(4-methyl-6-tertiary-butylphenol)	Antioxidant	X
TDAE	Treated distillates aromatic extract	Extender oil	
Si69	Bis (3-triethoxysilylpropyl) tetrasulfide	Coupling agent	
Si363	3-mercaptopropyltriethoxysilane reacted with ethoxylated C_{13}-alcohol	Coupling agent	
Octeo	Octyl-triethoxysilane	Coating agent	X
S	Sulfur	Cross-linking agent	
ZnO	Zinc oxide particles	Cure activator	
SA	Stearic acids	Cure activator	
MBTS	Bis(2-benzothiazole) disulfide	Cure accelerator	
TBBS	N-butyl-2-benzothiazole sulfonamide	Cure accelerator	
CBS	N-cyclohexyl-2-benzothiazole sulfonamide	Cure accelerator	
DPG	Diphenyl guanidine	Secondary cure accelerator	X

may provoke additional effects which have not been studied specifically as demonstrated in the next section for a coating agent with DPG. The association of these two chemicals commonly used in industrial formulations impacts the filler structure, which has been specifically studied by us using a new structural model discussed below [11].

In our recent studies, we have taken the approach opposite to the idea of making our systems resemble the complex real samples: nanocomposites are chosen to be as simple as possible, hence called simplified-industrial PNCs [12, 13]. In particular, care was taken not to introduce any trace of carbon black, ZnO nanoparticles, or crosslinking agents, which may render the interpretation of X-ray scattering experiments very difficult—just think of the strong scattering contribution of ZnO. Our samples are made of linear styrene-butadiene chains (SB, or SBR, if one wishes to emphasize absence or presence of crosslinking), of well-defined masses (typically between 100 and 200 kg/mol, PI below 1.1), possibly end-functionalized, in which silica NPs are dispersed. The fraction of functionalized chains among all chains is termed the F/NF fraction or called %D3 in some of our articles. Concerning the silica, we used Zeosil 1165 MP from Solvay (of specific surface 160 m^2/g) [14], which is dispersed by internal mixing (Haake) in the polymer. A silane coating agent termed "octeo", and DPG (see Table 1) were also added, thus, approaching the formulation of industrial systems, in particular in terms of the structural complexity of the filler. In the specific case of swelling studies, crosslinking has been introduced [15], as well as some cross-checks of its effect in [11]. Apart from this Zeosil-based system corresponding to our "simplified-industrial" one, we have set up an equivalent "model PNC" system using the same SB polymer but with colloidal silica NPs (of radius also ca. 10 nm). These model NPs are well-defined particles of low polydispersity, which are dispersed in the polymer by solvent casting, i.e., mixing in suspension with polymer solutions in a common solvent and drying. The main advantage of the colloidal NPs is that their scattering contribution is well-known, and one can build on it for a more complex analysis.

3 Overview Over Filler Structure in Industrially Relevant Nanocomposites

The filler structure of "simplified-industrial systems" as described above with ingredients limited to a strict minimum is nonetheless quite complex, due to the multi-scale structure of the industrial silica. Two methods have been combined in a quantitative way, TEM and SAXS. While the former provides an intuitive analysis but has limited representativity of local details, the latter is highly representative but technically complicated to analyze. As often in such cases, it is good practice to observe the evolution of the signals by varying a single control parameter. We have, thus, progressively studied the influence of silica volume fraction [12], chain functionalization [16], and chain mass [17]. In a second set of experiments, the influence of small molecules was studied [11], before extending our studies to dynamical properties (see Sect. 4).

In the methodological development described in the following paragraphs, different levels of the spatial organization need to be analyzed: at the lowest level, the primary silica particles interact to form aggregates. The aggregate size is rather easily determined from a Kratky plot, i.e., a presentation of $q^2I(q)$ *versus* q, giving the q-position of the maximum corresponding to a break in slope in the intensity I(q)—an example is given below [12]. Aggregate polydispersity has to be assumed, and we have always taken 30 %—modifying this number would slightly shift results, but not change the tendencies. The mass of the aggregates, i.e., the aggregation number, can only be extracted from an intensity argument, i.e., the height of I(q). It gives access to the internal silica volume fraction of aggregates, which we call the compacity κ, and which is a key parameter for later understanding their mechanical properties. In practice, the height of the signal is related both to the aggregate mass and to their interactions, expressed by the inter-aggregate structure factor. The latter depends in a crucial way on the aggregate volume fraction (related to the number of aggregates per unit volume), not to be confused with their (internal) κ, and for which a quantitative analysis of many TEM-images was necessary. It gave access to the large-scale filler structure in terms of branches and sheets densely filled with aggregates. Altogether, this opens the road to what we termed a correlation hole analysis giving the volume fraction of aggregates and, thus, their interactions, their mass, and finally compacity. The main ideas are outlined below, dividing the task into two parts. First, the interplay between the local NP density and the resulting interaction measured in a scattering experiment and expressed by the correlation hole is conceptually introduced for model particles. In the second step, it is applied to industrially relevant PNCs, where the interaction takes place on two structural levels, within aggregates and between aggregates.

3.1 Correlation Hole Analysis of Local NP Density in Model Nanocomposites

While the idea to take interactions into account was originally motivated by our first studies of simplified-industrial PNCs as a function of silica fraction [12], a general model approach has been proposed by us only a few years later [18]. The idea was taken up recently by Arleth and Pedersen within a generalized description of structure factors [19]. An empirical description of particle interactions within aggregates has also been included by Beaucage in his famous multi-scale fractal description of scattering of aggregates [20, 21]. In our studies, adding more filler led obviously to stronger local interactions, but they had never been described quantitatively in the literature before for complex filler NPs of industrial origin.

In order to understand the concept, one can start by looking at the hard-sphere structure factor S(q), i.e., the Fourier transform of the pair correlation function of spheres suspended in a solvent and which possess excluded volume corresponding to their size, and no other long-range interaction. Such a structure factor influences the scattering I(q) by introducing a multiplicative term to the form factor P(q), which describes the particle shape. For monodisperse particles—monodispersity is only assumed for the sake of a simplified discussion, our analysis always includes polydispersity— the intensity reads as follows:

$$\mathrm{I(q)} = \Phi_{\mathrm{NP}} \Delta\rho^2 \mathrm{V\ S(q)\ P(q)} \tag{1}$$

where Φ_{NP} is the volume fraction of nanoparticles, $\Delta\rho$ their scattering contrast (in cm^{-2}), and V their volume. P(q) is normed to one at low q for convenience, while S(q) is normed to one at high q by definition [22]. If one looks at the shape of S(q) for hard spheres, it has a low-q limit which is below one, and which is the normalized isothermal compressibility of the gas of spheres. S(q) can be calculated analytically in this case, and the so-called Percus–Yevick structure factor [23, 24] has the following low-q limit:

$$\mathrm{S_{PY}(q \to 0)} = \frac{(1 - \alpha\,\Phi_{\mathrm{NP}})^4}{(1 + 2\alpha\,\Phi_{\mathrm{NP}})^2} \tag{2}$$

where α is a prefactor of the volume fraction. It depends on particle polydispersity, and for monodisperse spheres, α equals 1. The conclusion is that measuring the low-q limit of the structure factor gives access to the particle volume fraction Φ_{NP}! And conversely, any intensity measured in this range is affected by S(q), and the particle (or aggregate) volume V in Eq. (1) can only be extracted quantitatively if a trustworthy estimate of S(q) is available.

The next step is to consider aggregates as depicted in the inset of Fig. 1. Note that all our calculations include particle polydispersity denoted as σ, and numerical results have been obtained by averaging over many such aggregates. The resulting average scattering function I(q) is seen to start at low q from a high number which is related to the total aggregate mass. It then decreases following what is called a Guinier regime, proportional to a Gaussian function of parameter R_g, i.e., the radius of gyration of the aggregate. It finally meets the particle form factor P(q) in the intermediate and high q-range, where I(q) and P(q) perfectly superimpose. Using Eq. (1), one can calculate and plot the apparent structure factor S(q) of the aggregates. Its shape has three different regimes: the same Guinier regime at low q related to the aggregate size and volume, a high-q regime where it tends to one, and a depression around q* which is the correlation hole we are interested in. This is the manifestation of Eq. (2) for a system (i.e., an aggregate) of finite size. By measuring its depth S(q*), one can extract the internal volume fraction κ of the aggregate, provided one knows α in Eq. (2) as a function of particle polydispersity.

The question to be answered is, thus, to calculate α for well-known particle densities within aggregates. We have, thus, generated and equilibrated a very big ("infinite"

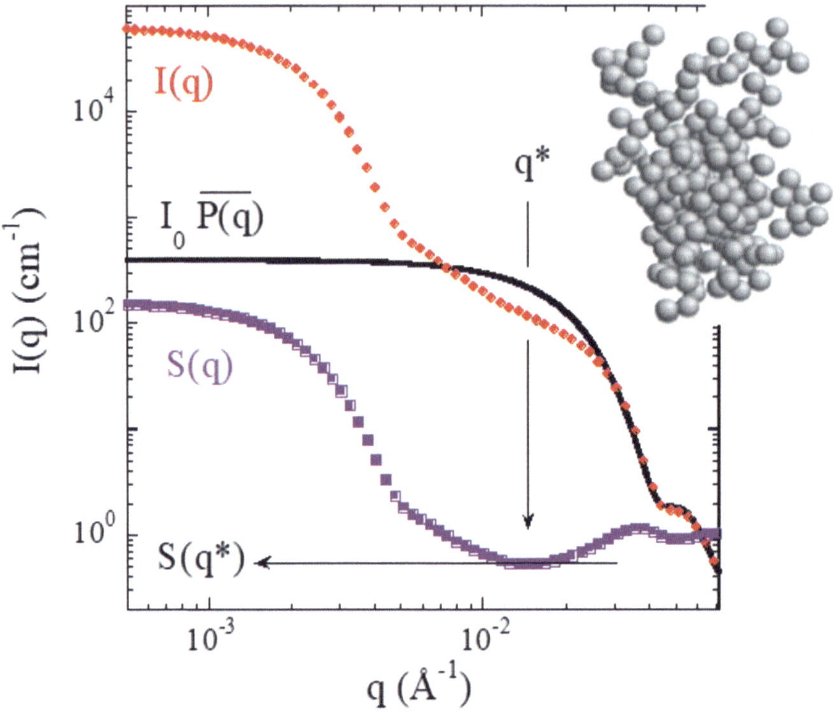

Fig. 1 Average intensity I(q) (circles), the average NP form factor P(q) (solid line), and the resulting apparent S(q) (squares) as a function of scattering vector q of simulated individual aggregates containing 200 polydisperse NPs ($R_0 = 10$ nm, $\sigma = 15$ %). Inset: example of aggregate. q* indicates the position of the correlation hole, and S(q*) its value. Adapted from Ref. [18] with permission from The Royal Society of Chemistry

on our scale) homogeneous system of repulsive spheres of average volume fraction Φ_{NP}, as shown in the cubic simulation box in the inset of Fig. 2a. The compressibility is a function of this global volume fraction, and if one cuts a spherical piece out of this representing an aggregate, then the local volume fraction $\kappa = \Phi_{NP}$, the compacity, is related to the depth of the correlation hole. Due to the finite size of such aggregates, the low-q structure factor increases up to a high value related to the number of spheres in the aggregate, i.e., to the aggregation number. Due to the aggregate of well-defined spherical shape, typical aggregate form factor oscillations are found in the intermediate q-range. The correlation hole, as shown in Fig. 2a, corresponds to the minimum of the envelope of the aggregate form factor oscillations. The higher the internal aggregate concentration, the deeper the correlation hole S(q*), as shown exemplarily for two Φ_{NP}-values, 20 % and 40 %. In Fig. 2b, S(q*) is plotted as a function of κ for different log-normal particle polydispersities. These functions can be fitted using Eq. (2), and $\alpha(\sigma)$ has been reported in [18]. As a result, it is, thus, possible to relate the observed correlation hole to the internal density and, thus, to

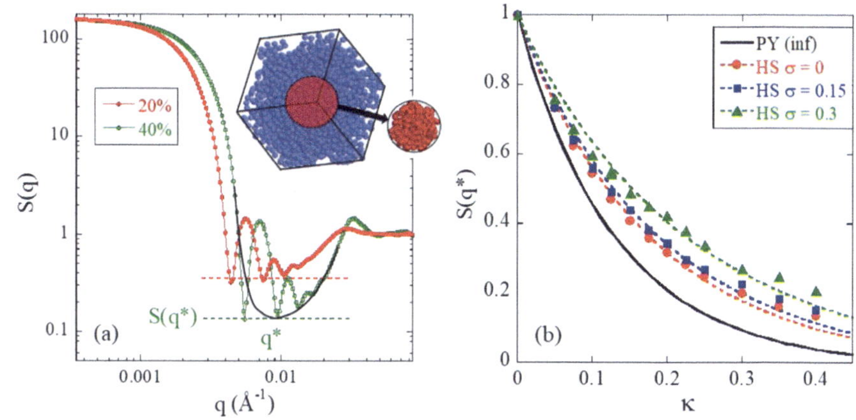

Fig. 2 **a** Structure factor S(q) of finite spherical systems made of hard spheres, for two local volume fractions, $\kappa = 20$ and 40 % (polydispersity $\sigma = 15$ %) generated by setting $\kappa = \Phi_{NP}$. q* represents the position of the correlation hole, the continuous black line illustrates its average shape. Inset: "Infinite" simulation box with equilibrated hard spheres and illustration of cutting out a finite spherical aggregate in the box center. **b** Low-q structure factor S(q*) of finite spherical systems as a function of κ based on simulations of hard spheres ($\sigma = 0$, 15, 30 %). Dotted lines are fits using Eq. (2) with $\alpha = 0.73$, 0.69, and 0.57, for $\sigma = 0$, 15, and 30 %, respectively. The solid line is the (infinite) Percus–Yevick prediction. Adapted from Ref. [18] with permission from The Royal Society of Chemistry

the average aggregate mass, if the size is known from a Kratky plot. Moreover, using simulated aggregates which we could characterize extensively in real space before extracting their compacity from their (calculated) scattering, we have been able to confirm the proposed "correlation hole" approach [18].

As the last point on this conceptual approach, the reader might wonder why the mass and density information is not extracted from the low-q Guinier scattering, where the aggregate radius of gyration and mass are measured. The reason is that in real systems, there is more than one aggregate, and the more interesting a system is, the more aggregates there are, in close vicinity, and thus interacting. This higher level of interaction modifies the low-q scattering such that the Guinier regime is eventually masked.

3.2 *Application of Correlation Hole Analysis to Simplified-Industrial PNCs*

In industrially relevant nanocomposites, the first level of interaction corresponds to the formation of aggregates by nanoparticles. It results in a correlation hole as described above. However, this is superimposed by a higher level of organization caused by the repulsive interaction of aggregates within dense branches or sheets.

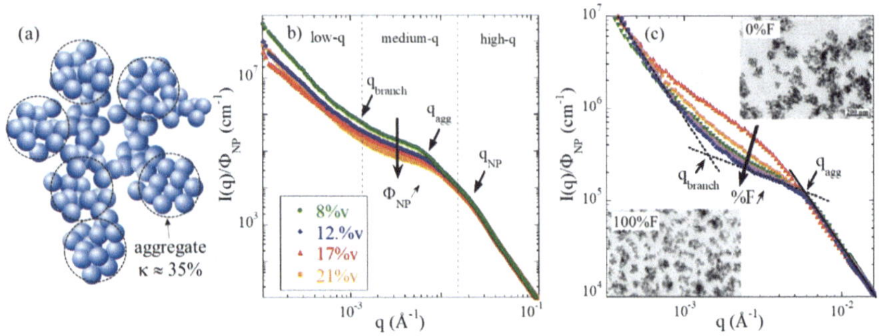

Fig. 3 **a** Schematic representation of the multi-scale filler structure in industrially relevant PNCs. Aggregates are highlighted with their local density κ. **b** Silica structure via the reduced SAXS intensity, $I(q)/\Phi_{NP}$, of 50%F-PNCs (140 kg/mol) for different silica contents Φ_{NP}, with indications of the different structural contributions. **c** Silica structure of 8.5 vol%-PNCs (140 kg/mol) with different matrix compositions from 0 to 100%F. In the inset, two TEM pictures are shown, for 0 % and 100%F, at $\Phi_{NP} \approx 8.5$ vol%. Adapted with permission from Refs. [12, 16] Copyright 2013 American Chemical Society

This second level of correlation hole analysis needs additional information on the presence of silica-free zones in the sample or, inversely, the volume of the branches and sheets, which are visible in TEM and have been quantified by image analysis and averages over many images. Aggregates are, thus, themselves concentrated in large-scale fractal branches or sheets of a lateral thickness of ca. 150 nm, extending over microns. This is illustrated in Fig. 3a.

The multi-scale filler structure of the Zeosil 1165 MP has been modeled starting with the primary silica beads as basic building units. Their scattering analysis reveals a log-normal filler size distribution ($R_0 = 8.55$ nm, $\sigma = 27$ %, giving an average radius $< R_{NP} >$ of 8.9 nm and an average bead volume of $V_{NP} = 3.6\ 10^3$ nm^3), in agreement with TEM studies. V_{NP} allows estimating aggregation numbers of silica NPs within aggregates. The typical radius of these aggregates found from Kratky plots ($q^2I(q)$ vs. q) is in the 40 nm range. These sizes are found in the scattered intensity in Fig. 3b, where the breaks in slope corresponding to the nanoparticles, the aggregates, and the branches are identified by arrows. Aggregates repel each other, thus, introducing a correlation hole effect as introduced in Sect. 3.1, which is illustrated by the arrow with increasing Φ_{NP} in Fig. 3b. In this normalized representation of $I(q)/\Phi_{NP}$, the increase of silica concentration leads to a decrease of the signal in the intermediate q-range, reinforcing the correlation hole. As shown above, its depth is directly related to the concentration of aggregates after subtracting a low-q power law associated with the surface scattering of branches. The aggregate content is higher than the nominal silica volume fraction due to the confinement within the fractal sheets, the remaining space being occupied by the polymer. Applying Eq. (2) with parameters given in [12] provides the aggregate volume fraction in the sheets/branches and, thus, together with the conservation law of silica, the aggregate mass and internal compacity κ. The resulting values are in the 30–40 % range, depending on the formulation, which corresponds to aggregation numbers of some 50 NPs. The latter does not depend strongly on the silica content, which was a surprising result given the highly different system viscosities during mixing. Such a result was not recognized before, probably because the decrease in the low-q intensity was attributed to a decrease in aggregate mass upon mixing at higher viscosity, neglecting the correlation hole effect due to S(q) in Eq. (1). It is also noteworthy that mechanical measurements could be described using the average aggregate compacity reported here [12].

In Fig. 3c, the influence of another important control parameter on the silica structure is shown. By varying the fraction of end-functionalized chains from 0 to 100 %, the inserted TEM pictures show a strong reorganization. Unfunctionalized chains induce more irregular structures, while the opposite is true for functionalized ones. In the reduced units of the X-ray scattering, the signal at intermediate q is depressed. If one takes into account the approximately constant correlation hole—these experiments are performed at fixed silica volume fraction—the aggregate mass is found to decrease with the functionalization. The latter is, thus, a convenient control parameter of filler structure in complex polymer nanocomposites. It also paved the road to a more global approach, based on the grafting density of the chains on the silica [25]. Indeed, it was found that the latter determined the silica structure in a unique way (within a restricted range). Originally, this finding was triggered by the observation of very similar TEM pictures for different compositions and gave the name to the corresponding article on "twins".

The analysis of the importance of grafting on final structure gave rise to a speculative sequence of processes during mixing—we are still hoping to study this in-situ one day. Depending on the sample viscosity defined by mainly chain mass and silica content—high viscosity favoring fragmentation—the silica pellets are broken up finely in the internal mixer. The aggregate size, thus, decreases in the first moment, before a reagglomeration process sets in. The latter is driven by the attractive interactions between bare silica particles and is impeded if chain grafting takes place. The higher the chain grafting, the lower the final aggregate masses, as described above. This mechanism is schematically shown in Fig. 4.

For the sake of completeness, it is also mentioned that the structural studies based on the correlation hole idea have been applied to industrial PNCs formulated with natural rubber and silica [26]. Quite some research is currently devoted to systems obtained by rapid coagulation [27], which seems to be an industrially promising way to return from synthetic to natural, bio-based, and, thus, sustainable materials. In the work by Boonsomwong et al. [26], the structure of such systems has been measured by SAXS, with specific attention paid to the mechanical rejuvenation of the samples. The remarkable result concerning structure is that the scattered intensity I(q) overlaps over a large q-range, including the correlation hole and the high-q scattering, both for all PNCs and the primary silica powder. Without being quantitative, this is a strong indication that the local correlations between the silica NPs are not affected by any mechanical work, in particular milling. On the other hand, the low-q structure evolves with concentration and milling, showing the break-up of large-scale NP assemblies, while the Payne effect weakens simultaneously.

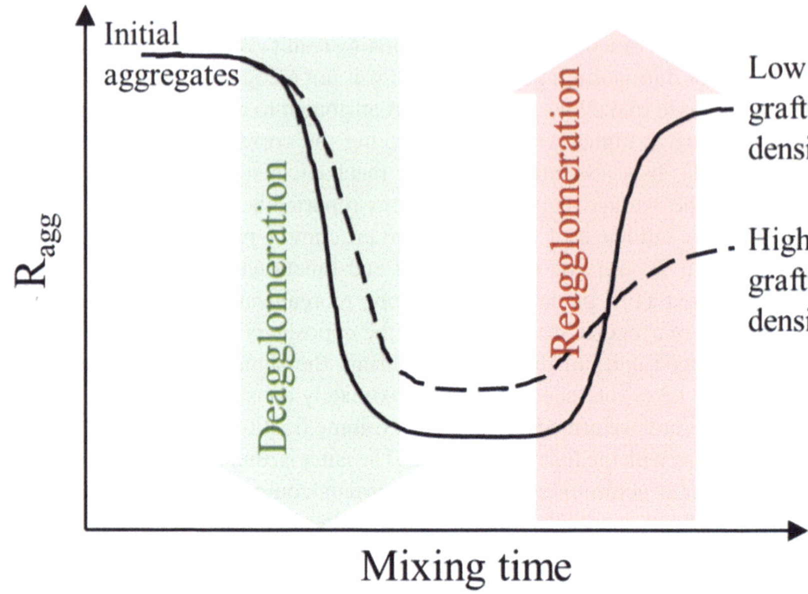

Fig. 4 Evolution of the aggregate size during the solid-state mixing process compatible with the result observed after mixing. Aggregates are thought to deagglomerate more or less depending on the viscosity of the polymer–particle blend, but reassemble in the "hot phase", in a way that is controlled by the grafting density. Adapted from Ref. [17] with permission from the Royal Society of Chemistry

3.3 *Impact of Small Molecules on Nanocomposite Microstructure*

The effect of so-called small molecules has been studied on virtually identical simplified-industrial PNC samples [11]. One of the most important molecules is a silane coating agent, commonly called "octeo" (see Table 1 and Sect. 2). In Fig. 5a, the reduced scattered intensity $I(q)/\Phi_{NP}$ is plotted for different amounts of octeo, the nominal value being 8 % in mass with respect to the silica filler, all other parameters remaining fixed.

The scattering in Fig. 5a is typical for our simplified-industrial PNCs as discussed in Fig. 3. The coating agent has a strong effect on the aggregate structure probed at intermediate q. The intensity levels vary considerably on this logarithmic scale. In absence of octeo, the dispersion is worst, with the highest intensity in Fig. 5a, and it improves with increasing octeo content as shown by the arrow. A quantitative analysis of these data shows a decrease in aggregation number N_{agg} from approximately 50 to 30, accompanied by a decrease in aggregate size R_{agg} from 34 to 30 nm as one can see from the Kratky plot shown as the inset. The decrease in N_{agg} indicates a higher compatibility of the filler with the matrix, as there is more polymer in contact with

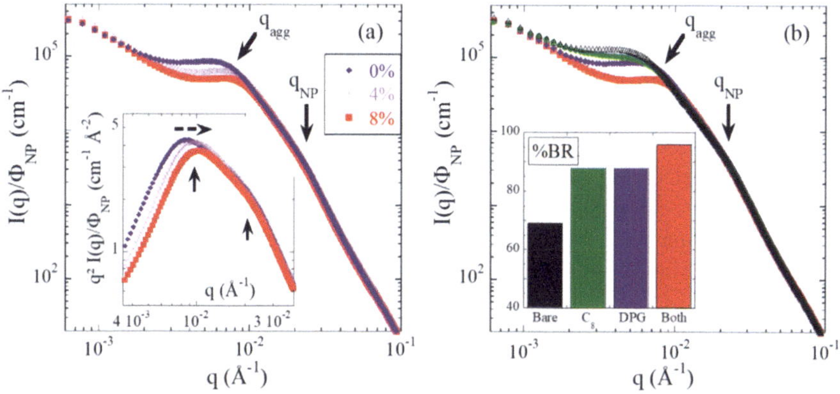

Fig. 5 Reduced SAXS intensities $I(q)/\Phi_{NP}$ allowing the structural comparison between F-SB PNCs with ca. 18 vol% of silica and different compositions in octeo and DPG. Arrows indicate the q-ranges of particle- and aggregate-scattering. **a** PNCs made with DPG and different amounts of octeo: 0 (diamonds), 4 (crosses), and 8 wt% (squares). The inset shows the corresponding Kratky plot $q^2I(q)$ *versus* q highlighting the structural evolution (dashed arrow). **b** PNCs made without small molecules (bare, triangles), with DPG only (diamonds), with octeo only (circles), and with both octeo and DPG (squares). The fraction of octeo (when present) is 8 wt%. The inset shows the bound rubber of the same PNCs. Adapted with permission from Ref. [11] Copyright 2017 American Chemical Society

silica surfaces for smaller aggregates. Incidentally, this induces a higher amount of bound rubber (not shown here, [11]).

We have also studied the simultaneous effect of both the coating agent octeo and DPG, as shown in Fig. 5b. A synergetic effect is found, in the sense that none or only one of the two molecules leads to a rather high scattering in the intermediate q-range, corresponding to a high aggregation number if one follows again the quantitative analysis based on the correlation hole outlined above. Only the simultaneous presence of both induces a low aggregation number and size ($N_{agg} \approx 30$, $R_{agg} = 30$ nm) and correspondingly a higher bound rubber value as shown in the inset of Fig. 5b.

3.4 Complete RMC Study of Model System

In the preceding paragraphs, we have described how one can extract quantitative information even in complex multi-scale filler systems, in particular on aggregate sizes and their internal densities, from small-angle scattering, with the help of large-scale image analyses based on TEM. This analysis is based on the implicit assumption that both primary particles and aggregates possess spherical symmetry, and polydispersity affects only the radii, not the shape. There is, however, no visual proof, say by imaging techniques, that this is indeed true, although such a simplified description is probably not too far from reality—after all, primary particles look somewhat "potatoe-like". If one wishes to go one step further in the calculations, one needs to step back (hopefully only momentarily) to the more ideal systems of colloidal nanoparticles. As their shape is essentially spherical and well-described by some average radius and polydispersity, such a model system is better suited for more involved calculations of scattering patterns. The reason behind this is again that Eq. (1) is strictly valid only for (monodisperse) objects of spherical symmetry. A more elaborate version of Eq. (1) for polydisperse spheres can be derived in a straightforward manner, as used by us in [12], but nonspherical shapes would include computationally intractable orientational dependencies of the form and structure factors.

We have designed model PNCs based on the same SB polymer, but introducing colloidal particles of a radius of 12.5 nm and log-normal polydispersity 0.12. Using a reverse Monte Carlo (RMC) approach [28, 29] which we have applied for the first time to individual PNC aggregates [30] and recently large-scale PNC systems [31], it is possible to explore particle configurations in space that have a scattering signal compatible with the experimentally observed intensity. Many details would need to be discussed here, but for lack of space (and adequacy of the topic), the interested reader is referred to the above-mentioned literature. The most important point is that there is no unique solution to this ill-posed problem, but all solutions define a family which can be investigated statistically. For illustration, a comparison of TEM pictures with (arbitrary) slices of the simulation box containing possible configurations is shown in Fig. 6a–d, for different silica fractions of the model system [31]. The "model" nature is immediately recognized in the TEM pictures, as beads are visible individually. The computer slices of the same thickness as the experimental ones

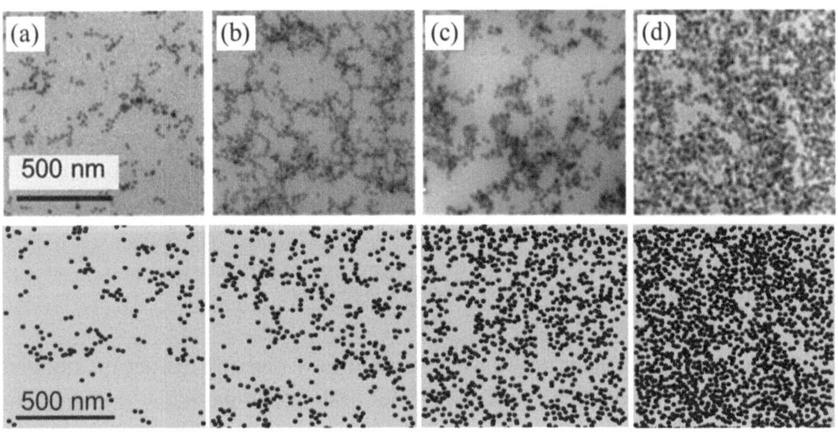

Fig. 6 Series of TEM pictures of colloidal silica-SB PNCs with increasing particle volume fraction Φ_{NP}. **a** 2.6 vol%, **b** 6.0 vol%, **c** 10 vol%, and **d** 20 vol%. For illustration, 70 nm-thick slices of the RMC-simulation box are shown below each picture, at the same nominal concentrations. Adapted from Ref. [31] with permission from The Royal Society of Chemistry

(70 nm) show strict monodispersity due to the plotting program. In reality, particles do have polydispersity.

In Fig. 6a–d, the crowding of the system, with the formation of strands and irregular aggregates, is nicely reproduced by the simulation. At higher concentrations, voids or pores seem to be formed, again in agreement with the experimental result.

Once one has obtained a set of particle positions that agree with the scattering—this agreement is shown in Fig. 7a where the simulation lines perfectly overlap with the experimental points—one can analyze these configurations quantitatively. An example of a simulation box is shown in the inset of Fig. 7a, where an aggregate (including periodic boundary conditions) is highlighted in red. It is then possible to evaluate the aggregation number and count the number of existing aggregates having this aggregation number. The result is an aggregate mass distribution plot as shown in Fig. 7b, where one can read off how many particles are single, doublets, etc. As the concentration is increased, the reduced scattering intensity at low q decreases again under the effect of the correlation hole. Bigger and bigger aggregates are formed until finally almost the entire box forms a giant aggregate. This can be seen in Fig. 7b by the formation of aggregates with around 10,000 NPs. It is, thus, interesting to see how the average aggregate mass evolves with concentration. We have proposed an indicator of aggregate mass, $<N_{agg}^2>/<N_{agg}>$, which is related to the low-q intensity in polydisperse systems.

Plotting this indicator in Fig. 7c shows the strong increase with particle volume fraction at a critical threshold, which is related to the percolation of the hard silica across the entire sample—or the simulation box on our scale. As a comparison, the

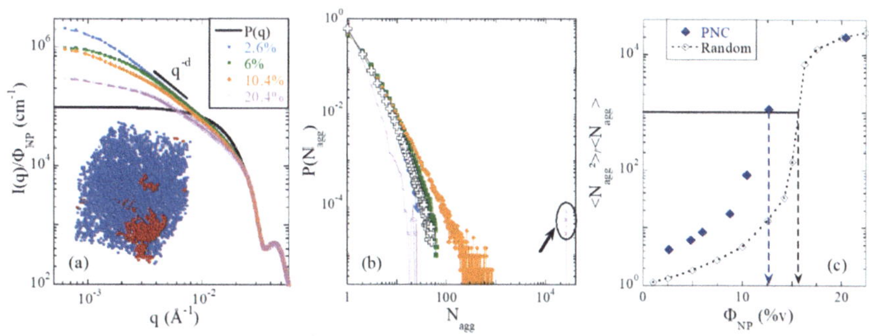

Fig. 7 **a** Reduced SAXS intensities $I(q)/\Phi_{NP}$ of colloidal silica-SB PNCs, for different NP volume fractions Φ_{NP}. The dotted lines superimposed to the symbols represent the result of the RMC fit. The NP form factor in the polymer is represented by a black line. **b** Examples of distribution functions of N_{agg} deduced from scattering normalized by the total number of aggregates in each configuration and averaged over 100 configurations. The distribution for a random dispersion of the same NPs ($\Phi_{NP} = 10.4$ vol% , $\delta = 0.6$) is represented by black crosses. **c** Aggregation indicator $< N_{agg}^2 > / < N_{agg} >$ as a function of Φ_{NP} for the silica-SB PNCs. The dotted line illustrates the aggregation of the same NPs with a random dispersion in the simulation box. The vertical arrows point toward the percolation threshold for both sets of NPs. Adapted from Ref. [31] with permission from The Royal Society of Chemistry

indicator calculated in the same manner for a random dispersion of the same polydisperse spheres in the box is shown to have a retarded increase. This demonstrates the propensity of the NPs to aggregate and eventually connect the entire sample, inducing mechanical percolation at lower filler content.

4 Dynamical Signatures in Industrially Relevant Nanocomposites Probed by BDS

BDS is a powerful tool with a straightforward principle: a sinusoidal electric field is applied to the sample sandwiched between two flat electrodes and one measures the complex impedance leading to the dielectric permittivity as a function of temperature and frequency as well as possibly pressure. The main benefit of BDS applied to polymeric materials is that it allows covering a broad frequency range (10^{-2}–10^{6} Hz in most cases). A complete dynamical process is, thus, captured in a single experiment, as opposed to mechanical measurements where it is generally needed to build master curves applying the time–temperature superposition principle. This is known to be problematic as soon as two processes with different T-dependence overlap. The two main polarization mechanisms responsible for the dielectric response of polymers and PNCs are the dipolar polarization due to the reorientation of permanent dipoles and the polarization resulting from the migration of charges. Any molecular motion involving fluctuations of the dipole moments is detectable by dielectric spectroscopy as detailed discussed in this section. At low frequency, the motion of ionic charges (here basically residual impurities) gives rise to conductivity but also to interfacial polarization, which is pronaunced in composite systems and may hinder a proper characterization of the segmental relaxation. Both contributions are schematically represented in Fig. 8.

4.1 Dielectric Response of Individual Components

We begin with an independent determination of the dynamics of the two main PNC components, namely the rubber matrix and the silica filler. Polymer melts display various dynamical processes depending on the probed length and time scales. Among them, the segmental dynamics (also named α-process) corresponds to the primary structural relaxation associated with the glass transition. It involves cooperative segmental motions of the polymer chain arising from conformational transitions. Figure 9 shows the typical evolution of the frequency-dependent dielectric loss for a pure styrene-butadiene random copolymer in the temperature range of the α-relaxation (typically 253–278 K). The position of this relaxation in the frequency window depends on the styrene content, a higher proportion of styrene units leads to a higher T_g [33]. In the frequency domain, the α-relaxation is commonly described

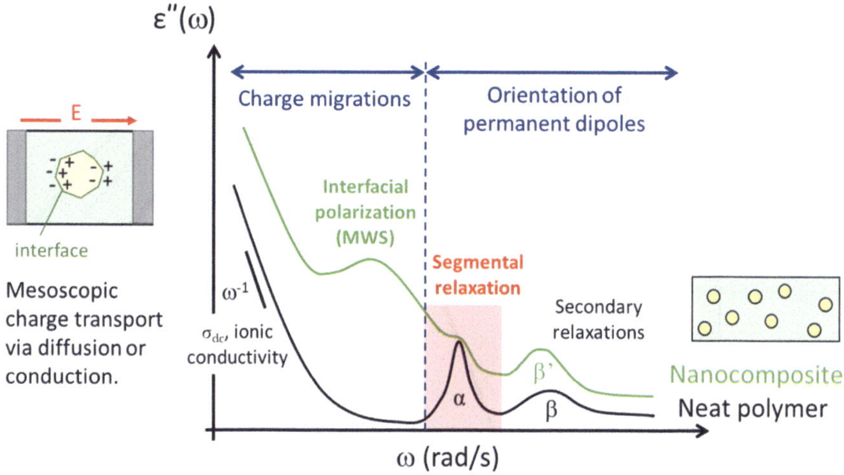

Fig. 8 Schematic overview of dynamical processes evidenced by isothermal dielectric loss spectra of polymer and polymer nanocomposites

by a Havriliak–Negami (HN) function with a simultaneous fit of the imaginary and real parts of the complex permittivity

$$\varepsilon^*(\omega) = \varepsilon_\infty + \frac{\Delta\varepsilon}{\left[1 + (\mathrm{i}\omega\tau_{HN})^\gamma\right]^\delta} \tag{3}$$

where $\Delta\varepsilon$ is the dielectric strength, τ_{HN} the characteristic relaxation time, and γ and δ are the width and asymmetry parameters of the HN distribution, respectively. As observed for many polymers [34], the α-process of the neat SB polymer has an asymmetric shape (δ in the range 0.4–0.6). We found that the shape of the distribution does not depend on the functionalization of the SB chains. However, it becomes symmetric in the vulcanized system, the low-frequency side of the relaxation peak being the only part impacted by crosslinking with a slowing-down of the cooperative segmental motions [35, 36].

On the high-frequency side of the low-temperature spectra in Fig. 9, the tail of a second process is discernible. It corresponds to the secondary β-relaxation observed below T_g with a typically weak, Arrhenius T-dependence (activation energy of 32 kJ/mol). This process is perfectly observable in the frequency window at 173 K. Among the different molecular units constituting the SB chain, it has been shown that local motions of the butadiene units are responsible for the β-relaxation [33], with distinct contributions of the 1,4-*cis* and 1,2-vinyl groups on the high- and low-frequency side of the dielectric process, respectively [37].

We now turn to the precipitated silica filler. Silica NPs are hydrophilic due to the presence of a large number of silanol groups on the surface [38]. These groups promote the adsorption of water via hydrogen bonding, and humidity levels of the

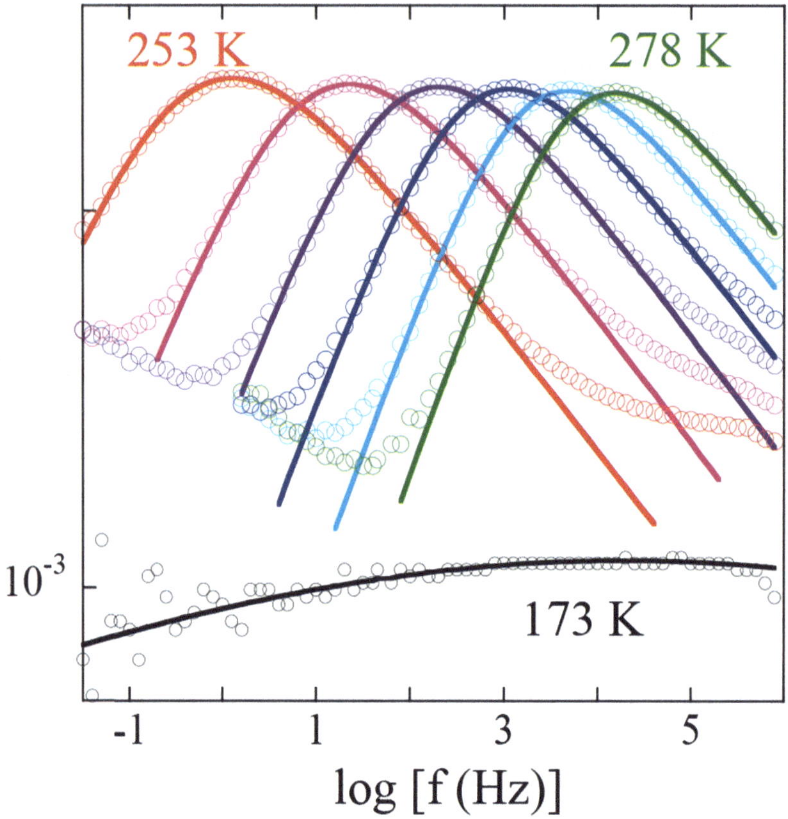

Fig. 9 Dielectric loss spectra as a function of frequency for pure F-SB between 253 and 278 K every 5 K and at T = 173 K (black circles). Solid lines represent fits with an HN function

order of 7 % have been reported for highly dispersible silica [14]. Cerveny et al. proposed a complete dielectric study of Zeosil 1165 MP produced by Solvay, highlighting the effect of the hydration level [39]. Silanols and water molecules contribute on a similar level due to their comparable and strong dipole moment. Figure 10a shows the evolution of the dielectric loss of silica particles with different water concentrations at low temperatures. Two main relaxation processes have been identified in the frequency window as illustrated in Fig. 10b: a low-frequency process (called process 2 in the figure) associated with the reorientation of hydrated, vicinal and geminal, silanol groups, and a well-defined high-frequency process (termed process 1) due to the relaxation of water molecules. The dielectric strength of the latter strongly depends on the water content with the formation of water clusters at high concentrations.

A similar identification of the dielectric relaxation processes of precipitated silica NPs has been made by Meier et al. in SBR nanocomposites [40], and an overall β'

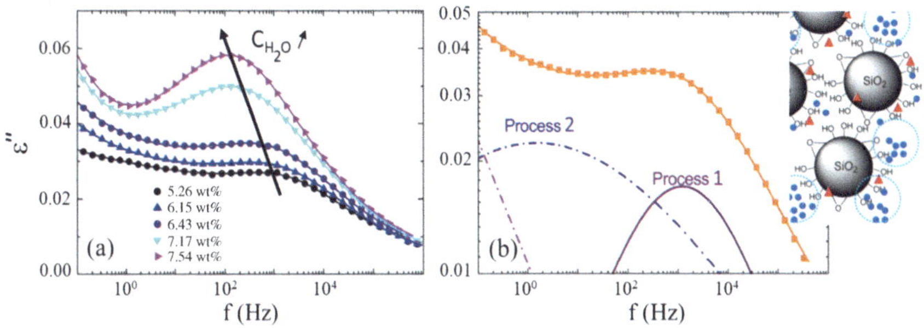

Fig. 10 Dielectric loss spectra of hydrated silica NPs at T = 146 K. The solid lines through the data points represent fits to the experimental data using the sum of two Cole–Cole functions ($\delta = 1$ in Eq. (3)) and a conductivity term. **a** NPs at different water concentrations, **b** decomposition of the dielectric response of NPs containing 6.4 wt% of water. At this temperature, processes 1 (solid line, water) and 2 (dashed-dotted line, silanols) are observed. Note that they are termed processes 2 and 3, respectively, in the original publication [39]. Inset: schematic drawing of the water adsorbed at the silica surface (blue dots: hydrogen bond water, red triangles irrotational water). Water molecules are distributed homogeneously or grouped to form clusters at high concentrations (dotted circles). Adapted with permission from Ref. [39] Copyright 2012 American Chemical Society

process is illustrated in Fig. 8 for the silica contribution. The authors studied the impact of surface modification, showing that grafting silane molecules at the silica surface leads to a reduction of the amplitude of the high-frequency silica process (process 1 in Fig. 10) and to an increase in the corresponding activation energy. Accordingly, the presence of nonpolar molecules at the silica/polymer interface reduces both the amount of interfacial water molecules and their mobility. Interestingly, the activation energy of this process is an indicator of the strength of the NP-polymer interaction: 0.60 eV (57.9 kJ/mol) for bare silica, 0.65 eV (62.7 kJ/mol) for the same silica NPs modified with a coating agent, and 0.69 eV (66.6 kJ/mol) when a coupling agent is used. On the opposite, it was found that there is nearly no impact of the dispersion state that can be varied by reducing the mixing time on the dielectric relaxation of silica.

4.2 *Structure-Dependent Charge Migration in PNCs*

The two different phases of nanocomposites—polymer and filler—are characterized by different electrical properties, and additionally, interfacial polarization known as the Maxwell–Wagner–Sillars effect is generally present in PNCs, including filled rubbers [34, 41]. The MWS process arises from the migration of charge carriers present from the stage of mixing, such as additives or impurities, at the filler/polymer interface. The trapped charge carriers form induced dipoles of large size leading to an additional contribution of the charges to measured spectra as illustrated in Fig. 8. In this section, we review BDS results focusing on the interfacial polarization processes in PNCs with multi-scale filler structure. These processes as well as ionic conductivity are affected by structural reorganizations, which are themselves closely related to the mechanical properties [42, 43]. In other words, the microstructure, i.e., the nanoparticle dispersion, sets the frame for the dynamics. Consequently, the evolution of the BDS signal can be used to detect changes in the dispersion state.

4.2.1 Influence of Water

A strong MWS polarization process has been reported in industrial rubber PNCs at the low-frequency side of the α-process (see Fig. 8 for illustration) [35, 42, 44]. The position of this process within the experimental frequency window depends strongly on the water content—it becomes slower upon in-situ drying within the dielectric cell—whereas both its intensity and spectral shape are nearly unmodified [35]. The activation energy for this process is of the order of 0.55 eV (53 kJ/mol), also without any significant variation with the water content. The dielectric response of the MWS polarization has been described by the interlayer model (ILM) in SBR PNCs [35]. This model proposed by Steeman et al. [45, 46] considers a two-phase system with an adsorbed water layer located at the interface. As evidenced in Fig. 11, the dielectric loss spectra at different hydration levels are well-reproduced by the ILM model using a single fit parameter, namely the water conductivity which depends on the water content and fixes the interfacial layer thickness. All these results indicate that MWS polarization in rubber PNCs is controlled by the amount of adsorbed water molecules at the silica surface, with a dielectric loss peak shifting reversibly within the frequency window.

4.2.2 Evidence of Filler Percolation

Electrical conductivity has been used for a long time as an indicator of the formation of percolating paths of conductive filler in non-conducting matrices [47]. For example, the relationship between electrical conductivity and dispersion in filled rubbers has been discussed for different NP types and shapes in a review article by Roland [48]. Here, we choose to highlight the original work by Le et al. on in-situ conductance measurements of SBR-carbon black PNCs [49]. A rather crude measurement of the dispersion by optical microscopy at selected moments during mixing gives access to the fraction of non-dispersed particle agglomerates bigger than a few microns and, thus, to the inverse parameter termed "dispersion" in Fig. 12. This parameter is found to increase with the mixing time of the compounds, while the conductance varies in a non-monotonous way. Its evolution has been separated into three stages by the authors: In the beginning, there is no infiltration of the CB by the polymer, and the electrical conductivity drops as the powder is incorporated. In stage 2 in Fig. 12, the dispersion value increases strongly, and so does the conductance: the CB powder disperses while building up a network. In the last stage, finally, the

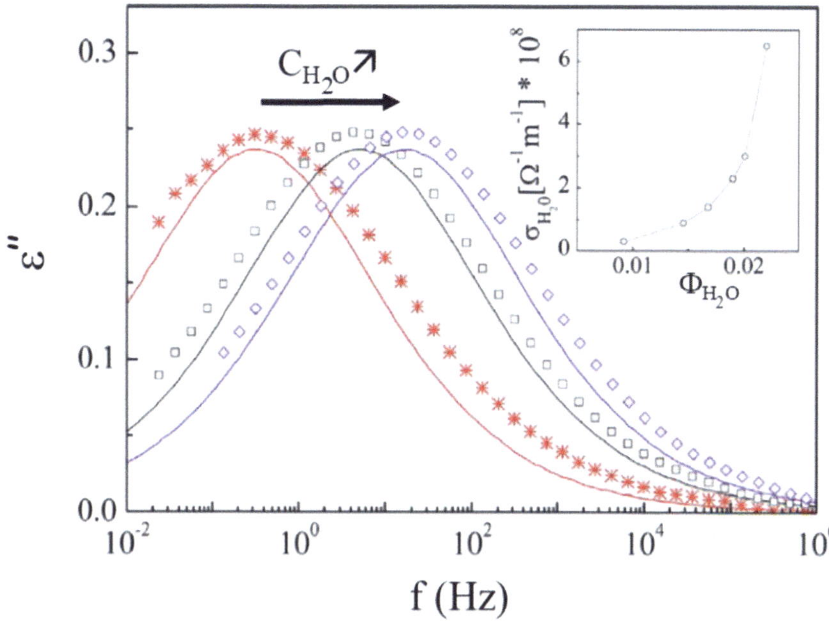

Fig. 11 Dielectric loss spectra at T = 295 K for SBR PNCs with 30 phr of silica and different water contents: C_{H2O} = 0.55 wt% (crosses), 0.95 wt% (squares), and 1.11 wt% (diamonds), after subtracting the α-relaxation and conductivity contributions. Solid lines represent the ILM descriptions. Inset: interlayer conductivity σ_{H2O} as a function of water volume fraction Φ_{H2O}. Adapted with permission from Ref. [35] Copyright 2013 American Chemical Society

best dispersion is obtained, corresponding to a break-up of any percolated network, and, thus, a slow decrease in conductivity.

In the preceding section, we have seen that conductivity is not the only process sensitive to charge migration. Besides MWS1 denominating the process discussed in Sect. 4.2.1, the second interfacial polarization process labeled MWS2 has been evidenced by our group a few years ago in simplified rubber PNCs in the high T-range, well above the glass transition [42, 43]. The MWS2 process is located at the low-frequency side of MWS1, and its evolution with the silica content in PNC is shown in Fig. 13a. It is visible at best for the lowest silica fraction with a broad dielectric loss peak in which two distinct contributions have been identified and described each by an HN function. For higher silica contents, the peak overlaps with the strong contribution of ionic conductivity, but two processes are always necessary to describe the dielectric spectra quantitatively.

For both MWS1 and MWS2 processes, the activation energies are found to be independent of the silica fraction. The activation energy of ca. 114 kJ/mol for MWS2 is similar to the one corresponding to the temperature dependence of the ionic conductivity. One may deduce that MWS2 is predominantly controlled by the dynamics of the polymer, as it was already the case for the ionic conductivity. On the other hand,

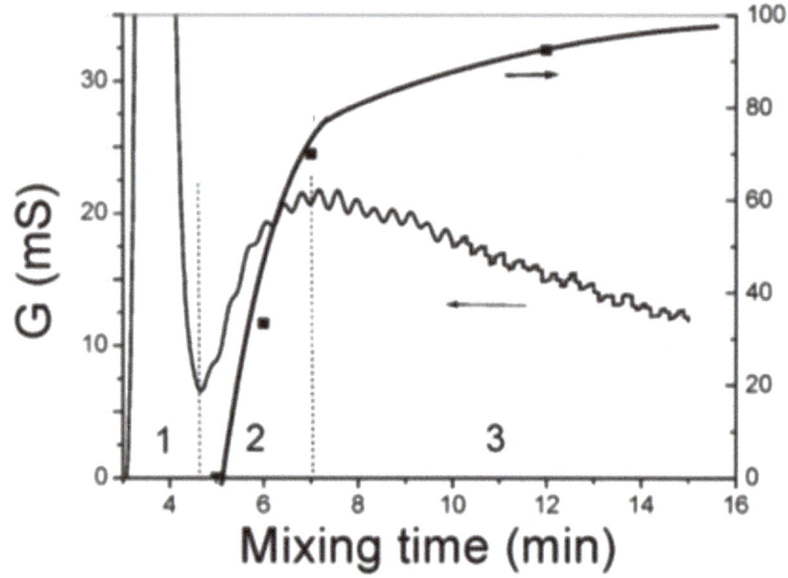

Fig. 12 Electrical conductance (left scale) measured in-situ during mixing of 50 phr N220 carbon black in SBR; corresponding filler dispersion (right scale) measured at selected times is indicated by the symbols with a line as a guide to the eye. Adapted with permission from Ref. [49] Copyright 2006 Rubber Division, American Chemical Society

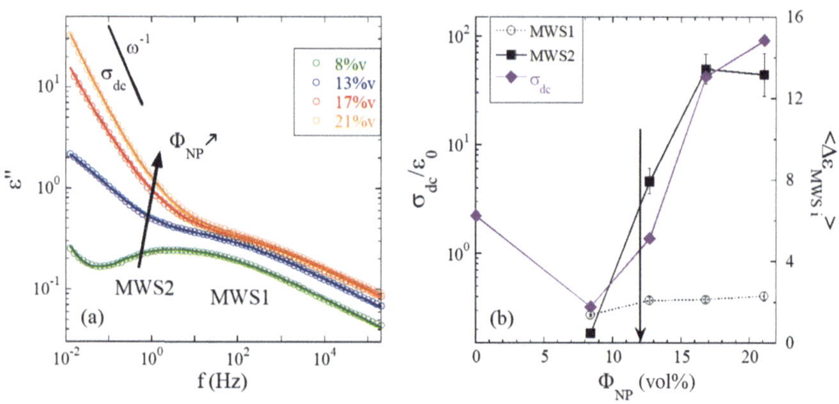

Fig. 13 **a** Dielectric loss spectra at 333 K for 50%F-SB PNCs with different Φ_{NP}. Solid lines are fits by the sum of two HN functions and a DC-conductivity term. **b** DC-conductivity at 393 K (diamonds) and the average value of the dielectric strengths over T for MWS1 (circles) and MWS2 (squares) as a function of filler fraction for the same PNCs. The percolation threshold as obtained from rheology is indicated by an arrow. Adapted from Ref. [42] with permission from the PCCP Owner Societies

the MWS2 time scales are controlled by the filler concentration but not by the hydration level, indicating different origins for both interfacial polarization processes. Whereas MWS1 is a local phenomenon based on charge diffusion in water layers along the surface of silica aggregates (which are the building units of the microstructure, see Sect. 3.2), the polarization mechanism of MWS2 consists of large-scale charge diffusion across the polymer matrix between aggregates. The evolution of the dielectric strengths of both MWS1 and MWS2 processes as a function of the silica fraction is reported in Fig. 13b together with the ionic conductivity. The latter exhibits first a decrease attributed to the reduction of free charges in the polymer matrix due to trapping at the silica surface. It is followed by a two-order of magnitude increase in the range of the percolation threshold as determined independently by means of rheology on the same samples. Here, it is found that MWS1 depends only marginally on the silica content and, thus, microstructure, whereas MWS2 is strongly impacted by large-scale structural reorganizations with the formation of a percolated network of aggregates. One may note here that a similar process to MWS2 has been partially identified in carbon black filled rubbers and linked to structural reorganization due to crosslinking or high-temperature treatments [50]. In these experiments, flocculation in SB-carbon black PNCs just below the percolation threshold has been induced by thermal treatments. The effect could be followed via the increase in mechanical strength, with the simultaneous development of a Payne effect indicative of filler networking. In parallel, BDS measurements have been interpreted in terms of charge migration across gaps between filler NPs, and closing of these gaps is consistent with the formation of filler networks. Innovative experiments coupling *simultaneous* dielectric and mechanical spectroscopy have been reported by Huang et al. in NR-carbon black PNCs [47]. These authors link the modification of interfacial polarization to mechanical properties. In PNCs of high filler content, a significant decrease of the real part of the permittivity under increasing strain is concomitant to the classical Payne effect on G'. A similar confrontation with mechanics has been followed by us in an above-mentioned article [26], however, without details on charge migration by BDS measurements indicative of filler contact. Scattering and mechanical Payne effect experiments show that thermal treatment may be used to rejuvenate a flocculated and possibly percolated aggregate structure, after having destroyed the filler network by milling.

It is, thus, possible to connect the dynamical dielectric response of these interfacial processes with the various structural length scales. Such an approach is original as BDS is not a spatially resolved experimental technique, which is why it is only rarely used [47, 51] to access structural information, beyond simple conductivity measurements. The latter are a common approach to the detection of percolation since the invention of the carbon microphone some 140 years ago.

4.2.3 Impact of Grafting Density

As described in Sect. 3.2, the use of graftable polymer chains is a powerful way of controlling the dispersion state which sets the frame for the mechanical properties of

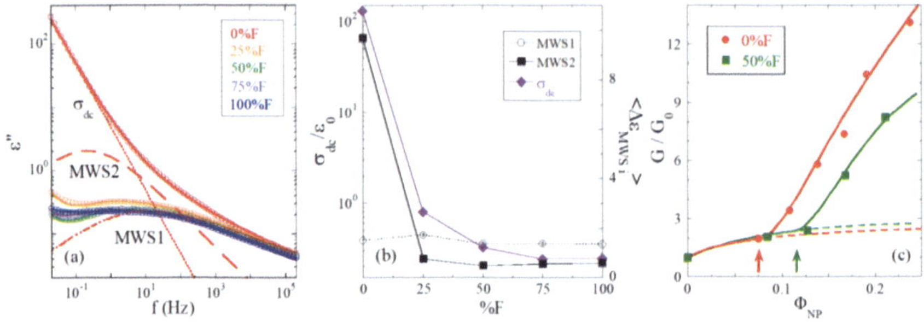

Fig. 14 **a** Dielectric loss spectra at 333 K for SB PNCs with ca. 8.5 vol% of silica and different amounts of graftable chains. Solid lines are fits by the sum of two HN functions and a DC-conductivity term. Individual contributions are included for the sample without graftable chains. **b** DC-conductivity at 393 K (diamonds) and the average value of the dielectric strengths over T for MWS1 (circles) and MWS2 (squares) as a function of matrix composition for the same PNCs. **c** Rheological response at 60 Hz in terms of the reinforcement factor G/G_0 as a function of silica fraction for SB PNCs with 0%F (circles) and 50%F (squares). Solid lines are fits discussed in [12], arrows indicate the percolation threshold. The purely hydrodynamic reinforcement, $1 + 2.5\Phi_{agg}$, is also included (dashed lines). Adapted with permission from Ref. [43] Copyright 2015 Elsevier

rubber PNCs. The dielectric response of a series of simplified-industrial nanocomposites with different amounts of functionalized SB chains is illustrated in Fig. 14a. The strong evolution of the dielectric loss with grafting is clearly visible, and the decomposition into the above-mentioned interfacial processes, MWS1 and MWS2, is also shown.

The evolution of the dielectric strengths as well as the ionic conductivity is reported as a function of the fraction of graftable chains in Fig. 14b in analogy to Fig. 13b. The strength of the local MWS1 process is mostly unaffected by grafting whereas the MWS2 strength is found to follow the same evolution as the ionic conductivity, as already observed in Fig. 13b. Here, a significant drop of the MWS2 intensity at ca. 25 % of functionalized chains is observed, which is interpreted as the break-up of the percolated aggregate network upon grafting at fixed silica content. This effect has been termed depercolation. It is directly related to the evolution of the filler structure with polymer grafting (Sect. 3.2). The formation of aggregates of lower mass at low grafting density notably shifts the mechanical percolation threshold as evidenced in Fig. 14c. There, the reinforcement factor defined as the ratio of the plateau moduli of PNC and the pure matrix is plotted as a function of the filler fraction for different grafting densities. These data have been fitted using a percolation model based on percolating aggregates within large-scale branches extending across the sample. It uses the aggregate volume fraction deduced from the combination of SAXS and TEM as the main variable instead of the silica fraction. The good description of the rheological data in Fig. 14c underlines the consistency of the proposed approach. Moreover, the volume fraction at percolation and its shift with functionalization agree nicely with the dielectric analysis of interfacial processes. Following the evolution of such processes is, thus, a sensitive tool to probe large-scale structural reorganization in rubber PNCs.

4.3 Segmental Dynamics of Rubber Compounds

The segmental relaxation of polymer matrices is the physical origin of dominating macroscopic rheological properties crucial for industrial applications. There is a large body of results related to fine-tuning these properties, and following our industrial application of car tire treads, we review here only effects of crosslinking-related additives (in particular activators), as well as blends, with dielectric spectroscopy as the main technique. Once the matrix-only part is understood, we will turn to polymer nanocomposites, with a critical discussion of the modification of the segmental dynamics close to nanoparticle interfaces in weakly interacting rubber systems like SBR-silica. Here, the BDS data are complicated to analyze due to overlapping processes, and some recent neutron scattering results are discussed.

4.3.1 Following Vulcanization Mechanisms by BDS in Unfilled Rubbers

The vulcanization of rubber leading to the formation of a permanent 3D network is required to enhance the macroscopic properties of the final product. This process involves a large number of additives beyond sulfur (e.g., ZnO, stearic acids, CBS, among others listed in Table 1), which makes it rather complex due to possible interactions between the different additives and also interactions with the silica filler. All additives are present in the formulation in small quantities with respect to the polymer (typically between 1 and 5 phr each) and quite a few of them possess a strong dipole moment (e.g., 2.90 D for CBS, 3.99 D for DPG) [52, 53]. It follows that a sensitive and selective technique such as dielectric spectroscopy is a highly appropriate tool to study the impact of the presence of vulcanization additives on the segmental relaxation. Modifications of the dynamics that are observed by BDS may be helpful for a better understanding of the vulcanization mechanism, which is critical for technological developments. These effects are reviewed here for unfilled rubber.

Ortiz-Serna et al. report dielectric measurements carried out on samples of natural rubber containing the common vulcanization additives but processed to avoid vulcanization [54]. The dielectric spectra reveal a broad loss peak that could be described by two contributions: a fast α-process due to the segmental motions of NR and a slower process assigned to the dynamics of stearic acids linked to the NR chains. The authors studied the impact of water and found, surprisingly, that the dielectric strength of the low-frequency contribution is higher for the dry rubber. This effect is interpreted as a consequence of different ordering of the lipid hydrocarbon chains which depends on the water content. High-pressure dielectric spectroscopy was also used to track the influence of temperature and density on the segmental dynamics. It shows that the relaxation processes detected here are driven by temperature variation instead of free volume. In a similar NR-based system, Hernandez et al. also report the second relaxation process, slower than the segmental dynamics of NR [55]. Going through a step-by-step study where the different vulcanizing additives are removed one by one, they rather attributed the slow process to ionic clusters of ZnO. These clusters act as crosslink precursors, which restrict the mobility of interfacial polymer segments, confirming the role of ZnO as an activator for sulfur vulcanization. This result is in agreement with a previous SANS study on vulcanized polyisoprene (the synthetic version of NR) swollen in a deuterated solvent [56]. The SANS results evidenced heterogeneous network structures formed by high-density network domains which surround ZnO clusters dispersed in the rubbery matrix. Stearic acids are nonetheless necessary to disperse ZnO and form zinc stearate after reaction, promoting crosslinking.

In the same spirit as Hernandez et al. [55], the effect of vulcanizing agents in unfilled SBR mixtures has been explored by Ortega et al. [52]. The same kind of slow contribution to the α-relaxation peak has been observed by BDS and it is evidenced in Fig. 15a. By comparing the dielectric response of the mixture without one or the other additives, the slow process could be assigned to the cure accelerator couple, DPG and CBS (see Fig. 15b). Furthermore, there is a dominant effect of the DPG content on the dielectric strength of this process. It was also demonstrated that DPG itself (and not byproducts from the vulcanization reaction) is related to the slow

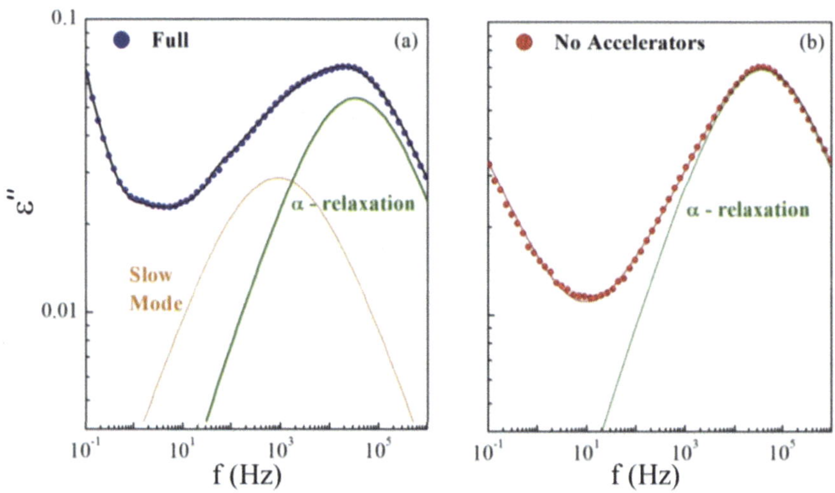

Fig. 15 Dielectric loss spectra at T = 289 K for F-SBR. **a** Full formulation, **b** the same formulation without DPG and CBS accelerators. Solid lines represent the fits to the experimental data using one or two Cole–Cole functions (δ = 1 in Eq. (3)) and a conductivity term. A single relaxation is observed in **b**. For the sake of clarity, the conductivity contribution is not shown. Adapted with permission from Ref. [52] Copyright 2019 Elsevier

process by comparing the dielectric response of a sample before and after in-situ vulcanization in the dielectric cell. BDS measurements under pressure showed that both contributions to the α-process display a similar activation volume (same shift of the relaxation times with pressure), indicating that the dynamics of DPG is coupled to the segmental dynamics. The latter is itself not modified by the presence of DPG. Lindemann et al. determined an activation energy of ca. 150 kJ/mol for the slow process in the same SBR system showing that it may eventually merge with the segmental relaxation in the high-temperature range [57]. Looking finally to the filled rubber mixture [52], the slow process is found to display a lower dielectric strength in presence of silica presumably due to the lower amount of DPG available in the matrix as this molecule tends to adsorb on the silica surface.

The examples discussed above highlight that a slower contribution to the segmental dynamics is systematically observed in rubber systems containing vulcanizing additives. This contribution may have different molecular origins depending on the polymer and the set of additives present in the recipe.

4.3.2 Studying Miscible Rubber Blends by Dielectric Spectroscopy

The dynamical properties of thermodynamically miscible polymer blends—even in absence of filler NPs—represent an entire research topic of their own [58]. This is due to the possibly wide dynamical asymmetry, where the interplay of the segmental dynamics of each compound can induce qualitatively new phenomena. Dynamical asymmetry refers to the fact that the α-relaxation of each polymer may be active at very different frequencies—or in other words, they display a large difference in individual T_g's. A representative example of asymmetric blends is the well-known poly(ethylene oxide)/poly(methyl methacrylate) blend [59]. Following the literature and in particular the groundbreaking work of the groups in San Sebastian [58] and Jülich, [59, 60] uncrosslinked blends usually show a common, intermediate macroscopic (calorimetric) T_g. If one resolves selectively the segmental dynamics with BDS, the surprising finding is that two distinct processes corresponding to the segmental dynamics of each component are reported with slight modifications due to the local composition in the blend. A rationalization of this effect based on the concept of self-concentration linked to chain connectivity has been proposed by Lodge and McLeish [61]. It was further refined by combining this concept with the Adam–Gibbs theory [62] linking structural relaxation time and configurational entropy [63]. If, however, one forces the interaction between the two chains by introducing specific connections (hydrogen bonds or crosslinks), then the local processes average out to a single one, i.e., dynamic heterogeneity may not persist on the probed length scale.

The segmental dynamics of unfilled vulcanized SBR/BR blends prepared by solid-phase mixing has been investigated recently. These are two miscible polymers, which are both dielectrically active, with a large difference in T_g of about 80 K. In this industrially relevant system, a single loss peak has been reported by BDS for the α-relaxation, indicating the same—or at least close—segmental relaxation times for the two polymers. In a recent contribution by the group of Blume, Rathi et al. [64] focus on the distribution and impact of extender oils (TDAE) on the mechanical performance of tire treads, namely wet-skid resistance, low wear, and low rolling resistance. These properties are difficult to predict presumably due to dynamical heterogeneities within the blend. They observe a systematic broadening of the glass transition in 50/50 vulcanized blends studied by calorimetry, dynamic mechanical analysis, and BDS. In dielectric spectroscopy, the authors decompose the single loss peak of the blend without TDAE (see Fig. 16) into a fast α- and a slow α_0-process, each described by an HN function: the fast α-process is associated with the contribution of dissociated polybutadiene segments, while the blended SBR and polybutadiene segments lead to the slower α_0-process. For the blends with extender oil, the general trend is that both the α- and α_0-processes show longer relaxation times with increasing amounts of oil, which was to be expected since the TDAE oil has slower relaxation dynamics compared to the blends. However, there is a larger effect on the fast α-relaxation coming from polybutadiene segments only, which is similar to the effect of the oil on pure BR. The dynamical heterogeneity is, thus, reinforced by the presence of the oil, contributing to the complex material response.

Starting from the individual components—BR and SBR of different microstructure—Schwartz et al. in collaboration with Goodyear quantitatively describe dielectric spectroscopy results of the vulcanized blends with a modified Adam–Gibbs

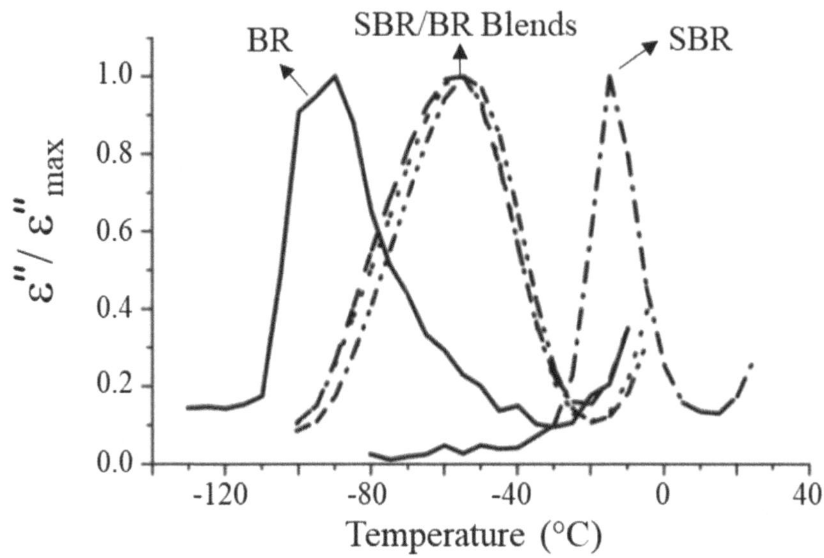

Fig. 16 Dielectric loss spectra normalized to their maxima versus temperature at 1 Hz for the pure polymers: SBR and BR, and for the SBR/BR blends with different contents of TDAE oil: ——0, - - 10, and – - 20 phr. Adapted with permission from Ref. [64] Copyright 2018 John Wiley and Sons

approach [36]. The latter relates the time scale of the segmental motion to the excess entropy of the mixture taking into account the interactions between components via an effective Flory-parameter, χ_{eff}. As outlined above, a single segmental relaxation process is observed due to the strong intermolecular interaction provided by the crosslinks, and the temperature dependence of the relaxation times is well-described using the proposed model. The resulting interaction parameter quantifies the T-dependent change in configurational entropy, and it can even be negative (lower entropy) for unfilled blends at the highest temperature, presumably implying a more efficient packing. The interaction is moreover shifted with the chain composition and also with silica-filling, using precipitated silica at a rather high load (120 phr). The model, thus, accounts for the presence of the filler NPs in terms of an increased interaction between chains.

4.3.3 The Complex Analysis of the Dynamics of Filled Rubbers

The segmental dynamics of SBR loaded with silica filler has attracted considerable attention over the past two decades. The question of modifications of the polymer dynamics close to the interface has been approached with various techniques, in particular mechanical measurements [65–68], and of course BDS [40, 69, 70]. The problem which has triggered considerable discussion in the literature is that most of these results—however well done on each specific system and for each experimental approach—are not mutually consistent: sometimes an interfacial layer of slowed-down dynamics is reported, and sometimes there is basically no impact of the presence of silica on the segmental dynamics. Depending on what one wishes to promote, the authors, thus, have the natural tendency to cite either one or the other of these "schools"... It is not the purpose of this modest chapter to review this entire discussion, but we attempt to replace some studies in this context. First of all, Arrighi et al. have shown the existence of the second high-temperature contribution in mechanical measurements of crosslinked SBR-silica systems represented as the loss tangent, tan(δ), *versus* temperature [65]. This additional process is enhanced by the addition of a coupling agent and it is attributed to a dynamical change in a nanometer-thick interface, assuming a given filler structure. The second tan(δ) peak in the viscoelastic response of SBR PNCs is also reported by Tsagaropoulos and Eisenberg [66] whereas it is not visible in the data of Mélé et al. [67] and Robertson et al. [68], though in both cases notable differences on the high-temperature side of the tan(δ) peak are observed. It has been pointed out by Robertson et al. that the observed differences in tan(δ) do not necessarily reflect modifications in the segmental dynamics of the polymer as the loss modulus is essentially not evolving with the polymer–filler interaction. They may rather be due to filler-network contributions and the variation of the elastic modulus (reinforcement), or restricted terminal flow in presence of NPs [71, 72]. It is, thus, not always necessary to invoke mobility-restricted polymer layers near the filler nanoparticles. The same SBR-silica system [65] has been studied by BDS by Pissis et al. [69], and they speculate on the influence of the surface treatment on structure—without having access to quantitative measurements—and finally on MWS processes that are more pronounced when untreated silica is used possibly due to aggregate formation. Fragiadakis et al. propose work on natural rubber filled with in-situ created silica [70]. Two states of dispersion have been prepared, unfortunately again without any quantitative characterization, and the authors showed that only the better dispersion leads to a small increase of the calorimetric T_g, with a silica-dependent low-frequency process observed at 2 to 3 orders of magnitude below the main α-relaxation. Such a strong effect—with a restricted mobility extending over a ca. 3 nm-thick interfacial layer—is difficult to understand taking into account the weak interaction between NR and silica and may result from very specific sample preparation. From these different examples, it is clear that the following points deserve further attention:

(a) System preparation: Any small surface-active molecule (e.g., coating agents) may also influence dynamical matrix properties and, consequently, also rheology. Complex systems are, thus, … complex to analyze.
(b) Robustness of the interpretation of complex BDS spectra: Given the very broad range of most dielectric processes, it is not easy to discard the influence of other processes on the position and shape of a given one (like MWS on the segmental relaxation).
(c) Crosscheck by other techniques: Which other microscopic probes can be used to analyze local polymer dynamics?
(d) Microstructure: Knowledge of the primary NP size is important, but not always sufficient to characterize the amount of polymer in interaction with filler structures. What about aggregation states and interlayer overlap?

In reply to these points, we have designed the simplified-industrial system defined in Sect. 2. Moreover, the adsorption isotherm of coating agents on filler NPs has been investigated in-situ and is discussed below. This work illustrates the differences between system design (all molecules should go to the interface) and reality (some do not!). Next, the overlap with MWS processes that have been detailed in the above subsection may or may not influence the shape of the α-relaxation process. Without a quantitative description of the latter, it is difficult to conclude on either a shift, a broadening, or the generation of an additional process for the segmental motions. As observed already by Meier et al., a broadening (without shift) of the dielectric loss peak with increasing silica content seems to be robust [40]—these authors do not identify any strong MWS process which could alter the interpretation, but this may be structure and loading dependent. The broadening is presumably related to the slowed-down dynamics of the chain molecules close to the interface, but the authors do not provide a detailed analysis of the α-relaxation and focus, as mentioned above, on the secondary relaxations. On the other hand, nano-dielectric measurements based on an AFM-type apparatus have shown that there is no detectable change in local dynamics across a sample, regardless of the distance to the next filler surface [73]. Despite locally different mechanical properties defining an interacting layer, the observation of bulk-like segmental dynamics in this layer may result from the use of a coupling agent acting as a flexible link between the polymer and the silica filler.

The question of cross-checking BDS results, in particular if peaks are ill-resolved, brought us to QENS outlined briefly below. These results clearly show that the change in dynamics in weakly interacting rubber systems is systematically present, but equally systematically small. Of course, structural investigations range also under "additional techniques", which is why we have opened the present chapter with a detailed review of recent advances in this field. If one focuses, e.g., on the thickness of the interfacial layer, the latter is usually deduced from some overall fraction of dynamically modified polymer. As a function of the filler landscape onto which this "slowed-down" volume is distributed, different thicknesses are obtained: thus, the thickness of the interfacial layer depends in a crucial way not only on particle size, but also on dispersion. As an example, Papon et al. have tackled this problem in silica-acrylate PNCs [74], making use of a RMC approach as described in Sect. 3.4. The result is a statistical representation of overlapping interfacial layers, which is related to the degree of vicinity, i.e., positional correlation, of the nanoparticles. Close-by NPs, thus, share an interfacial layer, thereby increasing their thickness with respect to the ideal, fully dispersed system having the same amount of slowed-down volume.

4.3.4 Dielectric and Neutron Spin Echo Spectroscopy Applied to Simplified-Industrial Systems

Figure 17a shows the dielectric data of a simplified-industrial PNC containing 20 vol% of silica. The dielectric loss exhibits a strong MWS contribution with a poorly defined high-frequency shoulder associated with the α-relaxation. The data in

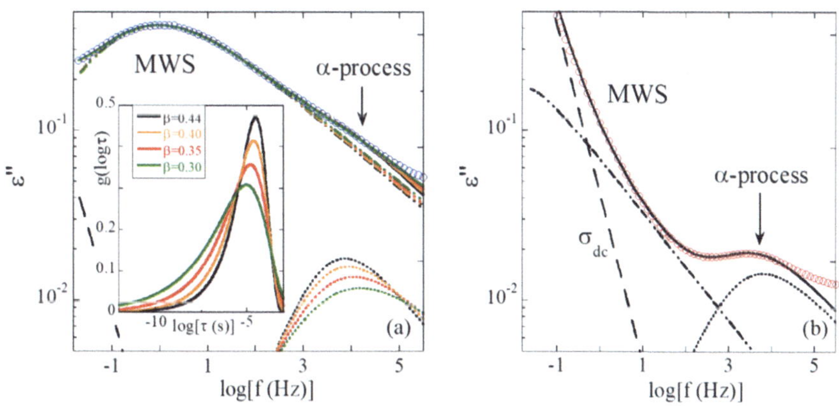

Fig. 17 **a** Dielectric loss spectra of SB PNCs with ca. 20 vol% silica at T = 273 K. Lines are fits by a purely dissipative DC-conductivity term (dashed line) and two HN functions: interfacial MWS process (dashed-dotted lines) and α-process (dotted lines). **a** Simplified-industrial PNC. Four different spectral shapes associated with the KWW distributions given in the inset are shown to reproduce the α-process. β = 0.44 corresponds to the stretching exponent obtained for the pure SB matrix (black curve). **b** Colloidal PNC. The KWW distribution of relaxation times associated with the α-process in this model PNC is given in the inset (red curve, β = 0.35). Adapted with permission from Ref. [32] Copyright 2020 American Chemical Society

Fig. 17a have been described using the sum of two HN functions plus a purely DC-conductivity term. For consistency with the following QENS data, where a stretched exponential function—known by the names of Kohlrausch, Williams, and Watts (KWW)—is used to describe the relaxation behavior in the time domain, we used coupled shape parameters in Eq. (3) to describe the α-process, following the connection between the frequency-domain HN function and the KWW function proposed by Alvarez et al. [75, 76]. It is clear from Fig. 17a that there are large uncertainties regarding the shape and position of the α-process. Indeed, an equivalently good description of the data is obtained using a selection of stretching exponents β ranging from 0.44 (the value obtained for neat SB) to 0.30. The corresponding—compatible—distributions of relaxation times for the segmental dynamics in PNC are, thus, possibly larger in presence of silica, as illustrated in the inset of Fig. 17a. Interestingly, the representative average characteristic times, $\langle\tau\rangle = \tau_W\, \Gamma(1/\beta)/\beta$, associated with these functions still lie in a very restricted range (−4.2 to −4.4 in log scale, including the pure matrix timescale), in agreement with the observation of a constant calorimetric T_g. Such results indicate that the segmental dynamics is only weakly affected by the presence of silica in the rubber system. This was confirmed by low-field NMR measurements on similar simplified-industrial samples [43], but also on their vulcanized counterparts including the whole ingredient list [77]. In both cases, the fraction of rigid polymer amounts only to a few percent, i.e., much lower than values reported for colloidal silica NPs dispersed in poly(ethyl acrylate) using either a coating or a coupling agent [78]. The small rigid fraction detected by NMR has nevertheless a critical role in the viscoelastic response of filled rubbers [77].

Here, one may wonder about the dynamical response of colloidal nanocomposites based on the same SB polymer as used in the simplified-industrial formulation but incorporating well-defined spherical NPs (see the model system defined in Sect. 2). In this case, the dispersion state as discussed in Sect. 3.4 is rather good, without the typical multi-scale structure of industrial PNCs which involves strongly interacting aggregates. Accordingly, one can see in Fig. 17b that the MWS process is less intense and is shifted toward a lower frequency. As a consequence, the α-process is now well-defined in the frequency window probed by BDS, and its description points toward a broader distribution of relaxation times ($\beta = 0.35$, see inset in Fig. 17a) with respect to the neat polymer. The position of the maximum of the distribution is not modified by the presence of silica, and only a negligible slowing-down of the average characteristic times ($\log \langle\tau\,(s)\rangle = -4.1$) is observed. There is, thus, a minor impact of the silica NPs on the segmental dynamics in the model PNC in spite of a larger range in relaxation times.

Given the impossibility to correctly resolve the α-relaxation in industrially simplified PNCs, QENS has been employed in a recent study [32]. Due to their rather low energy, which unlike the one of X-ray photons is comparable to k_BT, neutrons are often the only radiation probe capable of investigating spatially resolved dynamics. Neutron Spin Echo spectroscopy is one of these methods. It is based on the measurement of the changes in neutron speed after interacting with the sample. It is usually performed with deuterated samples (coherent scattering) in order to study the collective motions associated with the segmental relaxation, but this is difficult here for

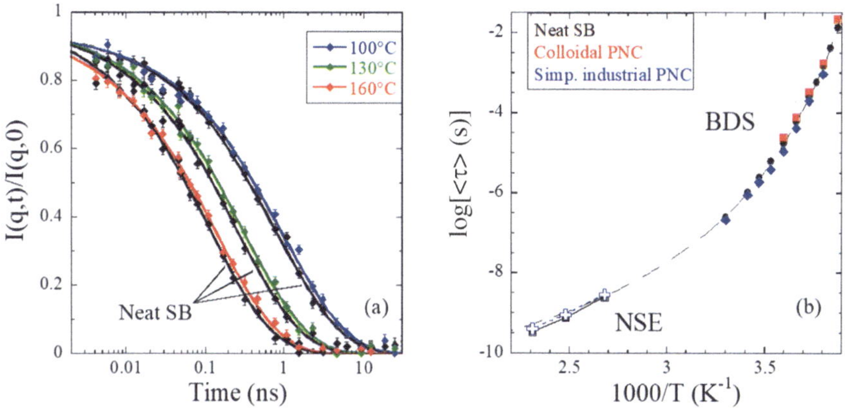

Fig. 18 **a** NSE data at q = 0.54 $Å^{-1}$ for the neat SB (black) and the simplified-industrial SB PNC with 20 vol% of silica measured at different temperatures. Lines are fits using a KWW function with $\beta = 0.55$. **b** Relaxation map with the characteristic average times for segmental relaxation from NSE data (q = 0.54 $Å^{-1}$) and BDS for the simplified-industrial and colloidal PNCs (same filler content) and the pure SB polymer. For the industrial PNC, the same shape parameter as for the matrix has been used in the analysis of the BDS spectra for consistency with NSE. The Vogel–Fulcher–Tammann (VFT) fit of the BDS times for the pure polymer matrix is represented by a dashed line. Adapted with permission from Ref. [32] Copyright 2020 American Chemical Society

practical reasons, in particular cost issues for large samples. NSE can also be used with hydrogenated samples to probe incoherent scattering, though the signal is strongly reduced in this case [79]. In our study, we have chosen the relevant q-range in order to avoid coherent scattering of the filler. We have then been able to demonstrate that incoherent NSE can be used to access small changes of segmental dynamics of simplified-industrial PNCs. The self-dynamics of the polymer chains in presence of the filler is shown in Fig. 18a where it is compared to the neat polymer. Due to the chosen q-vector, the spatial scale of these modifications is approximately 1 nm, and as they are caused by the NPs, they must take place close to the filler interface. As a result, our high-resolution NSE measurements show for the first time that the segmental motion is slightly but systematically slowed-down by the presence of the industrial filler NPs, without modification of the width of the time distribution in the high T-range well above T_g. Moreover, this is observed at all temperatures studied here (100–160 °C), and these results were included to the relaxation map in Fig. 18b. The NSE-relaxation times are in the continuation of the dielectric ones of both model and industrial systems, in a domain inaccessible for BDS as discussed above. They show that even in a strongly aggregated system, the shift in relaxation time is small—but nonetheless always present at high T.

4.3.5 Studying in-situ Adsorption of Small Molecules Onto Filler Particles by Dielectric Spectroscopy

We have seen in Sect. 4.3.1 that additives in the polymer matrix possibly modify its dynamical properties and that this can be probed by BDS. In a recent approach, we have made use of this effect in order to study the partition of silane coating molecules between the silica filler and the SB polymer matrix in a model PNC [80]. The idea behind this is that coating agents are part of the standard formulations of industrial PNCs, but it is difficult to know if they really go only to the filler interface. As we have seen in the NSE section above, it is challenging to isolate by BDS the segmental relaxation signal in the more complex industrial PNCs, and therefore, we have chosen to investigate the model system. Using the plasticization effect of the coating agents blended with the unfilled SB polymer, we have first established a BDS-calibration curve shown in Fig. 19a. This calibration is based on the shift of the α-process toward higher frequency when increasing the amount of coating agent in the unfilled matrix. In presence of silica in a PNC sample without grafting of the silanes, i.e., where silanes are free to go either in the bulk or at the silica surface, it is, thus, possible to determine the mass fraction of coating agent dispersed in the matrix, ζ_{free}, via its impact on the position of the dielectric loss peak. Then,

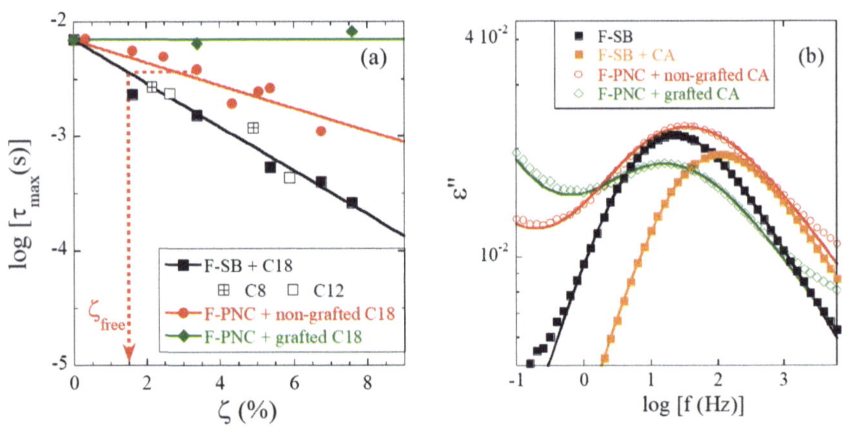

Fig. 19 **a** BDS segmental relaxation time at T = 258 K as a function of the nominal silane matrix concentration ζ in F-SB. Calibration curve for a polymer with various amounts of silane, C_8 (crossed empty squares), C_{12} (empty squares), and C_{18} (plain squares), in F-SB. The solid black line represents a linear fit to all silanes in log representation. The data for PNCs without (resp. with) activation of the grafting reaction are superimposed (red circles, resp. green diamonds) for nominal ζ in samples, together with a linear fit for the non-grafted PNCs. The construction (red dotted lines) illustrates the determination of the effective ζ_{free} within the matrix in presence of NPs. **b** Dielectric loss spectra at T = 258 K of F-SB (without and with C_{18} at $\zeta = 3.4$ %) and F-PNCs ($\Phi_{NP} \approx 10$ vol%, same amount of C_{18}). Possible silane grafting is indicated in the legend. Solid lines represent the fit to the experimental data based on Eq. (3) plus a purely dissipative dc-conductivity term. Adapted with permission from Ref. [80] Copyright 2020 American Chemical Society

by mass conservation, one can conclude on the quantity of coating agent actually covering the silica surface by adsorption. In Fig. 19b, the dielectric response for PNC systems is shown for different options of activation of grafting of the coating agents in PNCs. They are compared to the neat polymer (without any NPs) in the presence or absence of a coating agent. The result is that the loss peak for a PNC without grafting of the silanes is located in between those of the pure SB matrix blended with an equivalent amount of silanes and the same PNC where grafting has been activated (i.e., no free silane in the bulk). This demonstrates the partitioning of the coating agent between the functionalized matrix (via chemical compatibility) and the NP interfaces. A similar phenomenon is expected in industrial rubber PNCs in the first stage of mixing, before grafting reactions are activated. As a final result, we have determined the adsorption isotherm of the coating agent onto the filler within the PNC (not shown, [80]). The isotherm tells us that the maximum adsorbable amount is reached for the lowest silane concentration used in formulations, which also corresponds to industrial standards. It is concluded that this concentration must have resulted from optimization of properties of the nanocomposites while using only a minimal amount of agents.

5 Conclusion

We have reviewed recent studies of the filler structure and the polymer dynamics of nanocomposites of industrial relevance, including segmental relaxation and interfacial polarization processes at the polymer–nanoparticle interface. Structural studies have been shown to have progressed to a large extent through the quantitative analysis of the small-angle scattering, even for complex industrial filler nanoparticles with multi-scale structures. Interfacial polarization processes are not always visible in dielectric spectroscopy, as they depend on the large-scale filler structure of the samples (MWS2) or the water content (MWS1) and, thus, on details of the formulation. If MWS processes are dominant, which is typically the case for large filler assemblies or percolating networks, they may impact the analysis of segmental dynamics. While there has been a lively discussion in the literature about the effect of nanoparticle interfaces on the segmental dynamics of the polymer in close vicinity, we have chosen to broaden the view and include the effect of additives and blending. In presence of nanoparticles, several recent contributions including nano-dielectric measurements and neutron spin echo spectroscopy seem to indicate that the modification of the dynamical response of the polymer is only weakly affected in rubbery systems, in particular silica-SBR: in such elastomeric systems of weak polymer–particle interaction, there is a small but nonetheless detectable shift of the whole relaxation time distribution of segmental dynamics in the high T-range, in agreement with NMR studies. Our understanding of the weak interfacial layer in silica-based rubber system has, thus, made a few steps forward after patiently seeking and assembling information on the microstructure of such systems, characterizing the influence of additives, and finally focusing on segmental dynamics with combinations of dielectric and neutron scattering techniques. Weakly interacting rubber PNCs may be opposed to systems with a strong attraction like poly(2-vinyl pyridine) (P2VP) mixed with silica—reviewed by key players in the field in this collection [81]. In such systems, a distinct interfacial mobility extending over several nanometers is resolved by BDS. This second process is attributed to the slowed-down dynamics of the interfacial layer, and it shows a peculiar chain-mass dependence, which suggests that it is related to chain packing properties close to the interface [82]. On the other hand, bulk mobility is found not to be impacted. Interestingly, a measurable static interfacial layer formed by out-of-equilibrium chain packing leading to a lower density could be evidenced by SAXS [83]. It is tempting to identify the static and the dynamic interfacial layers in attractive PNCs, but many more questions need to be addressed beforehand: while the spatial extension of both layers is in the nanometer range, it is unclear why the decrease in local density as described by reverse Monte Carlo is accompanied by a slowing-down of this layer. In this context, the idea of chain stretching in the interfacial region has been put forward [81]. It is also consistent with measurable enhancement of mechanical moduli in the interfacial layer even in the glassy state. A balance between chain stretching and density variations may, thus, control the interfacial layer properties. There remain many open questions to be answered in the field of glassy or rubbery polymer nanocomposites, and it is concluded that the field of segmental dynamics in complex polymer–nanoparticle systems will stay a fascinating challenge for yet some years to come.

Acknowledgements Access to experimental installations and chemicals by Michelin, Solvay, and Synthos is gratefully acknowledged. Support by the European Soft Matter Infrastructure (EUSMI) for conducting the BDS experiments in San Sebastian is highly appreciated. Scattering experiments would not have been possible without powerful beamlines and support at large-scale facilities (ESRF, ILL, and Soleil).

References

1. Jancar J, Douglas JF, Starr FW, Kumar SK, Cassagnau P, Lesser AJ, Sternstein SS, Buehler MJ (2010) Polymer 51:3321–3343
2. Heinrich G, Klüppel M, Vilgis TA (2002) Curr Opin Solid State Mater Sci 6:195–203
3. Chazeau L, Brown JD, Yanyo LC, Sternstein SS (2000) Polym Compos 21:202–222
4. Payne AR (1962) J Appl Polym Sci 6:57–63
5. Mark JE, Erman B, Eirich FR (1994) Science and technology of rubber. Academic Press, San Diego
6. Banc A, Genix A-C, Dupas C, Sztucki M, Schweins R, Appavou MS, Oberdisse J (2015) Macromolecules 48:6596–6605
7. Glomann T, Schneider GJ, Allgaier J, Radulescu A, Lohstroh W, Farago B, Richter D (2013) Phys Rev Lett 110:178001
8. Papon A, Montes H, Hanafi M, Lequeux F, Guy L, Saalwächter K (2012) Phys Rev Lett 108:065702
9. Stephen R, Thomas S (2010) Rubber nanocomposites: preparation, properties, and applications. Wiley
10. Wolff S (1996) Rubber Chem Technol 69:325–346

11. Musino D, Genix A-C, Fayolle C, Papon A, Guy L, Meissner N, Kozak R, Weda P, Bizien T, Chaussée T, Oberdisse J (2017) Macromolecules 50:5138–5145
12. Baeza GP, Genix A-C, Degrandcourt C, Petitjean L, Gummel J, Couty M, Oberdisse J (2013) Macromolecules 46:317–329
13. Genix A-C, Baeza GP, Oberdisse J (2016) Eur Polym J 85:605–619
14. Guy L, Daudey S, Bomal Y (2009) Kautsch Gummi Kunstst 62:383–391
15. Baeza GP, Genix A-C, Paupy-Peyronnet N, Degrandcourt C, Couty M, Oberdisse J (2016) Farad Discuss 186:295–309
16. Baeza GP, Genix A-C, Degrandcourt C, Petitjean L, Gummel J, Schweins R, Couty M, Oberdisse J (2013) Macromolecules 46:6388–6394
17. Baeza GP, Genix A-C, Degrandcourt C, Gummel J, Couty M, Oberdisse J (2014) Soft Matter 10:6686–6695
18. Genix A-C, Oberdisse J (2017) Soft Matter 13:8144–8155
19. Larsen AH, Pedersen JS, Arleth L (2020) J Appl Cryst 53:991–1005
20. Beaucage G (2012) Polymer science: a comprehensive reference, vol 2. In: Möller MA (ed) Elsevier BV: Amsterdam, pp 399–409
21. Beaucage G, Ulibarri TA, Black EP, Schaefer DW (1995) Hybrid organic-inorganic composites. Am Chem Soc 585:97–111
22. Lindner P, Zemb T (2002) Neutrons, X-ray and light scattering. North Holland, Elsevier, Amsterdam
23. Percus JK, Yevick GJ (1958) Phys Rev 110:1–13
24. Ashcroft NW, Langreth DC (1967) Phys Rev 156:685–692
25. Baeza GP, Genix A-C, Degrandcourt C, Gummel J, Mujtaba A, Saalwächter K, Thurn-Albrecht T, Couty M, Oberdisse J (2014) ACS Macro Lett 3:448–452
26. Boonsomwong K, Genix A-C, Chauveau E, Fromental J-M, Dieudonné-George P, Sirisinha C, Oberdisse J (2020) Polymer 189:122168
27. Cattinari G, Steenkeste K, Le Bris C, Canette A, Gallopin M, Couty M, Fontaine-Aupart M-P (2021) J Appl Polym Sci 138:50221
28. McGreevy RL (2001) J Phys Condens Matter 13:R877–R913
29. McGreevy RL, Pusztai L (1988) Mol Simul 1:359–367
30. Oberdisse J, Hine P, Pyckhout-Hintzen W (2007) Soft Matter 2:476–485
31. Musino D, Genix A-C, Chauveau E, Bizien T, Oberdisse J (2020) Nanoscale 12:3907–3915
32. Musino D, Oberdisse J, Farago B, Alegria A, Genix A-C (2020) ACS Macro Lett 9:910–916
33. Cerveny S, Bergman R, Schwartz GA, Jacobsson P (2002) Macromolecules 35:4337–4342
34. Kremer F, Schönhals A (2003) Broadband Dielectric Spectroscopy. Springer-Verlag, Heidelberg
35. Otegui J, Schwartz GA, Cerveny S, Colmenero J, Loichen J, Westermann S (2013) Macromolecules 46:2407–2416
36. Schwartz GA, Ortega L, Meyer M, Isitman NA, Sill C, Westermann S, Cerveny S (2018) Macromolecules 51:1741–1747
37. Souillard C, Chazeau L, Cavaillé J-Y, Brun S, Schach R (2019) Polymer 168:236–245
38. Zhuravlev LT (2000) Colloids Surf A Physicochem Eng Asp 173:1–38
39. Cerveny S, Schwartz GA, Otegui J, Colmenero J, Loichen J, Westermann S (2012) J Phys Chem C 116:24340–24349
40. Meier JG, Fritzsche J, Guy L, Bomal Y, Kluppel M (2009) Macromolecules 42:2127–2134
41. Tsangaris GM, Psarras GC, Kouloumbi N (1998) J Mater Sci 33:2027–2037
42. Baeza GP, Oberdisse J, Alegria A, Couty M, Genix A-C (2015) Phys Chem Chem Phys 17:1660–1666
43. Baeza GP, Oberdisse J, Alegria A, Saalwächter K, Couty M, Genix A-C (2015) Polymer 73:131–138
44. Hernández M, Carretero-González J, Verdejo R, Ezquerra TA, López-Manchado MA (2010) Macromolecules 43:643–651
45. Steeman PAM, Maurer FHJ, Vanes MA (1991) Polymer 32:523–530
46. Steeman PAM, Baetsen JFH, Maurer FHJ (1992) Polym Eng Sci 32:351–356

47. Huang M, Tunnicliffe LB, Zhuang J, Ren W, Yan H, Busfield JJC (2016) Macromolecules 49:2339–2347
48. Roland CM (2016) Rubber Chem Technol 89:32–53
49. Le HH, Qamer Z, Ilisch S, Radusch H-J (2006) Rubber Chem Technol 79:621–630
50. Meier JG, Kluppel M (2008) Macromol Mater Eng 293:12–38
51. Böhning M, Goering H, Fritz A, Brzezinka KW, Turky G, Schönhals A, Schartel B (2005) Macromolecules 38:2764–2774
52. Ortega L, Cerveny S, Sill C, Isitman NA, Rodriguez-Garraza AL, Meyer M, Westermann S, Schwartz GA (2019) Polymer 172:205–212
53. Bocharov VN, Bureiko SF, Koll A, Rospenk M (1998) J Struct Chem 39:502–507
54. Ortiz-Serna P, Díaz-Calleja R, Sanchis MJ, Floudas G, Nunes RC, Martins AF, Visconte LL (2010) Macromolecules 43:5094–5102
55. Hernández M, Ezquerra TA, Verdejo R, López-Manchado MA (2012) Macromolecules 45:1070–1075
56. Ikeda Y, Higashitani N, Hijikata K, Kokubo Y, Morita Y, Shibayama M, Osaka N, Suzuki T, Endo H, Kohjiya S (2009) Macromolecules 42:2741–2748
57. Lindemann N, Schawe JEK, Lacayo-Pineda J (2021) J Appl Polym Sci 138:49769
58. Colmenero J, Arbe A (2007) Soft Matter 3:1474–1485
59. Genix A-C, Arbe A, Alvarez F, Colmenero J, Willner L, Richter D (2005) Phys Rev E72
60. Hoffmann S, Willner L, Richter D, Arbe A, Colmenero J, Farago B (2000) Phys Rev Lett 85:772–775
61. Lodge TP, McLeish TCB (2000) Macromolecules 33:5278–5284
62. Adam G, Gibbs JH (1965) J Chem Phys 43:139–146
63. Cangialosi D, Schwartz GA, Alegría A, Colmenero J (2005) J Chem Phys 123:144908
64. Rathi A, Hernández M, Garcia SJ, Dierkes WK, Noordermeer JWM, Bergmann C, Trimbach J, Blume A (2018) J Polym Sci, Part B: Polym Phys 56:842–854
65. Arrighi V, McEwen IJ, Qian H, Prieto MBS (2003) Polymer 44:6259–6266
66. Tsagaropoulos G, Eisenberg A (1995) Macromolecules 28:6067–6077
67. Mele P, Marceau S, Brown D, de Puydt Y, Alberola ND (2002) Polymer 43:5577–5586
68. Robertson CG, Lin CJ, Rackaitis M, Roland CM (2008) Macromolecules 41:2727–2731
69. Pissis P, Fragiadakis D, Kanapitsas A, Delides K (2008) Macromol Symp 265:12–20
70. Fragiadakis D, Bokobza L, Pissis P (2011) Polymer 52:3175–3182
71. Robertson CG, Rackaitis M (2011) Macromolecules 44:1177–1181
72. Warasitthinon N, Robertson CG (2018) Rubber Chem Technol 91:577–594
73. Kummali MM, Miccio LA, Schwartz GA, Alegría A, Colmenero J, Otegui J, Petzold A, Westermann S (2013) Polymer 54:4980–4986
74. Papon A, Montes H, Lequeux F, Oberdisse J, Saalwächter K, Guy L (2012) Soft Matter 8:4090–4096
75. Alvarez F, Alegra A, Colmenero J (1991) Phys Rev B 44:7306–7312
76. Alvarez F, Alegría A, Colmenero J (1993) Phys Rev B 47:125–130
77. Mujtaba A, Keller M, Ilisch S, Radusch HJ, Beiner M, Thurn-Albrecht T, Saalwächter K (2014) ACS Macro Lett 3:481–485
78. Papon A, Saalwächter K, Schäler K, Guy L, Lequeux F, Montes H (2011) Macromolecules 44:913–922
79. Richter D, Monkenbusch M, Arbe A, Colmenero J (2005) Neutron spin echo in polymer systems, vol 174. Springer, Berlin, pp 1–221
80. Musino D, Oberdisse J, Sztucki M, Alegria A, Genix A-C (2020) Macromolecules 53:8083–8094
81. Cheng S, Carroll B, Bocharova V, Carrillo J-M, Sumpter BG, Sokolov AP (2017) J Chem Phys 146:203201
82. Genix A-C, Bocharova V, Kisliuk A, Carroll B, Zhao S, Oberdisse J, Sokolov AP (2018) ACS Appl Mater Interfaces 10:33601–33610
83. Genix A-C, Bocharova V, Carroll B, Lehmann M, Saito T, Krueger S, He L, Dieudonné-George P, Sokolov AP, Oberdisse J (2019) ACS Appl Mater Interfaces 11:17863–17872

Filler Networks of Carbon Allotropes of Different Shapes and Dimensions in a Polymer Matrix

Dielectric and Rheo-Electric Investigations

Ingo Alig, Konrad Hilarius, Dirk Lellinger, and Petra Pötschke

Abstract Dielectric permittivity and electrical conductivity of percolating filler networks of different carbon allotropes such as carbon black (CB), multi-walled carbon nanotubes (MWCNT) , and graphite nanoplates (GNP) dispersed in a polycarbonate (PC) matrix were studied by broadband dielectric spectroscopy (BDS) and rheo-electrical measurements. CB, MWCNT, and GNP represent carbon allotropes of different shapes and dimensionalities, which build electrical conductive network structures within the polymer matrix. In a first part, the dielectric properties of solid samples with different CB, MWCNT and GNP contents were studied by BDS. The concentration-dependent dielectric properties are compared and discussed with regard to the particle shape. Since the structure of the filler network is extremely sensitive to deformations of the melt, the second part deals with rheo-electical measurements on composite melts. Under steady shear conditions, an interplay between build-up due to attractive interactions between the particles, and break-up of the network structures indicated by steady-state values of DC conductivity and viscosity can be observed. The increase of the DC conductivity in the quiescent melt after shear deformation is due to a recovery of the filler network. In order to study the orientation kinetics of the particles, additional rheo-electric measurements perpendicular to the shear flow and in flow direction were performed. The influence of particle shape and dimensions on the electrical, dielectric, and rheological properties are discussed in terms of agglomeration and break-up of the filler network and orientation kinetics of the particles in the polymer melt.

Keywords Carbon allotropes · Electrical conductivity · Dielectric properties · Rheology · Filler network · Nanocomposites

---

I. Alig (✉) · K. Hilarius · D. Lellinger
Fraunhofer Institute for Structural Durability and System Reliability (LBF), Darmstadt, Germany
e-mail: ingo.alig@lbf.fraunhofer.de

P. Pötschke
Leibniz Institute of Polymer Research Dresden (IPF), Dresden, Germany

A. Schönhals and P. Szymoniak (eds.), *Dynamics of Composite Materials*,
Advances in Dielectrics, https://doi.org/10.1007/978-3-030-89723-9_10

Abbreviations

b and d	Length along the axis of symmetry of an oblate spheroid and its diameter perpendicular to this axis
BDS	Broadband dielectric spectroscopy
CB	Carbon black
CNT	Carbon nanotubes
d and L	Diameter and length of a cylindrical particle
D_{trans}, D_{rot}	Diffusion coefficient for translation and rotation
DC	Direct current
$G^* = G' + iG''$	Complex shear modulus
GNP	Graphite nanoplates
k_B	Boltzmann constant
k_o	Kinetic coefficient for agglomeration in a quiescent melt
k_1 and k_2	Kinetic coefficients for build-up and destruction of agglomerates in a sheared melt
l	Lateral dimension of a particle, which for plates is equivalent to the plate diameter b, and for cylinders to the length L
MWCNT	Multi-walled carbon nanotubes
n	Exponent
p_a	Agglomerate concentration
$p_{a,0}$ and $p_{a,\infty}$	Agglomerate concentration at $t = 0$ and at long times ($t \to \infty$)
p'	Inverse of the aspect ratio
p_c	Concentration at percolation threshold
Pe_{trans} and Pe_{rot}	Péclet number for translational and rotational diffusion
PC	Polycarbonate
$q = b/d$	Axis ratio of an oblate spheroid
r	Radius of a pressed plate
R_S	Radius of the spherical particle
s, s' and t'	Critical exponents
t	Time
T	Absolute temperature
v, H and z	Pressing speed, sample height before and after pressing, respectively
β and ν	Critical exponents
η_0	Zero-shear viscosity
η_{stat}	Stationary value of the shear viscosity
$\eta^{T}(t)$	Transient viscosity
$\tau(t)$	Time-dependent shear stress
$\gamma(t) = \dot{\gamma}t$	Total shear
$\dot{\gamma}$	Shear rate
$\varepsilon^* = \varepsilon' - i\varepsilon''$	Complex relative dielectric permittivity

ε_0	Vacuum permittivity
$\varepsilon'_s = \varepsilon'(\omega \to 0)$	Static relative dielectric permittivity
$\sigma_{DC} = \sigma'(\omega \to 0)$	Direct current conductivity
$\sigma^* = \sigma' + i\sigma''$	Complex electrical conductivity
$\sigma_{stat} = \sigma_{DC}(t \to \infty)$	Stationary value of the electrical conductivity
Ψ	Average angle of orientation with respect to the flow direction
$\omega = 2\pi f$	Angular frequency
$\omega_c = 2\pi f_c$	Crossover angular frequency

1 Introduction

Solid-state properties and rheological behavior of polymers containing micro- or nanofillers are known to be sensitive to the size, shape, volume fraction, dimensionality, surface modification of the particles, and their distribution within the filler network (see e.g., [1, 2]). The arrangement of the fillers within the polymer composite is determined by the filler–filler and filler–polymer interactions as well as translational and rotational diffusion within the polymer melt. The characteristic times for formation or rearrangement of the filler network are closely linked to the relaxation time spectrum of the polymer melt. Accordingly, structure and dynamics of the filler network not only depend on size, shape, and volume fraction of the particles and the interparticle forces, but they are also strongly dependent on the thermo-rheological or processing prehistory.

In 1996, Schüler et al. [3, 4] found that the agglomeration of carbon black (CB) in an epoxy matrix has a tremendous influence on the final electrical conductivity. For clay particles in a thermoplastic melt under quiescent conditions, Heinrich et al. [5] reported in 2005 on the influence of particle agglomeration or flocculation on the dynamic shear modulus. Agglomeration and structural reorganization and their dependence on thermo-rheological or processing history have been probed by several authors for polymer–carbon nanotubes [6–17] or polymer–clay composites (see e.g., [5, 18] and references therein) by means of electric, dielectric, rheological , and rheo-electrical measurements. Due to the close coupling to the viscoelastic properties of the polymer melt, the arrangement of the nanofillers is extremely sensitive to mechanical deformations, such as shear or elongational flow. Vice versa there is a feedback of the generated filler network on the viscoelastic properties of the composite melt. Therefore, rheological methods are well suited for investigating the filler dispersion (see e.g., [1, 2, 18]).

In the case of anisometric particles such as rods or platelets, both, the positions of the particles with respect to each other *and* their orientation determine the properties and rheological behavior of polymer composites [1, 18, 19]. The orientation of anisometric particles in shear or elongational flow typically results in anisotropic properties such as electrical or thermal conductivity. In elongated blend structures, containing

CB in one of the phases, large differences in the electrical conductivity perpendicular and parallel to the melt flow direction were found by broadband dielectric spectroscopy (BDS) [20].

Above a certain concentration, the so-called percolation concentration (p_c), the filler particles form a space-filling network. For conductive filler particles, this results in conductive paths through the polymer matrix. Since the electrical conductivity changes at the percolation threshold by orders of magnitude, this provides an extremely sensitive tool to probe the structure of the filler network by using DC electrical conductivity measurements or BDS. By combining a rheometer (e.g. plate–plate geometry) with simultaneous measurements of the electrical properties, it is possible to study changes in the structure of the filler network for different filler particles under different deformation modes and in the quiescent melt. For carbon nanotubes in polymer melts the interplay between destruction and build-up of the filler network under steady shear and in the quiescent melt has been studied in combined rheological and electrical conductivity measurements (see e.g., [6–17] as well as rheo-optical measurements [15].

Over the years, nearly all types of nanofillers have been used for the preparation of so-called polymer nanocomposites. Examples are organoclays, carbon black, carbon nanotubes (CNT), graphite oxide and graphite nanoplates (GNPs), fumed and colloid silica, cellulose whiskers, metallic oxide, etc. Nanocomposites containing carbon allotropes such as CB [17, 21–23], CNT [6–17], GNP [22–25], and their mixtures [17, 22] have been studied by time-resolved rheo-electrical experiments using different shear flow protocols.

It is the aim of this chapter to review some of our work on filler networks of carbon allotropes of different shapes and dimensions in a polymer matrix by the use of dielectric and rheo-electrical measurements including new results on GNP [6–16] and more recent direction-dependent conductivity measurements. In order to be comparable and due to the good compatibility to carbon allotropes, polycarbonate (PC) was chosen as the matrix polymer for all composites. The carbon allotropes studied differ significantly in their geometrical structures:

i. two carbon black types of different structures (fractality) of the aggregates with (spherical-)averaged diameters of about hundred nanometers,
ii. multi-walled carbon nanotubes (MWCNT), which resemble the shape of a cylinder, with diameters of several nanometer and a length in the micrometer range, and
iii. graphite nanoplates with a thickness of a few nanometers and a lateral size in the micrometer range.

Figure 1 shows transmission electron microscopy (TEM) images and schemes of the dominant geometric shape of filler particles together with the number of dimensions in the nanometer range ("nanoscale dimension") and the corresponding dimension of the charge transport within the particles. The fractal dimensions of the filler networks assembled by the filler particles are expected to have a dimensionality between 2 and 3, which depends on the thermomechanical pre-history.

	Carbon black		Carbon nanotubes	Graphite nanoplates
Carbon Allotropes	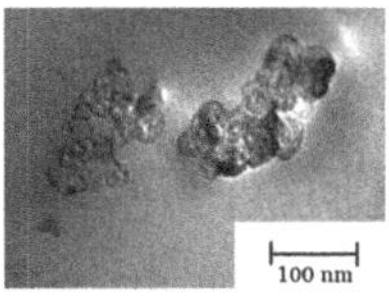		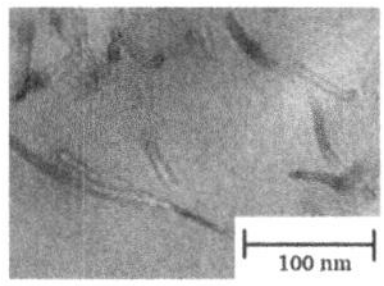	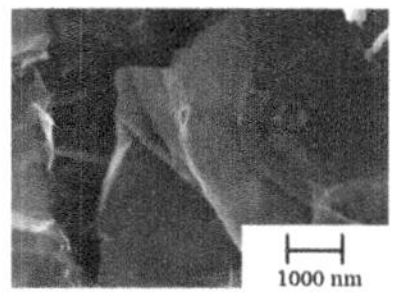
Geometrical shape	Sphere 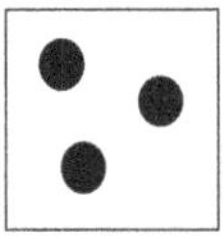	Fractal 	Cylinder 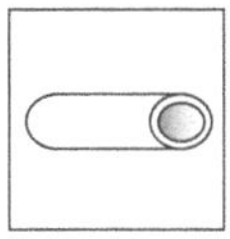	Plate 
Nanoscale dimensions of the particles	3		2	1
Dimension of charge transport in the particles	0		1	2

Fig. 1 Shape and number of dimensions in the nanoscale of different carbon allotropes

The chapter is organized as follow: After the introduction, some basics on agglomeration and orientation of nanofillers in the quiescent and sheared polymer melt are presented together with some basics of conductive filler networks. The following part deals with the materials, sample preparation as well as the dielectric and rheo-electric measurement methods. Then the test protocol for the rheo-electric experiments consisting on three time intervals for *annealing*, shear-induced *destruction*, and *recovery* of the filler network are described. The results of the measurements on the composites with different carbon allotropes are presented and discussed in terms of the influence of the filler dimensionality on structure and formation kinetics of the filler network as well as the filler orientation.

2 Polymer Nanocomposites: Particle Agglomeration and Orientation

2.1 Agglomeration

From a physical point of view, nanoparticles suspended in a fluid are exposed to particle–particle forces, particle–fluid interactions, viscous forces under flow, and Brownian forces [2]. The Brownian motion arises from a random force field representing the effect of the thermal fluctuations of the surrounding molecules on the particles. Without specific interactions, this would lead in a quiescent fluid at long times to a random distribution of the nanoparticles within the matrix. For agglomeration of nanoparticles, attractive interactions between the nanoparticles are needed to balance the tendency to randomization. In terms of phase separation between nanofillers and the macromolecules, a Flory-Huggins-type interaction parameter may be assumed. Similar to the phase separation of binary mixtures by spinodal or binodal decomposition, the segregation structures are temporary and develop with time. On the one hand, hydrodynamic forces during flow can fragment large droplets or agglomerates dispersed in a fluid, and, on the other hand, the flow field promote collisions, which are resulting in coagulation of small droplets or agglomerates. In Taylor's theory, the balance of these two effects leads to a preferred droplet size proportional to the interfacial tension between the droplet phase and the surrounding medium and inversely proportional to the shear rate, and the viscosity of the external fluid. In a generalized form, this can be transferred to the agglomeration and de-agglomeration of nanoparticles under deformation. For anisometric particles such as rods or platelets, the orientation has to be taken into account (see next section). In terms of transport processes, both, translational and rotational diffusion have to be considered.

Figure 2 show exemplarily transmission electron microscopy images of different arrangements of MWCNT (Nanocyl NC™ 7000 nanotubes from Nanocyl S.A., Sambreville, Belgium) in a PC matrix: (a) insufficiently dispersed *primary agglomerates* (1 wt%), (b) well-dispersed MWCNTs after proper melt mixing (1 wt%, random distribution, no preferred orientation) and (c) *secondary agglomerates* formed after isothermal annealing of well-dispersed MWCNT (0.825 wt%). For details of the sample preparation and prehistory see [12].

Figure 3 show the frequency dependence of the real part of the complex conductivity $\sigma'(f)$ of two samples with 0.875 wt% MWCNT in polycarbonate after melt pressing at 250 and 300 °C. The composite was prepared by master batch dilution

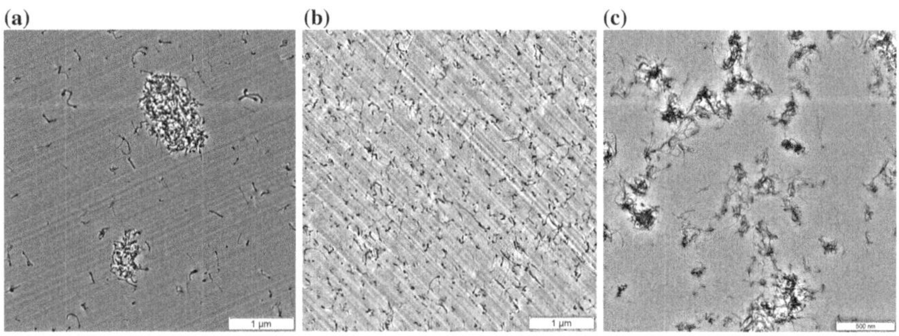

Fig. 2 Transmission electron microscopy (TEM) images of different arrangements of multi-walled carbon nanotubes in a polycarbonate matrix: **a** insufficiently dispersed *primary agglomerates,* **b** *well-dispersed* MWCNT after proper melt mixing, and **c** *secondary agglomerates* formed after annealing of well-dispersed MWCNT in polycarbonate. Adapted with permission from Elsevier from [12], in which the details of the sample preparation and pretreatment are described

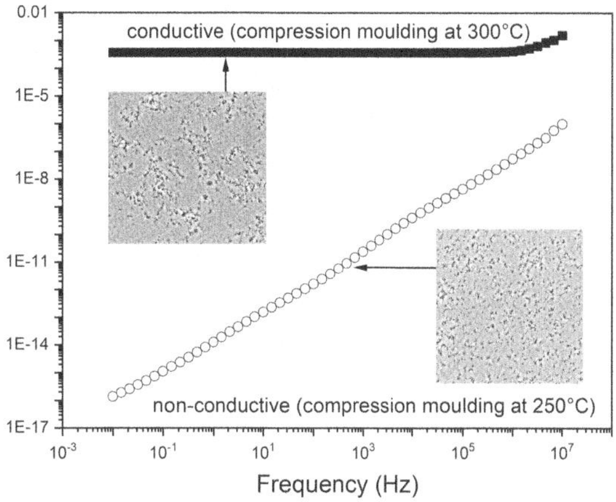

Fig. 3 **a** Frequency dependence of the real part of the complex conductivity (σ')and related TEM micrographs of compression-molded plates (0.875 wt% MWCNT), pressed at 300 °C (solid symbols) and at 250 °C (open symbols) with a pressing speed of 6.0 mm/min. The figure is adapted with permission from Elsevier from [12]. For details see there

at 250 °C. Due to the relatively high viscosity at 250 °C, the initially well-dispersed state after melt mixing is preserved during melt pressing, as indicated by the TEM image. This state of dispersion after melt mixing is expected to be frozen by vitrification. The frequency dependence of σ' and the DC conductivity (not shown) of this sample are similar to those of the neat polymer. This indicates that the nanotubes are separated and almost insulated by polymer chains. The distances between the nanotubes are expected to be above those for tunneling of electrons. The $\sigma'(f)$ curve of the sample pressed at 300 °C is several orders of magnitude above that of the sample pressed at 250 °C and shows a conductivity plateau, which is typical for an electrically conductive composite well above the percolation threshold. The TEM image (inset) of the sample pressed at 300 °C shows the formation of agglomerates. Such secondary agglomerates are expected to be formed at low viscosities (here due to the higher melt temperature) and/or longer annealing times. The pressing time at 300 °C is expected to be sufficient so that agglomeration can occur and a conductive pathway can be formed by percolation of the conductive agglomerates [6].

Applying shear or elongational flow to such agglomerated nanoparticles in a viscoelastic melt causes a break-up of the filler network into smaller agglomerates down to isolated particles (Fig. 4, upper row). On the other hand, the motion of the nanoparticles relative to each other in the deformation field promotes collisions and attractive interactions between them, and can than result in agglomeration or flocculation (Fig. 4, lower row).

In a steady flow under stationary conditions, build-up by shear-induced agglomeration and break up of agglomerates due to hydrodynamic forces are expected to be balanced and a pseudo-stationary (temporary) filler network develops. For a polycarbonate composite containing 1 wt% MWCNT this state is achieved after a total shear deformation ($\gamma = \dot{\gamma}t$) of about 40 rad [14].

Figure 5 exemplarily compares the DC conductivity during steady shear experiments with a shear rate of 0.02 s^{-1} in the melt at 230 °C of composites with initially *dispersed* and with initially *agglomerated nanotubes.* Both composites contain the

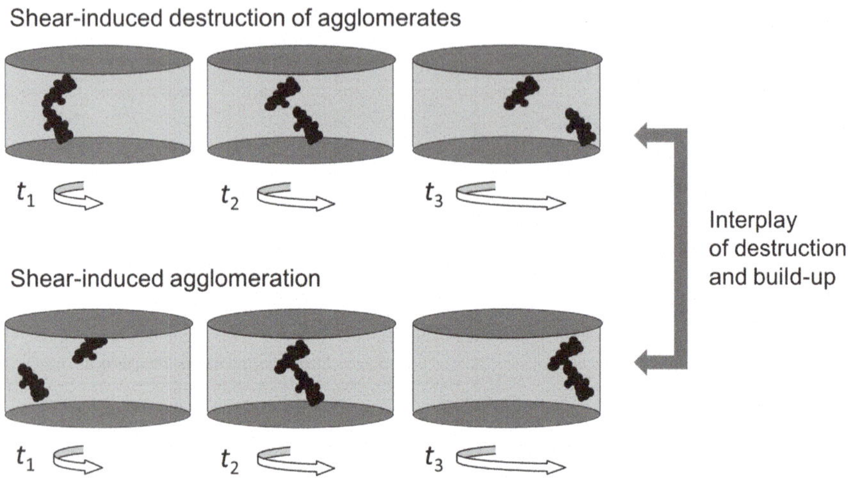

Fig. 4 Schematic representation of shear-induced agglomeration and deagglomeration of carbon black aggregates in a polymer melt under steady shear (plate–plate geometry)

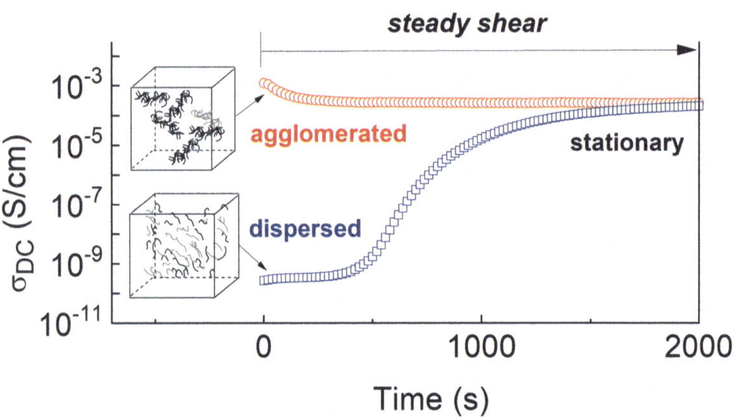

Fig. 5 Time dependence of the DC conductivity during steady shear for MWCNT composites with initially 'dispersed' (blue squares) and 'agglomerated' nanotubes (red circles). Both samples contain 1 wt% MWCNT in polycarbonate and differ only by the thermo-mechanical pre-treatment. The measurements are performed with a shear rate of 0.02 s^{-1} at 230 °C. The figure is adapted with permission from Elsevier from [14]

same amount of 1 wt% MWCNT in a polycarbonate matrix. The *agglomerated* composite was prepared by a thermo-mechanical pre-treatment as described in [14]. For description of the experimental setup see below. The DC conductivity of the sample with *dispersed* nanofillers shows an increase during steady shear, which is explained by dominance of shear-induced agglomeration as schematically shown in the lower part of Fig. 2. For the *agglomerated* sample, the DC conductivity decreases

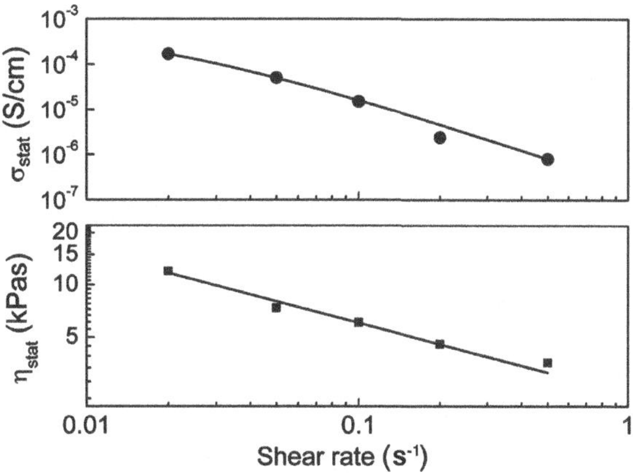

Fig. 6 Steady-state values (σ_{stat}) of the DC conductivity and viscosity (η_{stat}) after 600 s of steady shear versus shear rate (replotted with permission from Elsevier from [14])

during shear deformation, which can be explained by an at least partial destruction of the conductive paths. Since the real and imaginary parts G' and G'' of the complex shear modulus also decrease (not shown, see e.g., [14]) mechanical reinforcement undergoes a destruction as well. After a certain time, both samples regardless of the initial structure of the filler network (either 'dispersed' or 'agglomerated') approach constant values for the conductivity as well as for the rheological quantities, which indicate a stationary state of the dynamic filler network for a given shear flow. Figure 6 shows the shear rate dependence of the steady-state values of the DC conductivity and the viscosity. The DC conductivity changes by more than two orders of magnitude, whereas the viscosity changes only by a factor of about four. This indicates that the mechanisms of charge transport and reinforcement within a filler networks are quite different. However, the tendency that the DC conductivity and viscosity approach constant values under steady flow conditions supports the assumption that a *dynamic filler network* is formed by the interplay of shear-induced agglomeration due to attractive interactions between the nanoparticles and the destruction of the agglomerates in the shear field.

The increase in the agglomerate concentration in the quiescent melt can be described with a kinetic equation of nth order (see e.g., [5, 9, 10, 14, 16]):

$$\frac{dp_a(t)}{dt} = k_1\left(p_{a,\infty} - p_a(t)\right)^n, \qquad (1)$$

in which k_1 is the kinetic coefficient of the build-up in the agglomeration process, t the time, $p_a(t)$ is the agglomerate concentration, and $p_{a,\infty}$ is the agglomeration concentration at long time ($t \to \infty$). Measurements of the DC conductivity or the elastic modulus of thermoplastics with nanofiller [5, 9, 10, 16] or uncured rubber with carbon black [26] during agglomeration show that the agglomeration kinetics can be described within the error limits alternatively by first- or second-order kinetics. For agglomeration kinetics of second order, it is assumed that two filler particles combine to form a new agglomerate, whereas first-order kinetics is rather related to the attachment of a filler particle to an existing agglomerate. For sake of simplicity, we use the first-order description in the following. The related solution for agglomeration of first order is

$$p_a(t) = p_{a,0} + (p_{a,\infty} - p_{a,0})[1 - exp(-k_1 t)] \quad (2)$$

where $p_{a,0}$ is the agglomerate concentration at time t = 0. Agglomeration in the quiescent melt is assumed to be controlled by particle transport due to Brownian diffusion of the particles over a distance, which is needed to form a contact with another particle or agglomerate. If the particles (including aggregates and agglomerates) come close enough (within a "capture length"), attractive van der Waals forces pull them to an equilibrium distance, where they bond to each other.

As already discussed above, shear flow brings particles closer together, which then can agglomerate due to the attractive interaction. If, on the other hand, the shear forces are sufficiently high to destroy the existing agglomerates again, the agglomerate concentration decrease. The competition between shear-induced agglomeration and break-up can be described by a differential equation with an additional destruction term [10, 14]:

$$\frac{dp_a}{dt} = k_1(\dot{\gamma})(p_{A\infty} - p_A) - k_2(\dot{\gamma})p_A \quad (3)$$

where $k_1(\dot{\gamma})$, and $k_2(\dot{\gamma})$ are the kinetic coefficients for the build-up and destruction of the agglomerates. For a constant shear rate $\dot{\gamma}$, a dynamic equilibrium is established between the build-up and destruction. The solution of the differential Eq. (3) can be incorporated in an equation for electrical percolation or an effective medium model to describe the time evolution of the DC conductivity as a function of the agglomerate concentration during the shear. The balance between build-up and destruction then leads to a constant conductivity. Its steady-state value depends on the applied shear rate. Under steady shear conditions, this results in a steady-state value of the agglomerate concentration in the dynamic equilibrium:

$$p_{a,steady} = p_{a,\infty}\frac{k_1}{k_1 + k_2}. \quad (4)$$

However, the agglomeration model provides only an incomplete description, since only attachment and break-up of individual filler particles are included. The contribution of agglomerates of higher order and their size distribution is not included. Therefore, in [10] a hierarchical model is proposed, which takes the formation of agglomerates of different sizes by a set of differential equations into account [10, 14].

For the description of the DC conductivity, the solution of Eq. (2) for $p_a(t)$ has to be coupled with an equation for electrical percolation or an expression based on effective medium approach (see e.g., [16] and references).

Near the percolation threshold, the DC conductivity at constant current σ_{DC} of a composite with conductive fillers follows power laws above and below the percolation concentration p_c [27–32]:

$$\sigma_{DC} = \sigma(p) \propto \left\{ \begin{array}{l} \left(\frac{p_c - p}{p_c}\right)^{-s}, p < p_c \\ \left(\frac{p - p_c}{1 - p_c}\right)^{t}, p > p_c \end{array} \right\} \tag{5}$$

The usually accepted values of the critical exponents s and t [33–36] are $s = t \approx 1.3$ for two dimensions and $s \approx 0.73$, $t \approx 2.0$ for three dimensions (see e.g. [37] and references therein).

To study charge carrier diffusion in fractal conductive filler networks, both, the complex electrical conductivity $\sigma^*(p, \omega)$ and complex dielectric permittivity $\varepsilon^*(p, \omega) = \varepsilon'(p, \omega) - i\varepsilon''(p, \omega)$ can be used, with p being the content of the conductive filler and $\omega = 2\pi f$ the angular frequency (see e.g., [37] and references therein). The composition dependence of the static permittivity $\varepsilon'_s = \varepsilon'(\omega \to 0)$ near the percolation threshold p_c can be described for both cases, $p < p_c$ and $p > p_c$ by [31]:

$$\varepsilon'_S \sim |p - p_c|^{-s'}. \tag{6}$$

The critical exponent s' is defined by $s' = 2\beta - \nu$, where β is the percolation exponent, which characterizes the probability that a site belongs to the infinite cluster (for $p > p_c$), and ν is the exponent of the concentration dependence of the correlation length of the clusters. The numerical value for this critical exponent for 3D has been found to be $s' \cong 1.33$ [34–36], which agrees with experiments on MWCNT composites in [37] for $p < p_c$.

In a system in which conductive agglomerates form the percolating network, the filler content p in Eq. (6) has to be replaced by the time-dependent content of conductive agglomerates $p_a(t)$ and the concentration p_c at the percolation threshold of the 'percolating' conductive agglomerates. This has been assumed to be the case for CB and MWCNT composites (see e.g., [8–17]).

To describe the time-dependent insulator–conductor transition (sometimes mentioned as dynamic percolation), the solution of the differential equations for the concentration of conductive agglomerates $p_a(t)$, e.g., (1) or (3) can be combined with the effective medium theory [38–40] to calculate the time-dependent conductivity of the composite without any singularity:

$$(1 - p_a(t)) \frac{\sigma_m^{\frac{1}{s}} - \sigma_{DC}^{\frac{1}{s}}(t)}{\sigma_m^{\frac{1}{s}} + \frac{1 - p_{c,a}}{p_{c,a}} \sigma_{DC}^{\frac{1}{s}}(t)} + p_a(t) \frac{\sigma_a^{\frac{1}{\mu}} - \sigma_{DC}^{\frac{1}{\mu}}(t)}{\sigma_a^{\frac{1}{\mu}} + \frac{1 - p_{c,a}}{p_{c,a}} \sigma_{DC}^{\frac{1}{\mu}}(t)} = 0, \tag{7}$$

where σ_m and σ_a are the DC conductivities of the polymer matrix and the agglomerate $p_{c,a}$ is the percolation concentration of the agglomerates, s (for p $< p_{c,a}$) and μ

(for p > $p_{c,a}$) are the critical exponents and $\sigma_{DC}(t)$ is the conductivity at the agglomerate concentration $p_a(t)$. The critical exponents μ and s can be assumed to be 2 and 0.73, respectively, which are typical values for 3D filler systems (see [37] and references). Alternatively, Fournier et al. [41] proposed an empirical equation to describe the time-dependent conductivity at a percolation transition. It has been shown that both equations describe the behavior of MWCNT systems in polycarbonate in a comparable way [11].

2.2 Particle Orientation

For anisometric particles, the geometric shape of the diffusing particles has to be included into the description of the Brownian motion of the nanoparticles [42–48]. For the description of translational and rotational diffusion of the nanoparticles studied some approximations related to their shape have to be made. Neglecting their fractal structure, the carbon black agglomerates can be approximated by spheres. For a sphere, the diffusion coefficient for translation D_{trans} and rotation D_{rot} in a viscous medium are given by [42, 43]:

$$D_{trans}^{Sphere} = \frac{k_B T}{6\pi \eta_0 R_S}, \; D_{rot}^{Sphere} = \frac{k_B T}{8\pi \eta_0 R_S^3}. \tag{8}$$

R_S is the radius of the sphere embedded in a fluid with the zero-shear viscosity η_0, k_B the Boltzmann constant, and T the absolute temperature.

For nanotubes (CNT) the shape of a cylinder can be assumed. The translational and rotational diffusion coefficients of cylinders are [47, 48]

$$D_{trans}^{Cylinder} = \frac{k_B T}{6\pi \eta_0 \left(\frac{3}{2p'^2}\right)^{\frac{1}{3}} \left(\frac{L}{2}\right)} \cdot \frac{\ln(p' + \gamma')}{\left(\frac{2p'^2}{3}\right)^{\frac{1}{3}}} \tag{9}$$

and

$$D_{rot}^{Cylinder} = \frac{k_B T}{8\pi \eta_0 \left(\frac{3}{2p'^2}\right) \left(\frac{L}{2}\right)^3} \cdot \frac{9\ln(p + \delta)}{2p'^2}$$

with

$$p' = \frac{d}{L}, \delta \approx -0.662 + \frac{0.917}{p'} - \frac{0.05}{p'^2}$$

and

$$\gamma' \approx 0.312 + \frac{0.565}{p'} + \frac{0.100}{p'^2}$$

L is the length and d the diameter of the cylinder, p' is the inverse of the aspect ratio.

The shape of a graphite nanoplate (GNP) can be approximated by an oblate spheroid. The translational and rotational diffusion coefficients [44–46] are given by

$$D_{trans}^{Oblate} = \frac{k_B T}{6\pi\eta_0 \left(\mathrm{db}^2\right)^{\frac{1}{3}}} \cdot \frac{q^{\frac{2}{3}} \arctan\left(\sqrt{q^2 - 1}\right)}{\sqrt{q^2 - 1}}, \tag{10}$$

and

$$D_{rot}^{Oblate} = \frac{k_B T}{8\pi\eta_0 \left(\mathrm{db}^2\right)} \cdot \frac{3q^2\left(2 - q^2\right)}{2\left(1 - q^4\right)\sqrt{q^2 - 1}} \arctan\left(\sqrt{q^2 - 1} - 1\right),$$

in which b is the length of the particle along its axis of symmetry of the oblate spheroid and d its diameter perpendicular to this axis, which define an axis ratio $q = b/d$. The axis of rotation of the oblate spheroid is located within its plane. For plates, which are approximated by oblate spheroids, b corresponds to the thickness of the plate and d to its diameter. For an oblate spheroid, q can be considered to be the inverse aspect ratio.

The translational and rotational diffusion coefficients for spheres, cylinders, and oblates are given in Table 1 for typical geometrical values of the fillers studied. In addition, the values for nanotube agglomerates (agglom. CNT) assuming an agglomerate radius of 250 nm for the MWCNT studied here, are included in Table 1. For the zero-shear viscosity of the polycarbonate melt at 230 °C a value of 6409 Pa s was taken. Based on this, the translational diffusion of nanotubes and to a larger extend of GNP are expected to be considerably slower compared to that of CB. The translational diffusion coefficient is expected to determine the rate of network build-up due to agglomeration in the quiescent melt. Based on the values in Table 2, the build-up of the filler network between spheres, representing CB aggregates, and cylinders

Table 1 Diffusion coefficients for different geometric shapes at a temperature of 230 °C and a zero-shear viscosity of 6409 Pa s in polycarbonate calculated from Eqs. (8), (9), and (10)

Particle shape (approximated filler)	Geometrical parameter	$D_{trans} (m^2/s)$	$D_{rot} (s^{-1})$
Sphere (CB aggregates)	$R_s = 50$ nm	$1.15 \cdot 10^{-18}$	$3.45 \cdot 10^{-4}$
Sphere (CNT agglomerates)	$R_a = 250$ nm	$2.30 \cdot 10^{-19}$	$2.76 \cdot 10^{-6}$
Cylinder (CNT)	$L = 1\,\mu$m, $d = 10$ nm	$5.30 \cdot 10^{-19}$	$4.76 \cdot 10^{-6}$
Oblate ellipsoid (GNP)	$b = 15\,\mu$m, $d = 10$ nm	$6.02 \cdot 10^{-21}$	$3.01 \cdot 10^{-11}$

Table 2 Péclet numbers for translational and rotational diffusion in a shear field for shear rates of 0.01 s^{-1} and 1 s^{-1} for nanocomposites with particles of different shape representing CB, CNT agglomerates, CNT and GNP. The polymer melt (PC) has a viscosity of 6409 Pa s

Shape (approximated filler)	Pe_{trans}		Pe_{rot}	
	$\dot{\gamma} = 0.01\ s^{-1}$	$\dot{\gamma} = 1\ s^{-1}$	$\dot{\gamma} = 0.01\ s^{-1}$	$\dot{\gamma} = 1\ s^{-1}$
Sphere (CB, $R_s = 50$ nm)	$8.7 \cdot 10^1$	$8.7 \cdot 10^3$	$2.9 \cdot 10^1$	$2.9 \cdot 10^3$
Sphere (agglomerated CNT, $R_a = 250$ nm)	$1.1 \cdot 10^4$	$1.1 \cdot 10^6$	$3.6 \cdot 10^3$	$3.6 \cdot 10^5$
Cylinder (CNT, $L = 1\ \mu$m)	$1.9 \cdot 10^3$	$1.9 \cdot 10^5$	$2.1 \cdot 10^3$	$2.1 \cdot 10^5$
Oblate ellipsoid (GNP, $b = 15\ \mu$ m)	$7.6 \cdot 10^9$	$7.6 \cdot 10^{11}$	$3.3 \cdot 10^8$	$3.3 \cdot 10^{10}$

for MWCNT differ by a factor of 2. The translational diffusion of oblate ellipsoids, which are representing GNP, is even more than two orders of magnitude slower than that of spheres, which approximate the CB aggregates.

The influence of the anisotropy of particles on rotational diffusion is even more pronounced. The orientation of a cylinder, representing an individual nanotube, is two orders of magnitude slower, than that of a "spherical" CB agglomerate (see e.g., [10]). It should be noted that D_{rot} of oblate ellipsoids as a representative for GNP is by seven orders of magnitude smaller than for the sphere representing the CB. The difference to randomly distributed or agglomerated nanotubes is still five or four orders of magnitude, respectively. This suggests a fundamentally different orientation behavior of GNP than for the other nanofillers and even its agglomerates, at least for comparable melt viscosities.

The case of the shear-induced orientation of nanoplates in a polymer melt has been discussed for nanoclay in Ref. [18], and this argumentation can be adapted to GNP. It is generally accepted that anisometric particles tend to align toward the flow direction, when the hydrodynamic forces due to shear flow dominate the action of Brownian forces [49]. A dimensionless Péclet number can express the ratio of the timescale for Brownian motion to that for convective motion. The corresponding Péclet numbers for translational and rotational diffusion are

$$Pe_{trans} = \frac{\dot{\gamma} l^2}{D_{trans}(\dot{\gamma})}, \text{ and } Pe_{rot} = \frac{\dot{\gamma}}{D_{rot}(\dot{\gamma})}, \tag{11}$$

in which l is the lateral dimension of the particle in the shear field. For platelets, l is equivalent to the diameter of the plate b, whereas for nanotubes, l has to be replaced by the length L of the cylinder ($l = L$).

The Péclet numbers have been calculated for shear rates of 0.1 and 1 s^{-1}, for spheres, cylinders, and platelets (see Table 1) using the average dimensions of CB aggregates, MWCNTs, their agglomerates and GNP, and are given in Table 2.

For translational diffusion, it can be seen in Table 2 that for all types of nanofillers as well as for CNT agglomerates the hydrodynamic forces in the shear field dominate the Brownian ones ($Pe_{trans} \gg 1$). Interactions between the particles are expected to further suppress their diffusive motion.

Even at slow flow rate, the Péclet numbers for rotational diffusion are larger than unity ($Pe_{rot} > 1$) for all types of particles. This indicates the dominance of hydrodynamic forces compared to the Brownian forces. Interestingly, the Péclet numbers for rotational diffusion differ not more than about one order of magnitude from those of translational diffusion for all nanocomposites including the agglomerated CNT. The Péclet numbers for both translational and rotational diffusion of platelets differ by five and more orders of magnitude from those of the other nanoparticles. This suggests that GNP align already at very low shear rates and that rotational diffusion of shear-oriented GNP in a highly viscous melt is extremely suppressed. Already at moderate concentrations, the platelets are expected to become trapped by each other, which allows only minor orientation or translational rearrangements toward randomization. Consequently, it is difficult (if not impossible) to prepare samples with randomly oriented GNP by melt processing. The flow rate for squeeze flow due to pressing is in our case is in the order of $\dot{\gamma} = 0.1\ s^{-1}$ (see [22] and below). As we discuss in more detail later, the resulting orientation angles of GNP with respect to the flow direction oriented by squeeze flow during pressing are approximated to be between 13° and 18° for about 10 wt% of GNP in a polycarbonate.

Furthermore, the values in Table 2 support the finding for CNT in Sect. 2.1 that the competition between agglomeration and de-agglomeration is the dominating mechanism and that orientation is of second order at moderate flow conditions. Correspondingly, the electron microscopic image in Fig. 2 shows for compression-molded plates only agglomerates and no significant orientation of individual nanotubes. The same is expected for CB in a polymer melt and will also be shown later. As discussed in [16] the situation is different for melt spinning with take-up velocities of up to 800 m/s and in low viscous liquids or semi-dilute dispersion in which orientation of the MWCNT along the direction of flow has been found even at low shear stress. Shear-induced orientation was also found near to the walls in injection molded parts [16].

3 Experimental

3.1 Materials

For the matrix polymer, polycarbonate (PC) was chosen due to its low DC conductivity of $2 \cdot 10^{-9}$ S/m at 230 °C in the polymer melt. In the glassy state at room temperature, the DC conductivity of $2 \cdot 10^{-15}$ S/m is six orders of magnitude lower than in the melt, since the charge transport of ions is frozen. Furthermore, polycarbonate is suitable for the incorporation of carbon-based fillers by melt mixing and allows better filler distribution in the polymer by melt mixing compared to other thermoplastics such as polyolefins (see e.g., [16]). For all composites of this

Table 3 Types of carbon black (CB) and its characteristics according to the supplier

Type	Size of primary particles (nm)	Specific surface from BET (m^2/g)	Dibutyl phthalate absorption (ml/100 g)
Printex XE2	30	910	380
Printex L6G	18	250	123

study, Makrolon® 2600 [50] from Bayer MaterialScience (Leverkusen, Germany) was used. It is a short-chain type with a molecular weight of $3.3 \cdot 10^4$ g/mol, and a low melt viscosity suitable for injection molding. According to the manufacturer, the density at room temperature is 1.2 g/cm^3. The glass transition temperature T_g was determined by differential scanning calorimetry (DSC) to be 150 °C.

For the multi-walled carbon nanotubes, NC7000™ from Nanocyl S.A. (Sambreville, Belgium) is used [51]. The manufacturer specifies the average diameter (d) and average length (L) to be 9.5 nm and 1.5 μm, respectively. This corresponds to an inverse aspect ratio d $/L$ of about 1/150. The specific surface area is given by the supplier with 250–300 m^2/g.

For the carbon black, two different types with different specific surface areas are used. The notations and some of the characteristic properties are summarized in Table 3. Both CB types are supplied by Orion Engineered Carbons (Germany).

The specific surface area was determined by gas adsorption of nitrogen using the Brunauer–Emmett and Teller (BET) method. Another method for the characterization of the CB structure is the oil adsorption of the soot particles. Oil is added to the particles, usually dibutyl phthalate (DBP), and the amount of oil at which the viscosity becomes maximum under constant shear is determined as given in Table 3. Printex L6G has a lower primary particle size, surface area, and oil absorption than Printex XE2.

For the graphite nanoplates, a commercial grade xGnP®-Grade M-15 from XG Sciences (USA) was used. According to the manufacturer, the particles have an average diameter of 15 μm and a thickness of around 6–8 nm, which corresponds to around 20 graphene layers [52]. The type used is thus classified as GNP, even if named by the producer as graphene nanoplatelets. The specific surface area is 120–150 m^2/g (BET measurement).

3.2 *Sample Preparation*

All composites were produced by melt mixing using a micro-compounder DSM Xplore 15 (DSM Resolve, The Netherlands), representing a co-rotating conical twin-screw extruder with a filling volume of 15 cm^3. A mixing time of 5 min and a mixing temperature of 280 °C at a rotation speed of 250 rpm were selected for all composites. Figure 7 shows exemplarily TEM image of a PC composite with 4 wt% CB (Printex L6G), 3 wt% MWCNT, and 10 wt% GNP (xGnP-M15).

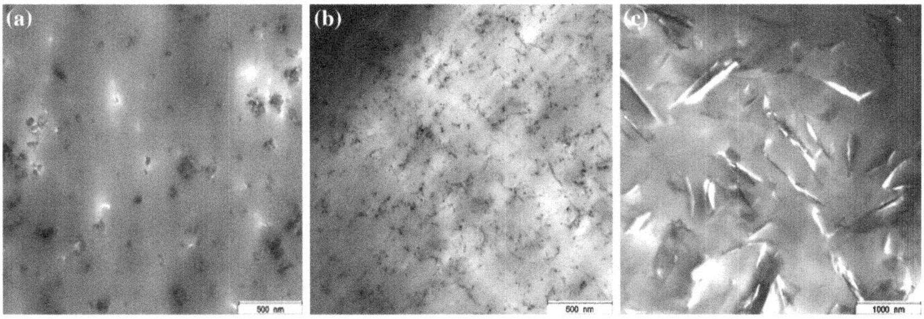

Fig. 7 TEM images of PC composites with **a** 4 wt% CB (Printex L6G), **b** 3 wt% MWCNT, and **c** 10 wt% GNP

For all allotropes, the dispersion was satisfactory. For the MWCNT these processing parameters provided a suitable dispersion with only a slight shortening of the nanotube length. The compounded strands of ca. 2 mm in diameter were melt pressed into circular plates using a PW40EH hot press (Paul-Otto Weber GmbH, Germany). For this, the strands were positioned in a preheated stainless-steel mold heated to 250 °C and covered with PTFE film as a separating layer. After 2 min heating the plates were pressed together with a speed of 6 mm/min and a maximum force of 50 kN. The resulting thickness of the pressed plates was between 500 μm and 1 mm. For DC conductivity and BDS measurements, samples with a diameter of 26 mm were punched out of the pressed plates. Gallium–indium–tin (Galinstan, Geratherm Medical AG) was applied to both sides of the sample to form circular electrodes of 20 mm diameter.

For the measurements in the rheoelectric setup, the samples were first dried for 90 min at 120 °C under reduced pressure. A dried sample was placed on top of the lower electrode of the rheometer, which was preheated to 230 °C, and the upper electrode was moved down in order to touch the surface of the sample. The thermal expansion of the sample was compensated by moving the sample holders slightly apart during heating. This procedure was completed after about 2 min before the recording started. A contact force between 0.08 N and 0.4 N was set, to guarantee sufficient contact between the electrodes and the composite melt over the course of the measurement.

3.3 Dielectric and Electric Measurements

The frequency-dependent real part of the conductivity (σ') and the real (ε') and imaginary part (ε'') of the complex permittivity ($\varepsilon^*(\omega) = \varepsilon'(\omega) - i\varepsilon''(\omega)$) were measured by BDS at standard climate (23 °C and 50 % ± 10 % relative humidity) at frequencies f between 10^{-2} Hz and 10^6 Hz. The angular frequency is $\omega = 2\pi f$.

The dielectric function $\varepsilon^*(\omega)$ and the complex electrical conductivity$\sigma^*(\omega)$ are related to each other by

$$\sigma^*(\omega) = \sigma'(\omega) + i\sigma''(\omega) = i\omega\varepsilon_0\varepsilon^*(\omega). \quad (12)$$

For the BDS measurements, a frequency response analyzer SI 1260 (Solartron) in combination with a broadband dielectric converter (Novocontrol, Germany) was used. The measurements were carried out on discs prepared from hot-pressed plates (see above). For some measurements, hot-pressed plates of 22 mm in diameter are used, with gold electrodes with 20 mm diameter sputtered on both sides of the discs. The measured data were analyzed by means of the software WinDeta (Novocontrol, Germany). Alternatively, measurements of the DC conductivity were performed on the plates at standard climate by means of a Keithley 6517A electrometer (Keithley Instruments, Solon, Ohio, USA) in combination with the resistivity test fixture Model 8009 (Keithley).

3.4 Rheo-Electric Measurements

For time-resolved rheo-electrical measurements with defined shear flow protocols, a commercial rheometer (ARES, TA Instruments, USA) was modified to allow simultaneous measurements of electrical and rheological properties (see e.g., [8, 16]]). The rheometer motor is controlled by an external NI PXI 8186 data acquisition system (National Instruments, USA), which also records the resulting torque from a torque transducer which is attached to the opposite plate.

For a sinusoidal shear deformation, torque and phase difference δ is used to calculate real and imaginary parts of the shear modulus ($G^*(\omega) = G'(\omega) + iG''(\omega)$). In order to measure $G'(\omega)$ and $G''(\omega)$ under steady shear conditions (see applied shear protocols below), a sinusoidal shear deformation can be superimposed to a continuous shear deformation, which results in a *modulated shear deformation*.

Measurements with continuous shear are an alternative to frequency-dependent measurements. The transient viscosity is related to the time-dependent shear stress $\tau(t)$ by

$$\eta^{\mathrm{T}}(t) = \frac{\tau(t)}{\dot{\gamma}}. \quad (13)$$

Here, $\dot{\gamma}$ is the constant shear rate. The total shear deformation $\gamma(t) = \dot{\gamma}t$ is increasing during shearing with time.

For the measurements of the DC conductivity, different electrode arrangements are used (Fig. 8): an electrode geometry with two insulated rings (*capacitor arrangement*), and a geometry where one ring electrode is replaced by an *interdigitating ring* structure (see Fig. 9). The ring geometry was chosen in order to obtain a nearly constant shear rate in the examined volume.

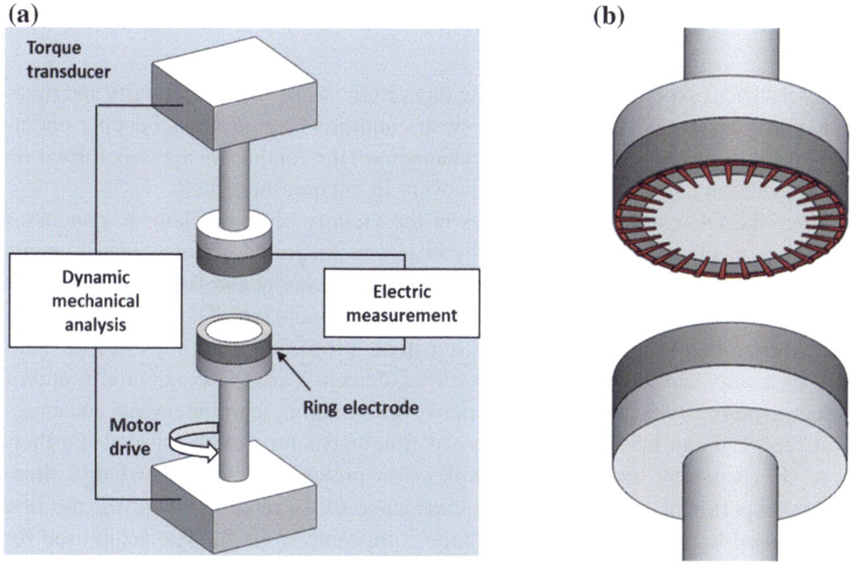

Fig. 8 Experimental setup for combined measurements of the dynamic shear modulus and DC conductivity (schematically) using ring electrodes for measurement at a defined shear rate. For measurement of the DC conductivity in the shear plane and perpendicular to the shear direction, the ring electrode can be replaced by a ring electrode with interdigitating electrodes (see Fig. 9)

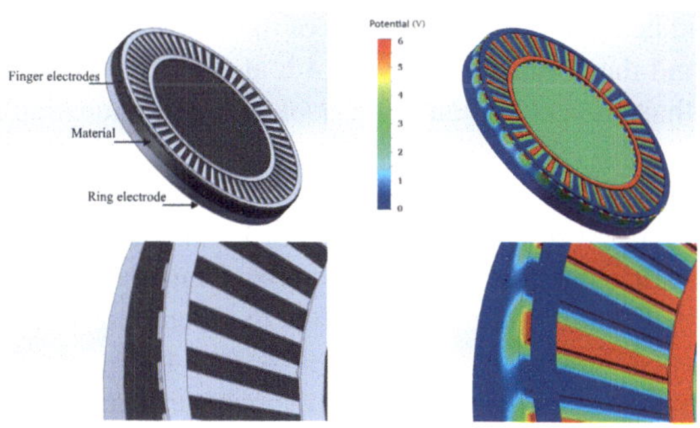

Fig. 9 Electrode arrangement for measurement of direction-dependent DC conductivity using a plate–ring geometry with interdigitating electrodes (left). Simulation of the electrical potential between the strips of the interdigitating electrodes (green) and between the interdigitating ring electrode and the opposite ring electrode (blue)

Replacing the Keithley 6514 electrometer by the frequency response analyzer SI 1260 in combination with a broadband dielectric converter (see above), rheo-BDS measurements were possible too. The geometrical factors needed to convert the measured impedances into specific conductivity and permittivity values were determined by FEM simulations.

3.5 Rheological Protocols

Rheo-electrical experiments combining measurements of DC conductivity and rheological properties under well-defined shear conditions can provide a deeper understanding of the underlying physical mechanisms of the conductive network formation and destruction as well as structural recovery in the quiescent melt.

Since the DC conductivity changes in the vicinity of an insulator-to-conductor transition (or electrical percolation) by orders of magnitude, those measurements provide an extremely sensitive tool to study small changes in the arrangement of conductive filler particles within a conductive filler network. The conductive filler particles can be considered as highly sensitive "probes" for the influence of shear flow on composite melts. By time-resolved detection of DC conductivity and/or dielectric permittivity, the kinetics of network formation, rearrangements and structural recovery can be studied in a broad time and temperature interval. Furthermore, the rheometer setup allows complex flow protocols. To our knowledge, rheo-electric experiment (also called rheo-impedance measurements) were for the first time applied to polymer–carbon nanotube composites in [6–8]. The setup used for this study is described in more detail in [16].

To obtain further insight into the effect of a weak shear flow on formation, destruction, and recovery of the arrangement of the different filler types, the same test protocol was applied to all composites at 230 °C in order to enable a comparison between the different materials. This temperature is well above the glass transition temperature of polycarbonate. Although after melt mixing a slight reduction of the molar mass could be observed, the changes of the molar mass during heat treatment of the neat PC and the compounds during 4 h at 230 °C in the rheometer are with about 4 % less than the experimental error of the gel permeation chromatography (GPC) [22].

The DC conductivity of the materials was determined within the rheometer setup in plate–plate geometry under nitrogen using different electrode geometries such as ring or interdigitating electrode geometries (for details see below). A rheological protocol assembled of three steps was applied to all composites. This protocol was originally developed for nanotube composites [8, 10, 14]. Figures 10a and b show the three step protocol together with the shear rate (input variable) as well as the measured DC conductivity and viscosity exemplarily for a polycarbonate composite containing 3 wt% carbon black at 230 °C. Instead of the DC conductivity, the AC conductivity and the complex permittivity can be measured by replacing the electrometer by an impedance analyzer.

In the first step (*annealing*) of the protocol, the sample is annealed at 230 °C for 1 h without shear in order to allow the network structure, which is influenced by the previous processing and sample preparation such as melt mixing and hot pressing, to equilibrate. The structure of the filler network resulting from such thermo-rheological history is usually not well defined and the annealing step should allow the network structure to clear or "erase" the process history at least partially. In the second step (*steady shear*), the sample is sheared with a shear rate of 1 s^{-1} for 120 s.

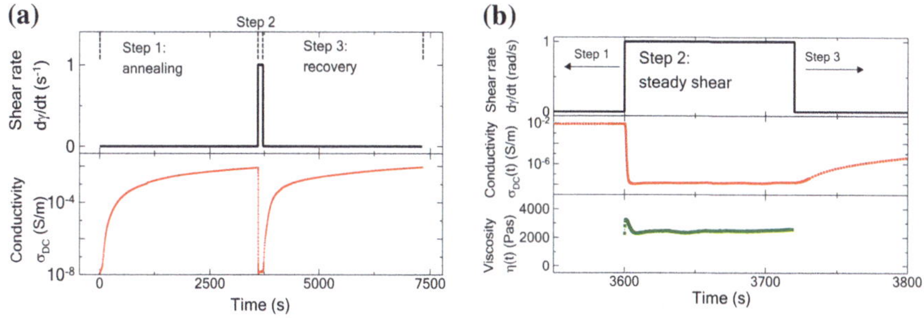

Fig. 10 **a** Three-step rheological protocol for rheo-electric measurements composed of an *annealing step* (step 1), a *steady shear* step (step 2) and a *recovery* step (step 3) for 3 wt% carbon black XE2 in PC at 230 °C. **b** Measurement the of DC conductivity and viscosity during steady shear (enlargement of step 2)

The DC conductivity and the transient viscosity are measured simultaneously. For the measurement of G' and G'', an oscillation can be superimposed to the steady shear deformation (see above). Step 2 is shown enlarged in Fig. 10b for all three quantities. It is assumed that by steps 1 and 2, a defined thermo-rheological pre-history can be "imprinted" to the composites by the defined shear and the previous thermal annealing. It is expected that, for a sufficient time of steady shear, a dynamic equilibrium structure of the filler network (e.g., by the interplay of destruction and agglomeration), which is represented by a constant conductivity and viscosity, can be achieved. The structure of the filler network generated by this procedure serves as a starting point for the third step (*recovery*), which is analogous to the first process step but with a better defined initial state. The sample is again held for 1 hour without shear at 230 °C and the DC conductivity is measured. Sometimes an oscillation with a very low shear amplitude is applied to monitor the changes of the complex shear modulus in parallel.

For measurements of the anisotropic DC conductivity of the GNP composites with interdigitating electrodes, the times of the steps were extended. The network formation process was prolonged to 2 h and the recovery time up to 14 h.

4 Results and Discussion

4.1 *Broadband Dielectric Spectroscopy of Solid Composites*

For sample characterization, in this paragraph, the dielectric properties of all carbon allotrope composites studied are summarized. The matrix polymer for all composites was polycarbonate. The BDS measurements were performed at 23 °C on compression-molded plates of 2 mm thickness as described above. The dielectric relaxation spectrum for the neat PC can be found in reference [22]. In the composites, the contribution of the PC to the dielectric relaxation spectrum is masked by the contribution from the conductive filler network and can be neglected for higher filler concentrations. It has to be noted that the dielectric spectra represent the structure of the filler network for the actual preparation conditions (melt mixing, squeeze flow during pressing and vitrification).

4.1.1 Carbon Black Composites

Figure 11 shows the real part $\varepsilon'(f)$ and imaginary part $\varepsilon''(f)$ of the complex dielectric permittivity and the real part $\sigma'(f)$ of the complex electrical conductivity for different concentrations of carbon black type XE2 (a) and type L6G (b). The conductivity spectra $\sigma'(f)$ are typical for conductive fillers in a non-conducting matrix below and above the insulator–conductor transition. The appearance of a low-frequency plateau in the real part of the electrical conductivity at carbon black concentrations of 4.5 wt% XE2 and at about 9 wt% L6G is related to the increase in the DC conductivity ($\sigma_{DC} = \sigma'(f \to 0)$. The increase in σ_{DC} by orders of magnitude represents an insulator–conductor transition, which is related to a percolation threshold, for which p_c is the critical filler content of the threshold.

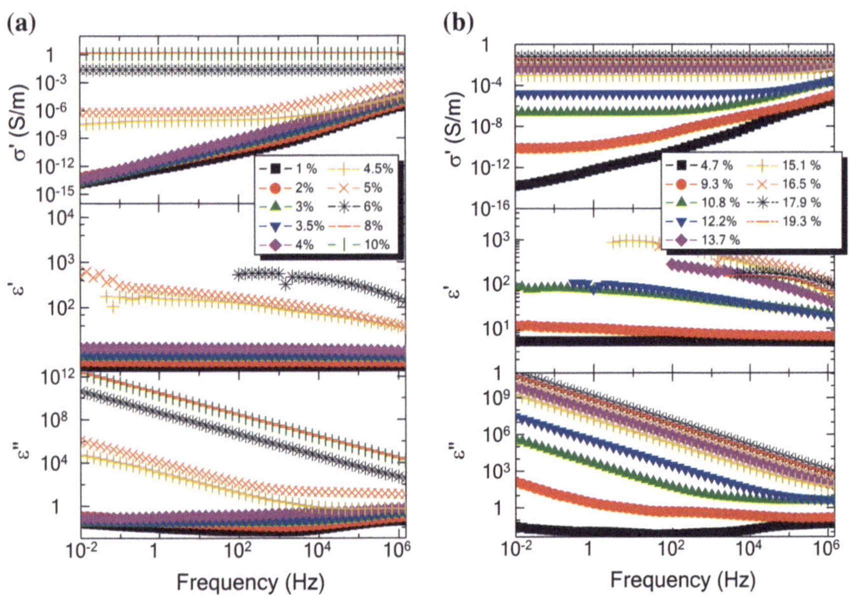

Fig. 11 Real part σ' of the complex electrical conductivity (above) and real ε' and imaginary part ε'' of the complex dielectric permittivity (below) as a function of the frequency for different filler concentrations (in wt%) of carbon black type XE2 (**a**) and type L6G (**b**) at 23 °C in PC

The percolation transition shown in σ_{DC} is coupled to an increase of the low-frequency contribution of the real part of the permittivity $\varepsilon'(f \to 0)$, which is discussed in more detail later in connection with Fig. 15. Above the percolation threshold p_c, the real part of the permittivity $\varepsilon'(f)$ decreases with increasing frequency as expected for charge carrier diffusion in fractal structures or RC networks. The curves for $\varepsilon''(f)$ represent the same dataset as the frequency-dependent conductivity, with $\sigma'(\omega) = \omega\varepsilon_0\varepsilon''$ and $\omega = 2\pi f$. This leads to the same interpretation. Thus, the low-frequency plateau in the conductivity "translates" to a power law $\varepsilon''(f < f_c) \sim f^{-1}$.

For concentrations above p_c, the frequency dependence of the electrical conductivity changes from the plateau value at low frequencies $\sigma'(f < f_c) = const.$) to a power law dependency at higher frequencies: $\sigma'(f > f_c) \sim f^b$, with $b \approx 1$. This change in the frequency dependency can be characterized by a crossover frequency $f_c = \omega_c/2\pi$. The values of f_c can be extracted from fitting of the $\sigma'(f)$ data (see [22]).

This type of frequency dependence of $\sigma'(f)$, with a cossover from a DC conductivity plateau to a power law behavior, is characteristic for ion conductivity in glass-forming liquids and can be related to the dynamic heterogeneity in those systems. This behavior can be explained by a spatially randomly varying (frequency-independent) conductivity [53]. In a discretized model, it can be described by an equivalent circuit (representing a simple cubic lattice) of resistors and capacitors (see [53]). Restricted or frustrated charge carrier hopping leads to similar results.

One may assume that, for conductive fillers embedded in an almost non-conducting polymer matrix, the crossover frequency f_c represents a transition from charge carrier transport on length scales larger than the correlation length of the conductive sites or clusters for $f < f_c$ to anomalous charge carrier diffusion at short length scales within a fractal filler network for $f > f_c$. Possibly, there is an analogy to the spatially varying electrical conductivity in glass-formers (see [53]). In composites with conductive fillers, the frequency dependence of the complex conductivity can be explained as well by charge carriers accumulated at the boundaries between conductive (here: carbon allotropes) and less conducting regions (here formed by polymer chains). Microcapacitors due to a polymer interlayer in the nanometer scale between the filler particles are discussed, e.g., for CB reinforced elastomers. However, the description of the frequency dependence of the complex conductivity and permittivity for conductive filler clusters in a polymer matrix is still under debate. In addition to classical assumptions, such as fractal resistor–capacitor arrays, tunneling conductivity have to be considerd.

In order to get an estimate for the DC conductivity $(\sigma_{DC} \approx \sigma'(\omega \to 0))$, the conductivity spectra for the composites are fitted for $p > p_c$ by an equation proposed by Jonscher [54], which is an extension of the Barton–Nakajima–Namikawa (BNN) equation [55–57]:

$$\sigma'(\omega) = \sigma'(0) + A\omega^b. \tag{14}$$

The results of the fits (not shown) to $\sigma'(f)$, with $f = \omega/2\pi$, for all allotropes are summarized in Fig. 14 and will be discussed later.

4.1.2 Carbon Nanotube Composites

In Fig. 12, the real part of the electrical conductivity as well as the real and imaginary parts of the complex permittivity are plotted as a function of frequency for different concentrations of MWCNT in PC below and above the percolation threshold.

The frequency-dependent dielectric behavior (σ', ε' and ε'') for the PC-MWCNT composites is similar to that of the CB composites in Fig. 11, with the typical features of a percolation threshold between 0.5 and 1 wt%. Below this transition (curves for 0.25 and 0.5 wt%), a typical behavior for non-conductive composites is indicated. The real and imaginary parts of the dielectric permittivity below the threshold are small and mainly due to the dielectric relaxations of the polycarbonate (see [22]). As in the case of the CB composites, with increasing filler concentration, a low-frequency plateau appears in the $\sigma'(\omega)$ curves. For concentrations above the percolation threshold, the typical crossover from the low-frequency

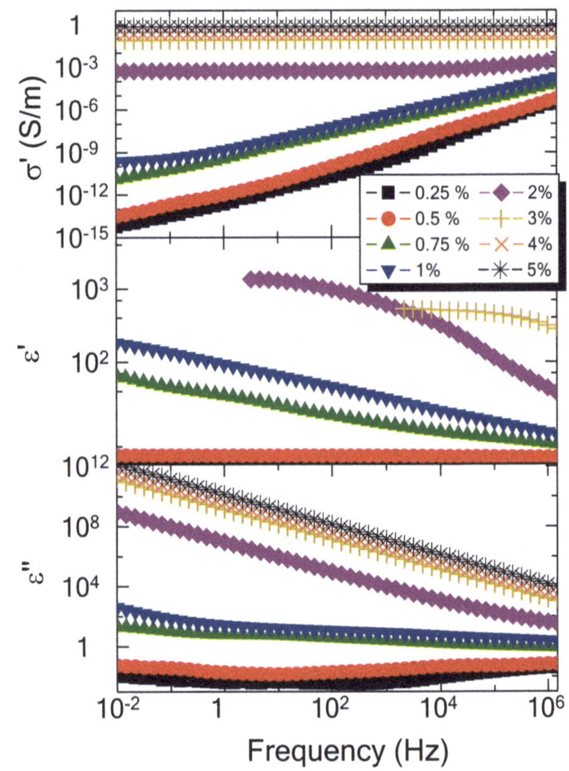

Fig. 12 Real part σ' of the complex electrical conductivity (above) and real ε' and imaginary parts ε'' of the complex dielectric permittivity (below) as a function of frequency for different concentrations (wt%) of MWCNT at 23 °C in PC

conductivity plateau ($\sigma'(\omega < \omega_c) = const.$) to a power law ($\sigma'(\omega > \omega_c) \sim \omega^b$) above a crossover frequency ω_c is obvious. As expected, for charge carrier diffusion in fractal clusters, the value of ω_c shifts toward higher values with increasing nanotube concentration. To estimate a value for the DC conductivity, the $\sigma'(\omega)$ for $p > p_c$ are fitted by Eq. (14) and compared in Fig. 14 to those for the CB composites.

The increase of the low-frequency values of ε´ with concentration up to values of 10^3 can be again explained by the emerging clusters formed by the filler particles (see discussion on Fig. 15) and has been observed before for similar MWCNT composites in PC [37].

4.1.3 Composites Containing Graphite Nanoplates

Figure 13 shows the real part of the complex electrical conductivity together with real and imaginary parts of the dielectric permittivity as a function of frequency for different concentrations of GNP in PC. The conductivity spectra $\sigma'(\omega)$ are similar to those of CB and MWCNT. A conductivity plateau appears at concentrations above 10 wt% with a value of only $1.4 \cdot 10^{-6}$ S/m at 20 wt%.

Compared to MWCNT composites, the percolation threshold of the GNP composites, which is indicated by the appearance of the low-frequency conductivity plateau,

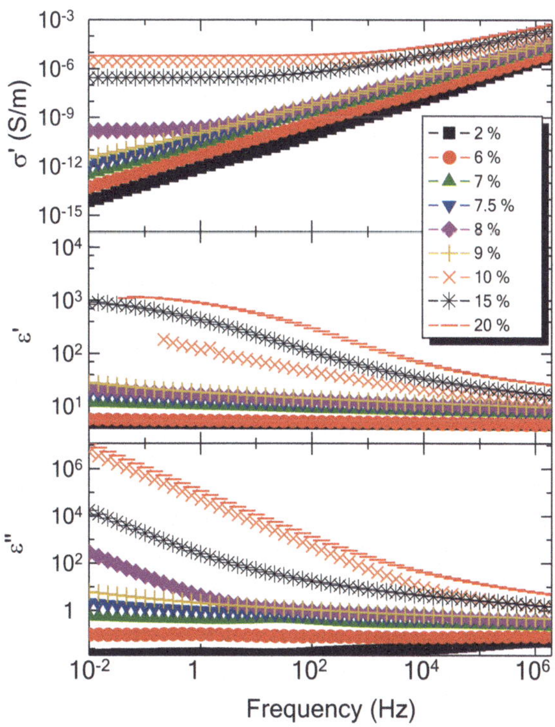

Fig. 13 Real part σ' of the complex electrical conductivity (above) and real ε' and imaginary parts ε'' of the complex dielectric permittivity (below) as a function of the frequency for different concentrations (wt%) of GNP at 23 °C in PC

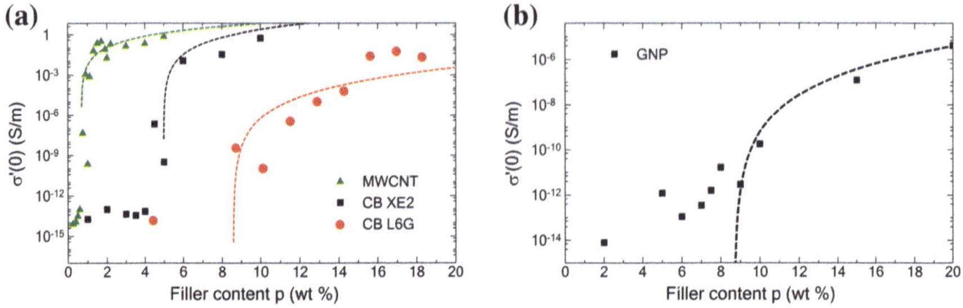

Fig. 14 Comparison of the extrapolated DC conductivity ($(\sigma_{DC} = \sigma'(f \to 0))$ as a function of filler content (in wt%) between two types of CB (XE2 and L6G) and MWCNT (**a**) as well as GNP (**b**) at 23 °C in PC. The dashed lines are fits using a power law Eq. (3) for concentrations above the percolation threshold

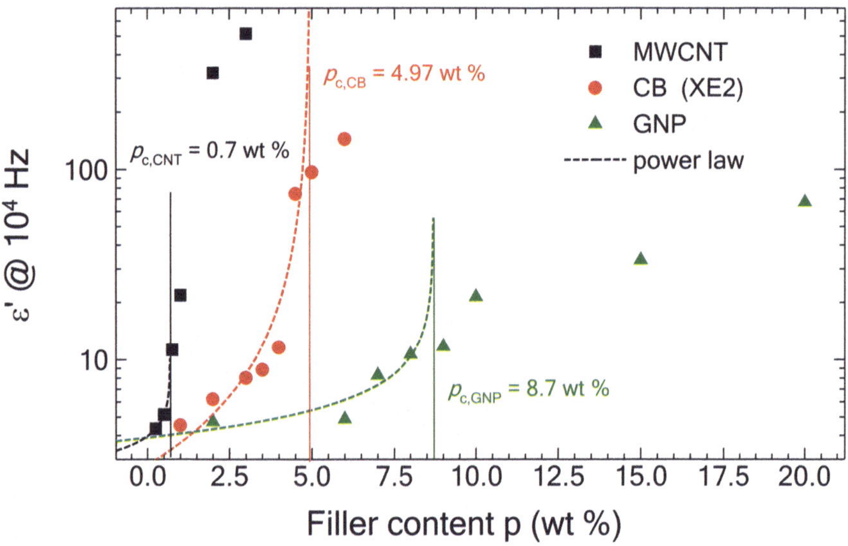

Fig. 15 Comparison of the real part of the dielectric permittivity at $f = 10^4$ Hz as a function of filler content between CB (XE2), MWCNT and GNP composites at 23 °C in PC. The dashed lines are fits with Eq. (6) for concentrations below the percolation threshold

occurs at considerably higher concentrations, and is only weakly pronounced. This is remarkable, since both, GNP and MWCNT, have a similar ratio between lateral size and thickness (or diameter), and therefore, a low percolation concentration was expected for both. The $\sigma'(\omega)$ curves for $p > p_c$ are also fitted by Eq. (14) and summarized in Fig. 14. The low-frequency values of the permittivity are discussed along with the corresponding values for CB and MWCNT composites in conjunction with Fig. 15.

4.1.4 Comparison of Composites with Different Carbon Allotropes

Figure 14 compares the low-frequency conductivities $\sigma'(\omega \to 0)$ of both types of CB, MWCNT (a) and GNP composites (b), which is a measure for the DC conductivity ($\sigma'(0) \approx \sigma_{DC}$). The low-frequency conductivity values are estimated by fitting the $\sigma'(\omega)$ curves for all composites by Eq. (14). The large standard deviations, indicated by the error bars, are most probably due to differences in the processing or preparation conditions. For instance, the filler dispersion and their agglomeration depend strongly on the melt viscosity, which varies not only with the filler type but also depends considerably on the filler content itself.

The dashed lines represent fits [22] with Eq. (5) for $p > p_c$ to determine p_c. For p_c values of (0.7 $\pm$ 0.2) wt%, (4.9 $\pm$ 0.5) wt%, and (8.6 wt% $\pm$ 0.5) wt% for MWCNT, CB type XE2, and CB type L6G, respectively, were evaluated. The corresponding values in volume content are 0.46, 3.3, and 5.6 vol%. As expected, the values for the MWCNT composites are considerably lower than for both CB types. However, for percolation of cylinders with the aspect ratio of the MWCNT, an order of magnitude lower values for p_c were expected [16]).

The lower value of p_c for the CB type XE2 compared to L6G can be explained by the higher structuration of the CB aggregates, with more primary particles per aggregate. The higher fractality of XE2 results in a lower percolation concentration. For a more detailed interpretation of the influence of the CB structure on the DC properties and parameters such as porosity or fractality, see [22] and references therein. The fit value of p_c for the GNP composites (Fig. 14b) was found to be (8.7 $\pm$ 1) wt%, which is equivalent to 4.9 vol%. Taking into account the similar ratio of lateral size to diameter or plate thickness for MWCNT and GNP, respectively, the value for the GNP is surprisingly higher than that for MWCNT. For other GNP types, Kim et al. [24] and Via et al. [58] reported values of 1.3 and 4.6 vol%, respectively. The reasons are not clear and may lie in the different GNP types or preparation routes such as granular coating [58]. The possibility of different orientations of the GNP in the shear flow is discussed later.

In Fig. 15, the real part of the dielectric permittivity at a frequency of 10^4 Hz, assumed to represent $\varepsilon'(\omega \to 0)$, are plotted as a function of filler content for CB (XE2), MWCNT, and GNP composites at 23 °C in PC. The dashed lines are fits with Eq. (6) for concentrations below the percolation threshold, which yield p_c values of 0.7, 5, and 8.7 wt% for MWCNT, CB (XE2), and GNP, respectively. For details of the fitting, see [22].

An increase of $\varepsilon'(f \to 0)$ with filler content is visible for all allotropes studied. It can be explained by the emerging clusters formed by the fillers and has been observed before for MWCNT composites [37]. One possible explanation is the time-delay (represented by a waiting time distribution) of charge carriers diffusing in fractal clusters, which are probing the size of the cluster. Alternatively, the capacitance in resistor–capacitor networks may explain this increase.

A decrease of $\varepsilon'(f \to 0)$ above the percolation threshold, as proposed by Eq. (6), could not be found for any of the composites. However, the origin of the remaining high values above the percolation threshold are not yet understood. It is assumed that permanent microcapacitors are formed by polymer chains that are trapped between the filler particles [37]. Such microcapacitors are expected to be similar to the so-called bound rubber in CB–rubber composites.

Although the general features for an insulator to conductor transition and the main assumption of the percolation theory are fulfilled for the frequency dependence of the complex conductivity and DC conductivity as well as for the permittivity (at least for $p < p_c$) for all three types of carbon allotropes, there are large deviations from the theoretical predictions. Based on our previous findings related to the strong influence of shear or elongational flow or the processing conditions during melt mixing (see e.g., [16]), deviations between the composites and even between samples within one series are therefore not surprising.

4.2 Rheo-Electric Experiments

This paragraph summarizes the rheo-electric experiments using the three-step shear flow protocol described in Fig. 8 on CB, MWCNT, and GNP composites in the PC melt at 230 °C. The three steps are the *annealing step* for 1 hour without shear, the *steady shear step,* with a shear rate of 1 s^{-1} for 120 s, and finally, the *recovery step* in the quiescent melt. For measurements of the anisotropic DC conductivity of the GNP composites, the protocol was partially modified. We focus on investigations during steady shear and the following structural recovery, since the structure of the filler network during annealing reflect the processing prehistory and is therefore not well defined.

4.2.1 Carbon Black Composites

Figure 16a shows the time dependence of the DC conductivity for CB (XE2) composites with different filler contents during steady shear. At the end of the annealing procedure, only samples with a CB concentration above 2 wt% show conductivity values which are significantly above those of the polymer matrix. The DC conductivity drops for all samples after the onset of shear deformation. After about 30 s, steady-state values are achieved. For samples with 3 and 3.5 wt% CB, the conductivity drops to values close to the conductivity of the neat melt. Both concentrations are below the percolation threshold as found in the BDS experiments on solid samples. It is therefore assumed that the CB agglomerates are almost destroyed by the applied shear and the shear-dispersed and -isolated CB aggregates do not build structures with significant conductivity. For CB concentrations above 4 wt%, the steady-state values of the *conductivity plateau* are significantly above the conductivity of the PC melt and rise with the filler concentration. This indicates a dynamic equilibrium between destruction and build-up (dynamic clustering) under steady shear conditions as discussed on Fig. 5 and in [14].

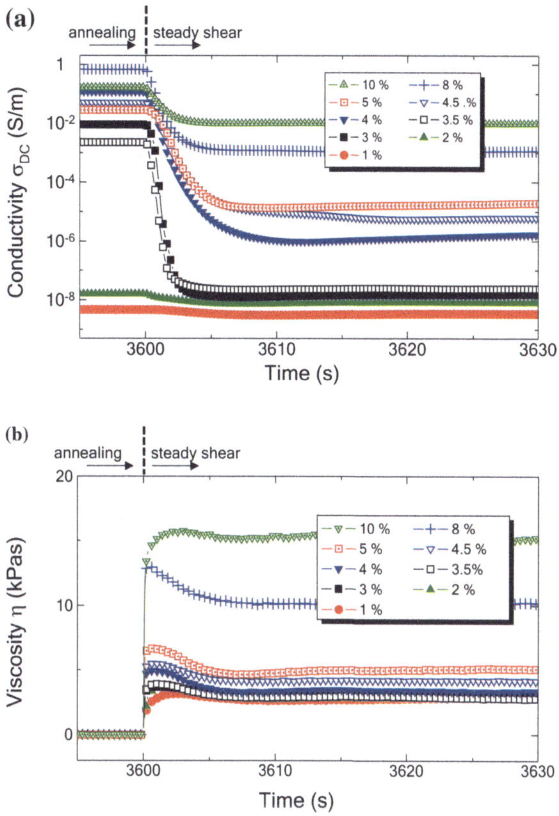

Fig. 16 Time dependence of the DC conductivity **a** and the viscosity **b** of composites with different concentrations (wt%) of the carbon black XE2 at 230 °C during shear deformation with a shear rate of 1 s^{-1} for 120 s.

Figure 16b shows the transient viscosity of the CB composites as a function of the filler concentration during the shear deformation. As for CNT composites in [14], at about 2 s, which is equivalent to a shear deformation of about 2 rad, a maximum viscosity can be recognized for all concentrations and an equilibrium value is obvious for longer shear deformations, which depends on the filler concentration. At a concentration of 5 wt%, the viscosity plateau rises to a value of 10.2 kPa s, which can be related to a rheological percolation. The rheological percolation behavior has been found to differ from the DC one [16], which is indicated by the linear concentration dependence of the rheological properties compared to changes of DC conductivity by orders of magnitude in a logarithmic scale. The rheological percolation threshold is further expected at somewhat higher concentrations than the electrical one [59]. The findings for CB type L6G [22] (not shown here) for the time-dependent conductivity and viscosity are similar to those for XE2. Differences are due to the higher percolation concentration of 9.3 wt% of L6G due to the lower fractality of its CB aggregates.

Figure 17 shows the *recovery* of the conductivity for different filler contents of XE2 (a) and L6G (b) in the quiescent melt at 230 °C after the end of shearing.

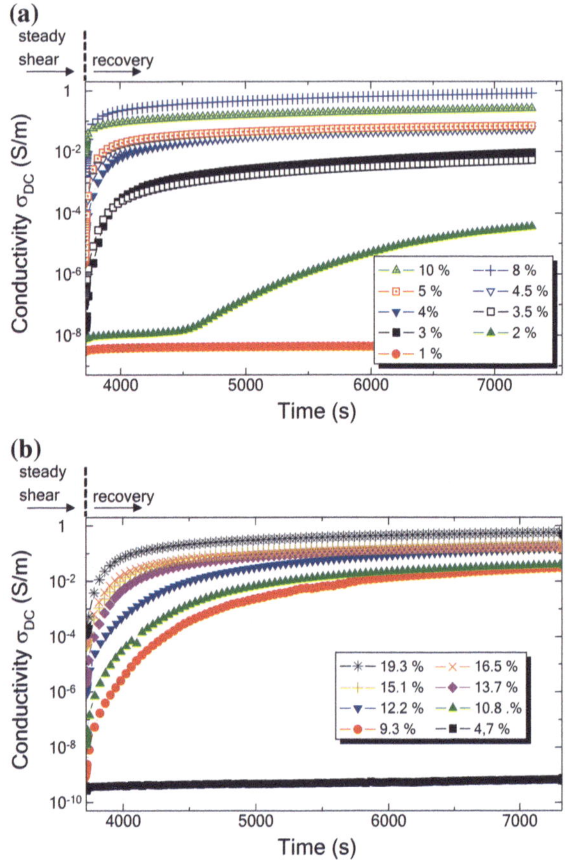

Fig. 17 Time dependence of the DC conductivity of PC composites with different concentrations (wt%) of CB XE2 **a** and L6G **b** in the quiescent melt at 230 °C after shearing for 120 s with a shear rate of 1 s^{-1}

After about 3000 s of recovery the conductivity reaches constant values. As for the steady-state values during shear, these values are also dependent on the filler concentration. Thus, quite different "percolation thresholds" exist for composites containing the same type of filler in dependence on the thermo-rheological history. Thus, the percolation concentrations are different for (i) compression-molded plates in the solid state, (ii) sheared molten composites in the dynamic equilibrium and (iii) structurally recovered melt composites as the three limiting cases of this study.

Interestingly, for the sample with 2 wt% CB XE2 the conductivity increases after a time delay. A similar delay was found for CB composites at concentrations close to the critical concentration with a CB of low fractality in [22]. The agglomeration model provides a possible explanation for the delay time. If the filler content is below a critical filler concentration, the agglomeration process is expected to lead first to the formation of agglomerates, which are electrically insulated from each other. The DC conductivity increases only slightly below the percolation concentration (see Eq. (6) with a small exponent $s \approx 0.73$). When after a certain time the critical agglomeration concentration has been reached, i.e., the agglomerates are grown and start to touch each other, the DC conductivity increases significantly with time.

4.2.2 Carbon Nanotube Composites

In Fig. 18 the DC conductivity (a) and viscosity (b) for MWCNT composites are plotted versus time during steady shear of 1 s^{-1} at 230 °C. Before this, the samples were annealed for 1 h at the same temperature.

After the onset of the shear deformation, the DC conductivity drops by up to five orders of magnitude within about 10 s, which can be explained by destruction of electrically conductive paths by the shear deformation. For concentrations below 1 wt% the conductivity decreases to the level of conductivity of the polymer matrix. For filler concentrations above 1 wt% the shear-induced agglomeration is sufficient to maintain an electrically conductive network. The relative conductivity decrease is stronger near the percolation threshold, since in this concentration region the build-up or destruction of a small number of electrical contacts cause large changes of the

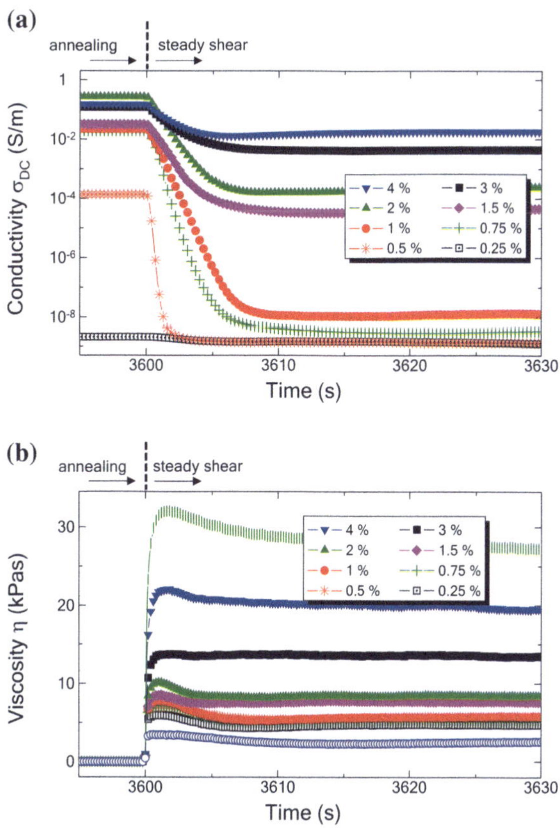

Fig. 18 Time dependence of DC conductivity **a** and viscosity **b** of PC-MWCNT composites of different concentrations (wt%) at 230 °C during shear deformation with a shear rate of 1 s^{-1}

conductivity. For filler concentrations above 3 wt% only a moderate conductivity decrease can be observed.

After about 30 s of shearing, equilibrium values for the DC conductivity (Fig. 18a) as well as the viscosity (Fig. 18b) are established at all MWCNT concentrations. The equilibrium values for both quantities depend on the filler concentration as already shown in Fig. 6. As for the CB composites, the steady-state values of the conductivity and viscosity are attributed to an equilibrium between the build-up and destruction of agglomerates [14, 16]. As also discussed before (see above for CB and [16] for CNT), the changes of rheological properties and of charge carrier transport couple differently to the structure of the filler network [16]. The increase in viscosity, with a maximum at 2 s for all filler concentrations, equivalent to a shear deformation of 2 rad, can be attributed to elongated chains (like "strain hardening") followed by a destruction of the filler network. Since the unfilled PC melt does not show such a maximum [14], the latter seems reasonable.

The recovery of the conductivity in the quiescent melt at 230 °C after shearing is shown in Fig. 19 for the MWCNT composites. The steady shear step of the flow protocol provides an almost defined initial state for the recovery experiments. As expected, the conductivity increases for samples above the percolation concentration with time, due to the dominance of filler agglomeration. For a filler concentration

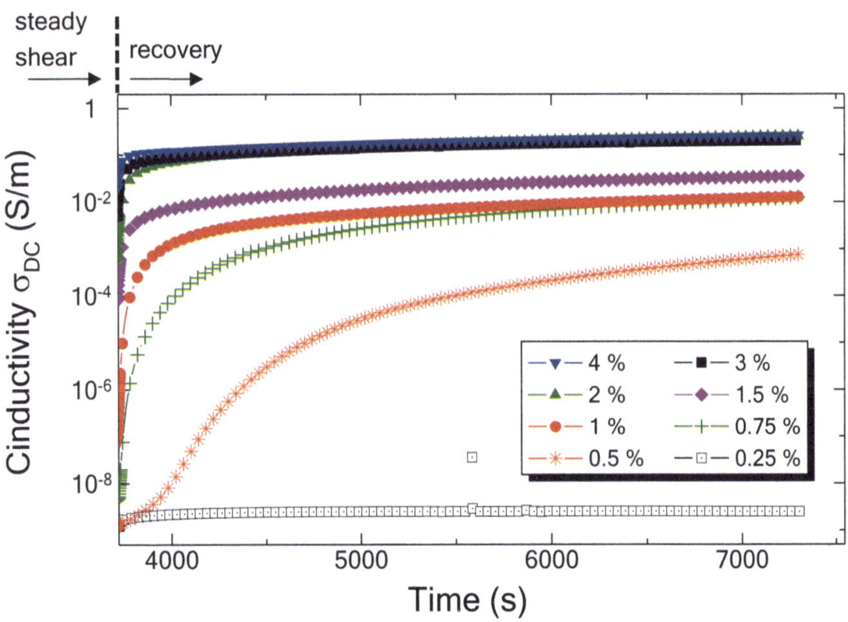

Fig. 19 Time dependence of DC conductivity of PC composites with different MWCNT concentrations (wt%) in the quiescent melt at 230 °C after shearing for 120 s with a shear rate of $1\ s^{-1}$

of 0.5 wt% the agglomeration process is somewhat stretched. As for CB composites, this can be explained by agglomerates, which are initially separated before a critical agglomeration concentration is reached after some time. For modeling of the conductivity with the agglomeration model in Sect. 2.1 see Ref. [22].

4.2.3 Composites Containing Graphite Nanoplates

For measurements on GNP composites the thermo-rheological protocol was prolonged since annealing and recovery in the melt were found to be significantly slower compared to CB and MWCNT composites. Figure 20 compares isothermal annealing of the pressed plates and the recovery after the steady shear step for a GNP composite with 10 wt% in the quiescent melt at 230 °C. The general features are similar to *annealing* and *recovery* of CB and MWCNT composites. However, the shape and time dependence of the conductivity curves differ significantly compared to those of the composites with CB and MWCNT.

Whereas the DC conductivity initially increases by about four orders of magnitude within about 2000 s, followed by a slow increase with less than one order of magnitude per hour, the recovery step after steady shear shows only a slow increase of the conductivity. The slope of this increase during the recovery is similar to that for the annealing at longer times. In the time of the experiment, no constant values are reached in both steps. The two different regimes in the annealing step are possibly

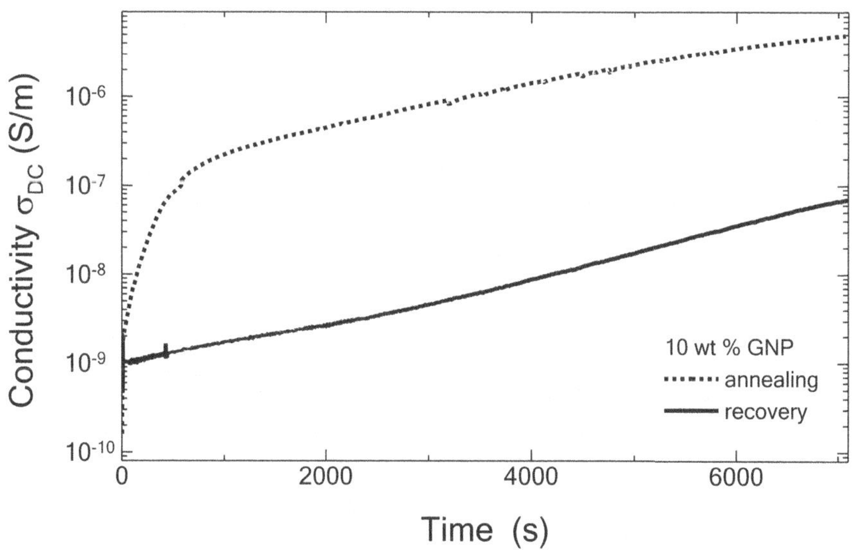

Fig. 20 Time dependence of the DC conductivity of a composite with 10 wt% GNP during annealing (dotted line) and recovery after steady shear (solid line) in the quiescent melt at 230 °C

related to different mechanisms. From the translational and rotational diffusion coefficients of oblate ellipsoidal in Table 2 it is expected that the rotational diffusion of GNP is considerably faster than the translational motion. Such initial increase during annealing is rather related to hindered rotation orientation diffusion of plates. However, due to the extremely low values of the rotation diffusion coefficient only very small rotation angles are expected in the time scale of the experiment. The low DC conductivity values and the high percolation concentrations in BDS measurements (perpendicular to the shear plane of compression molding) together with small-angle X-ray scattering results parallel and perpendicular to the main extension of the pressed plates, indicate that the GNP are already highly oriented as the result of the compression molding (for details see below and [22]). In addition, the limitations for orientation due to sterically hindrance at higher GNP contents has to be taken into consideration. Therefore, no considerable orientation randomization is expected to occur in a viscous polymer melt within the time scale of our experiments.

To check these assumptions, anisotropic rheo-electric conductivity measurements were performed using interdigitating electrodes as described in the experimental section. These measurements, together with FEM calculations of the electrical field of the electrode geometry, allow determining the DC conductivity parallel and perpendicular to the shear plane. The shear directions during the compression molding and the steady shear step are (almost) the same. The conductivity measurements perpendicular to the shear plane probe almost the same direction as used in the BDS and the rheo-electric measurements in plate–plate or plate–ring geometry described before.

Figure 21 shows exemplarily the measurements of the anisotropic conductivity during *steady shear* (step 2) and *recovery after shear* (step 3) for a composite containing 10 wt% GNP. In step 2 (Fig. 21a) the DC conductivity differs between parallel and perpendicular direction by two orders of magnitude. This difference is already evident when the shear deformation starts and results from the orientation of the graphite plates by the squeeze flow during compression molding. This orientation could not be "erased" by the isothermal *annealing* for two hours. Interestingly, the decrease in conductivity under shear in Fig. 21b occur within seconds. This indicates that already small changes in the positions of the plates by the shear flow distinctly

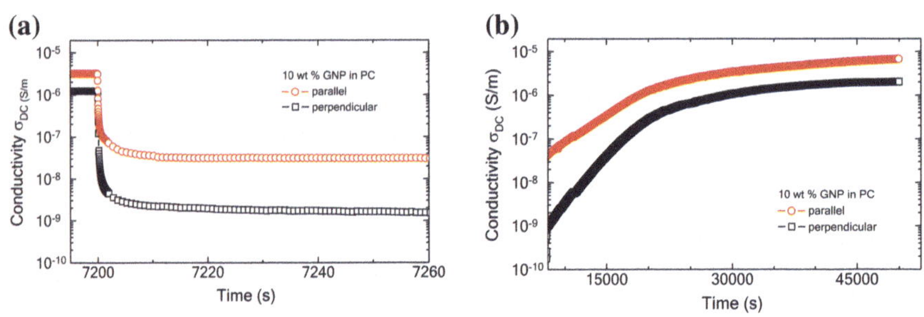

Fig. 21 Time dependence of the DC conductivity parallel and perpendicular to the direction of the shear flow (shear plane) of a PC composite with 10 wt% GNP during steady shear **a** and recovery in the quiescent melt after shear **b** at 230 °C

destroy the conductive paths. However, it is difficult to decide, whether orientation or translation of the plates with respect to each other is the dominant mechanism for the destruction of the conductive paths. The higher translational Péclet number Pe_{trans} (relation between hydrodynamic transport due to shear flow and translational diffusion of the center of mass) compared to Pe_{rot} for oblate ellipsoids, representing the GNP (see Table 2), seems to prefer the destruction of conductive paths by translational motion. D_{trans} and D_{rot} of the GNP are orders of magnitude smaller than for spheres or cylinders, representing CB aggregates and agglomerates, nanotube agglomerates or isolated MWCNT (see Table 1) This explains the extremely slow recovery of the conductivity of the composite with GNP after steady shear for both directions.

Similar to Fig. 20 the time dependence of the DC conductivity parallel and perpendicular to the shear plane is shown in Fig. 22 for a PC composite with 10 wt% GNP during the annealing step (dotted lines) and the recovery step (solid lines) in a double logarithmic scale. As shown in Fig. 21, the DC conductivity within the shear plane (noted as *parallel*) is about one order of magnitude higher than perpendicular to it. This supports the assumption of highly oriented GNP due to the compression molding step and that for long times of annealing or structural recovery no uniform orientation distribution can be achieved.

In the following, we try to get an estimate for the degree of GNP orientation by the compression molding step. For details, see Ref. [22]. Based on Abbasi et al. [60], the shear rate for compression molding assuming Newtonian flow can be estimated by:

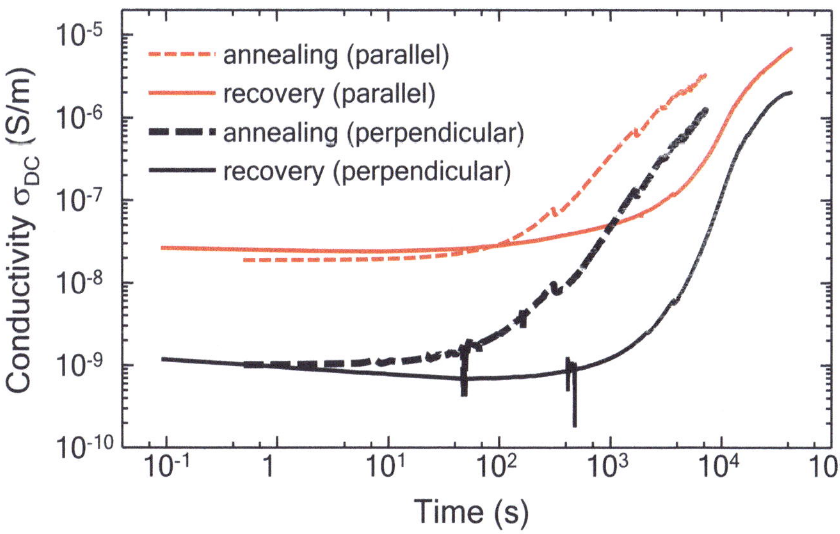

Fig. 22 Comparison of the time dependence of the DC conductivity parallel and perpendicular to the direction of the shear flow (shear plane) of a PC composite with 10 wt% GNP during annealing (dashed lines) and recovery (solid lines) in the quiescent melt after shear at 230 °C

$$\dot{\gamma} = \frac{3v}{H^3}\sqrt{4z^2(z-H)^2 + \frac{r^2}{2}(2z-H)^2} \tag{15}$$

in which r is the radius of the pressed plate, H the sample height before pressing, v the pressing speed and z the sample thickness after pressing. With a radius of the extruded strands of $r = 2$ mm, which are assembled between the plates of the press, $H = 1.5$ mm, $z = 1$ mm and $v = 0.6$ mm/s, a shear rate of $\dot{\gamma} = 0.1\ s^{-1}$ can be estimated. In a paper by Reddy et al. [61] from combined rheo-optical experiments, a power law relation between the average angle of orientation Ψ with respect to the flow direction and the Péclet number for GNP in a low viscous matrix is given:

$$\Psi = Pe_R^{-0.05} \cdot 45^\circ. \tag{16}$$

A mean angle Ψ of 0° corresponds to the full alignment of the GNP in the direction of the shear field. Using the rotational diffusion coefficients of oblate ellipsoids given in Table 1, for a GNP with a diameter of 15 μm and shear rates of $\dot{\gamma} = 0.1\ s^{-1}$ and $1\ s^{-1}$, orientation angles Ψ of 15° and 13°, respectively, are calculated. Based on the excluded volume of the plates and a percolation concentration of 4.9 vol% for the GNP [22], a value of Ψ between 15 and 17.5° can be estimated. This result, together with 2D small-angle X-ray scattering results on the pressed samples within the shear plane are shown in [22], is strongly supporting the assumption that the GNPs are oriented by squeeze flow.

4.2.4 Modeling of Agglomeration of Different Carbon Allotropes

Figure 23 shows a comparison of the time dependence of the DC conductivity (measured with the ring geometry perpendicular to the shear plane) of composites with MWCNT (1 wt%), CB XE2 (3 wt%) and GNP (10 wt%) during the steady

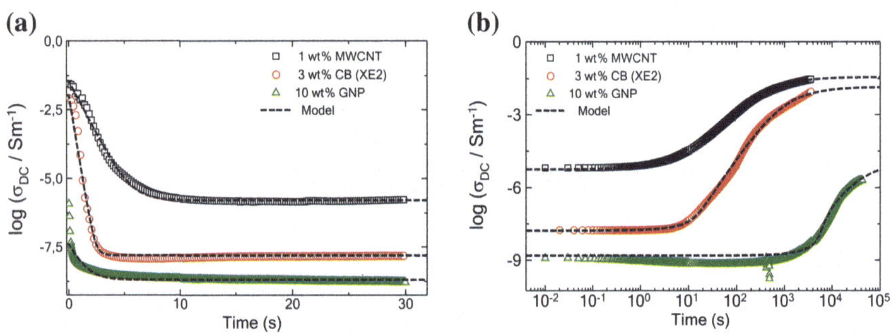

Fig. 23 Comparison of the time dependence of the DC conductivity for PC composites containing MWCNT (1 wt%), CB XE2 (3 wt%), and GNP 10 wt% during steady shear conditions (step 2 of the protocol) **a** and in the recovery step **b** in the quiescent melt after shearing (step 3) at 230 °C

Table 4 Parameters for the agglomeration model for the different carbon allotropes during recovery. The parameters for GNP represent the conductivity measured perpendicular to the shear flow

	$p_{a,\infty}$	$p_{c,a}$	$p_{a,0}$	σ_M(S/m)	σ_a(S/m)	$k_0(\mathrm{s}^{-1})$	μ	s
MWCNT (1 wt%)	0.3	0.2	0.201	$2 \cdot 10^{-10}$	2.31	$6.2 \cdot 10^{-3}$	2	0.73
CB XE2 (3 wt%)	0.3	0.2	0.199	$4 \cdot 10^{-10}$	1.6	$1.2 \cdot 10^{-3}$	2	0.73
GNP (10 wt%)	0.3	0.2	0.190	$2 \cdot 10^{-11}$	$5.1 \cdot 10^{-4}$	$2.4 \cdot 10^{-4}$	2	0.73

Table 5 Parameters for agglomeration under steady shear for different carbon allotropes during steady shear. For the GNP composite the parameters are for the conductivity measured perpendicular to the shear flow

	$p_{A,\infty}$	$p_{A,c}$	$p_{A,0}$	σ_M(S/m)	σ_A(S/m)	$k_1(\mathrm{s}^{-1})$	k_2 (s^{-1})
MWCNT (1 wt%)	0.3	0.2	0.3	$2 \cdot 10^{-10}$	2.31	0.59	0.29
CB XE2 (3 wt%)	0.3	0.2	0.3	$4 \cdot 10^{-10}$	1.6	1.43	0.73
GNP (10 wt%)	0.3	0.2	0,218	$2{\cdot}10^{-11}$	$5.1{\cdot}10^{-4}$	0.39	0.18

shear deformation (step 2) and the recovery step (step 3). The different filler concentrations were chosen to represent similar distances to the percolation concentration of the respective carbon allotropes. The time dependence of the conductivity can be modeled using the agglomeration approach described in Sect. 2.1 for both steps of the rheological protocol. The parameters for modeling the agglomeration without and under steady shear are summarized in Tables 4 and 5, respectively. From TEM images of agglomerated MWCNT composites [10], for $p_{a,\infty}$ a value of 0.3 is taken for long recovery times. The percolation concentration of the agglomerates was set to $p_{a,c} = 0.2$, which corresponds to the percolation concentration of randomly distributed spheres [62]. The coefficient for agglomerate build-up rate k_0, the intrinsic agglomerate conductivity σ_a and the initial agglomerate concentration $p_{a,0}$ are the remaining fit parameters. For a detailed description of the modeling and a discussion of the limitations of the model see [22].

As shown in Fig. 4, the model provides a reasonable description of the time development of the conductivity in the quiescent melt (recovery step) and under steady shear conditions for all three composites. During shear, shear-induced build-up and agglomerate destruction, which are two opposing processes, determine the concentration of agglomerates. On the one hand, the shear flow brings particles together, so that they can agglomerate by attractive interaction between them. The agglomeration under shear is described by the rate constant k_1. On the other hand, shear deformation leads to a destruction of already existing agglomerates, which is

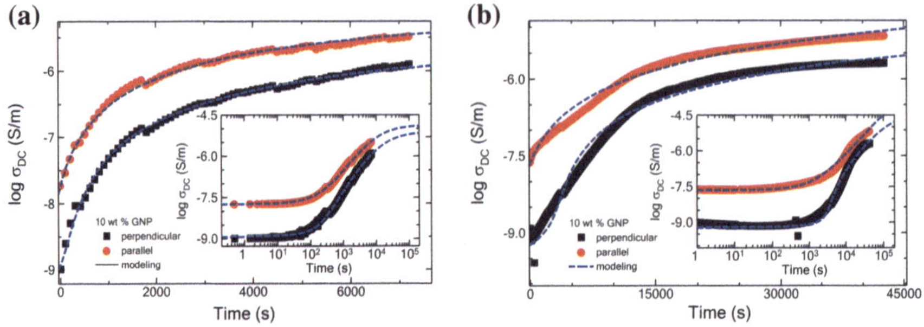

Fig. 24 Comparison of the time dependence of the DC conductivity parallel and perpendicular to the shear plane for PC composites containing GNP (10 wt%), during initial annealing (**a**) and recovery after steady shear in the quiescent melt (**b**). The dotted lines are fits with the agglomeration approach described in Sect. 2.1. The insets represent double logarithmic plots for the same data sets

modeled by the kinetic constant k_2 for agglomerate destruction. The agglomeration in the quiescent melt is described in the model by k_0. The interplay between build-up and destruction of agglomerates under continuous shear conditions leads to a transient network structure, which is reflected by a constant value of the DC conductivity at the given shear conditions and filler content. Since the shape, content and state of agglomeration and/or orientation are expected to be quite different for the three composites, a detailed interpretation of the parameters is not provided.

Finally, the conductivity parallel and perpendicular to the shear plane is modeled by using two parameter sets. Figure 24 shows the fits for both directions of the anisotropic DC conductivity measurements of a GNP composite containing 10 wt% GNP during isothermal annealing (step1) and recovery (step 3), both in the quiescent melt at 230 °C.

The data are the same as shown in Fig. 22 and the fit parameters are given in Table 6. All other parameters are the same as in Table 4.

The fits with the agglomeration model to the parallel and perpendicular components of the anisotropic conductivity during annealing (Fig. 24a) lead to different kinetic coefficients for both directions. Fitting of the conductivity of the recovery after shear deformation (Fig. 24b) results in lower coefficients k_0 compared to the annealing (step 1) for both directions. This leads to the assumption that the shear deformation in step 2 (see Fig. 21, not fitted) leads to a changed arrangement of the

Table 6 Rate constant k_0 for agglomerate build-up and the intrinsic DC conductivity of the agglomerates σ_a of a PC composite with 10 wt% GNP at different measurement directions in comparison between build-up and recovery

Direction	Annealing		Recovery	
	Perpendicular	Parallel	Perpendicular	Parallel
$k_0(\mathrm{s}^{-1})$	$2.5 \cdot 10^{-4}$	$4.0 \cdot 10^{-4}$	$2.3 \cdot 10^{-5}$	$1.0 \cdot 10^{-5}$
σ_a(S/m)	$5.1 \cdot 10^{-4}$	$9.0 \cdot 10^{-4}$	$3 \cdot 10^{-4}$	$3.1 \cdot 10^{-3}$

GNP even by weak shear. However, this arrangement does not recover within the time interval of our measurements.

It is important to point out that the agglomeration model in Sect. 2.1 has been developed for isotropically distributed, spherical particles and/or agglomerates. An extension to anisometric, highly oriented fillers goes far beyond these assumptions. Despite the relative good correspondence between measured data and modeled curves, the application of the agglomeration model to pre-oriented GNP is at best a very first attempt.

5 Concluding Remarks

In this study, electrical and dielectric properties of polycarbonate based composites containing carbon allotropes of different shapes and dimensions were measured on compression-molded plates at room temperature and in the melt. For all types of filler, the frequency-dependent complex conductivity and permittivity in solid samples show the typical features of a percolation phenomenon formed by conductive particles or clusters (agglomerates). The concentration values of the insulator to conductor transition could be related to shape, fractality, or aspect ratio of the particles. However, for carbon black and carbon nanotubes (here MWCNT) an agglomeration to clusters has to be assumed to explain the experimental findings. Despite the high aspect ratio of nanotubes, the percolation concentration was found to be higher than expected for this ratio and no significant anisotropy of the conductivity was found during and after weak shear flows. This lead to the assumption that the percolation of rather spherical nanotube agglomerates or clusters is the dominant mechanism for the formation of the conductive network.

To compare the kinetics of formation and destruction of the filler networks of carbon allotropes of different shapes and dimensionalities in the melt, a flow protocol combining isothermal annealing of the initial, melt pressed samples, weak steady shear, and, finally, isothermal recovery of the thus generated structures was applied to the composite melts. On the one hand, it has been found that rheological, electrical, and dielectric properties of polymer composites containing different carbon allotropes depend strongly on the micro- and nanostructure of the filler particles. In addition, the structure of the filler network was found to be not only dependent on the size, shape, interparticle forces, and volume fraction of the particles but is extremely sensitive to the thermo-mechanical prehistory. The interplay of build-up and destruction of the filler network in carbon allotrope nanocomposites was studied by measurements of conductivity during shear flow and under quiescent conditions. When applying a weak shear flow for a limited time to samples with well-dispersed nanofillers with concentrations below the percolation threshold, the conductivity and the viscosity were found to increase, whereas for samples, in which the filler network was established, a decrease of the conductivity was found. Both, conductivity and transient viscosity reached steady-state values due to the interplay of destruction and build-up of agglomerates in the melt flow. The graphite nanoplates (GNP) were found

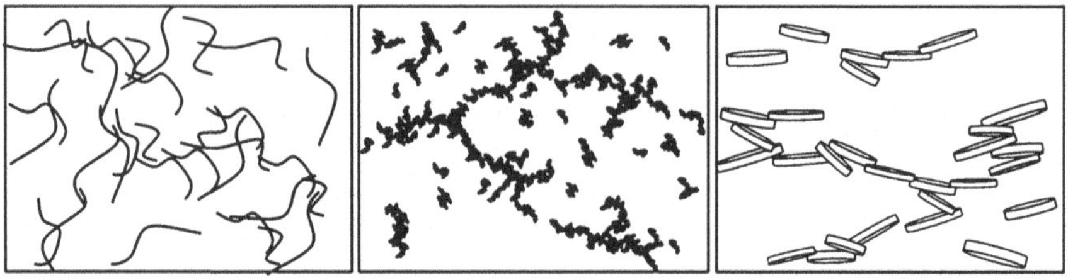

Fig. 25 Schematic representation of the agglomerated fillers for composites with different carbon allotropes: nanotubes (left), carbon black (middle), and graphite nanoplates (right)

to be strongly oriented already at weak shear flow, indicated by significantly different conductivities in shear direction and perpendicular to it, which were already existent after compression molding and remained during all steps of the flow protocol. This can be explained by the very large Péclet number, which compares the shear rate and the orientation diffusion in the melt. This orientation is expected to be arrested along all steps of the flow protocol.

In summary, it can be stated that agglomeration of the filler particles in the quiescent melt can be observed for all carbon allotropes (CB, MWCNT, and GNP) studied. A schematic representation of the structures formed by agglomeration for the different filler geometries with at least one dimension in the nanoscale, such as spherical shaped fractals, worm-like or cylindrical particles and platelets are shown in Fig. 25. The time-dependent evolution of the DC conductivities during annealing, shear deformation, and recovery in the quiescent melt can be described for the composites with different carbon allotropes by using the agglomeration approach. The MWCNT were found to form spherically shaped agglomerates in the quiescent melt and under a weak shear flow, showing a higher percolation threshold and less orientation than expected. In contrast to composites with CB and MWCNT, which form agglomerates, GNP orient already at low flow rates, although GNP have a similar aspect ratio as MWCNT. In the time scale of the experiments, the oriented platelets cannot recover to a randomized state by rotational diffusion.

Acknowledgements The authors thank the German Federal Ministry for Economic Affairs and the German Federal Ministry of Economics and Technology (BMWi) for financial support via the German Federation of Industrial Research Associations (AiF grants No. 122Z, 14454N, and 17546 BG). We also thank the researchers within these projects, Dr. Martin Engel, Sergej M. Dudkin, Dr. Sven Pegel, Wolfgang Böhm, Dr. Tetyana Skipa, Dr. Marina Saphiannikova, Dr. André Bormuth, Dr. Tobias Villmow, and Dr. Michael Thomas Müller for their contributions to the content of this book chapter. We thank Prof. Gert Heinrich for inspiring discussions.

References

1. Cassagnau P (2008) Melt rheology of clays and fumed silica nanocomposites. Polymer 49:2183–2196

2. Cassagnau P (2013) Linear viscoelasticity and dynamics of suspensions and molten polymers filled with nanoparticles of different aspect ratios. Polymer 54:4762–47759
3. Schüler R, Petermann J, Schulte K, Wentzel HP (1996) Percolation in carbon black filled epoxy resin. Macromol Symp 104:261–268
4. Schüler R, Petermann J, Schulte K, Wentzel HP (1996) Agglomeration and electrical percolation behavior of carbon black dispersed in epoxy resin. J Appl Polym Sci 63(13):1741–1746
5. Heinrich G, Costa FR, Abdel-Goad M, Wagenknecht U, Lauke B, Härtel V et al (2005) Structural kinetics in filled elastomers and PE/LDH composites. Kautsch Gummi Kunstst 58:163–167
6. Kharchenko SB, Douglas JF, Obrzut J, Grulke EA, Migler KB (2004) Flow-induced properties of nanotube-filled polymer. Mater Nat Mater 3(8):564–656
7. Obrzut J, Douglas JF, Kharchenko SB, Migler KB (2007) Shear-induced conductor-insulator transition in melt-mixed polypropylene-carbon nanotube dispersions. Phys Rev B Condens Matter Mater Phys 76 (19):195420
8. Alig I, Skipa T, Engel M, Lellinger D, Pegel S, Pötschke P (2007) Electrical conductivity recovery in carbon nanotube–polymer composites after transient shear. Physica Status Solidi B Basic Solid State Phys 244(11):4223–4226
9. Alig I, Lellinger D, Dudkin SM, Pötschke P (2007) Conductivity spectroscopy on melt processed polypropylene-multiwalled carbon nanotube composites: recovery after shear and crystallization. Polymer 48(4):1020–1029
10. Alig I, Skipa T, Lellinger L, Pötschke P (2008) Destruction and formation of a carbon nanotube network in polymer melts: Rheology and conductivity spectroscopy. Polymer 49(16):3524–3532
11. Alig I, Skipa T, Lellinger D, Bierdel M, Meyer H (2008) Dynamic percolation of carbon nanotube agglomerates in a polymer matrix: comparison of different model approaches. Phys Status Solidi B 245(10):2264–2267
12. Pegel S, Pötschke P, Petzold G, Alig I, Dudkin SM, Lellinger D (2008) Dispersion, agglomeration, and network formation of multiwalled carbon nanotubes in polycarbonate melts. Polymer 49(4):974–984
13. Lellinger D, Xu D, Ohneiser A, Skipa T, Alig I (2008) Influence of the injection moulding conditions on the in-line measured electrical conductivity of polymer-carbon nanotube composites. Phys Status Solidi B 245(10):2268–2271
14. Skipa T, Lellinger D, Böhm W, Saphiannikova M, Alig I (2010) Influence of shear deformation on carbon nanotube networks in polycarbonate melts: Interplay between build-up and destruction of agglomerates. Polymer 51:201–210
15. Bauhofer W, Schulz SC, Eken AE, Skipa T, Lellinger D, Alig I, Tozzi EJ, Klingenberg DJ (2010) Shear-controlled electrical conductivity of carbon nanotubes networks suspended in low and high molecular weight liquids. Polymer 51:5024–5027
16. Alig I, Pötschke P, Lellinger D, Skipa T, Pegel S, Kasaliwal GR, Villmow T (2012) Establishment, morphology and properties of carbon nanotube networks in polymer melts. Polymer 53:4–28
17. Hilarius K, Lellinger D, Alig I, Villmow T, Pegel S, Pötschke P (2013) Influence of shear deformation on the electrical and rheological properties of combined filler networks in polymer melts: Carbon nanotubes and carbon black in polycarbonate. Polymer 54:5865–5874
18. Cardinaels R., Zhu Y, Moldenaers P, Mewis J (2011) Shear-induced structural buildup in polymer-clay nanocomposites. In: Proceedings of 27th polymer processing society annual meeting 2011 (PPS-27), Marrakech, Morocco, OP-3–1124, Polymer Processing Society.
19. Krishnamoorti R, Yurekli K (2001) Rheology of polymer layered silicate nanocomposites. Curr Opin Colloid Interface Sci 6:464–470
20. Zoldan J, Siegmann A, Narkis M, Alig I (2006) Dielectric spectroscopy of anisotropic polypropylene/Nylon-66 blends containing carbon black. J Macromol Sci Part B 45:61–83
21. Bormuth A (2008) Anisotropie der elektrischen Leitfähigkeit, diploma thesis. Technische Universität Darmstadt, Germany

22. Hilarius K (2014) Elektrische und rheologische Eigenschaften von kohlenstoffbasierten Füllstoffnetzwerken in Polymeren (Electrical and rheological properties of carbon based filler networks in polymers), Ph.D. thesis, Technische Universität Darmstadt, Germany
23. Starý Z, Krückel J, Schubert DW (2014) Conductivity of carbon black-based polymer composites under creep in the molten state. Polymer 55:3980–3986
24. Kim H, Macosko CW (2009) Processing-property relationships of polycarbonate/graphene composites. Polymer 50:3797–3809
25. Boothroyd SC, Johnson DW, Weir MP, Reynolds CD, Hart JM, Smith AJ, Clarke N, Thompson RL, Coleman KS (2018) Controlled structure evolution of graphene networks in polymer composites. Chem Mater 30:1524–1531
26. Böhm GGA, Nguyen MN (1995) Flocculation of carbon black in filled rubber compounds. I. Flocculation occurring in unvulcanized compounds during annealing at elevated temperatures. J Appl Polym Sci 55:1041–1050
27. Kirkpatrick S (1973) Percolation and conduction. Rev Mod Phys 45:574–588
28. De Gennes PG (1976) On a relation between percolation theory and the elasticity of gels. J Phys Lett 37:1–2
29. Efros AL, Shklovskii BI (1976) Critical behaviour of conductivity and dielectric constant near the metal-non-metal transition threshold. Physica Status Solidi B Basic Res 76:475–485
30. Straley JP (1976) Critical phenomena in resistor networks. J Phys C Solid State Phys 9:783–795
31. Bergman DJ, Imry Y (1977) Critical behavior of the complex dielectric constant near the percolation threshold of a heterogeneous material. Phys Rev Lett 39:1222–1225
32. Webman I, Jortner J, Cohen MH (1977) Critical exponents for percolation conductivity in resistor networks. Phys Rev B 16(6):2593–2596
33. Clerc JP, Giraud G, Laugier JM, Luck JM (1990) The electrical conductivity of binary disordered systems, percolation clusters, fractals and related models. Adv Phys 39(3):191–308
34. Bunde A, Havlin S (1996) Fractals and disordered systems. Springer, Berlin-Heidelberg-New York
35. Stauffer D, Aharony A (1994) Introduction to percolation theory. Taylor & Francis Ltd., London
36. Sahimi M (1994) Applications of percolation theory. Taylor & Francis Ltd., London
37. Pötschke P, Dudkin SM, Alig I (2003) Dielectric spectroscopy on melt processed polycarbonate - multiwalled carbon nanotube composites. Polymer 44:5023–5030
38. McLachlan DS, Blaszkiewicz M, Newnham RE (1990) Electrical resistivity of composites. J Am Ceram Soc 73:2187–2203
39. Almond DP, Bowen CR (2004) Anomalous power law dispersions in ac conductivity and permittivity shown to be characteristics of microstructural electrical networks. Phys Rev Lett 92:157601
40. Andrews R, Jacques D, Qian D, Dickey EC (2001) Purification and structural annealing of multiwalled carbon nanotubes at graphitization temperatures. Carbon 39:1681–1687
41. Fournier J, Boiteux G, Seytre G, Marichy G (1997) Percolation network of polypyrrole in conducting polymer composites. Synth Met 84:839–840
42. Einstein A (1906) Eine neue Bestimmung der Moleküldimensionen. Ann Phys 324(2):289–306
43. Saffman P, Delbrück M (1975) Brownian motion in biological membranes. Proc Natl Acad Sci 72(8):3111–3113
44. Koenig SH (1975) Brownian motion of an ellipsoid. A correction to Perrin's results. Biopolymers 14:2421–2423
45. Perrin F (1936) Mouvement Brownien d'un ellipsoide (II). Rotation libre et dépolarisation des fluorescences. Translation et diffusion de molécules ellipsoidales. J Phys Radium 7(1):1–11
46. Perrin F (1934) Mouvement brownien d'un ellipsoide-I. Dispersion diélectrique pour des molécules ellipsoidales. J Phys Radium 5:497–511
47. Tirado MM, de La Torre JG (1980) Rotational dynamics of rigid, symmetric top macromolecules. Application to circular cylinders. J Chem Phys 73:1986
48. Tirado MM, de La Torre J (1979) Translational friction coefficients of rigid, symmetric top macromolecules. Application to circular cylinders. J Chem Phy 71:2581

49. Larson RG (1999) The structure and rheology of complex fluids. Oxford University Press, New York
50. Datasheet Makrolon 2600, Edition 28.09.2011, Bayer MaterialScience, Germany
51. Datasheet Nanocyl NC 7000, Edition 2009–03–10 V05. Sambreville, Belgium
52. xGnP Graphene Nanoplatelets Grade M Datasheet, 2012, XG Sciences, Lansing, MI, USA
53. Dyre JC (1993) Universal low-temperature ac conductivity of macroscopically disordered nonmetals. Phys Rev B 48:12511–12526
54. Jonscher AK (1977) The universal dielectric response. Nature 267:673–679
55. Barton J (1966) Dielectric relaxation of some ternary alkali-alkaline earth-silicate glasses. Verres Refract 20:328–335
56. Nakajima T (1971) Annual report conference on electric insulation and dielectric phenomena, 1972, p 168
57. Namikawa H (1975) Characterization of the diffusion process in oxide glasses based on the correlation between electric conduction and dielectric relaxation. J Non-Cryst Solids 18:173–195
58. Via MD, King JA, Keith JM, Bogucki GR (2012) Electrical conductivity modeling of carbon black/polycarbonate, carbon nanotube/polycarbonate, and exfoliated graphite nanoplatelet/polycarbonate composites. J Appl Polym Sci 124(1):182–189
59. Sumfleth J, Buschhorn ST, Schulte K (2011) Comparison of rheological and electrical percolation phenomena in carbon black and carbon nanotube filled epoxy polymers. J Mater Sci 46:659–669
60. Abbasi S, Carreau PJ, Derdouri A (2010) Flow induced orientation of multiwalled carbon nanotubes in polycarbonate nanocomposites: Rheology, conductivity and mechanical properties. Polymer 51:922–935
61. Reddy NK (2011) Suspensions and Gels Composed of Non-Spherical Nanoparticles: Rheology and Scattering (Suspensies en gels opgebouwd uit niet-sferische nanodeeltjes: Reologie en lichtverstrooiing), Ph.D. thesis, KU Leuven, Belgium
62. Munson-McGee SH (1991) Estimation of the critical concentration in an anisotropic percolation network. Phys Rev B 43:3331

Epoxy-Based Nanocomposites—What Can Be Learned from Dielectric and Calorimetric Investigations?

Paulina Szymoniak and Andreas Schönhals

Abstract Epoxy-based nanocomposites are promising materials for industrial applications (i.e., aerospace, marine, and automotive industries) due to their extraordinary mechanical and thermal properties. Regardless of the broad field of applications, there is still a considerable need to identify their structure–property relationships. Here, a detailed dielectric and calorimetric (DSC and fast scanning calorimetry) study on different epoxy-based nanocomposites was performed. Bisphenol A diglycidyl ether (DGEBA) cured with diethylenetriamine (DETA) was employed as the polymeric matrix, which was reinforced with three diverse nanofillers that exhibit different interaction strengths with the epoxy matrix (halloysite nanotubes, surface modified halloysite nanotubes, and taurine-modified layered double hydroxide). The structure, molecular mobility, and vitrification behavior are discussed in detail, focusing on the intrinsic structural and dynamic heterogeneity, as well as interfacial properties.

Keywords Nanocomposites · Epoxy nanocomposites · Dynamics · Interphase · Rigid amorphous fraction · Dielectric spectroscopy · Flash DSC · Temperature modulated Flash DSC · Temperature modulated TMDSC

Symbols and Abbreviations

A_T	Step height in a periodic temperature program
BDS	Broadband dielectric spectroscopy
CNT	Carbon nanotubes
C_p^*	Complex heat capacity
c_p^*	Complex specific heat capacity
CRR	Cooperatively rearranging region
Δc_p	Calorimetric strength
$\Delta\varepsilon$	Dielectric strength

P. Szymoniak (✉) · A. Schönhals
Bundesanstalt für Materialforschung und –prüfung (BAM), Unter den Eichen 87, 12205 Berlin, Germany
e-mail: Paulina.Szymoniak@bam.de

A. Schönhals and P. Szymoniak (eds.), *Dynamics of Composite Materials*, Advances in Dielectrics, https://doi.org/10.1007/978-3-030-89723-9_11

DETA	Diethylenetriamine
DGEBA	Bisphenol A diglycidyl ether
DMA	Dynamic mechanical analysis
DSC	Differential scanning calorimetry
δT	Temperature fluctuation
ΔT_g	Width of the glass transition
$\varepsilon', \varepsilon''$	Real and imaginary part of the complex dielectric function
E_A	Activation energy
f	Frequency
FD	Frequency domain
f_p	Relaxation rate
$f_{p,thermal}$	Thermal relaxation rates from static FSC
FSC	Fast scanning calorimetry
HF	Heat flow
HN	Havriliak–Negami
HNTs	Halloysite nanotubes
ξ	Cooperativity length scale
LDH	Layered double hydroxide
MAF	Mobile amorphous fraction
mHNTs	Surface modified halloysite nanotubes
MTHPA	Methyl tetrahydrophtalic anhydride
MWS	Maxwell/Wagner/Sillars
NP	Nanoparticle
PDA	Polydopamine
PNC	Polymer nanocomposite
RAF	Rigid amorphous fraction
SAXS	Small angle X-ray scattering
SHS	Specific heat spectroscopy
$\dot{T}$	Heating rate
τ	Relaxation time
T-LDH	Taurine-modified layered double hydroxide
TEM	Transmission electron microscopy
TD	Temperature domain
T_g	Calorimetric glass transition temperature
TMDSC	Temperature modulated DSC
TMFSC	Temperature modulated FSC
T_0	Vogel temperature
t_p	Length of the isothermal period in a periodic temperature program
VFT	Vogel–Fulcher–Tammann
ω	Radial frequency

1 Introduction

In the extremely broad field of polymer nanocomposites (PNCs), regardless of the number of the remaining open questions, some understanding of model systems has been already achieved. On the contrary, more complex polymer nanocomposites of higher industrial significance are often addressed in applied studies but only scarcely considered in fundamental investigations. One class of such materials, avoided in fundamental studies, is epoxy-based nanocomposites, due to their complexity.

Owing to the low production costs, lightweight, and extraordinary properties of epoxies that can be further improved by the introduction of nanofillers, these materials are promising for numerous industrial applications. Due to the superior mechanical properties, excellent adherence, chemical and corrosion resistance, high specific strength, and ease of processing [1, 2], epoxies serve as structural and metal bonding materials in aerospace, marine, and transportation industries. They are one of the most used adhesives because of their low shrinkage and a multitude of available curing options. Furthermore, because of their high corrosion resistance, strong adhesion, and good physical properties, epoxies are also excellent coating materials. Other applications range from laminates, electronic devices, encapsulants, packaging, and many more [3]. Regardless of the broad field of applications of epoxy-based materials, while the chemistry of epoxidation is generally understood, there is still a considerable need to identify their structure–property relationships and ways to engineer their properties for the desired applications.

Epoxies are thermosets formed via a curing process (crosslinking reaction) of an epoxy resin with a hardener (curing agent). They comprise oxirane rings (cyclic ether) as an "epoxy" functionality and can contain either aliphatic, aromatic, and/or heterocyclic structures in the molecule. The most common epoxy resin (85 % of the world's epoxy resin production [1]), also used in this work, is bisphenol A diglycidyl ether (DGEBA). Hardeners include polyfunctional amines, acids or anhydrides, phenols, alcohols, and thiols. The chemical structure of the hardener which is mixed with the resin in predetermined ratios controls the nature of the crosslinks, and, thus, the cure kinetics, and, therefore, the properties of the epoxy. The curing reaction is initiated most often by heating the reaction mixture to temperatures higher than the room temperature. The versatility of epoxy lies in the fact that there are multiple combinations of components available (also different resin to hardener ratios) and curing conditions (temperature and time), yielding different polymer structures with different thermo-mechanical properties (see Sect. 3).

Furthermore, the properties of epoxy-based materials are affected not only by the chosen components. For a specific resin–hardener combination, the properties can be tuned to a certain extent through modifying the molecular architecture by varying the crosslinking density. For instance, increasing the crosslinking density generates high stiffness and strength. Nevertheless, highly crosslinked materials often behave undesirably brittle because plastic deformation is strongly limited. The inherent brittleness of epoxies is known as a disadvantage of these materials. Other drawbacks include local stress concentrations, which might initiate cracks that lead to a spontaneous failure [4] and relatively high flammability [5]. Therefore, for decades, the primary aim in the field of epoxies is to provide these materials with higher toughness and lower flammability, allowing for a broader range of possible applications. Furthermore, these property improvements must take place without significantly waiving other important characteristics like thermo-mechanical properties or the modulus.

One successful approach employed by many researchers [6] is the incorporation of inorganic nanofillers (one dimension <100 nm) in the epoxy matrix. To increase the toughness, reduce local stress concentrations, generate high stiffness, and impact resistance, the employed nanoparticles (NPs) should have small spatial dimensions and, therefore, a high specific surface, a higher rigidity than the matrix, sufficient filler–matrix interactions, and incorporated water to reduce flammability, as well as other functionalities, selected for the desired application.

Novel multifunctional epoxy nanocomposites with improved mechanical properties and new functionalities such as electrical conductivity [7], anticorrosive [8], magnetic [9], and optical [10] properties have been already successfully prepared employing a broad range of nanofillers [11]: carbon nanofibers [12], carbon nanotubes [13], layered double hydroxides [14, 15], halloysite nanotubes [16, 17], iron and iron oxide nanoparticles [18], aluminum oxide [19], aluminum oxide hydroxide (boehmite) [20, 21], graphene [22], nanoclay [23], silica [24], and zinc oxide [25] just to mention a few.

Typically, for PNCs, the problems in the understanding of the structure–property relationships are related to the dispersion of NPs and the formation of an interfacial region. For epoxies, in addition, the NPs are present in the mixture during the curing reaction. Depending on the surface chemistry, some nanofillers may catalyze or take part in the network formation, as they could introduce additional functionalities and small molecules such as water to the crosslinking reaction. Furthermore, NPs might act as spatial barriers in the diffusion-controlled curing reaction, slowing down the diffusion coefficient and leading to a decrease in conversion. As a result, the crosslinking density of an epoxy-based PNC might be lower than that of the corresponding unfilled material [26]. Lastly, preferential adsorption of one of the components (resin/hardener) onto NPs might take place, altering the stoichiometry of the reaction mixture. In this scenario, not all reaction partners might be able to react with each other, which leads again to a decrease in the crosslinking density of the network compared to the unfilled material.

This chapter aims to provide an overview of the structure, dynamics, and vitrification behavior of three different DGEBA-based nanocomposite systems where diethylenetriamine (DETA) is the curing agent and the nanoparticles have different interaction properties with the crosslinked matrix (DGEBA/DETA). The employed techniques include conventional differential scanning calorimetry (DSC), fast scanning calorimetry (FSC), temperature modulated DSC (TMDSC), and FSC (TMFSC), as well as broadband dielectric spectroscopy (BDS). First, the properties of the pure epoxy will be addressed, which is necessary to understand the behavior of the nanocomposites. In the second part, the general properties of epoxy-based PNCs are discussed, whereas in the third part, the differences due to the different nanoparticle–matrix synergy are presented.

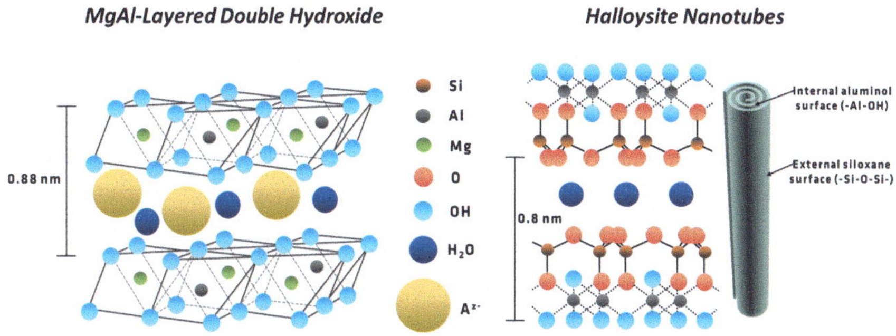

Fig. 1 Schematic representation of the structure of MgAl-layered double hydroxide (left) and halloysite nanotubes (right). Adapted with permission from Ref. [16]. Copyright (2021) MDPI

2 Experimental

2.1 *Materials*

In this chapter, three different DGEBA-based nanocomposites are considered. DGEBA was cured with the amine-based hardener DETA. The first system is reinforced with taurine-modified layered double hydroxide (T-LDH). For the second system, halloysite nanotubes (HNTs) were employed as nanofiller where surface modified halloysite nanotubes (mHNTs) are used for the third nanocomposite. The pure epoxy and the PNCs were prepared under identical curing conditions. The nanocomposites have different nanofiller contents, ranging from 3 to 18 wt%. The samples were named as Ep/T-LDHX, Ep/HNTX, and Ep/m-HNTX, where X represents the nominal weight concentration of the NPs. Details about the preparation are given in Refs. [17, 27, 28].

2.1.1 Nanofillers

Taurine-Modified MgAl-Layered Double Hydroxide (T-LDH)

Layered double hydroxides (LDHs) combine the features of conventional metal hydroxide-type fillers, like magnesium hydroxide, with the layered silicates, like montmorillonite. Due to the large amount of tightly bound water and their high-temperature endothermic decomposition, LDHs are commonly used to improve the flame retardancy and mechanical properties (i.e., tensile modulus, yield strength, and hardness) of polymers [30, 31].

LDHs are anionic clays with the chemical formula $[M^{2+}{}_{1-x}\ M^{3+}{}_{x}\ (OH)_2]^{x+}$ $(A^{z-})_{x/z}{\cdot}mH_2O]^{x-}$, where M^{2+} and M^{3+} are divalent and trivalent metal cations (here Mg^{2+} and Al^{3+}), and A^{z-} is an exchangeable interlayer anion. Their structure (Fig. 1) resembles brucite-like ($Mg(OH)_2$) layered crystals, where magnesium is octahedrally surrounded by hydroxyl groups. However, in LDHs, the divalent cation Mg^{2+} of the pristine brucite is partially replaced in an isomorphous way by a trivalent cation (here Al^{3+}), resulting in positively charged layers. The charge is compensated by interlayer anions (A^{z-}). Typically, $CO_3{}^{2-}$, $NO_3{}^{-}$, and OH^- are the intergalleries anions, which can be replaced by ion exchange. Therefore, LDHs can be modified by intercalation with functional ions [31]. The size of the gap between the layers (interlayer, gallery) depends on the employed anion. Layered materials can have an intercalated or exfoliated morphology in nanocomposites depending on the interfacial interaction between the polymer and the NPs and the preparation procedure (see Chapter "(Nano)Composite Materials–An Introduction" for more details). In most cases, intermediate morphologies between intercalation and exfoliation, and partial exfoliation and partial intercalation can also be found [32]. A modification of the LDH by exchanging the anion with, for instance, sodium dodecyl benzene sulfonate (SDBS) results in increased intergallery distance, enabling a further exfoliated and/or intercalated morphology of the nanofiller in the polymer matrix [33]. A further approach is to exchange A^{z-} with molecules containing functional groups to enhance the interaction of the filler with the polymer matrix. Here, the intergalleries of the LDH are modified by exchanging A^{z-} with taurine, which is a small molecule; therefore, no increase of the interlayer distance is observed [15]. Since taurine bears amine functionalities ($-NH_2$ groups), it is expected that it reacts with the oxirane rings of epoxy resin by forming covalent bonds.

Halloysite Nanotubes (HNTs)

Halloysite nanotubes (Fig. 1) are gaining increased attention as a cheaper, less toxic, and more environmentally friendly alternative for the commonly used carbon nanotubes (CNTs) [34] due to the similar aspect ratios of the two fillers. CNTs are employed as nanofillers due to their high aspect ratio, simple processing, and the surface modification possibilities. The incorporation of CNTs was found to be useful to improve the properties of epoxy-based PNCs leading to increased mechanical, electrical, and thermal characteristics [35]. HNTs were shown to improve, for instance, the thermal stability, mechanical properties, biocompatibility, flame retardancy, and waterproof behavior of a polymer matrix [17, 36–39]. Furthermore, HNTs can be incorporated without intercalation or exfoliation into the epoxy matrix.

Structurally, HNTs are aluminosilicates with the chemical formula $Al_2Si_2O_5(OH)_4{\cdot}nH_2O$, where if $n = 2$, the HNT is in a hydrated state with a layer of water molecules in the interlayer space [40]. These nanofillers have a hollow tubular structure with an outer surface chemically similar to the SiO_2 [41], containing siloxane groups at the exterior surface. The internal surface of HNTs has aluminol groups. The SPONGE program (scattering pattern calculator written in Python) [42] was used to assign the features observed in the SAXS patterns employing simulations of rolled cylinders to identify the dimension of the used halloysite nanotubes [16]. It was found that the length of the pristine HNTs is larger than 400 nm (exceeding the experimental accessible SAXS range), the outer diameter is ~125 nm, and the single sheet thickness is ~4 nm.

Modified Halloysite Nanotubes (mHNTs)

To investigate the effect of a surface modification on the structure and molecular mobility of the epoxy-based nanocomposite, the halloysite nanotubes were modified polydopamine (PDA) and ultrafine iron trihydroxide ($Fe(OH)_3$) nanoparticles (size <4 nm). Pristine HNTs, due to their external siloxane groups, have a weak interaction with the epoxy, leading to an increased formation of clusters and agglomerates [43]. Applying a 3 nm thick PDA coating on the external surface of the tube results in a large number of peripheral hydroxyl groups which can interact with the matrix and force a better dispersion of the nanofiller in the epoxy matrix. Iron nanoparticles were employed to endow PDA-coated HNT nanofillers with improved fire retardancy compared to unmodified HNT [16].

2.2 Methods

2.2.1 Calorimetry

Differential Scanning Calorimetry (DSC). DSC, employing a Perkin Elmer DSC 8500 instrument, was used to study the vitrification kinetics of the materials. The samples (5–8 mg) were measured in 50 μl aluminum pans in the temperature range from 298 to 455 K at a constant heating rate of $\dot{T} = 10\ \mathrm{K\ min^{-1}}$. The heat flow data of the second heating run were analyzed to ensure that the samples are fully cured.

Fast Scanning Calorimetry (FSC). FSC was used to extend the study on the glass transition behavior of the materials investigated by DSC to higher heating rates. A power compensated differential scanning calorimeter Mettler Toledo Flash DSC1 [44] was used which is based on non-adiabatic chip calorimetry [45]. The employed UFS1-sensor [46] allows for heating rates $\dot{T}$ in the range from 0.5 to 40 000 K s^{-1}. Because thermosetting materials cannot be melted on the sensor, the high viscosity silicon oil AK 60 000 from Wacker Chemie AG was used to improve the thermal contact between the sample (200–400 ng) and the sensor, as well as to minimize thermal lags.

Temperature Modulated DSC and FSC (TMDSC, TMFSC). Temperature modulated calorimetry is a class of techniques that belongs to a broader family of specific heat spectroscopy (SHS) and is a powerful tool to study segmental dynamics from the perspective of enthalpy fluctuations. Analogously to BDS, it can be described in the framework of the linear response theory. However, for SHS, the thermal processes are related to time-dependent enthalpy fluctuations that result from an external periodic temperature perturbation.

TMDSC is performed on a Perkin Elmer DSC 8500 device. Measurements were carried out on the same samples as the ones used for DSC. The temperature program consists of heating steps with a step height of $A_T = 2\,\mathrm{K}$ and a heating rate of

$\dot{T} = 50 \text{ K min}^{-1}$ as well as isothermal periods with length of $t_p = 18 - 120\,s$ corresponding to modulation frequencies of 5.6×10^{-2}–8.3×10^{-3} Hz.

The complex heat capacity (C_p^*) is obtained from the area under the heat flow ($HF(t)$) rate peaks:

$$C_p{}^*(T, t, \omega, \ldots) = \frac{1}{A_T} \int_0^{t_s} HF(t)dt \tag{1}$$

where the complex specific heat capacity (c_p^*) is calculated by

$$c_p^* = \frac{C_P^*}{m} \tag{2}$$

with m the sample mass.

Typically, the discussed quantity obtained from TMDSC measurements is the so-called reversing heat capacity, i.e., modulus of complex specific heat capacity $|c_p^*| = \sqrt{c_p'^2 + c_p''^2}$.

TMFSC is another technique to study the complex heat capacity of a material, or equivalently its modulus, however, extending the frequencies accessible by TMDSC [47, 48]. In principle, the same temperature program like for TMDSC is employed, with $A_T = 2\text{K}$, $\dot{T} = 200$ and 2000 K s^{-1}, and $t_p = 1,\ 0.1$ and 0.05 s corresponding to modulation base frequencies of 1, 10, and 20 Hz. Higher harmonics were also extracted from the base frequencies (from 1 to 11 Hz for 1 s, from 10 to 110 Hz for 0.1 s, and from 20 to 220 Hz for 0.05 s). Here, $C_P{}^*$ is calculated from the ratio of the Fourier transformations of the instantaneous heat flow (HF) and the instantaneous heating rate [49]:

$$C_p^* = \frac{\int_0^{t_p} HF(t)e^{-i\omega t}dt}{\int_0^{t_p} \dot{T}\ (t)e^{-i\omega t}dt} \tag{3}$$

Due to the experimental limitations, i.e., small sample mass which cannot be estimated easily, TMFSC results in the complex heat capacity and not in the complex specific heat capacity. For discussion, again the reversing heat capacity is used.

2.2.2 Broadband Dielectric Spectroscopy

Dielectric measurements were carried out in a frequency range of 10^{-1}–10^{7} Hz and a temperature range of 173–423 K for Ep/T-LDH and 173–503 K for Ep/HNT and Ep/mHNT. A Novocontrol high-resolution Alpha analyzer interfaced to an active sample cell was used. For temperature control, a Quatro Novocontrol cryosystem was employed with temperature stability better than 0.1 K.

Analysis of dielectric spectra

In general, the analysis of dielectric data is carried out by fitting the model function of Havriliak and Negami (HN-function) to the dielectric loss spectra [50, 51], which reads

$$\varepsilon_{HN}^{*}(\omega) = \varepsilon_{\infty} + \frac{\Delta\varepsilon}{(1 + (i\omega\tau_{HN})^{\beta})^{\gamma}} \tag{4}$$

where β and γ are the shape parameters ($0 < \beta \leq 1$ and $0 < \beta\gamma \leq 1$), $\Delta\varepsilon_{HN}$ is the dielectric strength, and τ_{HN} is the characteristic relaxation time, related to the maximum position of the corresponding dielectric relaxation process (relaxation rate $f_p = 1/(2\pi\tau)$). However, for epoxy-based materials, the HN-function can be used unambiguously only to analyze the localized processes at temperatures below the calorimetric glass transition T_g. Around and above T_g, the dielectric spectra are dominated by conductivity and polarization contributions, overlaying the structural relaxation. This indicates the presence of charge carriers in these materials, most likely ionic impurities (sodium and chloride ions), remaining from the synthesis [52]. The strong contribution of the ionic conduction in the measured spectra can be partly eliminated employing a "conduction-free loss" method, which is given for a Debye function by [53]

$$\varepsilon''_{deriv} = -\frac{\pi}{2}\frac{\partial\varepsilon'(\omega)}{\partial\ln\omega} = (\varepsilon'')^2 \tag{5}$$

In the resulting spectra, the Ohmic conductivity contribution ($\varepsilon''\sim 1/\omega$) is removed because for such a dependency, the real part ε' is independent of frequency. Moreover, the relaxation peak in ε''_{deriv} is narrower than the peak in ε'' due to the squared ε'' in Eq. 5.

For the HN-function, the derivative of the HN-function reads:

$$\frac{\partial\varepsilon'_{\mathrm{HN}}}{\partial\ln\omega} = \frac{\beta\gamma\,\Delta\varepsilon_{\mathrm{HN}}(\omega\tau_{\mathrm{HN}})^{\beta}\cos\left(\frac{\beta\pi}{2} - (1+\gamma)\Psi(\omega)\right)}{\left[1 + 2(\omega\tau_{\mathrm{HN}})^{\beta}\cos\left(\frac{\beta\pi}{2}\right) + (\omega\tau_{\mathrm{HN}})^{2\beta}\right]^{\frac{1+\gamma}{2}}} \tag{6}$$

$\Psi(\omega)$ is defined as

$$\Psi(\omega) = \arctan\left[\frac{\sin\left(\frac{\beta\pi}{2}\right)}{(\omega\tau_{HN})^{-\beta} + \cos\left(\frac{\beta\pi}{2}\right)}\right] \tag{7}$$

The power of the "conduction-free" loss approach is shown in Fig. 2a and b.

In the frequency domain (FD), the analysis of the dielectric spectra yields one segmental relaxation (α-relaxation). The relaxation rates of the α-relaxation are plotted in Fig. 3a in the Arrhenius diagram (relaxation map). Further, the conduction-

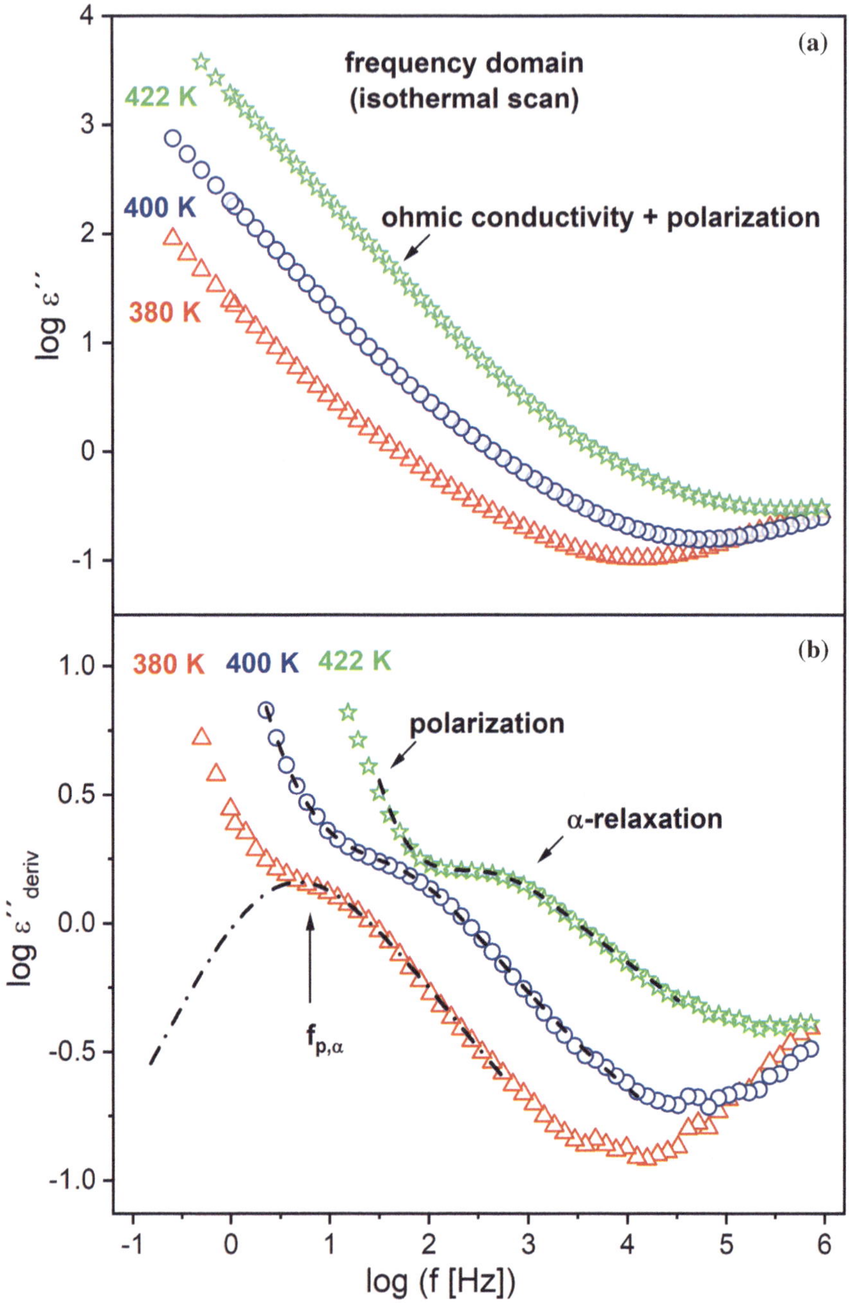

Fig. 2 Frequency dependence of **a** ε'' and **b** ε''_{deriv} for the pure epoxy for the second heating run at 380, 400, and 422 K. Black dashed lines in part b are the fits of Eq. 6 to the data including a power law for the contribution of the polarization. The dash-dotted line gives the contribution of the segmental relaxation at 380 K. Reproduced with permission from Ref. [15]. Copyright (2019) Elsevier

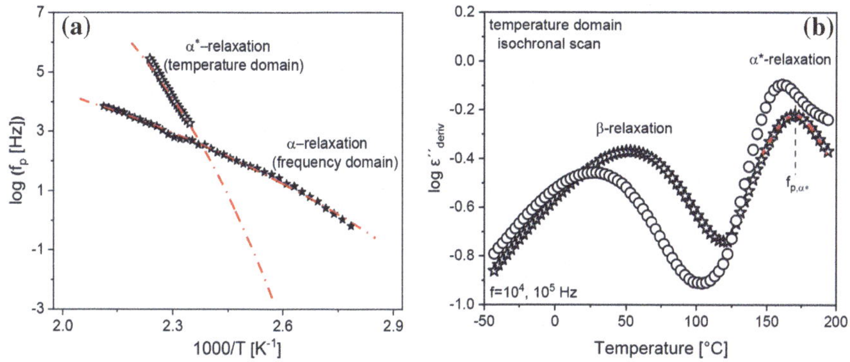

Fig. 3 **a** Relaxation map for the pure epoxy showing the relaxation rates for the α- (empty symbols) and α*-relaxation (filled symbols). The red dash-dotted lines indicate the Vogel–Fulcher–Tammann (VFT) fits to the data. **b** Temperature dependence of ε''_{deriv} for the pure epoxy for the second heating run at f = 10^4 Hz (circles) and f = 10^5 Hz (stars), where the red line is the Gaussian fit to the data. For the definition of the VFT function, see below. The data are taken from Ref. [15]

free dielectric loss can be also analyzed in the isochronal representation (temperature domain; TD). Typically for homopolymers, an identical temperature dependence of the relaxation rates of the processess in FD and TD is expected and observed. However, in some cases for heterogeneous materials, when multiple processes coexist in the same range of temperatures or frequencies, an additional analysis of TD could provide information about a process with a weaker intensity than the one observed in the FD [54]. The conduction-free dielectric loss in TD, the besides the β-process, shows one further process which is called α*-relaxation. It can be analyzed by a Gaussian fit, as shown in Fig. 3b. Nevertheless, a Lorentzian function or parabola can be also used to analyze relaxation processes observed in the temperature domain. The relaxation rates obtained from TD were plotted in Fig. 3a and compared with those obtained from the FD. The relaxation rates obtained from the TD domain have a different temperature dependence than those of the α-relaxation obtained obtained from the frequency domain.

3 Unfilled Epoxies

First, due to the complexity of these systems, unfilled epoxies must be considered before discussing the nanocomposites. Figure 4 depicts the calorimetric (TMFSC) and dielectric responses at 10 Hz of two fully cured DGEBA-based materials crosslinked with diethylenetriamine and methyl tetrahydrophtalic anhydride (MTHPA) for comparison. As shown in the inset, due to the different hardeners employed, the two investigated epoxy-based materials have significantly different network structures, and, therefore, a different behavior, evidenced, for instance, by

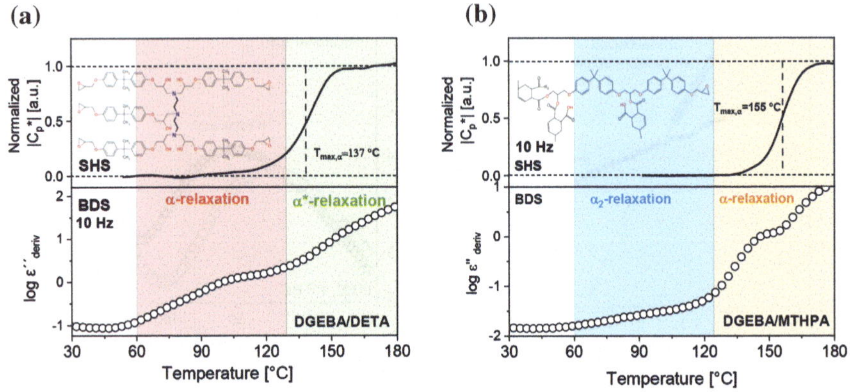

Fig. 4 Specific heat (temperature modulated fast scanning calorimetry) (top panel) and broadband dielectric spectroscopy (bottom panel) data at 10 Hz for bisphenol A diglycidyl ether cured with diethylenetriamine **a** and methyl tetrahydrophtalic anhydride. **b**. The heat capacity curves were normalized by the glassy and rubbery plateau. The insets show the crosslinked structures of DGEBA/DETA **a** and DGEBA/MTHPA **b**. The data are taken from Refs [20].

a ca. 18 K higher glass transition temperature probed by TMFSC for the anhydride-cured sample. The difference between amine- and anhydride-cured epoxies lies not only in their glass transition temperatures. The former materials tend to have better mechanical properties and water resistance, whereas the latter ones are cheaper, exhibit lower cure shrinkage and viscosity, high thermal stability, and, as shown, higher glass transition temperature [55].

As discussed in Ref. [20], the two materials show an intrinsic dynamic and spatial structural heterogeneity on a molecular scale in the network (Fig. 5a), related to regions with different crosslinking densities. For thermosetting materials formed via a curing reaction of two components, an existence of non-homogeneous nodular domains composed of regions of higher crosslinking density, surrounded by interstitial phase of lower crosslinking density, was found in the literature employing dynamic mechanical analysis (DMA) [21], atomic force microscopy [58], X-ray scattering techniques [20], and other methods [57]. Recently, it was discussed that one of the most important parameters is the temperature of the first curing step, the so-called pre-curing temperature (it should be noted that curing is usually a two- or three-step process performed with a different temperature for each step) [58]. As illustrated in Fig. 5b lower pre-curing temperatures result in a more homogeneous network structure, evidenced by Tanaka et al. [58]. An increasing heterogeneity of the network affects the macroscopic properties, for instance, by a decreased fracture behavior and solvent crack resistance [58].

For the two considered materials, the structural heterogeneity was evidenced by X-ray scattering, as well as by a combination of calorimetric and dielectric investigations, where two distinct segmental relaxation processes (dynamic glass transition) were found for each material. First, as discussed in Ref. [20], the X-ray scattering patterns show three distinct broad bumps for the epoxy-based materials, regardless

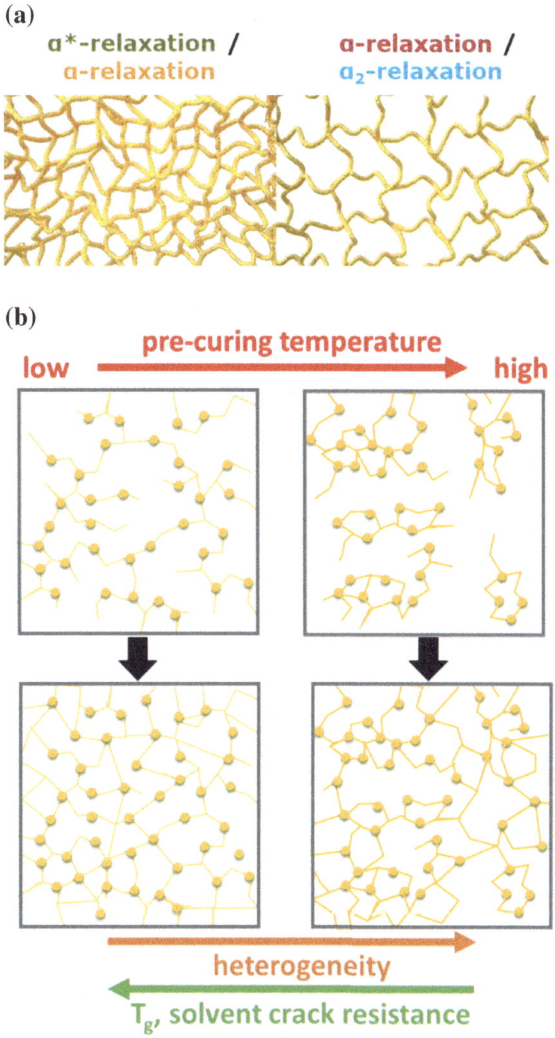

Fig. 5 Schematic representation of **a** structural and dynamic heterogeneity of an epoxy network and **b** different network structures expected for the pre-curing at lower and higher temperatures [58]

of the hardener used. In the X-ray investigations, broad peaks are a rough fingerprint of short-range intermolecular correlations of amorphous materials, where for most homogeneous samples, the most commonly occurring intermolecular distances result in one broad peak, denoted as the amorphous halo. For epoxies, in parallel to the amorphous halo, the first additional bump originates from the preferential packing arrangement of isopropylidene groups of DGEBA, forming elongated clusters [59]. The second bump is related to regions with lower than average crosslinking density. The assignment of that peak was done by correlating the averaged molecular distances calculated from the Bragg approximation with the length scale of the cooperatively rearranging region originating from regions with lower than average crosslinking density [14]. Furthermore, the two α-relaxation processes in Fig. 4 (α- and α*-relaxation for DGEBA/DETA and α_2- and α-relaxation for DGEBA/MTHPA) are related to cooperative segmental fluctuations within the epoxy network but they originate from regions with different crosslinking densities [14, 20, 60]. Nevertheless, the two segmental relaxation are evidenced by both the calorimetric and dielectric responses differently for the two materials. On the one hand, the high-temperature process (α*- and α-relaxation for DGEBA/DETA and DGEBA/MTHPA, respectively) shows a relatively narrow, well-pronounced peak for the anhydride-cured sample, whereas for DGEBA/DETA, this process was fully resolved only at high frequencies (~10^4 Hz). On the other hand, the low-temperature process (α- and α_2-relaxation for DGEBA/DETA and DGEBA/MTHPA, respectively) showed a higher intensity for the amine-cured material. For the sake of simplicity, this chapter focuses only on nanocomposites based on DGEBA/DETA. For more details on DGEBA/MTHPA nanocomposites, the reader is referred to Refs. [20, 21, 60].

4 Model Epoxy Nanocomposite (Ep/T-LDH)

The morphology of epoxy-based nanocomposites depends strongly on the synergy between the used components. Thus, considering the chemistry of epoxidation and the complex interaction between the matrix and the particles, it is difficult to draw general conclusions about the properties of these materials. Here, employing a combination of dielectric spectroscopy and calorimetry (DSC, FSC, TMDSC, and TMFSC), the structure, molecular mobility and the vitrification of a model epoxy-based PNC are investigated. The selected system is based on DGEBA crosslinked with DETA as a matrix and taurine-modified LDH as nanofiller.

The morphology of the Ep/T-LDH system was studied employing X-ray scattering techniques by fitting a model of stacked-disks as structure factor to the SAXS data [14, 61]. A reduction of the average number of layers of the nanofiller particles was found from 7 for pure T-LDH to 3 for the nanofiller embed in the nanocomposites. Additionally, a decrease of the interlayer distance (lamellar repeat distance) was observed for the nanocomposites. The layers with the largest interlayer distance at the edges of the T-LDH particles are most prone to exfoliate. Therefore, when exfoliated, these layers do not contribute to the average interlayer distance calculated from X-ray data. Therefore, a partial exfoliation of T-LDH is the morphology of the nanocomposites which was further confirmed by transmission electron microscopy (TEM). This morphology is most likely due to the amine groups of taurine which can form covalent bonds with the resin, leading to a disruption of the LDH stacks.

4.1 Localized Fluctuations

The dielectric behavior of the pure epoxy (Fig. 6a) and the nanocomposite with 15 wt% of the nanofiller (Ep/T-LDH15, Fig. 6b) is shown in Fig. 6. A β-relaxation is observed for the two materials at lower temperatures and higher frequency besides the α-relaxation and the above-mentioned conductivity (and polarization) contribution. Moreover, a comparison of Fig. 6a, b shows that the dielectric behavior of the nanocomposite is more complex than that of the pure thermoset. First, the localized processes will be discussed. As mentioned above, the secondary processes for epoxy-based materials are analyzed employing the HN-function. The temperature dependence of the relaxation rates of this process (not shown here) follows an Arrhenius behavior, which is typical for localized processes [15]:

$$\log f_p = f_\infty \exp\left[\frac{-E_A}{RT}\right] \tag{8}$$

where f_∞ is the relaxation rate at infinite temperatures, E_A is the activation energy, and R is the ideal gas constant.

The molecular origin of this process is controversially discussed in the literature [62–64] with the β-relaxation being assigned to crankshaft motions of small fragments of segments between crosslinks [65], localized fluctuations of dipoles formed during the crosslinking reaction (hydroxyl ether and secondary- or tertiary amine groups) [66], or local motions of dipoles of unreacted components during curing (specifically epoxide rings) [64]. Recent findings [14] for the system discussed here employing low field proton NMR spectroscopy revealed that due to the significant amplitude of protons at low temperatures, the β-relaxation is most probably associated with the rotational motions of para-substituted phenyl rings of DGEBA (phenyl rings with the adjacent ether linkages).

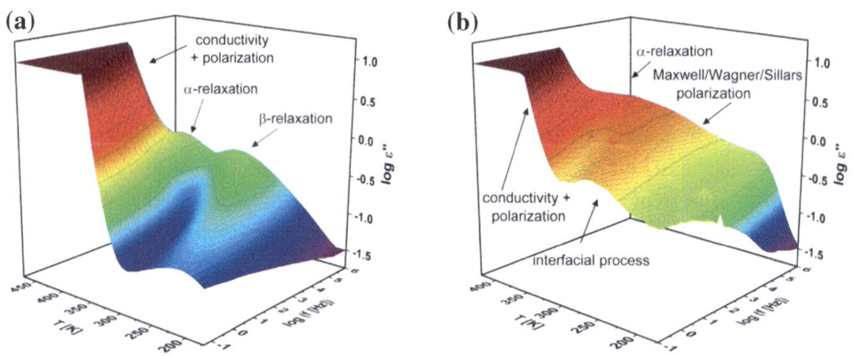

Fig. 6 3D representation of the dielectric loss spectra as a function of frequency and temperature for **a** the pure epoxy and **b** a nanocomposite with 15 wt% T-LDH content (Ep/T-LDH15). Reproduced with permission from Ref. [15]. Copyright (2019) Elsevier

For the discussed nanocomposite, the composition dependence of the β-relaxation cannot be considered due to the plateau (Fig. 6b) overlaying the dielectric spectra over a broad range of temperatures. This is different for the systems Ep/HNT and Ep/mHNT as discussed below.

Figure 6b reveals a (anomalous) plateau in the dielectric loss for Ep/T-LDH15, which extends over a broad temperature range of 120 K and hides the sub-T_g dielectric processes. The origin of this plateau cannot be associated with molecular dipoles because the absolute dielectric loss level is about one order of magnitude higher for PNC compared to the unfilled epoxy. This plateau is most likely related to interfacial polarization effects, known as the Maxwell/Wagner/Sillars (MWS) polarization, which is due to the blocking of charge carriers at inner boundary layers within an inhomogeneous system. Here, the source of the interfaces is the partially exfoliated nanofiller. Because the distances of the T-LDH layers in the matrix are subjected to a distribution, also a distribution of the interfacial polarization effects can be assumed, leading to a broad contribution in the dielectric loss. It should be noted that typically MWS polarization is observed at temperatures above the T_g, because the most common mechanism for the mobility of charge carriers is the slave mechanism, where the charge carrier mobility is coupled to segmental dynamics. However, for small enough charge carriers (like sodium ions present in the system), the conduction can be described by the Grotthuss mechanism [67], where the protons diffuse through a hydrogen bond network (i.e. between the hydroxyl groups of T-LDH and epoxide groups of DGEBA).

Moreover, the dielectric spectra given in Fig. 7 reveal an additional sub-T_g process at temperatures above the β-relaxation for all Ep/T-LDH samples. This process is not observed for the pure epoxy and its intensity increases with the increasing concentration of the nanofiller. Therefore, it was assigned to dipolar fluctuations of components of the epoxy systems adsorbed and/or covalently bonded with the nanofiller. In the former case, the resin or the hardener adsorb on the particle surface. In the latter scenario, covalent bonds between the $-NH_2$ groups of taurine and DGEBA might be formed. Considering that the layers of LDH become partly exfoliated during the preparation, the second scenario is more probable. This process was analyzed employing the "conduction-free loss". The temperature dependence of its relaxation rates is Arrhenius-like (inset Fig. 7). The activation energies of this process, estimated from the Arrhenius function (Eq. 8), range from 58 to 72 kJ mol^{-1}, depending on the filler content. The relatively low values of E_A further indicate the localized nature of this process. The dielectric strength ($\Delta\varepsilon$) for this process was estimated from the step height of the real part of the complex dielectric function ε' and it was found to increase with increasing temperatures, which is typical for localized processes [15]. This can be discussed in terms of the Debye theory of dielectric relaxation, which predicts that $\Delta\varepsilon$ is proportional to the number density of the involved dipoles [68]. Considering that the interaction strength between the adsorbed segments and the nanoparticle becomes weaker with increasing temperature, the number of mobile dipoles (or their fluctuation angle) increases with increasing temperature, resulting in

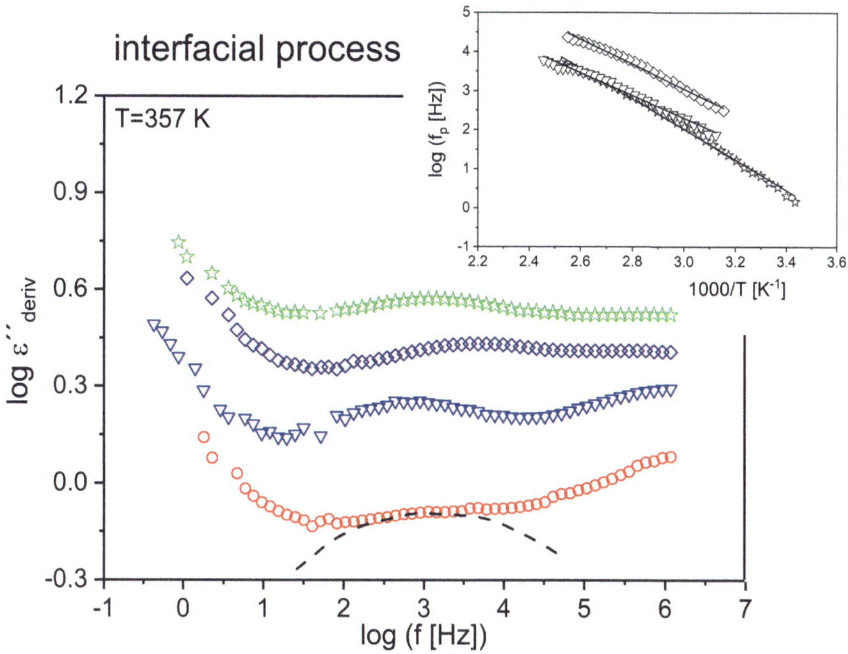

Fig. 7 Frequency dependence of the ε''_{deriv} for the Ep/T-LDH composites with various concentrations: 6 wt% (circles), 12 wt% (down-sided triangles), 15 wt% (diamonds), and 18 wt% (stars) at T = 357 K. The curves are shifted along the y-scale for sake of clarity. The dashed line is a guide for the eyes. The inset gives the relaxation map for the interfacial process. Reproduced with permission from Ref. [15]. Copyright (2019) Elsevier

an increase of the dielectric strength. It should be noted that the presence of dielectrically active localized fluctuations in the interfacial region depends on the investigated system and the particle–matrix synergy.

4.2 Segmental Dynamics

The segmental dynamics of these epoxy-based nanocomposites is now further considered. Similar to the pure epoxy discussed above, for all concentrations of the nanofiller, the dielectric measurements showed two relaxation processes, the α- and α*-relaxation, both related to structural relaxations. For both processes found in the frequency and temperature domain, the temperature dependencies of the relaxation rates are different like for the pure epoxy [14]. The segmental mobility (and corresponding vitrification behavior) was further studied by calorimetry besides dielectric spectroscopy. First, the calorimetric relaxation rates were estimated employing TMDSC and TMFSC. The calorimetric data and an example of its analysis can be found in Refs. [14, 15]. A combined relaxation map from the two calorimetric techniques and BDS is depicted in Fig. 8 for the pure epoxy and one nanocomposite (Ep/T-LDH15) as an example. It is observed that TMFSC and TMDSC data have the same temperature dependence as the dielectric α*-relaxation, indicating that these three techniques probe the same process. The temperature dependence of the relaxation rates of the α*-relaxation that is curved in this representation and can be described by the Vogel/Fulcher/Tammann (VFT) equation [69–71]:

$$\log f_p = \log f_\infty - \frac{A}{T - T_0} \tag{9}$$

where A is a constant and T_0 is the Vogel or the ideal glass transition temperature.

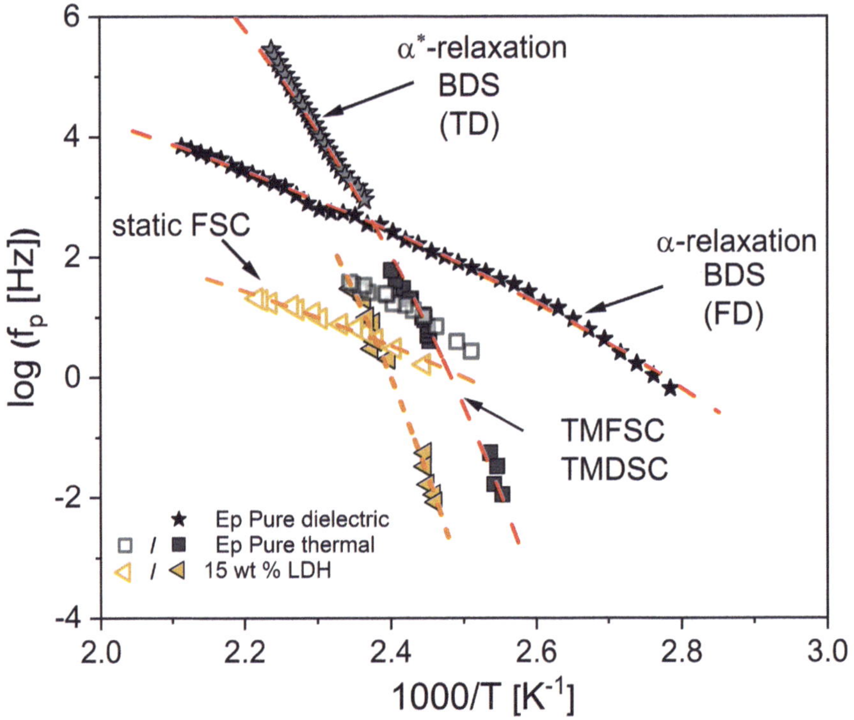

Fig. 8 Relaxation map for the pure epoxy (Ep—black squares) and for an epoxy/T-LDH nanocomposite with 15 wt% nanofiller (Ep/T-LDH15—orange left-sided triangles) obtained from static (open symbols) and dynamic fast scanning calorimetry measurements (filled symbols), as well as temperature modulated DSC data (filled symbols). The dashed lines are VFT fits to the corresponding data (color coded). The dielectric temperature domain, TMFSC, and TMDSC data are taken from Ref. [14], whereas the dielectric frequency domain and static FSC data are taken from Ref. [15]. Adapted with permission from Ref. [14]. Copyright (2021) The Royal Society of Chemistry

Moreover, to obtain a complete picture of the segmental dynamics of the materials, thermal relaxation rates were also calculated from fast scanning calorimetry measurements at constant heating rates (static measurements), as discussed in detail in Ref. [15]. According to the fluctuation approach model to the glass transition [72, 73], a relationship between the heating rate and a thermal relaxation rate can be derived according to

$$f_{p,thermal} = C\frac{\dot{\mathrm{T}}}{2\pi\,\Delta\mathrm{T}_g} \tag{10}$$

where ΔT_g is the width of the glass transition and C is a constant with a value of approximately one. The data obtained from static FSC measurements are also added to the relaxation map in Fig. 8. Surprisingly, the thermal relaxation rates estimated from the static FSC do not follow the same temperature dependence as the calorimetric relaxation rates obtained by TMDSC and TMDFSC, which coincide with the α*-relaxation. On the contrary, $f_{p,thermal}$ show a similar temperature dependence to the dielectric α-relaxation. The apparent activation energies obtained by approximating the dielectric α-relaxation and the FSC thermal relaxation rates with the Arrhenius equation are 108 kJ mol^{-1} for the dielectric α-relaxation and 127 kJ mol^{-1} for static FSC. This indicates that the two processes should have a similar molecular origin. It should be noted that the relatively high E_A values are indicating cooperative fluctuations. The shift between the data of approximately 30 K (at 10 Hz) results probably from the different sensitivities of the two techniques. Dielectric spectroscopy is sensitive to dipolar fluctuations, whereas calorimetry senses enthalpy fluctuations. The difference of approximately two decades in frequency can be explained by the cooperativity approach to the glass transition. It states that slowing down of the relaxation times with decreasing temperature is related to a correlation length, which increases with decreasing temperature. This correlation length defines the maximal number of segments fluctuating cooperatively; however, it allows also for less cooperative modes. Therefore, the shift between the slower calorimetric response is probably related to that this method senses more cooperative fluctuations than the faster dielectric response, but in principle with the same molecular origin.

Like the pure epoxy, it can be concluded that epoxy-based PNCs also exhibit a structural and dynamic heterogeneity. The two segmental relaxation processes (α- and α*-relaxation) originate both from cooperative segmental fluctuations of the epoxy network but from regions with different crosslinking densities. To correctly assign the two relaxation processes to regions with higher or lower crosslinking density, the Adam–Gibbs theory of the glass transition should be recalled [74, 75]. According to this theory with increasing chain stiffness (i.e., increasing crosslinking density), the configurational entropy decreases, which is related to the size of cooperatively rearranging regions (CRR) [76]. In addition, Sasaki et al. [77] proved experimentally that the configurational entropy decreases with an increasing degree of crosslinking, indicating that the number of configurations allowed in the smallest

CRR decreases as well. Considering that the temperature dependence of the relaxation rates of the α*-relaxation is highly fragile [78, 79], the α*-relaxation was assigned to regions with lower crosslinking density. Consequently, α-relaxation, with a weaker temperature dependence of the relaxation rates (strong behavior), corresponds to regions in the matrix with a more rigid structure, i.e., higher crosslinking density.

To further support this hypothesis, the cooperativity length scale (ξ), which is proportional to the volume of a CRR, was calculated from the TMDSC data for the α*-relaxation employing again the fluctuation approach to the glass transition [80]

$$\xi = \sqrt[3]{\frac{k_B T_g^2 \Delta \frac{1}{c_p}}{\rho(\delta T)^2}}, \tag{11}$$

where k_B is the Boltzmann constant, δT is the width of the glass transition, ρ is the mass density, and $\Delta(1/c_p) = 1/c_{p,glass} - 1/c_{p,liquid}$. Furthermore, $c_p \approx c_v$ is assumed. ξ for the pure epoxy and the nanocomposites was found to be 0.26–0.39 nm. Typically, the glass transition is considered as a cooperative process with an underlying length scale of 1–3 nm [81]. However, network-forming materials show a decrease of ξ with increasing crosslinking density [82] due to their rigid network structure. The cooperativity length scale for the α-relaxation could not be estimated from the static FSC measurements due to the unknown sample mass. Nevertheless, since $\xi \sim \frac{1}{\delta T}$ and considering that δT is larger for the FSC than for the TMDSC curves, it is concluded that ξ is smaller for the α- than for the α*-process. This further proves that the α-relaxation is due to regions with higher crosslinking density.

The two segmental relaxation processes observed for the pure epoxy and epoxy nanocomposites are also evidenced by two vitrification mechanisms observed in the heat flow curves obtained by conventional DSC measurements. Figure 9 shows the first derivative of the heat flow normalized by its maximum value $\left((dHF/dT)/(dHF/dT)_{max}\right)$ plotted versus temperature reduced by T_g. A step-like change at the glass transition in this representation shows a peak that is expected to be symmetric for dynamically homogeneous materials. Here, the peak corresponds to the vitrification related to the α*-process. Nevertheless, even for the pure EP, a low-temperature wing is observed related to segments with a higher molecular mobility than the average one. Therefore, the low-temperature wing is assigned to the α-process. For the nanocomposites, the second vitrification mechanism, related to α-relaxation, becomes more pronounced with increasing nanofiller concentration. Consequently, it can be concluded that for higher NPs contents, the relative contribution of the α-process to the whole response increases. As discussed above, curing is a diffusion-controlled reaction. In that respect, the nanoparticles will act as spatial obstacles affecting the degree of conversion and, thus, the crosslinking density. This will lead to more loose ends and/or higher heterogeneity and, therefore, more regions with a lower than average crosslinking density. In both cases, the mobility of the segments is higher than average that is represented by the main peak

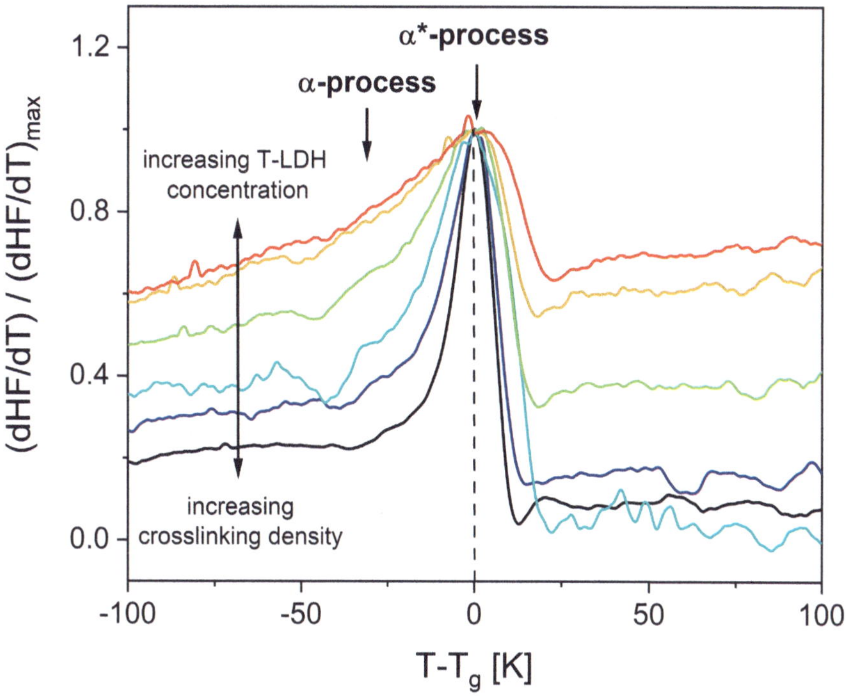

Fig. 9 Normalized derivative of the heat flow $(dHF/dT)/(dHF/dT)_{max}$ versus temperature normalized by the glass transition temperature for pure epoxy (EP—black) nanocomposites with various concentrations EP/T-LDH6—dark blue, EP/ T-LDH9—blue, EP/ T-LDH12—green, EP/ T-LDH15—orange, and EP/ T-LDH18—red. Reproduced with permission from Ref. [14]. Copyright (2021) The Royal Society of Chemistry

(α*-process). As expected for well-dispersed nanoparticles, the heterogeneity of the matrix increases with increasing nanofiller content, leading to a more pronounced low-temperature wing (α-process) in the derivative of the heat flow curves.

4.3 *Competition of Nanoparticle-Induced Mobilization and Immobilization of Segmental Dynamics*

An increase in the heterogeneity of the epoxy-matrix with increasing nanoparticle content discussed in the previous section (i.e., nanoparticle-induced mobilization of segmental dynamics) should be considered for all epoxy-based systems. This microscopic effect influences the macroscopic properties of the material, resulting, for instance, in a decrease of the calorimetric T_g and a decrease in the calorimetric strength (Δc_p), i.e., the step height of the specific heat capacity curves at the glass transition [see, for instance, Ref. 20]. Moreover, the matrix heterogeneities might deteriorate the mechanical properties of the PNCs, as the failure of the composite occurs initially through microscopic matrix cracking. Nevertheless, the macroscopic properties of an epoxy nanocomposite are also affected by a competing mechanism of the NP-related mobilization and immobilization of segments. The latter effect is related to the interfacial region, denoted as an adsorbed layer or rigid amorphous fraction (RAF), formed from polymer segments physically adsorbed and/or chemically bonded onto the nanoparticle surface. The behavior of the segments within RAF might differ from that of the bulk matrix, for instance, in their structure (i.e., crystallinity, packing density), conformation, and the molecular dynamics of the segments. These differences result from entropic (presence of a solid boundary) and energetic (dissimilarity between polymer–polymer and polymer–particle interaction) effects. As implied by the term "rigid", these segments are immobilized with respect to the cooperative segmental mobility. Localized fluctuations are possible within the interfacial region and can be detected by dielectric spectroscopy (Fig. 7).

The competition between the two mechanisms can have diverse consequences for different systems, depending on the reagents, nanoparticles, and curing conditions. For instance, for the considered system, the competition between the mobilization and immobilization effects is observed by the composition dependence of the calorimetric T_g. As shown in Fig. 10, T_g increases first with increasing concentration of the nanofiller till ca. 15 wt%, followed by a decrease for higher filler contents. The strong increase of T_g for the nanocomposites is probably due to the additional amine group of the taurine-modified nanofiller that might react with the DGEBA, leading to the formation of an immobilized interface. Additionally, also the adsorption of segments on the particle surface cannot be excluded. It should be noted that the formation of an interfacial region was confirmed for the discussed system by dielectric spectroscopy, as discussed in Sect. 4.1. Due to the particle-related confinement, the segments have lower molecular mobility and higher energy barriers for their conformational changes and, thus, higher T_g values. Nevertheless, with increasing nanofiller content, the hindering effect of NPs on the diffusion of the reaction partners will increase, weakening the increase of T_g and leading to the observed maximum at 15 wt%. For high NP concentration, the mobilization effect, i.e., lower crosslinking density, becomes the dominating effect over the immobilization of segments, which is still present for these samples. T_g is a rather ambiguous quantity because it is a complicated average for numerous different effects in the sample.

A more accurate property to discuss the competition between mobilization and immobilization effects is the heat capacity and calorimetric strength, which is proportional to the number of mobile segments. For the considered system, the Δc_p is dominated by the immobilization of the mobile segments, showing a systematic decrease with increasing T-LDH content [14]. However, for instance, a DGEBA/MTHPA material reinforced with boehmite nanoparticles (aluminum oxide hydroxide (γ-AlO(OH)) shows an interplay between the decrease in crosslinking density and the formation of a rigid amorphous fraction [20, 60]. A straight-forward separation of the mobilization and immobilization effects of the segmental dynamics is impossible employing calorimetric and dielectric methods only.

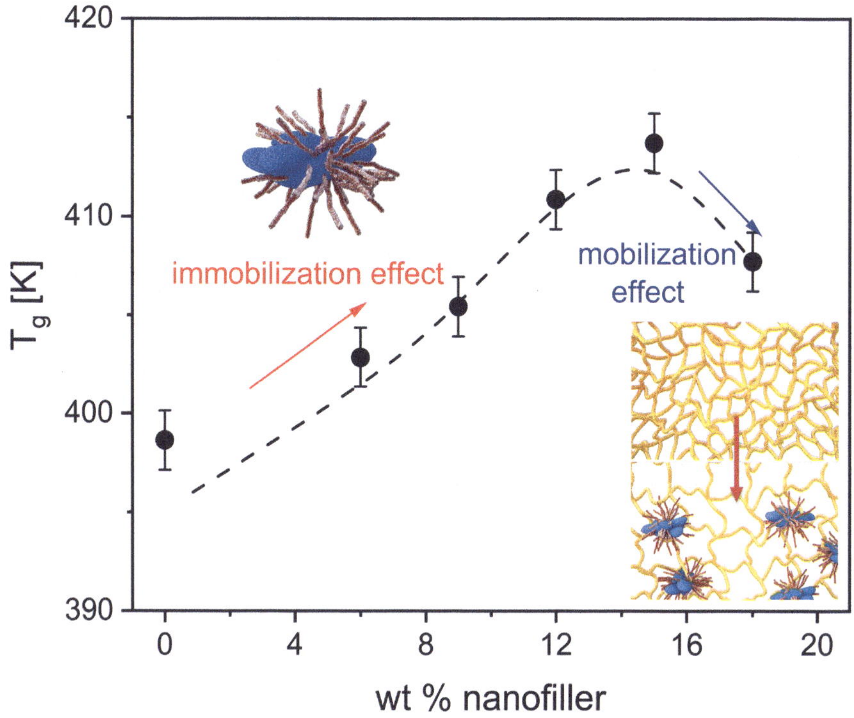

Fig. 10 T_g values obtained from static DSC (heating rate of 20 K min^{-1}) as a function of the filler content for Ep/T-LDH nanocomposite. The dashed line is a guide for the eyes. The data are taken from Ref. [14]

5 Comparison of Epoxy Nanocomposites with Different Nanofillers

The previous section provided a detailed discussion on a selected epoxy-based PNC as an example. It was discussed that epoxy-based nanocomposites are intrinsically heterogeneous, showing two distinct segmental relaxation processes related to regions with different crosslinking densities as discussed above. Nevertheless, other properties such as the interfacial behavior differ for all epoxy materials and each system must be considered separately. In this section, a comparison of three different epoxy-based PNC is provided. The matrix for each case is prepared with the same resin and hardener employing identical curing conditions. The focus lies on different alumina-based nanofillers which are expected to have different interactions with the matrix. In the first series of PNCs (Ep/HNT), halloysite nanotubes are employed as nanofiller. HNTs, as described in detail in Sect. 2.1.1., are aluminosilicate clays and their external surface consists mostly of SiO_2 groups, whereas the interior has aluminol groups. The epoxy and hydroxyl groups of the resin are chemically inert toward aluminols and siloxanes under the curing conditions. Therefore, the interfacial interaction between halloysite nanotubes and the matrix can be considered as weak. To improve the dispersion of the nanotubes and enhance the interaction with the matrix, HNTs were modified on the exterior surface employing a thin polydopamine coating and ultrafine $Fe(OH)_3$ nanoparticles. These modified nanoparticles yield the second type of epoxy nanocomposites (Ep/mHNT). On the one hand, the modification was employed due to enhanced catalytic abilities of iron trihydroxide NPs which result in a significant increase in both char yield and maximum degradation rate of the material, decreasing its flammability [16]. On the other hand, PDA, due to the large number of hydroxyl groups is expected to increase the interfacial interaction with the matrix and force a better dispersion of the nanofiller. The -OH groups of polydopamine form hydrogen bonds with the epoxide rings of DGEBA which are stronger than the van der Waals forces between the nanoparticles and the epoxy. Thus, the interfacial interaction of Ep/mHNT is considered to be stronger than that of Ep/HNT. Lastly, Ep/T-LDH, discussed in detail in the previous sections, is taken as a third epoxy-based PNC due to the strong interaction of NPs and the matrix via the formation of covalent bonds between taurine and DGEBA.

Figure 11 shows the calorimetric glass transition temperature as a function of the concentration of the NPs for the three nanocomposites Ep/HNT, Ep/mHNT, and

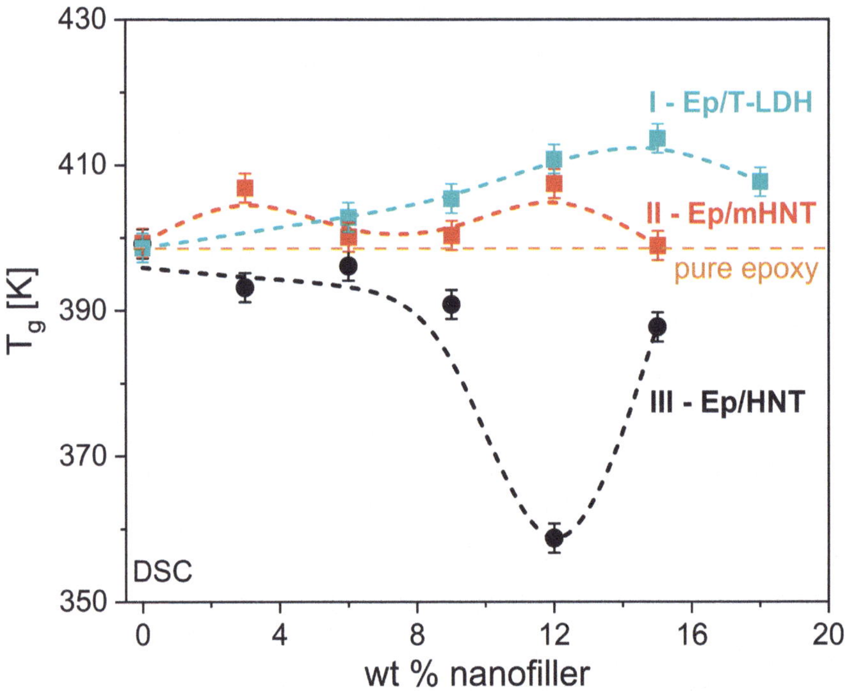

Fig. 11 Glass transition temperature as a function of nanofiller content estimated by DSC for Ep/HNT (black), Ep/mHNT (red), and Ep/T-LDH (blue). The orange dashed line indicates the T_g of the pure epoxy. The data are taken from Refs. [14, 16]

Ep/T-LDH (the interaction strength is in ascending order). The complex composition dependence of T_g for Ep/HNT and Ep/mHNT was discussed in detail elsewhere [16], whereas for the Ep/T-LDH, it was discussed in Sect. 4.3, as well as in Ref. [14]. Here, it is important to note that for each sample of the PNC with the weakest NP-matrix interaction, the T_g is the lowest of the three materials, having values below the T_g of the unfilled epoxy. This behavior indicates that for Ep/HNT, the dominating mechanism is an NP-induced mobilization effect due to the decreasing crosslinking density with increasing nanofiller content. Interestingly, this material is subjected to another kind of immobilization effect of the segmental dynamics. On the one hand, this results from the tubular shape and high aspect ratio of the nanoparticles. On the other hand, due to the weak NP-matrix interaction, the nanofiller is not well dispersed in the polymer matrix. Consequently, two phases are formed in the material, epoxy- and HNT-rich domains. In the former phase, HNT nanotubes are homogenously dispersed with large inter-tube distances. In the latter phase, a high HNT content is embedded in a continuous matrix phase, introducing an additional confinement to the segmental dynamics of the epoxy segments due to the small inter-tube distances. Therefore, even though Ep/HNT nanocomposite is dominated by the mobilization effect, for the highest NP loading (Ep/HNT15), an immobilization effect related to the inhomogeneous distribution of NPs is observed, which yields an increase of T_g.

Furthermore, Fig. 11 reveals that the surface modification of the halloysite nanotubes results in a higher glass transition temperature for Ep/mHNT than for Ep/HNT. Here, the T_g value for each sample is similar or higher than that of the pure epoxy, indicating that the modification of the HNT surface resulted in a stronger NP-matrix interaction via the formation of hydrogen bonds. Interestingly, because of the better dispersion of mHNT particles than of the HNTs, no immobilization of the segmental dynamics due to the two-phase structure of the material, i.e., epoxy- and NP-rich regions, is observed. Lastly, the Tg values for Ep/T-LDH are the highest of all three nanocomposites due to the strong immobilization effect for this system. These results show that the glass transition, which is one of the most important properties of polymeric materials, is strongly affected by the nanofiller—matrix synergy.

The molecular dynamics probed by dielectric spectroscopy is also different for the three systems. Figure 12a, b shows the dielectric loss spectra in 3D representations as a function of frequency and temperature for 15 wt% of the nanofiller for Ep/HNT and Ep/mHNT. As previously discussed, epoxy nanocomposites are intrinsically heterogeneous that leads to two distinct structural relaxations. This behavior is observed for all investigated systems and, therefore, will not be repeated in this section. The main difference between the three nanocomposites (Figs. 6 and 12a, b) is the absence of a dielectrically active process related to localized fluctuations in the interfacial region for Ep/HNT and Ep/mHNT. As discussed in Sect. 4.1, this process was observed for each composition of the Ep/T-LDH nanocomposite. The difference in dielectric responses of the three materials lies again in the different NP-polymer interactions because the matrix has the same chemical composition for each PNC. For the Ep/T-LDH, the interfacial process is fully resolved in the dielectric spectra because of the strong localization of a large number of dipoles via covalent bonds between taurine that bears $-NH_2$ groups and oxirane rings of epoxy resin. The fact

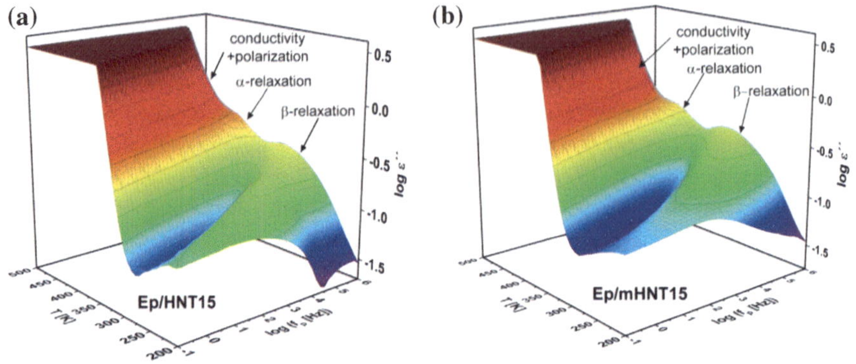

Fig. 12 3D representation of the dielectric loss as a function of frequency and temperature for **a** Ep/HNT15 and **b** Ep/mHNT15. Adapted with permission from Ref. [16]. Copyright (2021) MDPI.

that the interfacial region was not detected by dielectric spectroscopy for the two other PNCs does not exclude the formation of a rigid amorphous fraction for the two systems. Indeed, the behavior of the calorimetric T_g for Ep/mHNT implied immobilization effects in this system. A powerful method to study the interfacial properties qualitatively and quantitatively in nanocomposites is investigation of the calorimetric strength. The formalism, initially introduced by Wunderlich et al. for semicrystalline polymers [83], was extended to polymer nanocomposites [84]. It describes a PNC in terms of a three-phase model, where an amorphous nanocomposite is composed of a mobile amorphous fraction (MAF), a possible RAF, and the nanofiller. MAF and RAF can be calculated by

$$MAF = \frac{\Delta c_{p,PNC}}{\Delta c_{p,\ pure\ epoxy}} \tag{12a}$$

$$RAF = 1 - filler\ content - MAF, \tag{12b}$$

where $\Delta c_{p,PNC}$ and $\Delta c_{p,pure\,epoxy}$ are the step heights of the specific heat capacity curve at the glass transition for the nanocomposite sample and the unfilled epoxy, respectively.

Nevertheless, for epoxy-based nanocomposites, mobilization and immobilization effects usually occur simultaneously and cannot be disentangled from each other by neither calorimetric nor dielectric means. Therefore, for epoxy PNCs, RAF calculated employing Eqs. 12a and 12b is only a lower limit of the amount of the immobilized interface.

As shown in Fig. 13, the amount of RAF is different for the three nanocomposites including a different concentration dependence. For the PNC system with the weakest NP-matrix interaction (Ep/HNT), RAF can be calculated only for the lowest nanofiller concentrations, reaching up to 4 wt% for Ep/HNT6. For higher concentrations of HNT, RAF cannot be calculated because for these concentrations, Δc_p

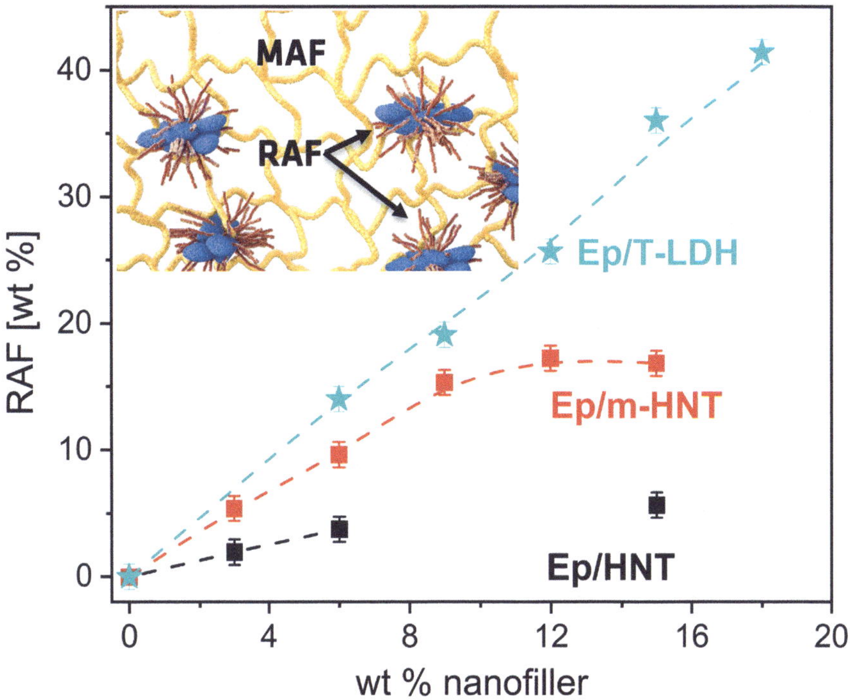

Fig. 13 Rigid amorphous fraction as a function of nanofiller content estimated by TMDSC with Eq. 12 for Ep/HNT (black), Ep/mHNT (red), and Ep/T-LDH (blue). The data are taken from Refs. [14, 16]

becomes larger than that of the bulk. This results from a mobilization effect strongly dominating the cooperative segmental mobility, as indicated also by the concentration dependence of the calorimetric glass transition temperature. For the system with the mHNT as nanofiller, the amount of RAF showed a systematic increase with increasing nanofiller concentration and reached values up to ~16 wt% for the highest loaded sample. Consequently, it can be concluded that for this PNC system, the interphase formation was dominant over matrix mobilization. The difference between Ep/HNT and Ep/mHNT lies in the surface modification of the nanofiller for the latter system, leading to a stronger matrix-NP interaction and a more homogenous distribution of the nanofiller in the epoxy matrix. Lastly, for the Ep/T-LDH system, where covalent bonds are formed between NPs and the matrix, the amount of RAF reaches values up to 40 wt%, indicating that the interface dominates the macroscopic material properties to a larger extent. For Ep/T-LDH, RAF was also estimated employing low field NMR measurements where the obtained results agree quantitatively with the calorimetric data. These results underline that, even though some similarities between different epoxy-based nanocomposites can be observed, as indicated in Sect. 4, each PNC must be investigated separately, and no general conclusions can be drawn for this class of materials. For reactive systems like epoxies, taking into consideration the different interactions and synergy between the reagents and the nanoparticles is important to understand their structure–property relationships and to successfully design materials with desirable properties. As indicated by the presented studies, good NP-matrix interaction must be ensured to achieve high values of the calorimetric glass transition temperature and large amounts of the interfacial region.

6 Conclusion

The aim of this chapter was to obtain a deeper understanding of epoxy-based nanocomposites, which are relevant for industrial applications, but rarely studied on the fundamental level. This work focuses on three systems, each based on bisphenol A diglycidyl ether and diethylenetriamine as crosslinking agents. The systems were cured under the same conditions. They are reinforced with halloysite nanotubes, surface modified halloysite nanotubes, and taurine-modified layered double hydroxide.

The latter system was taken as a model nanocomposite to discuss the general properties of this class of materials studied by dielectric spectroscopy and calorimetry. It was shown that epoxy-based nanocomposites, similar to the pure epoxy, show a structural and dynamic heterogeneity, indicated by two distinct segmental relaxation processes. This results from regions with different crosslinking densities. For thermosetting materials formed via curing reaction of two components, non-homogeneous nodular domains composed of regions of higher crosslinking density are surrounded by the interstitial phase of lower crosslinking density. Each of these regions exhibits inherent segmental mobility and, consequently, its own vitrification mechanism. Calorimetric investigations showed that with increasing nanofiller content, the material heterogeneity increases, leading to a more pronounced contribution of the segmental relaxation that originates from regions with lower crosslinking density.

Furthermore, the properties of epoxy-based nanocomposites are a result of the interplay of two simultaneous competing effects: NP-related mobilization and immobilization of the segmental dynamics. The former is due to NPs acting as spatial barriers in the diffusion-controlled curing reaction, slowing down the diffusion coefficient and leading to a decrease of conversion. In consequence, the average crosslinking density of a PNC will be lower than the corresponding unfilled epoxy, leading to higher segmental mobility. The latter results from the formation of an immobilized fraction at the particle/matrix interface (rigid amorphous fraction), where polymer segments are adsorbed and/or bonded to the surface of the NPs and, therefore, do not contribute to the cooperative segmental mobility.

Further, a comparative study on the three different epoxy nanocomposites was shown. The chosen PNCs, Ep/HNT, Ep/mHNT, and Ep/T-LDH exhibit different interactions with the matrix. HNTs have weak interaction with the epoxy, due to the external exterior siloxane groups. On the contrary, mHNTs because of the surface modification can form hydrogen bonds with the matrix. Lastly, for Ep/T-LDH due to the taurine modification of the nanofiller, it is expected that covalent bonds with the epoxy are formed. As a consequence of the different NP-matrix synergy, the three systems show different behavior of the calorimetric glass transition and interfacial properties. For the weakly interacting system, T_g for each concentration of the nanofiller is below the T_g of the unfilled epoxy. The value of RAF, which could be quantified only for low concentrations of HNT, is below 5 wt%. This indicates that because of the weak NP-matrix interaction, for Ep/HNT, the mobilization of the segmental dynamics due to the nanofiller is dominant. On the contrary, the surface modification of HNTs resulted for Ep/mHNT in an increase of T_g, compared to Ep/HNT and pure epoxy, as well as in higher value of RAF (ca. 16 wt% for the highest concentration) than for the unmodified system. Lastly, for the PNC where NPs and matrix are covalently bonded, the T_g and RAF are substantially higher than for the two other PNCs, where RAF reached ca. 40 wt% for the highest T-LDH concentration. Consequently, it can be concluded that the NP-matrix synergy significantly affects the material properties, and a sufficient bonding between them must be ensured to achieve good thermal properties and high amounts of the interfacial region.

References

1. Schab-Balcerzak E (2006) Modifications of Epoxy Resins' Properties. In: Bregg RK (ed) Frontal Polymer Research. Nova Science Publishers, New York
2. Srivastava I, Raiee MA, Yavari F, Raiee J, Koratkar N (2012) Epoxy Nanocomposites: Graphene a Promising Filler. In: Mukhopadhyay P, Gupta R (eds) Graphite, Graphene, and Their Polymer Nanocomposites. CRC Press, Bosta Roca
3. Azeez AA, Rhee KY, Park SJ, Hui D (2013) Epoxy clay nanocomposites—processing, properties and applications: a review. Compos B Eng 45(1):308–320
4. Wetzel B, Rosso P, Haupert F, Friedrich K (2006) Epoxy nanocomposites—fracture and toughening mechanisms. Eng Fract Mech 73:2375–2398
5. Rakotomalala M, Wagner S, Döring M (2010) Recent developments in halogen free flame retardants for epoxy resins for electrical and electronic applications. Materials 3:4300–4327
6. Kumar SK, Benicewicz BC, Vaia RA, Winey KI (2017) 50th anniversary perspective: are polymer nanocomposites practical for applications? Macromolecules 50:714–731
7. Guo J, Long J, Ding D, Wang Q, Shan Y, Umar A, Zhang X, Weeks BL, Wei S, Guo Z (2016) Significantly enhanced mechanical and electrical properties of epoxy nanocomposites reinforced with low loading of polyaniline nanoparticles. RSC Adv 6:21187–21192
8. Olad A, Barati M, Behboudi S (2012) Preparation of PANI/Epoxy/Zn nanocomposite using Zn nanoparticles and epoxy resin as additives and investigation of its corrosion protection behavior on iron. Prog Org Coat 74:221–227
9. Wang D, Kou R, Choi D, Yang Z, Nie Z, Li J, Saraf LV, Hu Z, Zhang J, Graff GL, Liu J, Pope MA, Aksay IA (2010) Ternary self-assembly of ordered metal oxide-graphene nanocomposites for electrochemical energy storage. ACS Nano 4:1587–1595
10. González-Campo A, Orchard K, Sato N, Shaffer M, Williams C (2009) One-pot, in situ synthesis of ZnO-carbon nanotube–epoxy resin hybrid nanocomposites. Chem Commun 4034–4036

11. Gu H, Zhang H, Ma C, Sun H, Liu C, Dai K, Zhang J, Wei R, Ding T, Guo Z (2019) Smart strain sensing organic–inorganic hybrid hydrogels with nano barium ferrite as the cross-linker. J Mater Chem C 7:2353–2360
12. Zhu J, Wei S, Ryu J, Budhathoki M, Liang G, Guo Z (2010) In situ stabilized carbon nanofiber (CNF) reinforced epoxy nanocomposites. J Mater Chem 20:4937–4948
13. Qing Y, Wang X, Zhou Y, Huang Z, Luo F, Zhou W (2014) Enhanced microwave absorption of multi-walled carbon nanotubes/epoxy composites incorporated with ceramic particles. Compos Sci Technol 102:161–168
14. Szymoniak P, Qu X, Abbasi M, Pauw B, Henning S, Li Z, Wang D, Schick C, Saalwächter K, Schönhals A (2021) Spatial inhomogeneity, interfaces and complex vitrification kinetics in a network forming nanocomposite. Soft Matter 17:2775–2790
15. Szymoniak P, Li Z, Wang D, Schönhals A (2019) Dielectric and flash DSC investigations on an epoxy based nanocomposite system with MgAl layered double hydroxide as nanofiller. Thermochim Acta 677:151–161
16. Omar H, Smales G, Henning S, Li Z, Wang D, Schönhals A, Szymoniak P (2021) Calorimetric and dielectric investigations of epoxy-based nanocomposites with halloysite nanotubes as nanofillers. Polymers 13:1634
17. Li Z, Liu L, Jiménez González A, Wang DY (2017) Bioinspired polydopamine-induced assembly of ultrafine Fe(OH)3 nanoparticles on halloysite toward highly efficient fire retardancy of epoxy resin via an action of interfacial catalysis. Polym Chem 8:3926–3936
18. Zhu J, Wei S, Ryu J, Sun L, Luo Z, Guo Z (2010) Magnetic epoxy resin nanocomposites reinforced with core−shell structured Fe@FeO nanoparticles: fabrication and property analysis. ACS Appl Mater Interfaces 2:2100–2107
19. McGrath L, Parnas R, King S, Schroeder J, Fischer D, Lenhart J (2008) Investigation of the thermal, mechanical, and fracture properties of alumina–epoxy composites. Polymer 49:999–1014
20. Szymoniak P, Pauw B, Qu X, Schönhals A (2020) Competition of nanoparticle-induced mobilization and immobilization effects on segmental dynamics of an epoxy-based nanocomposite. Soft Matter 16:5406–5421
21. Ghasem Zadeh Khorasani M, Silbernagl D, Szymoniak P, Hodoroaba V, Sturm H (2019) The effect of boehmite nanoparticles (γ-AlOOH) on nanomechanical and thermomechanical properties correlated to crosslinking density of epoxy. Polymer 164:174–182
22. Rafiee M, Rafiee J, Wang Z, Song H, Yu Z, Koratkar N (2009) Enhanced mechanical properties of nanocomposites at low graphene content. ACS Nano 3:3884–3890
23. Zaman I, Phan T, Kuan H, Meng Q, La Bao L, Luong L, Youssf O, Ma J (2011) Epoxy/graphene platelets nanocomposites with two levels of interface strength. Polymer 52:1603–1611
24. Park I, Peng H, Gidley D, Xue S, Pinnavaia T (2006) Epoxy−silica mesocomposites with enhanced tensile properties and oxygen permeability. Chem Mater 18:650–656
25. Liu Y, Lin Z, Lin W, Moon K, Wong C (2012) Reversible superhydrophobic–superhydrophilic transition of ZnO nanorod/epoxy composite films. ACS Appl Mater Interfaces 4:3959–3964
26. Rastin H, Saeb M, Nonahal M, Shabanian M, Vahabi H, Formela K, Gabrion X, Seidi F, Zarrintaj P, Sari M, Laheurte P (2017) Transparent nanocomposite coatings based on epoxy and layered double hydroxide: nonisothermal cure kinetics and viscoelastic behavior assessments. Prog Org Coat 113:126–135
27. Kalali E, Wang X, Wang D (2015) Functionalized layered double hydroxide-based epoxy nanocomposites with improved flame retardancy and mechanical properties. J Mater Chem 3:6819–6826
28. Wang D, Das A, Leuteritz A, Mahaling R, Jehnichen D, Wagenknecht U, Heinrich G (2012) Structural characteristics and flammability of fire retarding EPDM/layered double hydroxide (LDH) nanocomposites. RSC Adv 2:3927
29. Wang D, Leuteritz A, Wang Y, Wagenknecht U, Heinrich G (2010) Preparation and burning behaviors of flame retarding biodegradable poly(lactic acid) nanocomposite based on zinc aluminum layered double hydroxide. Polym Degrad Stab 95:2474–2480

30. Peng H, Tjiu W, Shen L, Huang S, He C, Liu T (2009) Preparation and mechanical properties of exfoliated CoAl layered double hydroxide (LDH)/polyamide 6 nanocomposites by in situ polymerization. Compos Sci Technol 69:991–996
31. Liu H, Zhao X, Zhu Y, Yan H (2020) DFT study on MgAl-layered double hydroxides with different interlayer anions: structure, anion exchange, host–guest interaction and basic sites. Phys Chem Chem Phys 22:2521–2529
32. Njuguna J, Ansari F, Zhu H, Rodriguez VM (2014) Nanomaterials, nanofillers, and nanocomposites: types and properties. In: Njuguna J, Pielichowski K, Zhu H (eds) Health and environmental safety of nanomaterials: polymer nanocomposites and other materials containing nanoparticles. Elsevier Science, Burlington, pp 3–27
33. Leng J, Szymoniak P, Kang N, Wang D, Wurm A, Schick C, Schönhals A (2019) Influence of interfaces on the crystallization behavior and the rigid amorphous phase of poly(l-lactide)-based nanocomposites with different layered doubled hydroxides as nanofiller. Polymer 184:121929
34. Deng S, Zhang J, Ye L (2009) Halloysite–epoxy nanocomposites with improved particle dispersion through ball mill homogenisation and chemical treatments. Compos Sci Technol 69:2497–2505
35. Cividanes L, Simonetti E, Moraes M, Fernandes F, Thim G (2013) Influence of carbon nanotubes on epoxy resin cure reaction using different techniques: a comprehensive review. Polym Eng Sci 54:2461–2469
36. Du M, Guo B, Jia D (2006) Thermal stability and flame retardant effects of halloysite nanotubes on poly(propylene). Eur Polym J 42:1362–1369
37. Ye Y, Chen H, Wu J, Ye L (2007) High impact strength epoxy nanocomposites with natural nanotubes. Polymer 48:6426–6433
38. Lvov Y, Abdullayev E (2013) Functional polymer–clay nanotube composites with sustained release of chemical agents. Prog Polym Sci 38:1690–1719
39. Vijayan PP, Hany El-Gawady Y, Al-Maadeed M (2016) Halloysite nanotube as multifunctional component in epoxy protective coating. Ind Eng Chem Res 55:11186–11192
40. Liu M, Jia Z, Jia D, Zhou C (2014) Recent advance in research on halloysite nanotubes-polymer nanocomposite. Prog Polym Sci 39:1498–1525
41. Du M, Guo B, Jia D (2010) Newly emerging applications of halloysite nanotubes: a review. Polym Int 59:574–582
42. Bressler I, Pauw B, Thünemann A (2015) McSAS: software for the retrieval of model parameter distributions from scattering patterns. J Appl Crystallogr 48:962–969
43. Liu M, Guo B, Du M, Cai X, Jia D (2007) Properties of halloysite nanotube–epoxy resin hybrids and the interfacial reactions in the systems. Nanotechnology 18(45):455703
44. Mathot V, Pyda M, Pijpers T et al (2011) The Flash DSC 1, a power compensation twin-type, chip-based fast scanning calorimeter (FSC): First findings on polymers. Thermochim Acta 522:36–45
45. Adamovsky S, Minakov A, Schick C (2003) Scanning microcalorimetry at high cooling rate. Thermochim Acta 403:55–63
46. van Herwaarden S, Iervolino E, van Herwaarden F et al (2011) Design, performance and analysis of thermal lag of the UFS1 twin-calorimeter chip for fast scanning calorimetry using the Mettler-Toledo Flash DSC 1. Thermochim Acta 522:46–52
47. Merzlyakov M, Schick C (2001) Step response analysis in DSC—a fast way to generate heat capacity spectra. Thermochim Acta 380:5–12
48. Shoifet E, Schulz G, Schick C (2015) Temperature modulated differential scanning calorimetry - extension to high and low frequencies. Thermochim Acta 603(227):236
49. Kamasa P, Merzlyakov M, Pyda M, Pak J, Schick C, Wunderlich B (2002) Multi-frequency heat capacity measured with different types of TMDSC. Thermochim Acta 392–393:195–207
50. Havriliak S, Negami S (1966) A complex plane analysis of α-dispersions in some polymer systems. J Polym Sci, Part C: Polym Symp 14:99–117
51. Schönhals A, Kremer F (2002) Analysis of dielectric spectra. In: Kremer F, Schönhals A (eds) Broadband dielectric spectroscopy. Springer, Berlin, pp 59–96

52. Sheppard NF, Senturia SD (1989) Dielectric properties of bisphenol-a epoxy resins. J Polym Sci Part B Poly Phys 27:753
53. Wübbenhorst M, van Turnhout J (2002) Analysis of complex dielectric spectra. I. One-dimensional derivative techniques and three-dimensional modelling. J Non-Cryst Solids 305:40
54. Szymoniak P, Madkour S, Schönhals A (2019) Molecular dynamics of the asymmetric blend PVME/PS revisited by broadband dielectric and specific heat spectroscopy: evidence of multiple glassy dynamics. Macromolecules 52:1620–1631
55. (2021) http://www.appliedpoleramic.com/wp-content/uploads/2015/05/API-TECH-NOTE-3-Anhydride-Cured-Epoxy-Matrices. Accessed 18 Jun 2021
56. Duchet J, Pascault J (2003) Do epoxy-amine networks become inhomogeneous at the nanometric scale? J Polym Sci Part B Polym Phys 41:2422–2432
57. Kreibich U, Schmid R (2007) Inhomogeneities in epoxy resin networks. J Polym Sci Polym Symp 53:177–185
58. Aoki M, Shundo A, Yamamoto S, Tanaka K (2020) Effect of a heterogeneous network on glass transition dynamics and solvent crack behavior of epoxy resins. Soft Matter 16:7470–7478
59. Kumar S, Adams W (1987) Structural studies of epoxy resins, acetylene terminated resins and polycarbonate. Polymer 28:1497–1504
60. Szymoniak P, Qu X, Schönhals A, Sturm H (2021) Characterization of Polymer Nanocomposites. In: Sinapius M, Ziegmann G (eds) Acting Principles of Nano-Scaled Matrix Additives for Composite. Springer, Cham, pp 55–77
61. (2021) http://www.sasview.org/docs/user/models/stacked_disks.html. Accessed 18 Jun 2021
62. Ochi M, Okazaki M, Shimbo M (1982) Mechanical relaxation mechanism of epoxide resins cured with aliphatic diamines. J Polym Sci Part B Polym Phys 20:689–699
63. Ochi M, Yoshizumi M, Shimbo, (1987) Mechanical and dielectric relaxations of epoxide resins containing the spiro-ring structure. II. Effect of the introduction of methoxy branches on low-temperature relaxations of epoxide resins. J Polym Sci Part B Polym Phys 25:1817–1827
64. Mangion MBM, Johari GP (1990) Relaxations of thermosets. III. Sub-Tg dielectric relaxations of bisphenol-A–based epoxide cured with different cross-linking agents. J Polym Sci Part B Polym Phys 28:71–83
65. Meier DJ (1967) Molecular basis of transitions and relaxation In: Meier DJ (ed) Midland macromolecular monographs vol 4. Gordon and Breach Science Publishers
66. Hassan MK, Tucker SJ, Abukmail A, Wiggins JS, Mauritz KA (2016) Polymer chain dynamics in epoxy based composites as investigated by broadband dielectric spectroscopy. Arab J Chem 9:305–315
67. Marx D (2006) Proton transfer 200 years after von Grotthuss: insights from ab initio simulations. Chem Phys Chem 7:1848
68. Schönhals A, Kremer F (2002) Broadband dielectric measurement techniques and theory of dielectric relaxation. In: Kremer F, Schönhals A (eds) Broadband dielectric spectroscopy. Springer, Berlin, pp 01–57
69. Vogel H (1921) The temperature dependence law of the viscosity of fluids. Phys Z 22:645
70. Fulcher GS (1925) Analysis of recent measurements of the viscosity of glasses. J Am Ceram Soc 8:339
71. Tammann G, Hesse G (1926) The dependency of viscosity on temperature in hypothermic liquids. Z Anorg Allg Chem 156:245
72. Donth E (1981) Glasübergang. Akademieverlag, Berlin
73. Donth E (1982) The size of cooperatively rearranging regions at the glass-transition. J Non Cryst Solids 53(3):325
74. Adam G, Gibbs JH (1965) On the temperature dependence of cooperative relaxation properties in glass-forming liquids. J Chem Phys 43:139
75. Gibbs JH, DiMarzio EA (1958) Theory of helix-coil transitions in polypeptides. J Chem Phys 28:373
76. Montserrat T (1995) Effect of crosslinking density on ΔCp(Tg) in an epoxy network. Polymer 36:435–436

77. Sasaki T, Uchida T, Sakurai K (2006) Effect of crosslink on the characteristic length of glass transition of network polymers. J Polym Sci Part B Polym Phys 44:1958–1966
78. Angell CA (1991) Relaxation in liquids, polymers and plastic crystals—strong/fragile patterns and problems. J Non-Cryst Solids 131–133:13–31
79. Angell CA (1997) Entropy and fragility in supercooling liquids. J Res Natl Inst Stand Technol 102:171–185
80. Donth E (1982) The size of cooperatively rearranging regions at the glass transition. J Non-Cryst Solids 53:325
81. Schönhals A, Kremer F (2012) Amorphous polymers in polymer science. In: Matyjaszewski K, Möller M (eds) A comprehensive reference, vol 1. Elsevier
82. Sasaki T, Uchida T, Sakurai K (1958) Effect of crosslink on the characteristic length of glasstransition of network polymers. J Polym Sci Part B Polym Phys 2006:44
83. Wunderlich B (2003) Reversible crystallization and the rigid–amorphous phase in semicrystalline macromolecules. Prog Polym Sci 28:383–450
84. Sargsyan A, Tonoyan A, Davtyan S, Schick C (2007) The amount of immobilized polymer in PMMA SiO2 nanocomposites determined from calorimetric data. Eur Polym J 43:3113–3127

Index

A
AC conductivity, 238, 310
Activation energy, 17, 18, 64, 76, 78–80, 97, 99, 135, 137, 145, 151, 164, 168, 172, 174, 180, 183, 193, 195, 197, 203, 204, 213, 269, 271–273, 279, 336, 349
Adsorbed layer, 22, 43, 44, 49, 51, 67, 356
Adsorption, 13, 40, 52, 68–70, 80, 92, 269, 286, 287, 306, 338, 356
Adsorption isotherm, 255, 282, 287
Agglomeration, 14, 23, 291–293, 295–300, 303, 305, 311, 317, 320–323, 327–330
Aggregation, 105, 110, 125, 164, 165, 180, 255, 267, 282
Aggregation number, 174, 258, 260, 262–265, 267
α-relaxation, 22, 273, 279, 280, 282, 284, 343, 345, 348, 349, 353, 354
α*-relaxation, 345, 348, 351–354
Anisometric particles, 293, 296, 302, 304
Anisotropic composites, 228
Arrhenius equation, 96, 137, 212, 353
Arrhenius plot, 96, 99, 100, 109, 111, 114, 116, 168, 174
Association/dissociation equilibrium, 171, 172, 174, 180, 182

B
Barbituric acid, 178, 179
Barium titanate, 7, 228
Barrier of dissociation, 163, 182
β-relaxation, 17, 21, 349, 350
Blends, 129, 251, 253, 254, 277, 279–281
Boltorn polymers, 190
Bond lifetime, 163, 171, 182
Bound layer, 13, 22, 36, 89
Break-up, 263, 273, 276, 291, 297, 300
Broadband dielectric Spectroscopy, 22, 37, 65, 87, 90, 96, 103, 251, 254, 291, 292, 294, 311, 338, 342, 346
Brominated butyl rubber, 174
Build-up, 177, 178, 291, 292, 294, 297, 299, 300, 303, 318, 321, 322, 327–329
Butyl rubber, 163

C
Calorimetric strength, 355, 356, 360
Carbon black (CB), 2, 13, 230, 235, 243, 247, 252, 253, 257, 274, 275, 291, 293, 294, 298, 302, 306, 310–312, 318, 319, 329, 330
Carbon nanotubes (CNT), 4–7, 14, 88, 94, 104, 108, 110, 112, 116, 188, 292, 294, 302–305, 319, 322, 329, 335, 338, 340
Ceramic nanoparticles, 242
Characteristic time, 144, 284, 293
Charge carrier diffusion, 301, 313, 315
Charge migration, 255, 271, 273, 275
Chemical relaxation, 170
Clay, 2–4, 7, 10–12, 14, 15, 109, 188, 198, 293, 339, 357
Clusters, 5, 6, 8, 63, 65, 70, 71, 80, 164–166, 168, 172–174, 177, 178, 180–183, 235, 236, 238, 241, 242, 270, 271, 278, 301, 313, 315, 317, 329, 341, 347

A. Schönhals and P. Szymoniak (eds.), *Dynamics of Composite Materials*, Advances in Dielectrics, https://doi.org/10.1007/978-3-030-89723-9

Coating agents, 253, 255–257, 264, 265, 271, 282, 286, 287
Colloidal silica, 13, 130, 251, 253, 257, 284
Complex dielectric function, 1, 132–134, 147, 163, 336, 350
Complex permittivity, 21, 48, 132, 135, 145, 192, 200, 208, 210, 226, 229, 269, 307, 310, 314
Complex shear modulus, 131, 143, 144, 292, 299, 311
Composite, Composites, 1, 2, 5, 14, 22, 37, 124, 132, 225, 227, 231–233, 235, 236, 241–243, 245–247, 268, 291, 293, 296–298, 301, 306, 307, 310, 323–325, 327–329, 356
Composites with clustered nanofillers, 235
Compressibility, 259, 260
Conduction free loss, 343, 350
Conductive particles, 329
Confinement, 37, 39, 40, 54, 78, 90, 91, 101, 115, 152, 187–190, 198, 200–208, 212–214, 262, 356, 359
Correlation hole, 251, 258–263, 265, 267
Coupling agent, 7, 13, 255, 256, 271, 281, 282, 284
Critical exponents, 301, 302
Crosslinking density, 337, 338, 346–348, 353, 354, 356, 357, 359, 362
Crystallization, 87, 90–92, 94, 97–101, 103–108, 110–113, 115–117

D

DC conductivity, 1, 5, 6, 11, 41, 136, 193, 194, 201, 240, 241, 291, 297–301, 305, 311–313, 315–326, 328
Debye theory, 350
Depercolation, 276
Dielectric loss, 12, 19, 38, 40, 41, 43, 69, 71, 72, 95, 96, 134, 136, 140, 146, 147, 167, 169, 181, 227, 230, 268–274, 276, 278, 280, 282, 283, 286, 343, 345, 349, 350, 359, 360
Dielectric properties, 19, 47, 48, 180, 227, 228, 236, 238, 239, 242, 243, 245, 291, 311, 329
Dielectric permittivity, 19, 38–40, 46, 95, 96, 134, 192–195, 200, 202, 210, 211, 227, 235, 236, 243–246, 268, 291, 292, 301, 310, 312, 314–317
Dielectric strength, 35, 44–50, 52, 53, 103, 109, 113, 139, 163, 177, 194, 201, 202, 204, 209–211, 213, 230, 269, 270, 274–279, 335, 343, 350, 351
Differential scanning calorimetry (DSC), 22, 35–37, 41, 42, 94, 95, 101, 103, 104, 107, 108, 110, 112, 187, 196, 201, 204, 254, 306, 335, 336, 338, 341, 348, 354, 357, 358
Diffusion, 6, 14, 15, 17, 90, 91, 106, 116, 117, 127, 165, 167, 171, 172, 226, 238, 239, 241, 275, 292, 293, 300, 305, 324, 330, 356
Diffusion coefficient, 15–18, 226, 239, 240, 292, 302, 303, 324, 338, 362
Diffusion controlled, 172, 338, 354, 362
Dimensionality, 90, 110, 291, 293–295, 329
Dispersion, 4–8, 11–13, 36, 41, 65, 73, 89, 91, 111, 123, 128, 152, 189, 227, 229–231, 238, 254, 264, 271–275, 282–284, 297, 305, 307, 338, 341, 358, 359
Dissociation, 90, 163–165, 167, 169–172, 174–176, 178, 180, 182, 183
Distribution, 1, 5–10, 15, 16, 20, 21, 23, 38, 39, 44, 78, 91, 92, 96, 101, 106, 112, 130, 139, 141, 147, 170, 194, 225, 227, 228, 230, 251, 262, 267, 269, 280, 283–285, 287, 293, 296, 300, 305, 317, 325, 350, 359, 361
Domain polarization, 246
Double-quantum NMR, 177
Dynamical-mechanical analysis, 232, 280, 346
Dynamic heterogeneity, 55, 279, 313, 335, 347, 353, 362
Dynamics, 18, 20–22, 35–37, 41, 43, 50, 52, 55, 56, 63, 65, 67–69, 71–73, 78, 80, 87, 90, 91, 99, 100, 103, 104, 106, 110, 111, 113–117, 123, 125, 127, 138, 141–144, 149, 151, 153, 163, 164, 166, 167, 174–176, 178, 183, 187–190, 192, 193, 197, 198, 200, 201, 203, 206–208, 213, 214, 230, 233, 251, 253–255, 268, 272, 273, 277, 279–284, 287, 288, 293, 338
Dynamic shear modulus, 293, 309

E

Elastomer polymer composites, 227
Elastomers, 4, 166, 168, 182, 225, 227–230, 233, 234, 242, 243, 246, 313
Electrical conductivity, 3, 4, 21, 272, 291, 293, 294, 301, 308, 312–315, 338
Electrode polarization, 149, 166, 175
Electron microscopy, 110, 143, 183, 251

Elongational flow, 293, 297, 318
Energy-resolved elastically scattered intensity, 205
Epoxy, 8, 9, 14, 293, 335, 337–341, 344, 345, 347–363
Epoxy resin, 337, 340, 359
Equilibrium constant, 165, 169, 174
Ethylene-propylene-diene elastomer (EPDM), 225, 232, 236, 238, 241–244, 246
Exfoliation, 2, 340, 348

F
Fast scanning calorimetry, 335, 338, 341, 346, 352, 353
Filler dispersion, 254, 293, 317
Filler-Filler interactions, 255
Filler network, 291, 293–295, 297, 299, 301, 303, 310–313, 318, 322, 329
Filler-polymer interactions, 91, 293
Fillers, 2–5, 13, 14, 17–21, 36, 55, 87, 88, 91, 92, 95, 105–110, 114, 115, 117, 180, 189, 225, 227–235, 238, 241–247, 251–255, 257, 258, 262–265, 268, 269, 271, 272, 274–277, 279, 281–283, 285–287, 293, 295, 299–301, 303–305, 310–322, 328–330, 339, 340, 350, 356, 357
Flame retardancy, 1–3, 12–14, 19, 339, 340
Flocculations, 275, 293, 297
Flow, 127, 130, 132, 143, 167, 169, 178–180, 192, 252, 281, 291, 293–296, 299, 300, 304, 305, 308, 310, 312, 317, 322, 324–327, 329, 330
Fractal, 8, 63, 70, 80, 226, 235, 236, 239, 240, 258, 262, 294, 301, 302, 313, 315, 317
Frequency, 1, 5, 6, 12, 19, 21, 37, 38, 40, 41, 43, 50, 55, 64, 76, 77, 95–97, 99, 103, 115, 132–134, 136, 140, 142, 144–146, 149, 150, 164, 167, 175, 176, 181, 192, 193, 200–202, 204, 205, 207, 208, 210, 211, 226, 230, 236, 238, 239, 243–246, 252, 268–270, 272, 284, 286, 293, 296, 297, 301, 307–309, 312–318, 329, 336, 342–344, 349, 351, 353, 359, 360
Frequency domain, 96, 252, 268, 336, 343, 345, 352

G
γ-relaxation, 195
Glass transition, 22, 87, 90, 91, 94, 95, 97, 99, 106, 108, 110, 111, 125, 127, 130, 132, 137, 139, 144, 145, 147, 148, 164, 181, 192, 194, 196, 210, 255, 268, 273, 280, 336, 341, 343, 346, 353–355, 359, 360, 363
Glass transition temperature, 2, 5, 6, 13, 14, 64, 68, 91, 98, 188, 190, 192, 193, 195, 197, 201–207, 209, 210, 212, 233, 306, 310, 336, 346, 352, 355, 358, 359, 361, 362
Glassy dynamics, 22, 42, 63
Grafting density, 6, 130, 146, 263, 264, 275, 276
Graphene, 3, 4, 14, 21, 93, 112–114, 188, 208, 306, 338
Graphene oxide, 3, 21, 93, 108, 208, 225, 226, 230
Graphite nanoplates, 291, 292, 294, 303, 306, 315, 323, 329, 330
Graphite oxide, 187, 189, 190, 208, 213, 214, 294
Guinier regime, 259, 261

H
Halloysite nanotubes (HNT), 4, 9, 93, 335, 336, 338–342, 350, 357–363
Havriliak-Negami, 39, 95, 96, 99, 187, 194, 201, 209, 229, 269, 336
Heat flow, 336, 341, 342, 354, 355
Heterogeneity, 139, 141, 280, 346, 354, 355, 362
Hybrane polymers, 190
Hydrogen bond network, 196, 197, 203, 213, 350
Hydrogen bonds, 165, 180, 190, 191, 196, 197, 203, 204, 207, 212, 214, 271, 279, 358, 359, 363
Hydroxyl rotation, 196, 197, 212
Hyperbranched polymers, 187, 189, 190, 192, 193, 195, 197–199, 203, 208, 214

I
Image analysis, 262
Imidazolium ion, 174
Incoherent structure factor, 206, 207
Interaction, 3, 6–8, 17, 20, 22, 36, 37, 39–41, 54, 65, 67–69, 87, 89–92, 105, 106, 108, 109, 112, 116, 117,

125, 137, 139, 141–144, 147, 152, 164, 165, 168, 171, 181, 187, 189, 190, 198, 200, 207, 208, 213, 214, 230, 253, 255, 258, 259, 261, 263, 271, 277, 279, 281, 282, 287, 291, 295–297, 299, 300, 305, 327, 335, 338, 340, 341, 348, 350, 356–363
Intercalation, 7, 200, 208, 340
Interdigitating electrodes, 309–311, 324
Interfacial dynamics, 117, 125, 127, 225
Interfacial effects, 225, 227, 246
Interfacial layer, 16, 18, 22, 35–39, 41–51, 54–56, 65, 68, 69, 72, 80, 89–91, 117, 164, 181, 182, 272, 281–283, 287, 288
Interfacial polarization, 139, 164, 235, 251, 254, 268, 271, 273, 275, 287, 350
Interfacial water, 271
Interlayer distance, 3, 7, 113, 198, 200, 208, 214, 340, 348
Interparticle forces, 293, 329
Ionic conductivity, 132, 134, 136, 148, 149, 194, 210, 211, 272, 275, 276
Ionic liquid, 149, 177
Ionomers, 165–167, 177

K
Kirkwood-Fröhlich factor, 50, 51

L
Layered Double hydroxide (LDH), 1, 3, 7–9, 335, 336, 338–340, 348, 350, 362
Lifetime renormalization, 171, 182
Loss modulus, 175, 176, 181, 226, 281

M
Magic-angle spinning NMR, 178
Master curve, 73, 75–77, 175, 268
Matrix polymer, 63, 65, 90, 253, 305, 311
Maxwell-Wagner-Sillars (MWS), 10, 11, 36, 38, 39, 187, 202, 209, 210, 213, 230, 237, 246, 251, 254, 271, 272, 282–284, 350
Mean squared displacement, 205
Mechanical deformation, 293
Melting, 94, 100, 173
Micelle, 178
Modulus, 2, 36, 54, 71, 73, 78, 126, 144, 145, 175, 178, 181, 182, 232, 281, 299, 308, 338, 339, 342
Molecular dynamics, 16, 22, 132, 197, 205, 356, 359
Monodispersity, 267
Multi-filler composites, 225, 227
Multi-walled carbon nanotubes (MWCNT), 88, 93, 291, 292, 294, 296–298, 301–307, 314–318, 321–323, 325–327, 329, 330
MWS polarization, 10

N
Nanocomposites, 1–16, 19–23, 36, 40, 42, 43, 47, 50, 56, 71–74, 78, 87, 92, 107–109, 123–126, 129, 135, 138, 141–144, 147, 152, 153, 164, 181, 187, 189, 190, 192, 198–210, 212–214, 225, 230, 235, 254, 257, 258, 261, 264, 266, 270, 271, 276, 284, 287, 294, 304, 305, 329, 335, 337–341, 345, 348–352, 354–362
Nanohybrids, 198
Nanofiller, 3–10, 13–16, 20, 22, 94, 139, 141, 293–296, 298, 304, 329, 335, 337–341, 348–352, 354–363
Nanoparticle, 1–11, 13–23, 35–37, 39–41, 44–46, 48, 49, 51–54, 56, 63–68, 87, 89, 105, 123–131, 133, 134, 137–147, 150, 152, 153, 163, 166, 181, 188, 232, 251–253, 257, 259, 261, 262, 266, 272, 277, 281, 283, 287, 295–297, 299, 302, 305, 336, 338, 341, 350, 354–356, 358, 359, 362
Network formation, 310, 311, 338
Neutron scattering, 8, 36, 37, 65, 188, 190, 254, 255, 277, 288
NMR spectroscopy (NMR), 37, 65, 126, 173, 254, 284, 287, 349, 361

O
Onsager equation, 170
Order parameter, 178
Orientation, 15, 176, 178, 192, 230, 291, 293, 295, 296, 304, 305, 317, 324–326, 328, 330

P
Particle shape, 259, 291, 303
Particle size, 228, 232, 283, 306
Payne effect, 13, 253, 263, 275
Péclet number, 292, 304, 305, 325, 326, 330

Percolation, 5, 20, 70, 80, 235, 251, 254, 267, 268, 275, 276, 297, 300–302, 310, 313, 317, 319, 329
Percolation concentration, 294, 301, 316, 317, 319, 320, 322, 324, 326, 327, 329
Percolation threshold, 238, 267, 274–276, 292, 294, 297, 301, 312–321, 329, 330
Percus-Yevick, 8, 259, 261
Phenylurazole, 165, 168, 169, 171
π-π interaction, 165
Plate-plate geometry, 294, 298, 310
Poly(alkylene furanoate), 87, 92
Polarization, 10, 20, 21, 38, 39, 55, 149, 177, 202, 209, 225, 226, 236, 237, 241, 242, 246, 247, 268, 272, 343, 344, 349, 350
Polybutadiene, 168, 169, 280
Polycarbonate (PC), 5, 6, 17, 18, 291, 292, 294, 296–298, 302–307, 310–312, 314–318, 320, 322, 324–326, 328, 329
Poly(ε-caprolactone), 87, 88, 92, 110
Polydimethylsiloxane, 101, 173, 180, 225, 228, 230
Polydispersity, 124, 129, 130, 189, 253, 257–259, 261, 265–267
Polyhedral Oligomeric Silsesquioxane (POSS), 2, 3, 17, 18
Polyisobutylene, 173, 174
Polyisoprene, 39, 171, 252, 255, 278
Polylactide, 8, 87, 92, 97
Polymer, 1–9, 13–22, 35–56, 63–80, 87–95, 97, 99–101, 103–106, 109, 110, 112–117, 123–131, 135, 137–139, 141–151, 153, 163–166, 168, 172–174, 177–179, 181, 182, 187–214, 227, 235, 237–239, 241, 246, 251–257, 262–264, 266–269, 271–273, 275–288, 291, 293–295, 297, 298, 301, 304, 305, 310, 311, 313, 318, 321, 324, 329, 337, 339, 340, 356, 359, 360, 362
Polymer bridge, 63–68, 70–72, 78–80
Polymer composites, 163, 235, 237–239, 241, 293, 329
Polymer matrix, 3, 4, 6, 8, 14, 15, 18, 21, 36, 38, 46, 64, 68, 69, 71, 89–91, 95, 117, 147, 188, 189, 227, 252, 253, 255, 275, 285, 286, 291, 294, 301, 313, 318, 321, 340, 359
Polymer melt, 268, 291, 293–295, 298, 304, 305, 324
Polymer nanocomposites, 2, 20, 35, 55, 63, 64, 68–70, 87, 89, 123, 125, 128, 130, 137, 188, 198, 208, 214, 251, 252, 263, 269, 277, 288, 294, 336, 337, 360
Poly(methyl methacrylate), 22, 125, 127, 279
Poly(vinyl acetate) (PVA), 38–40, 42, 44, 47, 68, 71–78, 125
Power law, 41, 167, 229, 262, 301, 313, 315, 316, 326, 344
Precipitated silica, 269, 270, 281
Processing history, 293

Q

Quantum dot, 3
Quasielastic neutron scattering (QENS), 188, 190, 204, 205, 283, 284
Quiescent melt, 291, 292, 294, 299, 300, 303, 310, 318, 320, 322, 323, 326–328, 330

R

Random dispersion, 228, 267, 268
Reinforcement, 13, 73, 113, 163–166, 173, 180–182, 253, 276, 281, 299
Relaxation map, 168, 169, 175, 176, 179, 180, 196, 203, 212, 213, 285, 343, 345, 351–353
Relaxation rate, 135, 151, 178, 180, 336, 343, 345, 349–354
Relaxation strength, 134, 139, 146, 148, 169, 170, 172, 175, 176, 194, 196, 201, 203, 210, 212, 213,, 226, 237, 238, 240, 241, 285, 345, 351–353
Relaxation time, 22, 38, 41, 43–45, 48, 52, 55, 56, 64, 65, 70–72, 74, 78, 96, 99, 101, 109, 131, 133, 135, 137–139, 141, 144–146, 148, 151–153, 164, 174, 188, 194, 195, 197, 198, 201–204, 206, 207, 211–213, 226, 237, 238, 240, 241, 269, 279–281, 283–287, 293, 336, 343, 345, 353
Renewable polymers, 92
Reptation, 165, 167, 169, 171, 252
Reversing heat capacity, 342
Rheo-electrical measurements, 291, 293, 294, 308
Rheological behavior, 183, 293

Rheology, 73, 74, 80, 131, 144, 145, 148, 153, 167, 173, 175, 176, 178, 179, 251, 252, 254, 274, 275, 282
Rigid amorphous fraction (RAF), 22, 90–92, 95, 106–109, 116, 117, 336, 356, 360–363
Rotational diffusion, 296, 302–305, 324, 326, 330
Rubber, 1, 2, 12–14, 21, 22, 110, 112, 163, 174, 180, 230, 251–256, 263, 265, 268, 271–273, 275–277, 279, 282–284, 287, 288, 318

S
Secondary relaxation, 211, 282
Segmental dynamics, 22, 36–39, 41, 44, 50–52, 54–56, 65, 68, 72, 74, 80, 91, 114, 115, 132, 165, 168, 188, 193, 213, 214, 251, 252, 254, 255, 268, 277–279, 281, 282, 284, 285, 287, 288, 341, 350, 351, 353, 355, 356, 359, 362, 363
Segmental relaxation, 35, 37, 38, 40, 41, 43–45, 47–50, 52–56, 64, 69, 72, 74, 78, 88, 89, 95, 101, 111, 137, 139, 148, 174, 175, 179, 190, 193, 194, 197, 198, 201, 204, 205, 207, 210, 213, 214, 268, 277, 279–282, 284–287, 343, 344, 346, 348, 353, 354, 357, 362
Self-healing, 165, 174, 176
Shear deformations, 291, 297, 299, 308, 311, 318, 319, 321, 322, 324, 327, 328, 330
Shift factor, 72–75, 77, 78, 131, 143–145, 148, 175
σ_{DC}, 5, 6, 11, 194, 240, 242, 301, 312, 313
Silica, 3, 7, 17, 19, 20, 22, 40–44, 47, 50, 53, 56, 93, 94, 96, 104, 105, 108–110, 114, 115, 125–130, 132–134, 136, 137, 139, 142–145, 147, 150, 152, 153, 188, 192, 198, 230, 232, 246, 251–255, 257, 258, 262–273, 275–277, 279, 281–288, 294, 338
Small angle X-ray scattering (SAXS)s, 7–11, 36, 37, 48, 54, 101, 143, 174, 236, 251, 254, 257, 262, 263, 265, 267, 276, 288, 336, 340, 348
Sodium montmorillonite, 187, 188, 190, 198, 208, 213, 214
Solid-phase mixing, 256, 279
Solid state NMR, 173
Specific heat capacity, 94, 130, 342, 355, 360
Star polymers, 179
Steady flow, 297, 299
Steady shear, 291, 294, 297–300, 308, 310, 311, 318, 321–329
Sticky reptation, 165, 167, 169, 171
Storage modulus, 63, 64, 71, 75, 144, 175, 226, 232–234
Structure factor, 8, 9, 205, 206, 258–261, 266, 348
Styrene-butadiene rubber (SBR), 225, 226, 230–234, 251–253, 255, 257, 270, 272–274, 278–281
Synthetic elastomers, 227
Supramolecular bonds, 180

T
Telechelic polymers, 164, 172
Temperature domain, 336, 345, 351, 352
Temperature modulated DSC (TMDSC), 22, 48, 338, 341, 342, 348, 351–354, 361
Terminal relaxation, 64, 70, 71, 175, 180
Thermoplastic, 1, 2, 14, 20, 22, 166, 293, 305
Thermorheological complexity, 176
Thermo-rheological history, 310, 320
Thymine, 178, 179
Tire, 4, 12, 13, 22, 23, 251–253, 256, 277, 280
Transient networks, 165, 328
Translational diffusion, 303–305, 325
Transmission electron microscopy (TEM), 8, 9, 12, 73, 226, 232, 234–236, 254, 257, 262, 263, 265, 266, 276, 294, 296, 297, 306, 307, 327, 336, 348

U
Unfilled polymer, 4, 16, 95

V
Viscoelasticity, 281
Viscoelastic properties, 36, 65, 71, 73, 163, 293
Vogel-Fulcher-Tammann (VFT), 97, 99, 111, 188, 196–198, 204, 212–214, 285, 345, 352

W

Water, 3, 16, 19, 126, 131, 132, 180, 192, 208, 251, 269–273, 275, 277, 287, 338–340, 346

Wide angle X-ray scattering (WAXS), 8–10, 101, 112, 131, 143

X

X-ray diffraction, 196, 198, 208, 214

X-ray scattering, 8, 131, 174, 257, 263, 324, 326, 346, 348

Z

Zeolitic-Imidazolate Frameworks (ZIF), 18

The manufacturer's authorised representative in the EU is Springer Nature Customer Service Centre GmbH, Europaplatz 3, 69115 Heidelberg, Germany. If you have any concerns regarding our products, please contact ProductSafety@springernature.com

Printed and bound by CPI Group (UK) Ltd, Croydon, CR0 4YY
30/09/2026
02225496-0014